Fundamentals of Weather and Clima

D0243787

You have spread out the heavens like a tent
and built your home on the waters above.
You use the clouds as your chariot
and ride on the wings of the wind.
You use the winds as your messengers
and flashes of lightning as your servants.

PSALM 104, vv. 2–4
Good News Bible

Fundamentals of
Weather and Climate

ROBIN McILVEEN
Environmental Science Division
Institute of Environmental and Biological Sciences
Lancaster University

CHAPMAN & HALL
London · Glasgow · Weinheim · New York · Tokyo · Melbourne · Madras

Published by Chapman & Hall, 2-6 Boundary Row, London SE1 8HN, UK

Chapman & Hall, 2-6 Boundary Row, London SE1 8HN, UK

Blackie Academic & Professional, Wester Cleddens Road, Bishopbriggs, Glasgow G64 2NZ, UK

Chapman & Hall GmbH, Pappelallee 3, 69469 Weinheim, Germany

Chapman & Hall USA., 115 Fifth Avenue, New York, NY 10003, USA

Chapman & Hall Japan, ITP-Japan, Kyowa Building, 3F, 2-2-1 Hirakawacho, Chiyoda-ku, Tokyo 102, Japan

Chapman & Hall Australia, 102 Dodds Street, South Melbourne, Victoria 3205, Australia

Chapman & Hall India, R. Seshadri, 32 Second Main Road, CIT East, Madras 600 035, India

First edition 1992
Reprinted 1992, 1995 (twice)

© 1986, 1992 J.F.R. McIlveen

First published in 1986 as Basic Metereology: A physical outline

Typeset in 10/11pt Times Roman by Colset Private Ltd, Singapore

Printed in Singapore by Kin Keong

ISBN 0 412 41160 1

A Catalogue record for this book is available from the British Library

Library of Congress Cataloging-in-Publication Data available

Contents

Preface to this edition

I have taken the opportunity of a second edition to revise the original text from beginning to end. There were some errors of course, and I hope I have spotted most; and there were sections of text which were unclear when I read them again in the cool light of reappraisal. And of course there were sections which cried out for amendment (mainly extension) because of international and personal changes in perceived importance. Once I had begun to consider these, I became increasingly aware that the title of the book, and of many of its sections, reflected a traditional perception of the scope of meteorology somewhat at odds with the actual contents, and overtaken by the pace of work on climate, climate change and human interactions with the atmosphere. Accordingly I have retitled the book and many of its sections. The underlying rationale of the book (to provide a mechanistic introduction to the physical behaviour of the lower atmosphere) is unaltered, having been justified by positive reaction to the first edition.

Alterations include the addition of separate sections on the concept of chaos, stratospheric ozone, the long-term evolution of the atmosphere, human society and climatic change, diurnal variations of radiative heat fluxes, and summaries of mid-latitude of low-latitude climates. The particular problem of anthropogenic carbon dioxide and its effect on the atmospheric greenhouse is treated in several sections now, as befits its profound significance for the future of our planet. Previously separate chapters on radiation and global climate are amalgamated, and the large-scale weather systems in low and middle latitudes are now given separate chapters. To help instructors with students still using the first edition, there is a final appendix listing section equivalences and major errata. Smaller changes are a simplified derivation of the Coriolis acceleration, and a new appendix on geopotential height. Associated with all of this is the appearance of 40 new photographs and diagrams, all in addition to those used in the first edition. I have tried to include more examples from outside my immediate maritime, mid-latitude experience, but hope that the inevitable residual bias will be as interesting to those who live in other climates as their accounts are to me.

I am very grateful to those who have provided valuable suggestions and comments on the first edition, many of which have been assimilated in the second; to the publishers who have tactfully galvanized me toward completion; to my cartographer Jane Rushton who has once again translated impressionistic scribbles into legible diagrams; and to my wife and daughters Annwen and Heather who cheerfully combed the text for all the alterations to figure and section numbers. I hope the text continues to be interesting and useful, and will gladly receive comments from readers.

In the time between starting on the first edition (roughly 1982) and the present, public perception of aspects of meteorology and climatology has changed remarkably. To all of us involved in the study of our magnificent, and yet bewildering atmosphere, it is exciting to see our subject centre stage. It is also a little worrying, since however little else we know, we know how uneasily its stubborn subtleties are likely to fit into the simplistic schemes of politicians and the media. I will be well content with my several thousand hours of effort if I have helped more people to appreciate the simplicities and the subtleties of our atmosphere, and to begin to distinguish between them.

J.F.R. McIlveen
Lancaster University

Preface to the first edition

I have written this book for undergraduates and serious amateurs seeking introductory information about the behaviour of the lower atmosphere (especially the troposphere), and an outline of the physical mechanisms involved. There are many good texts which outline behaviour without giving any coherent account of mechanism; and there are several good texts which outline the theoretical framework of meteorology, but assume the reader to be fairly familiar already with the atmospheric behaviour being discussed. A few books bridge the gap, but I sense a tendency in them to concentrate on the many essential nuts and bolts at the expense of a more connected view of the vast and detailed atmospheric activity which daily surrounds us.

The reasons for the presence of such a gap in the literature become obvious as soon as you try to fill it. I have found it in practice very difficult indeed to balance the need to outline a minimum of actual behaviour against the need to offer some accompanying discussion of mechanism, without causing mental indigestion in the reader and writer, and without falling into undue emphasis on either description or mechanism. I know that I have succeeded only patchily, but I hope that my aim in the process will become sufficiently clear to the sympathetic reader that he or she will be able to read a personally more suitable rendering between the lines. The deliberately discursive style is adopted for the same reason, for although it expands the sheer volume of the text, it reveals the personal attitude of the author to the matters in hand, in ways which can be useful to the new student, anxious for clues about the relative importance of different parts of the total mass of information. Since I have also tried to maintain an open style of discussion, minimizing the diplomatic niceties which can so easily blunt the impact of basic ideas, I know that I have probably rushed into several places where more experienced angels fear to tread. I have never done so wittingly, though a patient critic has already pointed out that I may have erred in overplaying the concept of the highest opaque layer in Chapter 8.

The inclusion of both familiarization and mechanistic understanding in an introductory text raises the danger of saying a little about everything and not enough about any one thing. Whenever I have been conscious of this danger I have deliberately tried to err on the side of being selective, and this being the case it might be useful to read this book in conjunction with one from the first of the three groups mentioned above, although I have tried to make the book largely self-contained.

The balance and level of the text has risen largely, but not exclusively, from my experience of teaching introductory meteorology to students in the Department of Environmental Science of the University of Lancaster over a period of years. In that context it covers the material presented over at least the first two years of a three-year undergraduate course which covers a wide range of aspects of the environment. In the context of a more specialized degree, it probably covers the outline of the first year of meteorological studies, and provides a mechanistic core for later years.

A significant complication when planning courses or texts in meteorology is that people come to the subject with a wide variety of academic backgrounds. I have assumed that readers are familiar with the terminology and some of the elementary manipulations of the differential calculus, going little beyond the coverage of the Ordinary Level syllabuses of the English General Certificate of Education. This standard of background would not suffice in Physics, however, but rather than assume competence to the Advanced Level GCE standard (which anyway is seriously deficient in elements of classical physics, especially thermodynamics, and has virtually no coverage of fluids), I have chosen to outline the necessary basic relationships as they arise in the text. Sometimes a physical concept seems to me to be so fundamental to the meteorology in hand that I have chosen to describe it in some detail; and I have done the same thing when experience has taught me that a particular matter is difficult to grasp (the Coriolis effect is a notorious case in point). In other places I have included only a brief exposure in the body of the text, reserving formal treatments to appendices. I have always tried to outline only the essential basis of mechanism, and have been interested to note that this seems to involve less formal detail in some regions (such as thermodynamics and radiation) than other authors have felt necessary, and more in others (such as dynamics, energetics and surface behaviour). I have deliberately avoided vectors, partly because they are not widely encountered before first-year University level (at least not to a useful extent), and partly because their very success in pulling dynamical formulations into very neat packages can actually obscure the physical realities being described. It would be a useful step towards more advanced studies to transcribe the expressions in Chapter 7 (the dynamics chapter) into vector form.

I have taken considerable pains to assemble problems which will test the reader's familiarity and understanding of each chapter. They are graded into three levels as follows: the first level consists of short questions testing knowledge of concepts and nomenclature, often at a very basic level. The second consists of questions which test the application of concepts, usually in the form of equations, to simple problems. Where the procedure is at all complex or subtle it is clearly cued. The third level tests a more assured understanding of the material, either by setting simple questions without clues about procedure, or by setting more complex or subtle questions, or by seeking a discursive summary in the style of a short scientific essay.

On looking over the text and comparing it with the aims stated initially, I feel that I have probably been most successful in those parts where I have managed to recapture (at least internally in the process of writing) the spirit of enthusiastic and almost innocent inquiry which I encountered as a graduate student in the Department of Meteorology at Imperial College London, particularly at the hands of Dr J.S.A. Green and the late Professor F.H. Ludlam. I began to recall the latter's infectious enthusiasm for clouds at several points in the text, and hope that it has helped to integrate a subject which is too often left to specialist texts. Further back I feel a debt, especially when engaged in the deceptively simple business of trying to convey my sense of the logical way to approach complex reality, to two former physics masters at the Royal Belfast Academical Institution, W.L. Brown and H.F. Henderson, both dead now but remembered with affection and respect by two generations of more or less dedicated pupils.

I wish to thank Dr R.S. Harwood of Edinburgh University for urgently and carefully reading about half the text, and making many wise and useful comments. Grateful thanks also go to Jane Rushton of the Department of Environmental Science, for working with skill and enthusiasm on the numerous line drawings. I am grateful to colleagues who tolerated a considerable degree of distraction during the period of maximum effort, and am very grateful to my wife and family for putting up with considerable domestic disruption over the same period, usually with gentle forbearance. My wife also helped crucially with the considerable editorial load which accumulates in the later stages of a work of this scale. Finally I wish to say to the youngest members of my family that I am sorry I could not begin 'Once upon a time', but there is a kind of 'happily ever after' at the end.

Acknowledgements

I am very grateful to the following people and institutions for their kind permission to use copyright material, and in many cases for the supply of prints, etc. Those entries including NML were traced through the National Meteorological Library, Meteorological Office, London Road, Bracknell, Berks. All photographs not acknowledged below are by the author.

Fig. 1.1   National Remote Sensing Centre, Royal Aircraft Establishment, Farnborough, Hants.

Fig. 1.2   National Aeronautical and Space Administration, Washington D.C.

Figs. 1.3 and 12.10   Hong Kong Royal Observatory, Nathan Road, Kowloon.

Fig. 2.6   Meteorological Office, Bracknell, Berks. Crown Copyright.

Fig. 2.9   From 'The cyclone problem'—Professor F.H. Ludlam's inaugural lecture, Imperial College London, 1966. With permission from Mrs Ludlam.

Figs 2.11 and 10.23   Vaisala Oy, Helsinki, Finland.

Figs. 2.12 and 2.13   P.E. Bayliss. Department of Electrical Engineering and Electronics, University of Dundee, and the Natural Environmental Research Council.

Fig. 3.2   Copyright Dr J.E. Harries, Rutherford Appleton Laboratory, Chilton, Didcot, Oxfordshire, UK.

Fig. 3.8   From Dr J.N. Cape. Copyright Institute of Terrestrial Ecology, Penicuik, Midlothian, Scotland.

Fig. 3.10   Dr D.G. McGarvey, Department of Earth Sciences, The Open University, Milton Keynes, Bucks.

Fig. 3.11   Dr D.A.R. Simmons, 21 Dougalston Avenue, Milngavie, Glasgow.

Fig. 3.13   Dr R.I. Jones, Institute of Environmental and Biological Sciences, Lancaster.

Figs. 4.3 and 6.9   Copyright Robert Nutbrown, University of Warwick, UK.

Fig. 4.16   Chief Engineer, Lancaster City Council.

Fig. 5.3   Copyright BP Educational Service.

Figs. 5.5, 5.6, 5.7, 5.11, 6.6   Tephigrams after Metform 2810, Meteorological Office, Bracknell, Berkshire. Crown Copyright.

Fig. 6.14   Photograph by Sergeant Al Roberts, Royal Air Force. Copyright British Aerospace, Warton, Lancashire, UK.

Fig. 6.22   We have been unable to trace the owner of the copyright to this picture but would welcome any information relating to this matter.

Acknowledgements

Fig. 6.25   British Aerospace, Filton, Bristol.

Fig. 6.26   Dr K.A. Browning, from *Reports on Progress in Physics*, **41** (1978), The Institute of Physics, London.

Fig. 6.27   J.W.F. Goddard, Radio Communications Research Unit, Rutherford Appleton Laboratory, Chilton, Didcot, Oxfordshire.

Fig. 7.4   Copyright Paul Hodgkins.

Fig. 7.10   John Bowman, Department of Environmental Science, University of Lancaster.

Fig. 7.22   Dr R. Hide, Meteorological Office, Bracknell, Berks. From [45], the Royal Meteorological Society, James Glaisher House, Grenville Place, Bracknell, Berks.

Fig. 7.24   We have been unable to trace the owner of the copyright to this picture but would welcome any information relating to this matter.

Figs. 8.10 and 11.12   Rod Stainer, New Zealand Meteorological Service, Wellington, New Zealand.

Fig. 8.13   Captain J.R.C. Young, Brackenhill, Cotswold Road, Cumnor Hill, Oxford, in *Weather*, **20**: 18 (1965). The Royal Meteorological Society, as in Fig. 7.22.

Fig. 9.10   P. Larsson and M. Beadle, Department of Environmental Science, University of Lancaster, Lancaster, UK.

Fig. 9.11   Dr N.E. Holmes, 25–134, Crimea Road, Marsfield, New South Wales, Australia.

Fig. 9.24   Doug Allan, Underwater Images, 19 Armscroft Road, Gloucester.

Fig. 9.31   Taken in 1938 by Flt Lieut. G.R. Shepley (killed in action 1940). The Dean and Chapter of Salisbury Cathedral.

Fig. 10.14   Copyright NOAA, Rockville, MD 20852, USA.

Fig. 10.21   We have been unable to trace the owner of the copyright to this picture but would welcome any information relating to this matter.

Fig. 10.22   R. Cunningham, AFGL, Hanscom, Mass. 01731, USA. From *Weather*, **33**: 62 (1978). The Royal Meteorological Society.

Fig. 10.24   D.J. Tritton in *Physical Fluid Dynamics*, Van Nostrand Reinhold (UK) Ltd, Molly Millars Lane, Wokingham, Berks.

Fig. 10.30   Copyright B. Goddard, via Ross Reynolds, Reading University, Berkshire, UK.

Fig. 10.31   As Fig. 2.12.

Fig. 10.32   As on the figure caption

Fig. 11.1   S. Petterssen (1956) *Weather Analysis and Forecasting*, Vol. 1, Motion and Motion Systems. McGraw-Hill Book Company, New York.

Fig. 11.7   As Fig. 2.12.

Fig. 11.9   R.K. Pilsbury, Nyetimber, Uplands Road, Totland, Isle of Wight.

Fig. 11.13   Chief Engineer, Lancaster City Council.

Fig. 11.26   P.G. Wickham, in [85], Her Majesty's Stationery Office, London. Crown Copyright.

Fig. 12.1   Daily Weather Report of the Meteorological Office, 1200Z 4 August 1972. Crown Copyright.

Fig. 12.4   Copyright Professor Bruce Atkinson, Queen Mary Westfield College, University of London, UK.

Fig. 12.10   As Fig. 1.3.

Fig. 12.12   Dr Neil Frank, National Hurricane Center, Coral Gables, Florida.

Useful information

Commonly used prefixes for powers of ten

Name	Symbol	
nano	n	10^{-9}
micro	μ	10^{-6}
milli	m	10^{-3}
centi	c	10^{-2}
kilo	k	10^{3}
mega	M	10^{6}
giga	G	10^{9}

Spherical properties (radius R)

Diameter	$2R$
Circumference	$2\pi R$
Maximum cross-section area	πR^2
Surface area	$4\pi R^2$
Surface area between latitudes 30°	$2\pi R^2$
Volume	$\frac{4}{3}\pi R^3$

SI units and common equivalents

PRIMARY

Quantity	Name	Units	Equivalence
mass	kilogram	kg	
	tonne		10^3 kg
length	metre	m	
	angstrom	Å	10^{-10} m
	nautical mile		1852 m
time	second	s	
temperature	kelvin	K	
	Celsius	°C	same unit size but $T°C \equiv (T + 273.2)$ K
quantity of pure element or compound	mole	mol (formerly called gram molecule)	
angle	radian	rad	
	degree	°	0.017 45 rad ($360° = 2\pi$ rad)
electrical current	ampere	A	

SOME SECONDARY UNITS USED IN METEOROLOGY

Quantity	Name	Units	Equivalence
frequency	hertz	Hz	one cycle per second
volume	cubic metre	m^3	
	litre	l	10^{-3} m^3
electric charge	coulomb	C or A s	
specific mass	kilogram per kilogram		
	gram per kilogram	g kg^{-1}	10^{-3}
	parts per million	ppm	10^{-6}
velocity	m per s	m s^{-1}	
	knot (nautical mile per hour)	kt	0.514 m s^{-1}
angular velocity		rad s^{-1}	57.3° s^{-1}
temperature gradient		K m^{-1} or °C m^{-1}	

Quantity	Name	Units	Equivalence
velocity gradient, divergence, vorticity		s^{-1}	
density	kg per cubic m	kg m^{-3}	
acceleration	m per s per s	m s^{-2}	
momentum		kg m s^{-1}	
force	newton	N or kg m s^{-2}	
pressure	pascal	Pa or N m^{-2}	
	millibar	mbar or hPa	100 Pa
pressure gradient		Pa m^{-1}	
		mbar per 100 km	10^{-3} Pa m^{-1}
angular momentum (or moment of momentum)		kg m^2 s^{-1}	
moment of force		N m	
energy	joule	J or N m	
	gram calorie	cal	4.18 J
power	watt	W or J s^{-1}	
energy flux density		W m^{-2}	
	langley (cal per cm) per minute	Ly min^{-1}	697 W m^{-2} (radiation)
heat capacity		J K^{-1} or J $^\circ$C^{-1}	
specific heat capacity		J kg^{-1} K^{-1}	
		cal g^{-1} $^\circ$C^{-1}	4180 J kg^{-1} K^{-1}

(See Appendix 7.1 for further treatment of units)

Universal constants

Speed of light *in vacuo*	c	3.00×10^8 m s^{-1}	
Planck constant	h	6.626×10^{-34} J s	
Avogadro number	N	6.022×10^{23} mol^{-1} (molecules per mole)	
Universal gas constant	R_*	8.314 J K^{-1} mol^{-1}	
Gravitational constant	G	6.672×10^{-11} N m^2 kg^{-2}	
Stefan–Boltzmann constant	σ	5.670×10^{-8} W m^{-2} K^{-4}	

Useful values in units normally used

SOLAR AND TERRESTRIAL

Solar diameter	1.37×10^6 km
Solar–terrestrial separation	$150 \, (\pm 5) \times 10^6$ km
Effective blackbody temperature of solar surface	5800 K
Solar constant	1380 W m^{-2}
Sidereal day (period of terrestrial rotation)	86 040 s (23.93 hours)
Angular velocity of terrestrial rotation	7.303×10^{-5} rad s^{-1}
Coriolis parameter	1.0×10^{-4} rad s^{-1} (or s^{-1}) at latitude 43.2°
Solar day (mean period of apparent terrestrial rotation relative to the Sun)	86 400 s (24 hours)
Year (period of terrestrial orbit of Sun)	3.156×10^7 s (365.3 days)
Calendar year	3.154×10^7 s (365 days)
Angle between equatorial and ecliptic planes	23.4° (latitudes of tropics)
Effective blackbody temperature of planet Earth	255 K
Earth's planetary albedo in solar wavelengths	about 0.33
Terrestrial mean radius	6370 km
Average gravitational acceleration at mean sea level	9.81 m s^{-2}
Terrestrial surface area	5.10×10^{14} m^2
Meridional length subtending latitude degree	111.2 km or 60 nautical miles
Elevation of highest surface above mean sea level (Mt Everest)	8848 m
Depression of lowest surface below mean sea level (Dead Sea surface)	392 m

ATMOSPHERIC

Standard temperature	0 °C	⎫ Standard temperature
Standard atmospheric pressure	1013.2 mbar	⎭ and pressure (STP)

Effective molecular weight of dry air	29.0
Specific gas constant for dry air	$287 \text{ J kg}^{-1} \text{ K}^{-1}$
Specific heat capacity for dry air at constant volume	$717 \text{ J kg}^{-1} \text{ K}^{-1}$
Specific heat capacity for dry air at constant pressure	$1004 \text{ J kg}^{-1} \text{ K}^{-1}$
Density of dry air at STP	1.292 kg m^{-3}
Decadal scale height of isothermal dry atmosphere at $-20\ ^\circ\text{C}$	17.1 km
Velocity of sound in dry air at $0\ ^\circ\text{C}$	331.5 m s^{-1}
Thermal conductivity of dry air at $0\ ^\circ\text{C}$	$2.43 \times 10^{-2} \text{ W m}^{-1} \text{ K}^{-1}$
Dynamic viscosity of dry air at $0\ ^\circ\text{C}$	$1.71 \times 10^{-5} \text{ kg m}^{-1} \text{ s}^{-1}$
Saturated vapour at $0\ ^\circ\text{C}$	
density	$4.85 \times 10^{-3} \text{ kg m}^{-3}$
pressure	6.11 mbar
Specific latent heat of vaporization of water at $0\ ^\circ\text{C}$	2.50 MJ kg^{-1}
Specific latent heat of fusion of ice at $0\ ^\circ\text{C}$	0.334 MJ kg^{-1}
Density of water at $0\ ^\circ\text{C}$	1000 kg m^{-3}
Density of ice at $0\ ^\circ\text{C}$	917 kg m^{-3}
Density of mercury at $0\ ^\circ\text{C}$	$13\ 595 \text{ kg m}^{-3}$

Selection from International Civil Aviation Organization standard temperate atmosphere

Height/m	Pressure/mbar	Temperature/K	Density/kg m^{-3}
0	1013.3	288.2	1.23
5 000	540.5	255.7	0.736
10 000	265.0	223.2	0.414
15 000	121.1	216.7	0.195
20 000	55.3	216.7	0.0889

Typical magnitudes

Solar horizontal irradiance at Earth's surface	0–1000 W m^{-2} low latitudes and mid-latitude summer 0–200 W m^{-2} mid-latitude winter
Average tropospheric wind speed	10 m s^{-1}
Wind speed in jet core	50 m s^{-1}
Height of jet core above sea level	9 km polar front jet stream 12 km subtropical jet stream
Height of tropopause above sea level	10 km high latitudes 15 km low latitudes
Height of freezing level above sea level	4 km low latitudes 0–3 km mid latitudes
Depth of sub-cloud layer	1 km
Depth of precipitable water	30 mm
Average annual precipitation	1000 mm
Specific humidity	10 g kg^{-1} low troposphere 1 g kg^{-1} mid troposphere
Sea-level pressure gradient associated with gale force wind in mid latitudes	5 mbar per 100 km
Equivalent tilt of isobaric surface	40 m per 100 km

WEATHER SYSTEMS

Type	Breadth	Lifetime	Surface wind	Updraught	Rainfall
cumulonimbus	10 km	1 hr	15 m s^{-1}	5 m s^{-1}	10 mm hr^{-1}
extratropical cyclone	1000 km	1 week	15 m s^{-1}	5 cm s^{-1}	1 mm hr^{-1} (background)
hurricane	500 km	1 week	40 m s^{-1}	10 m s^{-1}	40 mm hr^{-1} (active annulus around the eye)

Introduction

1.1 The film of gas

From our earliest childhood we are aware of the skies above us, but daily familiarity blunts this awareness so much in most of us that it requires an especially striking sunset, or a particularly massive and ominous cloud overhead, to rekindle interest. Even then, the bias naturally introduced by the relatively tiny size of the human frame, and its usually surface-based viewpoint, can produce very misleading impressions of the scale of atmosphere at the bottom of which we spend almost all of our lives. In fact the Earth's atmosphere has two very different scales.

Horizontally the atmosphere is enormously larger than we can perceive from a point on the Earth's surface, just as the earth itself is larger. The *order of magnitude* (a term used throughout this book to describe magnitudes to the nearest power of ten) of the maximum horizontal scale of the atmosphere is of course that of the Earth itself, which is tens of thousands of kilometres, written $\sim 10\,000$ km. Photographs such as Fig. 1.1 allow us to appreciate this relatively vast scale much more directly than was possible before the advent of the meteorological satellite in the 1960s.

By contrast the vertical scale of the atmosphere is very much smaller than the radius of the Earth. In Fig. 1.2 you can barely see the thin limb of air illuminated on the horizon by the hidden sun. In fact on the scale of the Earth, the atmosphere is the merest film of gas clinging to the surface by gravitational attraction. As we shall see, distances which a person could walk in a few hours in the horizontal are a significant fraction of the vertical depth of the atmosphere. Our impression of the great height of a towering thunder cloud for example is greatly enhanced by our awareness of the difficulty we would encounter in climbing the cloud if it were a solid mountain. But although the atmosphere is really quite modest in vertical scale, its influence on the conditions in which we live is enormous, as will be detailed at many points throughout this book.

Unlike the ocean, the atmosphere has no definite upper surface. For example its density falls continuously with increasing height, from values close to 1 kg m^{-3} at the surface, to values more than ten thousand million (10^{10}) times smaller in the tenuous interplanetary gas found more than a few hundred kilometres above the surface. Despite this apparent lack of definition, the vertical extent of the atmosphere can be usefully quantified by a so-called *scale height*, which is the

Fig. 1.1 The face of the Earth centred on the equator at 0° longitude as seen in visible light by the European Space Agency geostationary satellite Meteostat 2. The picture was taken at 1130 Z on 25 September 1983 and shows large cloud patches in low latitudes associated with tropical weather systems in and around the Intertropical Convergence Zone (see section 4.7). Cloudy vortices and streaks at higher latitudes are associated with extratropical cyclones and their fronts and jet streams, and are especially extensive in the southern hemisphere on this occasion.

height interval in which a particular property, such as air density, changes by a particular numerical factor. Consider for example the case of atmospheric pressure. It is observed that throughout the lowest 100 km of the Earth's atmosphere the *decadal scale height* for pressure (the interval separating levels with pressures differing by a factor of 10) remains close to 16 km. Thus pressure at mean sea level (MSL) and at 16, 32 and 48 km above MSL, are about 1000, 100, 10 and 1 mbar respectively. Since the pressure is proportional to the weight of the overlying atmosphere, it follows that about 90% of the weight of the atmosphere is concentrated in the lowest 16 km — a layer whose depth is only about 0.2% of the Earth's radius. This concentration is produced in the easily compressible atmosphere because the air is squeezed against the Earth's surface by its own weight. Partly because of this concentration, but more importantly because the weather-giving cloudy activity of the atmosphere is almost entirely confined within the same layer, this shallow layer constitutes the bulk of the lower atmosphere discussed in this book.

1.2 The stratified atmosphere

Since much of the material of the atmosphere is squeezed into such a shallow layer overlying the surface, the distributions of temperature, humidity and indeed of almost every observable property are strongly *anisotropic*, in the sense that their gross vertical and horizontal distributions are very different — vertical structure being usually much more closely packed and vertical gradients correspondingly larger. For example, away from the complicating close proximity of the surface, temperature decreases with height at a typical rate of about 6°C km^{-1} up to heights of between 10 and 15 km above MSL, whereas the strongest extensive horizontal temperature gradients (those associated with vigorous fronts in middle latitudes) rarely exceed 0.05°C km^{-1}. The result is that *isopleths* (lines or surfaces of equal quantity, such as isobars and isotherms) are usually nearly horizontal, and vertical sections such as Fig. 4.6 usually have a horizontally layered or stratified appearance.

Corresponding to this marked stratification of the structure of the atmosphere, there is an equally marked stratification of its behaviour. Immediately overlying the Earth's surface is a region dominated by turbulent interaction with the surface — the *planetary boundary layer*. Its depth varies greatly with time and location, and even with the particular atmospheric property under discussion, but is often ~ 500 m. By contrast, the rest of the atmosphere (about 95% of the total mass) is much less directly influenced by the surface and is accordingly termed the *free atmosphere*. The planetary boundary layer together with the first 10–15 km of the free atmosphere (the larger value applying to lower latitudes) comprises the *troposphere*, so-called after the Greek word for turning because it is the site of the commotion and overturning associated with so much significant surface weather. The active and cloudy troposphere is separated from the relatively quiet and cloud-free *stratosphere* above by the nearly horizontal and often very sharply defined *tropopause*. This book deals mainly with the troposphere.

The anisotropy of the atmosphere is well marked only on scales which are at least not insignificant in comparison with the decadal scale height of the atmosphere. On much smaller scales, for example ~ 1 m, the atmosphere does not seem to

Fig. 1.2 The Earth's limb, photographed manually by Gemini astronauts in visible light from the setting sun. The thin atmospheric layer on the horizon is visible because of light scattered from air molecules, dust and some cloud. Molecular scattering produced a blue cast in the original colour photograph, while the others gave no colour bias (section 8.6). The clouds on the left are anvils spreading from the tops of shower clouds over the Andes.

'know' that it is so flattened by gravity and compressibility, and disturbances are much more isotropic. Such small scales are particularly important in the planetary boundary layer, which is heavily influenced by a hierarchy of turbulent eddies ranging in spatial scale from ~ 500 m to ~ 5 mm. Though the free atmosphere too is turbulent, from many points of view it is not essentially so. In fact it is dominated by much larger-scale disturbances, many of which share the flattened configuration of the lower atmosphere to such an extent that vertical sections have to be drawn with grossly expanded vertical scales. Though this practice is essential for clarity, it is useful to recall when examining diagrams such as Fig. 4.6 that the lower atmosphere and many of its common larger-scale disturbances are in fact two orders of magnitude 'flatter' than conventionally depicted.

1.3 The disturbed atmosphere

The atmosphere, and particularly the troposphere, is the site of incessant commotion. *Cumulus* clouds as big as the mountains they resemble (the name means hill in Latin) erupt and dissolve in periods of tens of minutes as the updraughts within them wax and wane. In middle latitudes vortices of continental scale (Fig. 1.1) grow and decay in periods of a few days. During their lifetime they are streaked with great bands of cloud which produce rain, snow and locally hail, and are associated with large areas of depressed atmospheric pressure at the surface, from which they take their popular title. The technical name for such depressions is *extratropical cyclones* — the prefix extratropical serving to distinguish them from the *tropical cyclones* which are a family of disturbances confined to low latitudes. The most intense of the latter are the *hurricanes* (Fig. 1.3) which seasonally haunt the Caribbean and other similar regions (where they are known locally by other names), and which constitute the most energetic type of tropospheric disturbance. Each of these weather systems contains and interacts with activity on relatively smaller scales: for example all types of cyclone contain cumulus, and cumulus contain large turbulent elements which give the characteristically knobbly appearance to their tops and sides (Fig. 1.4), and which can make flying in them uncomfortably or even dangerously bumpy.

Superficially it is a scene of huge chaos, but the existence of characteristic types suggests an underlying order. There is method in the apparent madness which it is the object of atmospheric science to identify and describe in the most simple and concise manner. According to its etymology, *meteorology* (which comes from the Greek for lofty) should be the study of everything above the Earth's surface, but in practice it has come to mean the study of the physical rather than the chemical nature of the lower atmosphere, especially the troposphere. And its meaning has been narrowed further to focus on the dynamic, changeable aspects of this nature, leaving the term *climatology* to cover the typical and average aspects. In fact most significant physical structures and behaviour of the lower atmosphere have their meterological and climatological aspects, which often overlap. And the confusion is compounded by the relatively recent explosion of interest in *climate change*, which of course focuses on the changeable nature of climate.

Curiosity is the mainspring of meteorology, as it is of every other branch of science, but so much atmospheric activity affects man so vitally, occasionally even disastrously, that for over a century meteorology has been financed with the aim of

Fig. 1.3 Low latitudes centred on longitude 140°E, seen in visible light from the Japanese Meteorological Agency's geostationary satellite GMS-3 at 0300 Z 16 October 1985. Among the substantial areas of cloud there are two which have the dense swirl appearance typical of severe tropical storms: Typhoon Dot lies east of the Philippines and Typhoon Cecil lies over Vietnam.

Fig. 1.4 Several swelling cumulus showing flat bases, knobbly sides and tops, and the mountainous bulk typical of rapidly growing convective clouds. Notice the 'silver lining' produced by strong forward scatter of sunlight, and the cloud shadow cast on haze nearer the camera.

routinely forecasting atmospheric behaviour, especially that affecting surface conditions, for as little as six hours ahead of time. In this period forecasting has developed into a complex and highly specialized technique practised by nationally organized networks of observers and analysts cooperating internationally. In the last few years forecasting of climate change has become very important too, attracting international effort. Though forecasting as such is considered only incidentally in this book, its bias towards those types of tropospheric commotion which strongly affect human activity naturally pervades meteorology and is reflected here. For such reasons in this book we will largely ignore the many beautiful but complex optical phenomena such as the rainbow, and will be highly selective in treatment of the otherwise very important chemical behaviour of the atmosphere, though each is a large subject in its own right. Atmospheric chemistry in particular has attracted increasing attention with the recent realization that man's industry is capable of altering natural balances substantially for the worse.

Much of the weather-producing activity of the atmosphere can be arranged in a spectrum according to scale, as in Fig. 1.5. There is a range of over eight orders of magnitude in space scales and a range of time scales which is only slightly smaller. Space and time scales are related so that big events happen relatively slowly while small events happen quickly. In fact the ratio of space to time scales is ~ 1 m s^{-1} for all the items tabulated. This value is a measure of the intensity of the activity of the atmosphere, and can be compared with the much larger value for the much more violently disturbed visible surface of the Sun, where *granulations* (a type of solar cumulus) with diameters ~ 100 km appear and disappear within minutes. The list in Fig. 1.5 is necessarily incomplete; known types of lesser importance are omitted for simplicity, and no doubt others are missing because they have yet to be identified among the welter of observations.

1.4 Physical laws

We believe that the observed structure and behaviour of the atmosphere can be related to the operation of physical and chemical laws, and the forecasting bias of

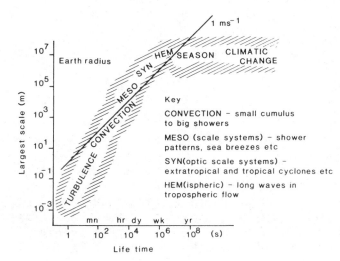

Fig. 1.5 Types of weather-related atmospheric disturbance displayed according to lifetime and maximum space scale. Each type is discussed at some length later in the text. Note that logarithmic scales (equal steps along an axis corresponding to equal multiples of value) cover wide ranges of value very compactly, but require careful interpolation.

meteorology directs attention to the former in particular. Fortunately there is no reason to suspect the operation of any fundamental physical law not already well established by the development of physical science during the last three centuries. In fact most of the laws which play important roles in the lower atmosphere appear at least in outline in school science courses: the laws of statics and dynamics (complicated in expression but not in essence by the fluid nature of the atmosphere), matter, heat and radiation. However, it is emphasized repeatedly in this book that even when considering some quite specific feature of the atmosphere, no single law suffices; instead there is an interaction of many, though fortunately it is often possible to pick out a few laws which between them dominate most of what is observed in any particular instance.

Consider for example the forces acting on a small volume of air, curiously but conventionally entitled an *air parcel*. In the lower atmosphere there are three such forces, arising respectively from pressure gradients, gravity and friction. At a given time and place there is a resultant F of these forces which is related to the acceleration a of an air parcel of mass M by a form of Newton's second law of motion.

$$\frac{F}{M} = a$$

(1.1)

Note that F and a are written in bold type to show that they have direction as well as magnitude, i.e. that they are *vector* rather than *scalar* quantities like mass, which have magnitude only. In different zones of the atmosphere, atmospheric motion (represented here by the acceleration a) differs considerably in type. For example, in the planetary boundary layer there are quite large frictional forces associated with the relatively intense turbulence there, especially near the surface, whereas in the free atmosphere such friction is relatively unimportant in comparison with other forces and can often be neglected without serious error. Indeed much of the progress made in this century by detailed application of relations such as eqn (1.1) to the atmosphere has depended on judicious separation of primary and secondary effects, since only then are the relations mathematically simple enough to be solved. The skeleton theory outlined in this book contains a number of simple examples of such separation, but in more advanced work the process is often elaborate and subtle. Note in passing that the eqn (1.1) (often called the *equation of motion*) is insufficient on its own to describe any physical situation in the atmosphere at all comprehensively; formal expressions of other laws such as the conservation of energy and of mass must be combined with eqn (1.1) before a definite solution is possible even in principle.

1.5 Cause and effect

It is difficult to avoid considering F and a in eqn (1.1) as cause and effect respectively, probably because in homes, factories and laboratories we arrange forces to achieve desired effects; but it is important to realize that the natural laws which man has identified and labelled indicate neither cause nor effect but simply relationship. For example, even if we were to say that a certain pressure gradient force caused air to accelerate, the terminology would be misleading since the resulting redistribution of air could be said with as much justification to cause a change in the pressure gradient force. This is a quite general point, but it is particularly relevant to the largely self-activating lower atmosphere, so let us consider it a little further in that context.

The temptation to identify cause and effect is strong because it satisfies our innate desire for order, but it is probably enhanced in meteorology by the nature of much weather-giving activity. Above the planetary boundary layer the dominant types of disturbance are in the cumulus and larger scales (Fig. 1.5), and are such that individual examples rest uneasily between statistical uniformity and individual uniqueness. In their combination of individuality, regularity and transience, cumulus, thunderstorms, depressions, hurricanes etc. resemble living things, and you will detect in forecasts and books (including this one) a tendency to treat them as such. This does not matter so long as it is recognized merely as a device to simplify description, but unfortunately it reinforces the tendency to identify imaginary causes and effects. The spurious animism implied by a statement such as 'the weakening of the anticyclone over the British Isles is allowing a belt of rain to move into north-west Ireland' is descriptively very convenient, but it is quite misleading if it is taken to imply any will, purpose or even priority in the action. In fact the weakening no more allows the advance than the latter causes the weakening; the two events are related. Such relationships are very much the concern of meteorology, but long experience has shown that it is positively unhelpful to search for 'chicken and egg' amidst the interminable flux of atmospheric events. However, there is a resounding exception to this rule in the conspicuous action of the Sun as prime mover of all atmospheric activity, as we now briefly consider.

1.6 The Sun and the atmospheric engine

It is obvious on any sunny day that the Sun has a direct and important influence on the atmosphere. Solar energy, much of it in the form of visible sunlight, pours continually onto the Earth, affecting both the surface and the overlying air. In fact the role of the Sun in determining the condition and behaviour of the Earth's atmosphere is even greater than our experience of its warmth, or observation of the growth of cumulus clouds on sunny mornings, might lead us to suppose. The atmosphere is quite literally solar-powered, and the Sun can be considered as the prime mover of all atmospheric activity, and as a cause of atmospheric effects in the restricted sense that it influences our atmosphere but is not itself significantly influenced in return.

Sir Napier Shaw, a distinguished pioneer of modern meteorology, wrote in the early years of the twentieth century that 'the weather is a series of incidents in the working of a vast natural engine'. As in man-made heat engines, heat is taken in at a *heat source*, and exhausted at a *heat sink* (Fig. 1.6) and *mechanical energy* is generated, i.e. massive bodies are made to move. The source is provided by the absorption of solar energy, mainly at the Earth's surface, and the sink by the emission of infra-red *terrestrial radiation* to space, mainly from the upper troposphere. The motion and unstable distribution of the atmospheric mass is the mechanical energy produced. Unlike man-made engines however, the atmospheric engine is not constrained by a rigidly pre-ordained structure of turbines, gear trains etc.: it generates and constantly maintains its own structure, which includes the hierarchy of disturbances associated with surface weather. The atmospheric engine

9

also controls its intake of solar energy by producing cloud masses which reflect a very significant proportion of sunlight to space, and rather less obviously regulates the output of terrestrial radiation to space. Between them these two effects act like a governor which keeps the activity of the engine close to its normal level. Obviously the behaviour of an engine which continually regenerates and regulates itself is much more complex and subtle than that of man-made engines, but there is no difference in underlying principle. The description of the atmosphere as a heat engine is detailed at many points throughout this book, and is discussed at length in the final chapter, but for the moment it is enough to accept that the sun is its furnace.

Fig. 1.6 Schematic diagram of the atmospheric heat engine with heat source, sink and mechanical energy flow. The mechanical energy is subsequently transformed as discussed in Chapter 13.

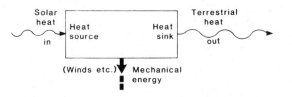

1.7 Determinism and chaos [1, 2]

Although equations like (1.1) do not identify cause and effect, they do show that present conditions are inescapably linked to all past and future conditions. Of course from the outset, people have been more interested in predicting future conditions; and it was quickly assumed that present conditions, together with the laws of behaviour, somehow 'determined' all future conditions (neatly side-stepping the question of what determines the laws). This assumption was stated most confidently by the French natural philosopher and mathematician Laplace when he was writing his *Celestial Mechanics* in the late eighteenth and early nineteenth centuries. Having recast Newton's laws into an extremely compact form for application to the motion of the planets, he claimed that perfect knowledge of the present state of everything in the universe, together with perfect understanding of all its laws, would give total knowledge of the future. Such Laplacian determinism, as it came to be known, underpinned the confidence of the scientific community as it grew and developed during the nineteenth century, investigating subjects (like thermodynamics and fluid mechanics) vastly more complex than Laplace's billiard-ball model of the the solar system. And meteorologists too came to share this confidence in the twentieth century, as they began to apply established physical laws to understanding and forecasting atmospheric behaviour.

The first obvious breach of Laplacian determinism came with increasing understanding of atomic-scale behaviour in the 1920s, when quantum theory postulated an observational uncertainty (named after Heisenberg) which is inescapable in practice and principle. Indeed it is now accepted that individual events on the atomic scale are blurred by an uncertainty which can only be reduced by considering the statistics of large numbers of such events. Fortunately, atoms are so small and numerous that this is exactly what happens in the study of events on scales much larger than the atomic. Nevertheless many scientists were unsettled by the apparent failure of a principle which had come to be seen as a watershed separating modern science from primitive capricious superstition, and to the end of

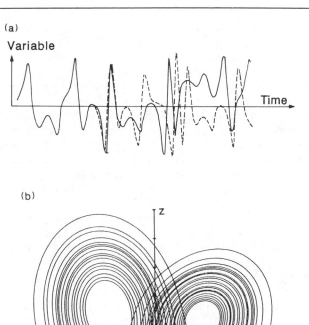

(a)

Variable

Time

(b)

z

y

x

Fig. 1.7 Portraits of chaos. (a) The butterfly effect: two runs of the same Lorenz atmospheric model, starting from minutely different initial conditions. Resemblance between behaviour of a chosen variable (displacement along the vertical axis) virtually disappears about two thirds the way through the time period (the horizontal) of the otherwise identical simulations.

(b) Loose repetition with unexpected switches: the line representing the continuous sequence of states of the Lorenz convection model turns in upon itself as the convection cell rolls about a horizontal axis. However the rolling changes slightly and progressively until it switches to slowly varying rotation in the opposite direction. (After E.N. Lorenz in [1] and [2]).

his life Albert Einstein resisted the idea of a purely statistical basis to physical law in his famous dictum 'God does not play dice'.

Despite this revolution on the atomic scale, larger-scale behaviour was still assumed to be forecastable in the Laplacian sense, until the spread of the electronic computer in the 1960s allowed people to begin to investigate numerically the *full* range of behaviour described by equations representing established natural laws, rather than the very limited range covered by analytical (e.g. algebraic) solutions. Quite unexpected results emerged separately from a wide variety of scientific studies: physical, chemical and biological. When brought together in the 1970s, recognition of common ground triggered one of those profound reviews of accepted wisdom which seem to agitate the scientific community every 50 years or so. Interestingly, what is now recognized as one of the most revealing of those early separate studies was made by the meteorologist Edward Lorenz in the early 1960s [2].

Lorenz was using a primitive computer to model global climate by allowing a numerical weather-forecasting model to run for long periods, simulating years of behaviour from assumed initial conditions. He found by accident that the model was so very sensitive that minuscule differences in some initial condition of the model atmosphere could give rise to completely different global conditions some

11

finite simulated time later. Lorenz guessed that this was a basic property of the real atmosphere, rather than an unrealistic property of his simple model. To express the implied limits to the scope of useful weather forecasting, meteorologists coined the graphic image of 'the butterfly effect', in which the flapping (or not) of a butterfly's wings at some place and time could lead to a completely different disposition of weather systems over the whole Earth some months later.

A little later Lorenz found similarly extreme sensitivity in an even more crude model of atmospheric convection; and subsequent work by others has found it in even simpler systems, such as the simple pendulum and the annual growth of animal populations. In every case the sensitivity arises from non-linear relations between parts of a system (i.e. sensitivity greater than simple direct proportionality), and the result is a type of irregular, effectively unpredictable behaviour which has been named *chaos*. It now seems that virtually every natural system has a capacity for such chaotic behaviour, and that the atmosphere in particular is probably chaotic on most of the scales depicted in Fig. 1.5.

Chaos is not random, despite superficial appearances. The present state of any chaotic system is still related to ('determines') all future states through the operation of natural laws, but the future behaviour of a chaotic system is predictable in only a limited sense, even in principle. Sensitivity to initial conditions means that uncertainties there (arising from inaccurate or inadequate observation, or even ultimately from atomic-scale uncertainty) grow more and more rapidly as the system evolves in time, and the same applies to any model simulation. As a result, the doubling of accuracy in the initial conditions of even the most perfect weather-forecasting model (for example) produces much less than a doubling in its useful forecasting period. Although weather-forecasting models are still imperfect in physical principle and computational practice, there is a growing consensus that it may never be possible to forecast weather with current confidence more than a couple of weeks in advance (current practice manages one week), confirming Lorenz's doubts of 20 years ago. Of course less specific seasonal forecasts may become possible, as understanding of climatic systems improves (section 4.9), but they too will be limited if those systems are chaotic, as seems likely.

Another property of chaotic systems is that they never repeat themselves precisely, although they often behave in a loosely periodic fashion. Quite simply, if they did repeat exactly they would not be chaotic, since the 'deterministic' nature of natural law would ensure that they repeated the cycle for ever, or until disturbed from the outside. This capacity for loose rather than precise repetition, with broadly similar things repeating, but always slightly differently, is of course just what we see in the atmosphere: depressions form every few days in preferred parts of the middle latitudes, but the time intervals vary considerably, and the depressions are never precisely the same; seasonal variations are broadly but never precisely the same; and no two cumulus, turbulent eddies or snowflakes are ever identical. Indeed the combination of individuality, regularity and transience mentioned in section 1.5 is now seen to represent the fingerprints of chaos.

More subtly, it is known from the study of simple chaotic systems that details of typical behaviour are closely related to the precise nature of the non-linear relations at work in them. Although the chaotic nature of such a large and complex system as the atmosphere is only beginning to be examined in detail, it is likely that typical atmospheric structure (for example the sizes and lifetimes of depressions) may be explicable to this extent some day. The doddering Polonius saw dimly through Hamlet's simulated insanity in one of Shakespeare's most seminal quotations: 'Though this be madness, yet there is method in 't'. Until the recognition of chaos as a type of behaviour in its own right, meteorology and climatology (in common with many other studies of simple and complex systems) concentrated on the method and shunned the madness: now it seems that the madness itself is subtly structured, and likely to reward careful study with increased understanding of our marvellous and yet bewildering atmospheric environment.

Problems

LEVEL 1

1. Find the order of magnitude of the depth/breadth ratio for the troposphere.
2. What is the atmospheric pressure two decadal scale heights above sea level?
3. Place the following in order of ascending height: troposphere, planetary boundary layer, stratosphere, free atmosphere, stratopause, tropopause. Deal with overlap using brackets.
4. What is the ratio (in m s^{-1}) of the space and time scales of solar granulations?
5. Divide the following atmospheric quantities into vectors and scalars: wind velocity, temperature, gravitational force, wind speed, pressure. (Note that experiment shows that the pressure at any point in a fluid is independent of direction.)
6. Identify the effective heat source for the atmospheric engine: the upper troposphere, tops of clouds, the Earth's surface, the stratosphere.

LEVEL 2

7. Figure 12.11 represents a vertical section through a tropical cyclone. If the vertical scale of the diagram were redrawn to be equal to the horizontal, what would be the length of the new vertical axis?
8. On a certain weather map point A has a temperature of 23 °C, and point B has a temperature of 15 °C. If isotherms are drawn at intervals of 2 °C for even values, at least how many isotherms must there be between A and B?
9. Horizontal accelerations of air in large weather systems are observed to be $\sim 10^{-4}$ m s^{-2}; what is the implied net force on an air parcel of mass 1 tonne?
10. Estimate the volume of the air parcel in problem 9 near sea level.

LEVEL 3

11. Sketch an example of a temperature field for problem 8 with more than the minimum number of isotherms.
12. Actually the spuriously animistic statement in section 1.5 about the anticyclone and the rain belt can be rewritten just as simply without implying animism, cause or effect. Try it.
13. A Hollywood film director is threatening to make a disaster movie in which

a hurricane develops from nothing to maturity in 12 hours. Sketch an argument demonstrating the unreality of the event. And what would you say to another proposing to have a rogue tornado scour the countryside for several days?

14. What ultimately happens to the mechanical energy generated by the atmospheric engine?

Observations

2.1 Care in observation

At an early stage in his development, man must have learnt to scan the skies and to consider his sensations of warmth, wind strength etc. for evidence of impending change in the weather, and in this sense observational meteorology may be over one million years old. The need for such awareness increased greatly with the onset of the agricultural revolution about ten thousand years ago, since farmers are understandably very sensitive to weather and climate at certain phases of the crop cycle. The rapid development of long-range ocean voyaging in the last millenium encouraged an even more critical sensitivity among mariners, and it is hardly surprising that modern meteorology originated largely in their need for reliable warning of heavy weather, and in its early stages was often assisted by their observations and perceptions. The development of aviation in the present century has produced a demand for foreknowledge of winds, cloud and hazardous conditions along flightpaths and at departure and destination airports.

OBSERVATIONAL AIDS

If sufficiently motivated, people can make themselves aware of a wide range of meteorological conditions, but the often weak and ambiguous reaction of many of the human senses severely limits the quality and quantity of our observations, even when directed by an anxious and enquiring mind. It is true that our three-dimensional hunter's vision is very acute, but this asset is largely wasted in the absence of other similarly accurate observations, and in the absence of even minimal understanding of what is observed. To resolve the apparent confusion of atmospheric behaviour, data must be as unambiguous and definite as possible, and therefore mechanical aids are needed to refine and extend our basic human sensations. Most of the more detailed observational information mentioned in this book was gleaned only after basic observational instruments had been developed. You can gain some idea of the crucial role played by such artificial aids by examining the inspired but erratic speculations about the atmosphere made by the

15

Roman poet and natural philosopher Lucretius about 55 BC [3], who relied completely on his unaided senses.

Given that mechanical aids to observation are needed, several general points arise. First, although it may be clear that temperature, for example, should be measured by thermometer rather than by bodily sensation of warmth, it may not be clear initially exactly how the thermometer should be exposed. Is atmospheric temperature more truly represented by a thermometer exposed to available sunlight, whose warmth is such a significant component of what we perceive to be the weather, or by a shaded thermometer? No doubt you know that the standard practice now is to use a carefully shaded thermometer, but this was not generally accepted for nearly a century after thermometers were first used meteorologically. Nor was the delay due to stupidity; the best argument for using shaded thermometers is that they give the most coherent patterns of temperature across time and geographical space, but this emerged only after extensive trial and (as we now know) error. There are theoretical justifications too, but they were unlikely to be understood until correct observations began to build a body of meteorological understanding in the nineteenth century.

The problem of how best to use an observing instrument is particularly acute when it does not correspond to any existing human sense. The measurement of atmospheric pressure by barometer is an early example which is discussed further in section 2.5, but the present hectic pace of technical innovation produces new examples frequently, the meteorological satellite being a conspicuous case. Novel data are at once challenging and baffling, and there is normally a substantial time period in which our understanding is struggling to assimilate the new data, and in which data can easily be misused. The satellite may still be in this phase.

Last, it is important to bear in mind the simple fact that all measurements are subject to limitation and error. The glossy trappings of modern instrumentation may tend to inspire blind confidence, especially if the data emerge from a computer, but this should be firmly resisted. Every meteorological observation is less than perfect by an amount which should be known when it is being interpreted. Imperfections in individual observations are compounded by the problems of representativity and consistency which arise when data are assembled from widely scattered observation points, as is normal in meteorology.

The following sections exemplify and enlarge on these points, beginning with a consideration of a few basic and familiar instruments used for making observations near the surface. The outline is highly selective, and much more comprehensive descriptions are found in manuals dedicated to observing instruments and techniques [4, 5]. The previous remarks about the time taken to assimilate new data will apply more or less forcefully to you depending on your previous knowledge of meteorology. It may not be at all clear to you why certain measurements are made the way they are, or are made at all. Clarification should increase with knowledge of the atmosphere, but this cannot be compressed into a single chapter, or indeed a single book.

2.2 Temperature measurement

Air temperature is one of the most important of atmospheric properties, but though our bodies are sensitive to it, they are quite unreliably so. We feel hot or

Fig. 2.1 Stevenson screen with door open for thermometer reading, showing the normal dry- and wet-bulb thermometers inside (vertical left and right respectively). The horizontal thermometers are maximum and minimum dry-bulb thermometers.

cold depending on whether our bodies are having to waste or conserve heat to maintain their internal temperatures at about 37 °C, and which of these is happening at any moment depends on exposure to sunlight, terrestrial radiation, wind, humidity (which affects our ability to lose heat by sweating), health, and the time, size and temperature of our latest meal, as well as on the temperature of the ambient air [6]. Clearly a less ambiguous sensor is required.

The mercury-in-glass thermometer was perfected by Fahrenheit in the early eighteenth century, but though it was used almost immediately to measure air temperature, over a century elapsed before the need for careful exposure was generally recognized. Such care is necessary because, like the human body, the thermometer is sensitive to solar and terrestrial radiation, from which it must be shielded to isolate the effect of air temperature. However, the shield must allow sufficient ventilation for the temperature of the air in contact with the thermometer to be at the same temperature as the air outside the shield. The familiar *Stevenson screen* (Fig. 2.1) is a simple, robust and reasonably satisfactory compromise solution. The white surfaces absorb little sunlight, and the thick wooden walls insulate the interior from the warming effect of residual solar absorption, and from the warming and cooling effects of terrestrial radiation. The louvred walls and floor allow natural ventilation of the interior by available wind. The access door is placed on the poleward side of the screen so that direct sunlight does not enter when readings are taken. The ventilation is the least satisfactory part of the compromise: on still, sunny days it is so poor that the temperature inside the screen may rise a degree or more above the external air temperature. The ventilation can be improved by using a fan to force a constant draught through the interior of the screen, but the instrument then requires a source of power, and is consequently more elaborate and prone to breakdown.

The mercury-in-glass thermometer is still in widespread use, though it has two major disadvantages: it cannot be used at very low temperatures (such as occur at high latitudes and altitudes), since mercury freezes at about −40 °C, and it cannot give a continuous record of temperature automatically. Alcohol-in-glass thermometers can register much lower temperatures, and are widely used in *minimum thermometers*, in which the alcohol meniscus moves a metal index inside the bore

17

of the thermometer, leaving it behind at the minimum value for subsequent reading. Simple continuously recording thermometers (*thermographs*) contain a bimetallic strip which moves a pen over a clockwork-powered chart drum; and in recent years a wide range of thermometers has been developed making use of the temperature sensitivity of electrical resistance, often to produce continuous records.

INSTRUMENTAL LAG

No measurement can be made instantaneously: in principle each is an average over a finite time interval, however short. The length of this interval is most simply assessed by observing the reaction of the measuring instrument to an abrupt change in the variable being measured (e.g. temperature). The *lag* of the instrument is conventionally defined to be the time taken to register 63% of the change, this value being appropriate to an exponential type of response (Appendix 2.1), such as is observed in many instruments. The lag is clearly a measure of the sluggishness of instrumental response, and of the degree of averaging therefore inherent in all its data. As a rule of thumb, an instrument cannot reliably measure variations occurring with time periods less than ten times the lag.

In thermometers the chief cause of lag is the thermal capacity of the actual sensor, and the speed of heat transfer between it and the air. The heat capacity of the mercury reservoir and glass bulb, and the slow conduction of heat through the glass, are limiting factors in the mercury-in-glass thermometer, and large specimens have in consequence lags of about a minute in a ventilation draught of 5 m s^{-1}. They are even more sluggish in slower draughts because heat flow between bulb and air is reduced. Such lags are quite adequate for routine hourly measurements near the Earth's surface, but they are much too long to allow measurement of the rapid temperature fluctuations which accompany turbulence in the air, each of which may last for only a fraction of a second. Thermometers for such purposes must have very small thermal capacities, and are therefore physically very small, having fine wires or beads as sensors.

2.3 Humidity measurement

The amount of water vapour in the air is measured by *hygrometers*, and is expressed in several different but related ways.

When evaporation or condensation is the main interest, the most relevant measure is the *relative humidity*, which is the ratio of the actual vapour density (or pressure) to the value which would produce *saturation* at the same temperature, usually expressed as a percentage. (The concept of saturation is examined more closely in section 6.2.) Since the saturation vapour density (or pressure) depends only on temperature — a dependence well-known from laboratory measurement (Fig. 6.2) — simultaneous measurement of relative humidity and air temperature determines the actual vapour density. Animal tissues, such as hair or skin, respond directly to relative humidity, and this response is the basis for several simple hygrometers. For example, in the common *hair hygrometer*, a hair is kept under slight tension so that the decrease in its length with increasing relative humidity is easily registered. The length variation is sufficiently regular and repeatable to allow any particular hair or bunch of hairs to be calibrated in different known relative humidities (determined by other methods) to produce a quantitative instrument. Hair and skin hygrometers have lags ~ 10 seconds at room temperature, but are much more sluggish at lower temperature. Unfortunately they suffer so severely from hysteresis and ageing that they are best used to register changes rather than absolute values. However, their simplicity and ability to produce a continuous record (in a *hygrograph*) make them very popular for semi-quantitative purposes. Some hygrometers measure relative humidity through its effect on the electrical resistance of a hygroscopic surface, and can be made somewhat more reliable than the hair and skin types.

Psychrometers measure the humidity of the air from the cooling effect of water evaporating into it, and their simplicity and reliability make them the most widely used instruments not requiring calibration. The simplest types consist of a *wet-bulb* thermometer mounted beside a normal thermometer (i.e. one with a dry bulb) inside a Stevenson screen (Fig. 2.1). Natural ventilation allows water to evaporate from the saturated wick of the wet bulb, cooling it below the temperature of the dry bulb by an amount which varies with the humidity of the ventilating air. In dry air, such as is found in desert regions, this *wet-bulb depression* is very large (~ 10 °C), whereas it is zero in saturated air, for example in fog. Given the wet and dry bulb temperatures, a special book of hygrometric tables [7] or a special slide rule gives the associated vapour pressure (or density, though pressure is conventionally preferred), relative humidity and dew-point temperature (see below). When the Stevenson screen is inadequately ventilated, the actual humidity of the air outside the screen is overestimated somewhat, since the air inside the screen is significantly moistened by evaporation from the wet bulb. Mechanical ventilation of both bulbs gives more accuracy in this respect, and is provided in the *aspirated* type of psychrometer. The wet bulb has much the same lag as the corresponding dry bulb, and of course requires similar shielding to minimize radiation error (if it is not already in a screen). Electrical psychrometers with tiny wet and dry bulbs can detect the rapid humidity fluctuations associated with atmospheric turbulence.

A further measure of the humidity of the air is the *dew-point* temperature, or simply dew point, which is defined to be the temperature of a chilled surface just cold enough to collect dew from the adjacent moist air. In a *dew-point meter* a polished metal surface is chilled progressively below the temperature of the ambient air, and the surface temperature is noted at the moment when the first dew appears. In this form it is an elaborate, sluggish but potentially very accurate instrument which requires no calibration. Dew points can be converted to other humidity measures by using the hygrometric tables mentioned previously. Very roughly, the *dew-point depression* below the ambient air temperature is twice the wet-bulb depression, ranging from tens of degrees in arid air to zero in saturated air.

The theoretical relationships between vapour pressure, density, relative humidity, wet-bulb temperature and dew-point temperature, and other measures of humidity, are not straightforward. They are outlined in sections 5.4 and 5.5 and discussed fully in advanced texts [8].

2.4 Wind measurement

Horizontal wind direction is indicated by *vanes* which are in principle the same as those exposed on church steeples and old buildings. An electromechanical device can be used to give a remote and recordable readout of the vane direction. The standard meteorological vane has a lag of several seconds in normal winds but is more sluggish and ultimately unresponsive in very light winds, which are anyway very variable in direction. Small, light and aerodynamically optimized vanes are needed to register the rapidly varying directions associated with turbulence.

The meteorological convention is to describe the direction of a horizontal wind by the direction from which it is blowing, as in everyday speech. Thus a north-westerly wind blows from the north-west, or equivalently from an angle of 315° (the *azimuth* measured clockwise from zero north). This opposes the convention used in other branches of physical science, where the direction of any motion is considered to be the direction towards which the body of air is moving, but arises naturally from the age-old practice of sailors and farmers of looking upwind to see impending weather, which in turn reflects the fact that many weather systems travel more or less with the wind near the surface (but not always; see section 10.4 for example). Because of its rapid variations, wind direction is seldom specified to better than the nearest 5°.

Wind speed is measured by various types of *anemometer*, but because these are surprisingly difficult to make usefully accurate and reliable, the earliest successful procedure used instead a classification of wind strength in terms of observable effects of wind on the sea and sailing ships. This famous *Beaufort scale* was soon extended to include the effects of wind over land and is still used very widely to describe the strength of wind near the surface. Subsequent measurement by anemometer has associated a range of wind speeds with each *Beaufort number* (appendix 2.2). Another maritime relic is the continued use of the *knot* (1 nautical mile per hour) as a unit of wind speed in parallel with the proper SI unit, the metre per second. The precise equivalence is given in the secondary table on p. xviii, but for rough purposes the metre per second can be approximated to 2 knots (1 m s^{-1} $\cong$ 2 kt).

The common *cup* anemometer (Fig. 2.2) is insensitive to the horizontal wind direction or azimuth, unlike the *propellor* type which has to be kept pointed into the wind by a steering vane. The cups of the former type whirl continually because the drag of the wind is greater when blowing into the mouth rather than into the back of each cup, but this results in greater lags in dropping rather than in rising gusts, with the result that wind speeds are significantly overestimated in gusty conditions. With careful design the rate of cup rotation can be made almost directly proportional to wind speed over a usefully wide range of speeds. In many observations the *run of wind* is found by converting the number of cup axle revolutions in a certain time period into a length of wind. Divided by the time period used, which ranges from ten minutes to a day depending on the type of observation, the run of wind yields the average wind speed in this period. Ten-minute averages, together with maximum speeds in gusts, are used to define wind speeds near the surface in routine observations. The instrumental lag associated with the robust standard type of cup anemometer ensures that gusts are themselves smoothed over periods of several seconds.

The only anemometer not requiring prior calibration in a wind tunnel (at least in principle) is the *pitot tube* type, which measures the excess pressure developed in air as it rams into the mouth of a small tube steered into the wind by a vane. Since the

Fig. 2.2 A standard British Meteorological Office pattern cup anemometer. Its robust construction enables it to cope with stormy conditions with minimum maintenance. Revolutions are counted and displayed on the instrument (as here), or remotely by electrical signal.

rammed air is more or less completely halted in the process, the air loses all its kinetic energy, with the result that the excess pressure is proportional to the square of the wind speed (by Bernoulli's equation) and the instrument is insensitive to low wind speeds.

So far it has been assumed that the horizontal component of wind is the only one of interest. This is largely true because the flattened shape of the atmosphere ensures that the horizontal wind speeds are usually more than 100 times larger than vertical ones, and almost always more than 10 times larger. When the vertical component is to be included, additional vanes and anemometers are used, such as a propellor on a vertical axle. Small-scale turbulence in particular has an important vertical component which is comparable with the horizontal ones because the turbulence is essentially isotropic. Updraughts in shower clouds are known to be quite significant, but cannot be measured by fixed anemometers for obvious reasons. They can however be measured from aircraft traversing through them by using routine and specialized aircraft instrumentations. The very weak but persistent updraughts which maintain the great sheets of frontal cloud (nimbostratus) are too small to be directly measurable by direct observation, though they can be inferred from other measurements, sometimes by way of quite sophisticated theory (section 7.13).

2.5 Pressure measurement

Man is insensitive to naturally occurring pressure variations below the audible range (say 20 Hz), with the result that the existence of substantial but much slower variations in atmospheric pressure was quite unsuspected before the development of the *barometer* in the mid seventeenth century. However, the relation between pressure variations and weather sequences was then almost immediately noted, and

21

(b)

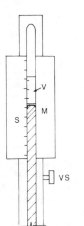

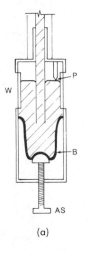

(a)

Fig. 2.3 (a) A Fortin-type mercury baromoter. The mercury reservoir in the soft bag B is adjusted by the adjusting screw AS so that the exposed mercury surface just touches the point P at the bottom of the fixed vertical scale S (visible through the window W). The Vernier slide V is moved by VS until it appears to rest on top of the mercury meniscus M. (b) A barograph with glass case removed. The aneroid capsule is visible on the right, and the irregular atmospheric pressure trace on the left shows variations of about 10 mbar in 24 hours.

the barometer has occupied an important place among meteorological instruments ever since. The relationship between pressure and weather is by no means as simple as the legends on domestic barometers suggest, but it is nevertheless interesting and important in ways which are apparent in Chapter 11.

The basic absolute instrument is still the *mercury-in-glass* barometer, working on the same principle as the ones first made by Torricelli in 1644 (Fig. 2.3). Atmospheric pressure acts on an exposed surface of mercury to maintain a column of mercury in a vertical glass tube whose top end is sealed and evacuated. The vertical height h of the top of the mercury column above the exposed surface is related to the atmospheric pressure p by the barometric equation eqn (4.16)

$$p = g \rho h \qquad (2.1)$$

where g is the *gravitational acceleration* and ρ is the density of mercury (see table on p. 183). Such barometers are heavy and frail, but if they are read carefully and corrected for variations in g and temperature (which affects ρ), they are accurate to better than 0.1 mbar. Marine barometers are mounted in gimbals so that they remain nearly vertical despite the movement of the ship.

The reading procedure imposes an effective lag of $\sim$ 10 s on the mercury barometer. However, most of the significant variations of atmospheric pressure associated with weather systems occur over much longer time periods, and in fact the standard hourly reading is usually quite adequate. Recently the more compact and robust *precision aneroid* barometer has largely replaced the mercury-in-glass type in the meteorological network. A partly evacuated (aneroid) flexible metal capsule expands as the ambient air pressure decreases and vice versa. The movement of the capsule is detected by a manually operated probe, and the instrument is calibrated against a master mercury-in-glass barometer. The common domestic aneroid barometer operates on the same basic principle, but the capsule movement is displayed by mechanical linkages whose friction and looseness make the indicated pressure liable to errors of several millibars. Even so this simple instrument can give a useful impression of the big slow pressure variations which are associated with weather systems in middle and high latitudes.

A major shortcoming of these barometers is their inability to yield a continuous record. This is overcome in the traditional *barograph* (Fig. 2.3), in which the movement of the aneroid capsule is communicated mechanically to a pen writing on a graph fixed to a drum rotating by clockwork. Even when very well made, the problems associated with mechanical linkages are even worse than in the aneroid barometer, but the pressure trends (the *pressure tendencies*) are graphically depicted, and are very useful for weather analysis and forecasting. At the moment electromechanical linkages are beginning to be produced which should give much better accuracy, and instruments using the pressure sensitivity of electrical properties of crystals may eventually give rise to barographs as accurate as the best mercury-in-glass barometers.

Note that all such instruments measure the ambient atmospheric pressure of the barometer, usually known as the *station pressure*. As discussed in section 2.9 these may need substantial correction to produce values which are consistent across a network of station at various altitudes.

2.6 Precipitation measurement

Most precipitation reaches the surface as rain, but strictly the term includes all the forms in which water and ice fall to the surface, which include the various types of snow and hail, as well as rain and drizzle. Wherever possible, amounts of snow and hail are expressed as rainfall equivalents. Rainfall measurements are of great interest to hydrologists concerned with the management of rivers and reservoirs, as well as to meteorologists.

The quantity of rainfall in an observation period is specified by the depth of water collected on a horizontal area, but because this may be as little as 0.1 mm in a typical collecting period (12 or 24 hours), *rain gauges* always use funnels to concentrate the precipitation after collection. The commonest standard gauge has a circular mouth about 130 mm in diameter, and a much narrower glass measuring cylinder with graduations adjusted to allow for the concentrating effect of the funnel. Automatic gauges with wider mouths operate by registering electrically the times taken to fill small cups beneath the spout of the collecting funnel (e.g. the *tipping bucket* gauge in Fig. 2.4).

The ability of the gauge to resolve temporal variations of rainfall usually depends on the sampling rate rather than on instrumental lag. For example, manual gauges are read once or twice per day, which is adequate for many purposes but obliterates all the interesting variations occurring in the intervals between successive readings. These are revealed by automatic gauges, many of which register the collection of 0.05 mm of rain. The time charts of rainfall revealed by automatic gauges are often extremely spiky, and show that large fractions of total rainfall occur in relatively short bursts of heavier rain (Fig. 2.5).

The great weakness of all these types of gauge is that their spatial representativity is always in doubt: X mm of rain falls at gauge A in a certain storm, but what would have been collected by identical gauges nearby if they had been in operation? From

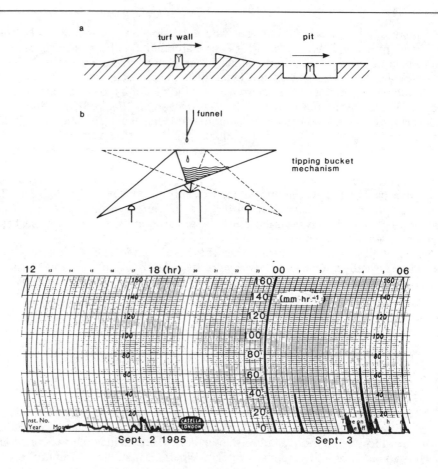

Fig. 2.4 (a) An idealized vertical section through rain gauges set inside a turf wall and in a pit, to minimize interference by airflow over the gauges. (b) A vertical section through the mechanism of a tipping bucket type of automatic gauge. The bucket see-saw tilts into the dashed position when the filling bucket is full.

Fig. 2.5 Rate of rainfall recorded on a Jardi-type gauge. The continuous rain (with embedded heavier bursts around 1800 Z) fell from a passing active occluded front. The isolated spikes early next day correspond to showers in the following colder north-westerly flow.

experimental work with dense networks of gauges it is known that measured rain-fall can be very strongly and closely patterned in space - the patterning arising from details of gauge exposure, topographical influence as well as the small-scale structure inherent in precipitating weather systems.

1. *Gauge exposure.* The efficiency of collection and therefore the accuracy of a gauge is quite sensitive to the way in which the collecting mouth is exposed. Until recently gauge mouths were always raised above the surrounding surface to avoid flooding or splash-in. But this deforms the flow of air over the gauge and may significantly reduce the amounts collected, especially in the higher winds which often accompany substantial rainfalls. Placing the gauge mouth about 30 cm above the surface has been found to be an adequate compromise over land. Recently flush-mounted gauges surrounded by a dummy splash-free surface have begun to come into increasing use (Fig. 2.4), and where this is not desirable the aerodynamic 'waisting' of the body of the normally elevated gauge has been found to reduce error.

    Problems of exposure are especially severe in ship-borne gauges, where spray and the effects of the ship's large bulk and motion are effectively unavoidable. Some might be reduced by mounting gauges on carefully designed unmanned observation buoys which are kept on station automatically, but these are sophisticated devices whose design is still being optimized. So far the distribution of rainfall over the seas is much more uncertain than it is over land,

and is probably the most inadequately observed major element of the world's weather.

2. *Topographical influence.* Rainfall can be very strongly enhanced over the windward side of high ground and reduced in its lee (forming a *rain shadow*). Since such patterns may have an important effect on agriculture and water supply, for example, and yet be too finely detailed to be properly represented by any feasible gauge network, quite subtle techniques have been devised by hydrologists to interpolate between gauges as realistically as possible [9].

3. *Inherent rainfall patterns.* These range in scale all the way from the fine streaks apparent in curtains of rain backlit by the sun, through showers, to the extensive swathes associated with major weather systems such as fronts. Showers in particular are difficult to allow for since they often contribute significantly to local totals, but yet are so small that they tend to slip between the inevitably sizeable gaps in the gauge network. This results in overall underestimation of the total amount of precipitation (since there is almost no chance of the few showers coinciding well enough with the few gauges to produce overestimation) and a wide variability between adjacent gauges.

Since about 1950 increasing use has been made of radar to observe the distribution and intensity of precipitation over wide areas (Fig. 2.6). The technique

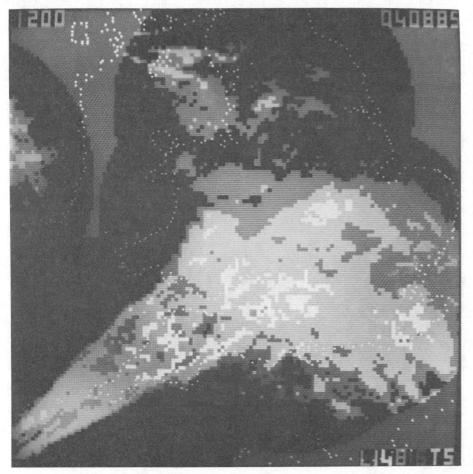

Fig. 2.6 A composite display from six meteorological radars in the southern British Isles at 1200 Z on 4 August 1985. The echo strengths correspond to rainfall rates of between about 0.5 and 8 mm hr^{-1}, and the rainy areas clearly outline the warm front (centre right) and the cold front (trailing left) associated with a small but vigorous depression centred west of Ireland. The rain over the Irish Sea (top centre) is associated with a relatively weak occluded front bending westwards toward the low-pressure centre.

originated in the Second World War when it was discovered that heavy precipitation produced radar echoes whose strengths were afterwards found to increase with the average size of the raindrops, snowflakes and hailstones, and with the numbers of these per unit volume of air. Since these in turn are fairly directly related to the rate of precipitation toward the underlying surface, carefully calibrated radars can be used to measure precipitation rates in a way which avoids many of the snags of conventional point measurement. Although echo strength varies somewhat depending on whether the precipitation is rain, hail or snow, the radar echo also improves on the conventional gauge in its ability to assess snowfall, which is very difficult to measure accurately otherwise because of crippling exposure problems and the choking of gauges, even when heated to measure equivalent rainfall.

2.7 Other surface observations

Several of the routine types of observation not mentioned so far are made by the unaided human senses. These are ones which have been found by long experience to be good indicators of the current state of the atmosphere from the point of view of forecasting, and are complex in ways which make them difficult to instrument. For example the amounts, types and heights of clouds are judged by eye, using internationally agreed categories of cloud which have been perfected over the last 150 years and are now enshrined in the *International Cloud Atlas* and derived handbooks [10]. The categories of type and height have been chosen so that cloud reports emphasize factors which are especially telling evidence of the presence of particular weather systems: for example, the high clouds which are characteristic forerunners of a warm front (Fig. 2.7). Further examples appear in Chapters 6, 10, 11 and 12, and the *Cloud Atlas* types and names are listed in Appendix 2.3.

Fig. 2.7 Cirrus plumes (cirrus uncinus) ahead of a warm front. Patches of ice cloud in the high troposphere are producing trails of larger crystals which are drawn out as they fall through the strong wind shear there. Note that the wind shears in direction as well as speed.

In a similar way the current or *present weather* is described by selecting one from a series of 99 types arranged in order of increasing forecasting significance, only the most significant type present being reported on any occasion. For example, all types of fog are less significant than rain, light rain is less significant than moderate rain, and the most significant weather type of all is a heavy hail shower with thunder (indicating extreme convective instability). *Past weather*, meaning weather in the period (often 6 hours) immediately before the time of observation, is similarly described using a list of 10 types. Present and past weather types are briefly summarized in appendix 2.4.

Of the many other types of observation which could be mentioned, the most important concern solar and terrestrial radiation. For nearly a century, periods of bright sunshine have been recorded by the elegant and simple *Campbell–Stokes* sunshine recorder (Fig. 2.8). Indeed its very success makes it difficult to replace by a comparable instrument with an output more convenient than its scorched paper strip. In recent decades various types of *solarimeter* have been devised to measure the intensity of direct and diffuse solar radiation by registering the electrical effects of its absorption. The simplest types use a thermopile to measure the warming of a

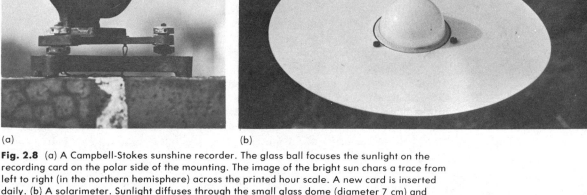

(a) (b)

Fig. 2.8 (a) A Campbell-Stokes sunshine recorder. The glass ball focuses the sunlight on the recording card on the polar side of the mounting. The image of the bright sun chars a trace from left to right (in the northern hemisphere) across the printed hour scale. A new card is inserted daily. (b) A solarimeter. Sunlight diffuses through the small glass dome (diameter 7 cm) and warms the blackened top end of a thermopile whose shielded lower end stays at ambient temperature. The white collar provides a radiatively uniform surround.

27

blackened surface beneath a glass dome (Fig. 2.8), and these can be made sensitive to terrestrial radiation as well by using a shield which is also transparent to infrared radiation. Since glass is opaque to terrestrial radiation, like most terrestrial solids and liquids, the dome formerly had to be made of quartz, but fortunately polythene now provides a cheaper transparent alternative. A pair of these, one facing up and the other down, with their electrical outputs arranged to give the algebraic sum of the downward fluxes, constitutes a *net radiometer*, directly measuring the resultant of all solar and terrestrial radiant fluxes passing downward (conventionally) through a horizontal surface. Values of the net flux measured near the Earth's surface are of great importance in quantifying the factors determining local climate.

Temperatures at and beneath the ground surface are particularly useful in agriculture. The daily minimum surface temperature is especially significant when it is near 0 °C, and is usually measured by an alcohol-in-glass minimum thermometer exposed just above short grass. This *grass minimum temperature* is often used as a measure of the degree of ground frost, though minimum temperatures actually vary considerably with the type of surface (section 9.4). Thermometers are also buried at various shallow depths in soil to measure the temperatures experienced by the roots of crops, etc.

2.8 Observation networks

By the early nineteenth century it was realized that weather in middle latitudes was organized in moving patterns with horizontal scales of 1000 km or more. Given their speed of movement and development, any analysis of such patterns could only be retrospective while the maximum speed of messages was that of a galloping horse. The development of the electric telegraph by Morse in 1844 transformed the situation. A network of meteorological observing stations connected by telegraph to a centre for analysis and forecasting was established in France in 1863, and the USA, Britain and other technically developed countries quickly followed suit. Thus began the network which now covers the land areas of the globe, albeit very unevenly (coverage increasing with local material wealth), and which is formally organized by the World Meteorological Organization (WMO) acting to coordinate the many national networks. The complete observing, communicating, and analyzing and forecasting system is now entitled *World Weather Watch* (WWW). We will consider only the observing part of WWW, under the separate headings of surface, upper air and satellite networks. First consider some quite general points.

WWW is geared to record the present state of the atmosphere with just enough detail and accuracy to permit useful forecasting. Initially, when only the surface network existed, the forecasting was primarily for shipping near coasts, but it is now aimed at all sea and air transport, and land transport in hazardous conditions, as well as agriculture and industry, and the public at large. In some parts of the globe, special hazards such as hurricanes or tornadoes receive special attention.

Just how much detail and accuracy will suffice for forecasting is a crucial question which is answerable only by experience. Given the obvious complexity of the weather and our limited understanding of it, it might seem that we should try to observe as much as possible, everywhere and all the time. However, since personnel and finance are obviously severely limited in practice, the question is always 'what

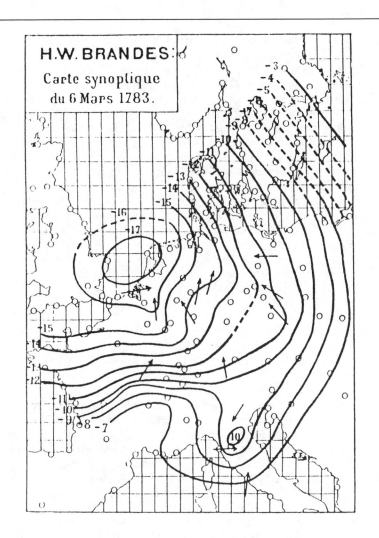

Fig. 2.9 An early synoptic chart reconstructed by the pioneer Brandes some decades after the event. The isobars (of pressure difference from normal, to allow for the effects of variable station altitude) clearly reveal the south-eastern quadrant of a depression centred between England and Holland, with cyclonic surface winds angled toward low pressure by friction (section 7.11).

is the minimum network of people and instruments which will provide data good enough for useful forecasting?' Even this question does not have a definitive answer, since the amount of coverage considered adequate depends on the level of our understanding of the systems to be forecast (in principle, better understanding permits more efficient observing), and with our needs. Up to now, forecasters have been content with networks only just good enough to define large-scale weather patterns, such as extratropical cyclones and their somewhat smaller-scale frontal structures, so that the networks of surface and upper air stations making *synoptic* (simultaneous) observations have evolved to fit this role, to the extent that this scale of weather phenomena is now called *synoptic scale*.

Even though the density of a reasonable compromise network is widely agreed by forecasters, the fact remains that this has been implemented by only the most wealthy nations: others have networks which are more or less inadequate, and the oceans are effectively data deserts. This patchiness of observational coverage in the face of the need for homogeneous, minimal coverage, is the fundamental problem facing WWW, and will require decades to solve.

Even if the synoptic network were to be completed overnight, it would still be much too open to permit more than random observations of sub-synoptic-scale

29

systems such as shower clouds and the quite substantial clusters which they sometimes form. Forecasting such events therefore is and will probably remain statistical, and predictions such as 'heavy showers are likely near the west coast' will no doubt remain normal. Weather radar can give more specific warning of all substantial showers within range, but showers are so short-lived that individual forecasts could be given only by interrupting local radio or television broadcasts. This is precisely what is done in parts of the United States in the tornado season (tornadoes being associated with very heavy showers), but it is worthwhile only because tornadoes are so damaging and dangerous. Continuously broadcast televised 'pages' of local forecasts could no doubt be updated frequently to provide similar though less arresting advice.

Lastly, in considering networks of surface and upper air observations in particular, note the need for standardization of observational practice: unless great care is taken, data from different stations may well differ more on account of differences in observational practice than atmospheric variations. This is obvious in principle, but its effective implementation requires surprising rigour in even the smallest matters.

2.9 The surface network

Ideally, observations of wind, temperature, humidity, pressure, visibility, and cloud and weather types are made at each station of the network every hour just before the hour (Greenwich Mean Time or Universal Time, usually denoted by Z), so that it is possible in principle to have a synopsis of world-wide data at any hour. In regions where such regularity cannot be relied on, efforts are made to ensure that the observations at 0000, 0600, 1200 and 1800 Z are made and reported. The treatment of precipitation totals is rather different, since manual gauges are traditionally read at 0900 and 2100 local time to reveal the systematic differences between night and day which are so conspicuous over land. Obviously all instruments are designed to conform to international standards, and are operated by staff trained to use them in a consistent and proper manner.

As far as is possible the observations are made in sites which are standard in construction and exposure, but yet are representative of the surrounding terrain. Full details, even down to the length of grass in the fenced enclosure, are specified in handbooks such as [5]. However, sites vary considerably in altitude, and since the atmospheric pressure decreases by about 1 mbar (0.1%) for every 10 m rise, substantial corrections must be made to station pressures to prevent the pressure variations associated with differences in station altitude from swamping those arising from weather systems. These corrections are made by estimating the additional pressure which would have arisen had the atmosphere continued down to mean sea level (the agreed datum level) at the station. The correction p is calculated from a form of eqn (2.1), where h is the height of the station above MSL and ρ is the fictional air density, which is equal to or closely related to the air density at the station. This is a satisfactory procedure for the many stations which lie within a few hundred metres of MSL, but it gives trouble at mountain stations where a kilometre or more of fictional atmosphere may have to be invoked.

No attempt is made to correct air temperatures for station altitude, but the height of the Stevenson screen above the local ground surface must be carefully chosen. Daily temperature variations decrease sharply with increasing height above the ground surface, because heating and cooling are concentrated at the surface. Ideally we might consider measuring air temperature above the surface layer, but since the screen would then have to be at least several tens of metres above the surface, there would be severe practical problems and the data would be largely irrelevant to the immediate human environment on many occasions. The compromise solution is to place all Stevenson screens in the network at the same height above the local surface, so that they are subject to similar daily cycles. The standard height of the thermometer bulbs (often known as *screen level* because it is in the middle of the screen interior) is 1.5 m, with slight national variations, and is obviously related to normal human stature.

Wind speed increases with height above the ground surface, as well as with distance from upwind obstructions. The latter effect is minimized by judicious siting, but the former again requires an arbitrarily chosen measurement level, since the frictional drag of the surface on the air only really disappears some 500 m above the surface. The actual level chosen is 10 m, which is about the highest level which can be reached using a small mast. The wind speed there is typically about three times the speed at 1 m height. At many secondary observation stations even this modest height is unattainable, and so the anemometer has to be sited at a lower level and its data multiplied by a standard factor to give even a reasonable approximation for the 10 m wind speed. Wind direction is measured at the same height as speed.

The distribution of surface synoptic stations throughout the Northern Hemisphere and in the British Isles in particular is shown in Fig. 2.10. With an average separation of 50 km, this is typical of the surface networks of wealthy, densely populated countries. In poorer and more sparsely populated areas the networks are much more open. Fully automatic stations, requiring manning only for maintenance and repair, are being developed to plug gaps and reduce high staffing costs. Over the sea there is increasing but still very inadequate coverage through the voluntary participation of merchant shipping. This is extremely valuable, though data from shipping suffer from the effects of non-standard exposure (since there is no such thing as a standard ship) and crowding along popular shipping lanes at the expense of intervening expanses. Because of their great capital and running cost, there are very few special weather ships, and so unmanned instrumented buoys with radio transmitters are being designed and tested for eventual inclusion in WWW. The worldwide total number of reporting land-based synoptic surface stations is static at about 2500, whereas the average number of ships reporting at the important synoptic hours has trebled in the ten years since 1975, and is approaching 1500.

In addition to the standard synoptic network, there is in Britain, for example, a much denser network of voluntarily manned stations, many of them at schools, where limited standard observations are made each day at 0900 Z. These data are not available to forecasters, but they are collected, scrutinized and published in monthly compilations by the British Meteorological Office (the national forecasting service); these compilations serve to define the British climate and its variations on a much finer scale than is possible with the synoptic network. Other countries have similar non-standard systems.

Networks of weather radars are as yet very incomplete, and confined to the wealthiest nations. There are quite large numbers of radars in North America, not all coordinated effectively; there is a small coordinated network in the southern British Isles, and a fairly comprehensive network is envisaged for Europe. The large useful range of radars means that adjacent sets can be separated by over 100 km in non-mountainous terrain. Their data are analysed, improved (for

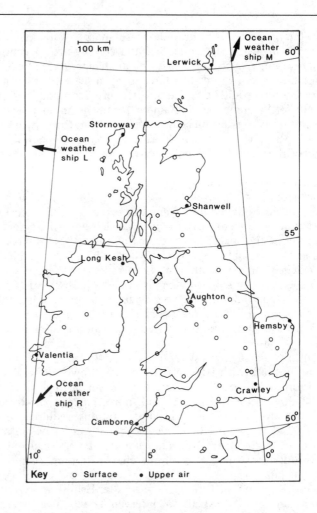

(a)

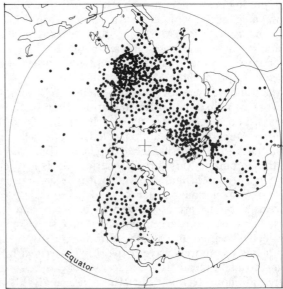

Fig. 2.10 (a) Map of British Isles surface and upper air stations. Weather ships M, L and R are far beyond the map edges in the direction shown. (b) Radiosonde stations in the Northern Hemisphere supplying upper air data on 21 October 1979. Note that the British Isles network shown in (a) is one of the densest parts of the network. The Southern Hemisphere is relatively empty being largely oceanic.

(b)

example by the removal of ground echo), amalgamated and displayed in real time largely by computer.

2.10 The upper air network

Observations of wind, temperature, relative humidity and pressure are made by *radiosondes* — free-flying balloons released from the upper air stations of the synoptic network (Figs. 2.10 and 2.11). The sondes are released at 0000 and 1200 Z daily, and climb at about 5 m s^{-1} until they burst between 20 and 30 km above MSL, whereupon the instrument package returns to the surface by parachute for possible re-use. While in flight the temperature, humidity and pressure data are sent by radio to the ground station, and the sonde's position is monitored by automatically tracking radar. The sonde has so little inertia that it moves with the

(a)   (b)

Fig. 2.11 (a) A radiosonde at launch. After launch the reel at the base of the balloon (now in the launcher's right hand) pays out a 30 m cable to keep the instrument package (now in his left hand) clear of local balloon effects. This particular sonde (a Vaisala RS-80) can be tracked for wind finding by Navaid systems, radio theodolite or radar (with a reflector immediately below the balloon). (b) Close-up of the instrument package of the British Meterological Office's Mark 3 radiosonde, sitting as it hangs in flight. The plastic ring on top supports the very fine wire of the electrical resistance thermometer; the open shield just beneath contains the skin hygrometer, and the main body houses the aneroid pressure capsule, electrical battery, converting, switching and timing devices, and radio transmitter.

33

horizontal wind of each layer it passes through. Successive horizontal positions of the sonde (for example at minute intervals) can therefore be made to yield a vertical profile of horizontal wind. Since the analysis of the radio and radar data proceeds while the sonde is in flight, the staff of the ground station can assemble and send to the local analysis centre within a few minutes of the end of the flight the profiles of wind, temperature and pressure up to the bursting level. Humidity data are usually ignored above the 10 km level because sensors are unreliable in the very low temperatures prevailing at these and higher levels. Strictly speaking, radiosondes do not measure a vertical profile, since the sonde's flight path can be a rather gradual slope in strong winds, but in the grossly flattened configuration of all synoptic scale systems, even a slope of 1:10 is effectively vertical. We therefore make no serious error in assuming the profiles to be vertical in most data analysis.

Midway in time between the full radiosonde ascents (i.e. at 0600 and 1800 Z), balloons bearing only a radar reflector are flown and tracked to give profiles of wind only. The alternation of these *wind sondes* with the radiosondes at six-hourly intervals provides sufficient resolution in time to define the structure of the troposphere and low stratosphere associated with synoptic-scale weather and northern hemisphere systems. Figure 2.10 shows the network of upper air stations in the British Isles, and this again is typical of such networks in wealthy countries. In other regions networks are much more open, and indeed they hardly exist over substantial land areas. Upper air data from sea areas are obtained only from the few weather ships remaining in the North Atlantic and North Pacific after increasing cost enforced a sharp decline in numbers in the last twenty years. These provide an invaluable but totally inadequate sampling of the 70% of the lower atmosphere which lies over the oceans, and it is a major task of WWW to improve the synoptic upper air network, or devise complementary systems. A direct improvement is envisaged through the development of completely automatic systems for launching radiosondes and telemetering data from suitable merchant ships. A parallel approach which is already in extensive use is to fit large civil aircraft with automatic sensors which will transmit pressure, temperature and wind data while in flight. However, the temperature and wind data are not very reliable, and are heavily concentrated at about the 300 mbar level along the major air routes. There are currently about 500 upper air stations reporting regularly, and the daily total of aircraft reports is well in excess of 2000.

Note that sampling intervals in space and time are much greater in the upper air network than in the surface network. You may wonder therefore why the upper air network is not totally inadequate given that the much finer mesh of the surface network is only just adequate. The answer is that atmospheric structure is much smoother and larger in scale aloft than it is near the surface, where surface inhomogeneities generate significantly smaller and more transient structures, especially over land. A much more open network is therefore adequate aloft, which is just as well given the much greater cost of making regular observations there.

2.11 The satellite network

The first meteorological satellite was launched in 1960 and immediately provided extremely interesting and useful data. First, the huge panoramic views of the atmosphere directly revealed and confirmed the structures of large cloudy weather systems which had previously emerged only after painstaking assembly and

Fig. 2.12 Visible image from polar orbiting Tiros N, showing a vast panorama of sea, snow and cloud in the north Atlantic at 1600 Z, 11 March 1980. Cellular patterns of shower clouds spread across a westerly flow, north and west of dense frontal cloud associated with a depression centred near Iceland, whose north-west tip is just visible upper centre. Oblique sunlight highlights a profusion of cloud structure on the small and mesoscales.

analysis of synoptic data. I suspect all meteorologists who were active before 1960 still have a sense of relief and awe when they see the work of the pre-satellite generations of analysts so vividly confirmed in pictures such as Fig. 1.1. Secondly the relatively high resolution of the satellite pictures (a few kilometres compared with the tens of kilometres of the synoptic surface network and the much coarser scale of the upper air network) revealed an almost bewildering range of sub-synoptic-scale structures in cloudy systems, many of which have yet to be understood and incorporated in observational models. For example, the quasi-regular patterning of shower clouds in vigorously convecting zones (Fig. 2.12) is quite obvious on satellite pictures but is completely lost in the wide gaps between stations in the synoptic network, and has yet to be understood in useful detail. Thirdly the

35

satellite data were, even from the outset, much more uniformly spread over the Earth's surface, covering the enormous oceanic and low latitude land areas which had virtually no synoptic network.

Meteorological satellites have multiplied and developed considerably since 1960, and the broad outline of a permanent network of satellites is probably now established, although details may continue changing for years to come.

Basically, meteorological satellites are platforms for electromagnetic scanning of the atmosphere from above — *top-side* observation in the vivid jargon. The scanning is passive in the sense that satellites merely make use of existing radiation emitted or reflected from the atmosphere, without adding to it as radar does — though special radars are being tried on account of their ability to penetrate cloud. The radiometers used are sensitive to one or more wavelength bands in the visible and infra-red ranges. The field of view from the satellite is scanned line by line, either by using the equivalent of a television camera, or by physically sweeping a very narrow-field radiometer across the view. In either case the data are sent in sequence to a receiving station on Earth for reconstitution of the whole picture. The width of a scan line on the Earth's surface fixes the ultimate limit of resolution since details less than a few lines across are unresolved. Most satellites operate so that this limit is a few kilometres at a point vertically beneath the satellite; it obviously increases with increasing obliqueness of view.

If visible wavelengths are used, the reflected sunlight picks out cloud vistas which are slightly blurred versions of those photographed conventionally by astronauts, and may be in full colour if at least three distinct bands of wavelengths are used and synthesized. A major disadvantage is the inevitable blindness on the night side of the Earth.

Radiometers sensitive to wavelengths in the far infra-red (i.e. >3 μm) are independent of solar radiation and respond instead to the terrestrial radiation emitted continually by the Earth's surface and atmosphere. This increases in intensity as the temperature of the emitting materials rises (Chapter 8) so that the resulting picture has a brightness scale which corresponds to a temperature scale in the original panorama. Sometimes the brightness scale is replaced by an arbitrary colour scale for ease of subsequent analysis by eye. Because the cloud-free atmosphere is nearly opaque in some parts of the infra-red spectrum, and nearly transparent in others (Fig. 8.4), quite different pictures emerge in different wavelengths. For example, wavelengths to which air is transparent yield pictures of the temperatures of the top surfaces of clouds (which are quite opaque in the infra-red), or of the ground or sea surface if unshielded by cloud. Monochrome cloud vistas produced in this way resemble visible pictures if the brightness–temperature correspondence is suitably chosen (Fig. 2.13). By contrast, pictures taken in wavelengths in which the air is opaque show the distribution of air temperatures in the middle and upper troposphere. Careful selection of wavelengths covering a range of opaqueness, together with subsequent complex analysis, can produce vertical temperature profiles of cloud-free air with low but nevertheless useful vertical resolution. These are potentially very valuable additions to the high resolution (in the vertical) but otherwise inadequate traditional radiosonde measurements, and are currently under active development.

Two distinct types of satellite orbit are in use. In the *sun-synchronous* type, the satellite orbits about 860 km (i.e. one seventh of an Earth radius) above the surface, passing near the poles but making an angle to the meridians which is just enough to allow the orbit to remain effectively fixed relative to the sun (Fig. 2.14). The satellite takes about 102 minutes between successive passes near one pole, while its radiometers scan the swath of planet passing continually below it. The radiometer data can be received in real time by any suitably equipped station as it is

Fig. 2.13 Infra-red image of frontal clouds over the north-eastern Atlantic (from polar orbiting NOAA-7 at 0853 Z, 18 October 1979). Though at this time only the eastern part of the panorama is sunlit, the image is unaffected since its brightness does not depend on visible light. The false brightness scale makes the coldest parts (the high cloud tops) look brilliant white, the low clouds and rather cool land surfaces grey, and the warmest water black, much as they would in visible light. A narrow ridge of high pressure over Britain is associated with clear skies there. An occluded front is approaching from the west, while a wave develops on an old cold front to the east, the two systems being connected by a loop of low or middle-level cloud associated with a weak cold front south of England. There are patches of vigorous convection east of England where cold air is pushing eastward south of the developing wave.

overflown, and any particular geographical location is most nearly overflown once every twelve hours at predictably varying clock times. In addition, data are accumulated over twelve-hour periods and transmitted very rapidly to a master station to provide a complete data set. Radiometers using visible wavelengths of course provide only half the coverage because of data black-out on the night side of the planet.

In the *geosynchronous* or geostationary type of orbit, the satellite orbits about 36 000 km (over five Earth radii) above the equator, moving in the same direction as the rotating Earth. Since the orbital period at this distance is exactly one sidereal day, the satellite hangs vertically above a fixed point on the equator (Fig. 2.14). The field of view is very nearly a full hemisphere, though of course the view of the polar regions and the western and eastern limbs is very oblique (see Fig. 1.1). A very high-resolution radiometer scans the view like a very slow television set, completing a scanning cycle in about 20 minutes and providing resolution which is nearly as good as in the very much lower altitude sun-synchronous type. The data are transmitted to one or more stations on the surface far below. In visible wavelengths the view disappears during the night, so that infra-red radiometers are always included. For World Weather Watch the global coverage is made usefully complete by having a network of five geosynchronous satellites equally spaced along the equator. The bias of good coverage toward low latitudes is welcomed because of the very poor synoptic network in those vast areas (remember half of the

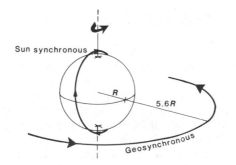

Fig. 2.14 Satellites in geosynchronous and sun-synchronous orbits. For ease of portrayal, the geosynchronous orbit has been drawn at a reduced scale relative to the earth. Note that the retrograde sun-synchronous orbit misses the poles by about 10° latitude.

37

surface area of the Earth lies between 30° N and 30° S). Geosynchronous satellites are very well placed to detect the appearance and movement of tropical cyclones, since these form over the warm oceans of low latitudes. Before the advent of the meteorological satellite, even the most vigorous of these systems (the hurricanes — section 12.4) were often detected only through chance encounter with ships, making estimation of storm paths and landfall very unreliable.

2.12 Epilogue

A very large scale of individual and collective effort is needed to maintain World Weather Watch, even in its present incomplete state. Shortly before every major synoptic hour (0000, 0600, 1200, 1800 Z) tens of thousands of observers all over the Earth are making their routine observations from the surface, and smaller numbers are launching and tracking radiosondes. Many more people are active around the clock in the global networks of associated telecommunication and data collection, and at the numerous forecasting offices. The completely international directorate and administration of the World Meteorological Organization coordinates the activities of the national networks and deals with matters ranging from the standardization of equipment and procedure to the planning and monitoring of improvements to WWW.

In the latter part of a century which has seen much of the optimism of our forefathers overwhelmed by confusion and strife, and a comprehensive approach to the human predicament frustrated by unenlightened sectional interest, it is heartwarming to see what people can do globally when they really put their minds to it. Of course the cooperation arises in the face of a fairly well-defined problem — the need to forecast the short-term future behaviour of the lower atmosphere — and the atmosphere is inescapably a global entity which is mercifully independent of Man's random subdivisions of the Earth's surface. Nevertheless we should give credit where credit is due, if only because the future of civilization depends on this kind of low-key, practical and sincere cooperation rather than on the strident rhetoric and divisive mayhem which is now so widely current.

Appendix 2.1 Instrumental response times

Suppose that an accurate and properly exposed thermometer has been exposed to air of a certain steady temperature for so long that the thermometer reading θ is steady at the same value. Now suppose that the air temperature suddenly increases to a different steady value T (Fig. 2.15a). The reading θ will begin to move toward the new value as heat flows into the thermometer. According to Newton's law of cooling (and heating), the rate of warming reduces in proportion to the temperature difference Δ between air and thermometer. Formally

$$\frac{d\Delta}{dt} = -k\,\Delta \tag{2.2}$$

where k is the constant of proportionality which describes the speed of response of this particular thermometer, and Δ is $T - \theta$. Equation (2.2) is the equation for exponential decay, whose solution is given in all basic mathematical texts.

$$\Delta = \Delta_0 \exp(-k\,t) \tag{2.3}$$

where t is the time elapsed since the sudden change Δ_0 in air temperature. This exponential response is graphed in Fig. 2.15a.

The lag coefficient λ, or simply lag, of the thermometer is defined to be the time taken for Δ to fall from Δ_0 to $\Delta_0 \exp(-1)$, which is $0.3679\,\Delta_0$. At this stage the thermometer has registered just over 63% of the initial jump in temperature. By inspection of eqn (2.3) it follows that λ is equal to $1/k$. The time taken for a larger or smaller response can also be found from eqn (2.3). For example, the time for Δ to fall to $\Delta_0/10$ is $2.3026\,\lambda$. Such relationships are common to all exponential behaviour, however it arises. Appendix 4.4 contains further examples.

The response of a thermometer to an oscillating air temperature is particularly relevant to meteorology. The analysis begins with eqn (2.2) as before, but is complicated by the continuous variation of T. Figure 2.15(b) depicts the response to a regular wavy oscillation in T. The sluggishness of the thermometer has two effects: the oscillation of the thermometer reading lags behind that of the air temperature, and its amplitude is reduced. The magnitudes of these effects depend on the ratio of λ to the time period τ of the air temperature oscillation. For example, detailed analysis of a pure sinusoidal oscillation shows that the ratio of amplitudes (Fig. 2.15(b)) is given by

$$\frac{A_\theta}{A_T} = \left[1 + \left(\frac{2\pi\lambda}{\tau}\right)^2\right]^{-\frac{1}{2}} \tag{2.4}$$

It follows that when τ is equal to 10λ the amplitude ratio is 0.85, which means that the thermometer underestimates the true amplitude by 15%. The corresponding phase lag Φ (Fig. 2.15b) is about $\tau/10$. The distortion of the thermometer's temperature pattern increases sharply with λ/τ, so that when this ratio is unity the amplitude ratio is 0.16 and the phase lag is 0.22τ.

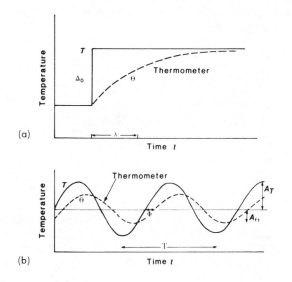

(a)

(b)

Fig. 2.15 Thermometer response to (a) a sudden rise in ambient temperature, and (b) a regular sinusoidal oscillation of ambient temperature.

Although we have considered only thermometers so far, the concepts and equations we have used apply to any instrument which responds exponentially to changes in the quantity it measures. Even when the response is not strictly exponential, when for example the instrument responds at first more quickly and then more slowly than the exponential curve allows, the lag is usually still defined as the time for 63% response. The lags of anemometers and wind vanes tend to decrease as average wind speed increases, since changes in speed and direction are communicated more efficiently by higher winds, and so it is customary to express their lags in terms of the runs of wind needed to produce 63% response. For example, the standard surface anemometer shown in Fig. 2.2 requires about 15 m of air to pass its cups in order to register 63% of a sudden change in wind speed. At the lowest wind speed to which this anemometer responds (about 2 m s^{-1}), this run corresponds to a lag in time of about 8 s, whereas the lag in response to gusts in a gale is roughly 1 s.

Appendix 2.2 The Beaufort scale

By about 1800 AD it was becoming obvious that it would be useful to categorize wind strength in terms of its effects on the sea and sailing ships. This was done so effectively in 1806 by Francis Beaufort while surveying for the British Royal Navy that his scale became the standard for all wind observations until 1946, and is still used in weather forecasts for shipping. The scale of effects in the Beaufort scale (Table 2.1) was capable of being used so consistently by trained seamen that assessment differences between observers of the same sea were believed to be substantially less than one Beaufort number. Beaufort also devised a concise code of *Beaufort letters* for weather description, which is still useful for weather diaries. He is recognized as a pioneer of the movement to make meteorological observations as consistent and concise as possible.

Table 2.1 lists simplified descriptions of effects on fairly open land surfaces, together with the equivalent ten-minutes average wind speeds measured by anemometer 10 m above the land surface — an equivalence established by anemometry long after the scale was initially devised [5].

Appendix 2.3 International cloud classification [10]

The basic division is into distinguishable types or *genera*, each with a latin name and standard abbreviation. Each genus occurs so often in one of three bands of heights above sea level, that it is assigned to that band, even though exceptions are quite common. The larger types of cumuliform cloud, in particular the cumulonimbus, are vertically so extensive that they extend to medium and high levels from their low-level origins and bases, but are ascribed to low levels for simplicity.

Table 2.1 Beaufort scale of wind force

Force	Specifications for use on land	Equivalent mean wind speed 10 m above ground	
0	Calm; smoke rises vertically.	0 kt	0 m s^{-1}
1	Light air; wind direction shown by smoke drift, not by vanes.	2	0.8
2	Light breeze; wind felt on face; leaves rustle; vanes move.	5	2.4
3	Gentle breeze; leaves and small twigs moving; light flags lift.	9	4.3
4	Moderate breeze; dust and loose paper lift; small branches move.	13	6.7
5	Fresh breeze; small leafy trees sway; crested wavelets on lakes.	19	9.3
6	Strong breeze; large branches sway; telegraph wires whistle; umbrellas difficult to use.	24	12.3
7	Near gale; whole trees move; inconvenient to walk against.	30	15.5
8	Gale; small twigs break off; impedes all walking.	37	18.9
9	Strong gale; slight structural damage.	44	22.6
10	Storm; seldom experienced on land; considerable structural damage; trees uprooted.	52	26.4
11	Violent storm; rarely experienced; widespread damage.	60	30.5
12	Hurricane; at sea, visibility is badly affected by driving foam and spray; sea surface completely white.	>64	>32.7

Nimbostratus often extend to both low and high levels from their middle-level centres of gravity.

LOW-LEVEL CLOUDS (0–2 km above sea level)

Stratus (St) — Low, structureless, extensive sheet of cloud, often just deep enough to produce drizzle or snow grains at the surface.
Stratocumulus (Sc) — Similarly shallow cloud sheet, broken into more or less regular cumuliform masses, giving drizzle or snow grains at the surface if cloud base is not too high.
Cumulus (Cu) — Detached hill-shaped clouds with flat bases which may be at a uniform level across any particular population.
Cumulonimbus (Cb) — Large cumulus, usually showering. Their considerable vertical extent ensures that bases are quite dark and upper parts often reach to high levels. The lower and middle levels of any particular Cb look like a conglomeration of cumulus, but the high levels often consist largely of ice cloud and as a result look like dense cirrus (see below).

MEDIUM-LEVEL CLOUDS (2–4, 7, 8 km in polar, temperate and tropical regions respectively

Altocumulus (Ac) — Shallow sheet broken into more or less regular, rounded patches or rolls, sometimes with virga (see below). Distinguished from Sc by smaller apparent size (<5), and from cirrocumulus (see below) by the presence of shading.

Altostratus (As) — Largely featureless extensive sheet of watery looking cloud, often thick enough to make the Sun appear as if seen through thick ground glass.

Nimbostratus (Ns) — Extensive sheet of cloud, usually precipitating. The considerable depth blots out the sun, moon and stars, making days dull and nights pitch black.

HIGH-LEVEL CLOUDS (3–8 km in polar, 5–14 in temperate and 6–18 in tropical regions)

Cirrus (Ci) — Detached, white fibrous clouds, often with a silky sheen.

Cirrocumulus (Cc) — Shallow patch or sheet, broken into more or less regular, unshaded and apparently very small blobs or ripples, which are usually fibrous in places.

Cirrostratus (Cs) — Shallow extensive sheet of white cloud, thin enough to be largely transparent and produce haloes round the sun and moon. Cs may look fibrous or smooth or both.

The genera may be amplified by adding a description name quite misleadingly called a *species*, exemplified by the following selection.

Fractus (fra) — Broken or ragged (of St and Cu).

Lenticularis (len) — Elements shaped like a lens or almond (of Sc, Ac and Cc).

Humilis (hum) — Of only slight vertical extent (of Cu).

Congestus (con) — Growing markedly, often by bulging like the heart of a cauliflower (of Cu).

Capillatus (cap) — Distinctly fibrous upper parts (of Cb).

The addition of a list of *varieties*, such as *radiatus* (banded), *supplementary features* such as *virga* (fibrous precipitation trails tapering down from cloud base to extinction above the surface), *accessory clouds* such as *pileus* (thin, smooth webs stretched over the tops of Cu con or Cb), and the new category of *mother-cloud* (producing other clouds also present) completes a scheme some of whose details suggest an unhappy marriage of observable reality and Linnaean taxonomy.

Appendix 2.4 Present and past weather

Table 2.2 lists the categories of present weather used internationally by observers. Each is given a number, and in any observation only the highest applicable number is reported. The descriptions have been simplified for present purposes, since the full version is rather legalistically worded to avoid ambiguity. Note that 'freezing' means freezing on impact with solid surfaces.

Table 2.3 similarly lists the categories of past weather, which by definition refers to the 6-hour period before each of the principal synoptic hours: 0000, 0600, 1200 and 1800 Z.

Table 2.2   Present weather categories   Appendix 2.4

00–19 No precipitation at station at time of observation
00   Cloud development unobserved during past hour
01   Clouds generally dissolving during past hour
02   Clouds generally unchanged during past hour
03   Clouds generally developing during past hour
04   Visibility reduced by smoke
05   Haze
06   Widespread dust
07   Dust or sand raised by local winds, but not by dust or sandstorm or whirls
08   Dust or sand whirls seen in last hour, but no dust or sandstorm
09   Dust or sandstorm seen at or near station in past hour
10   Mist
11   Shallow patchy fog or ice fog
12   Shallow continuous fog or ice fog
13   Lightning but no thunder
14   Precipitation within sight not reaching surface
15   Precipitation reaching the surface in distance
16   Precipitation reaching surface near but not at station
17   Thunderstorm but no observed precipitation
18   Squalls seen at time of observation or during past hour
19   Funnel cloud(s) seen at time of observation or during past hour

20–29 precipitation, fog or thunderstorm at station in past hour but not at time of observation
21   Drizzle (not freezing) or snow grains not in showers
22   Rain (not freezing) not in showers
23   Rain and snow or ice pellets, not in showers
24   Freezing drizzle or freezing rain
25   Rain showers
26   Rain and snow showers or snow showers
27   Hail and rain showers or hail showers
28   Fog or ice fog
29   Thunderstorm

30–39 duststorms, sandstorms, drifting or blowing snow

40–49 fog or ice fog at time of observation

50–59 drizzle at time of observation
50   Not freezing, intermittent, slight at time of observation
51   As 50 but continuous
52   As 50 but moderate at time of observation
53   As 52 but continuous
54   As 50 but heavy at time of observation
55   As 54 but continuous
56   Slight freezing drizzle
57   Moderate or heavy freezing drizzle
58   Slight drizzle and rain
59   Moderate or heavy drizzle and rain

60–69 rain at time of observation: as 50–59 but with drizzle replaced by rain, and rain (in 58 and 59) replaced by snow

70–79 solid precipitation not in showers
70   Snow flakes, intermittent, slight at time of observation
71–75 Inclusive: pattern is in 51–55
76   Ice prisms with or without fog
77   Snow grains with or without fog
78   Isolated starlike snow crystals with or without fog
79   Ice pellets

80–90 showery precipitation
80   Rain showers, slight

43

Table 2.2 continued

81	Rain showers, moderate or heavy
82	Rain showers, violent
83	Rain and snow showers, slight
84	Rain and snow showers, moderate or heavy
85	Snow showers, slight
86	Snow showers, moderate or heavy
87	Slight showers of snow pellets, whether or not encased in ice, with or without showers of rain or rain and snow
88	As 87 but moderate or heavy
89	Slight hail showers, with or without rain or rain and snow, without thunder
90	As 89 but moderate or heavy

91–94 current precipitation with thunderstorm in past hour

91	Slight rain
92	Moderate or heavy rain
93	Slight snow, or rain and snow, or hail
94	Moderate or heavy snow, or rain and snow, or hail

95–99 current precipitation and thunderstorm

95	Slight or moderate storm without hail but with rain and/or snow
96	Slight or moderate storm with hail
97	As 95 but heavy storm
98	Storm with sandstorm or duststorm
99	Heavy storm with hail

Table 2.3 Past weather categories

0	cloud covering half or less of the sky throughout the period
1	cloud covering more than half of the sky for part of the period and less than this for the rest
2	cloud covering more than half of the sky throughout the period
3	sandstorm, duststorm or blowing snow
4	visibility less than 1 km because of fog, ice fog or thick haze
5	drizzle
6	rain
7	snow or mixed rain and snow
8	showers
9	thunderstorm with or without precipitation

Problems

LEVEL 1

1. A manufacturer seeks your advice on plans to build a transparent Stevenson screen. Advise him rationally and politely.
2. Why should a Stevenson screen have a floor?
3. The vapour content of a certain room full of air is rising, though its air temperature and pressure are constant. Consider in turn whether each of the following should be rising or falling: vapour density, relative humidity, wet-bulb depression, dew-point depression.
4. What are the azimuths of the following winds: E, SE, S, NNW (i.e. half-way between NW and N)?

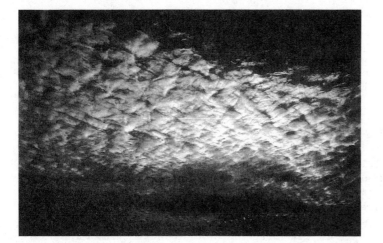

Fig. 2.16 Altocumulus — visually distinguishable from stratocumulus because its greater height makes the patchiness (caused by overturning within the layer) look smaller.

5. If you read a mercury barometer without correcting for a too high mercury temperature, will you over- or underestimate the atmospheric pressure? Mercury density decreases with increasing temperature.

6. Use appendices 2.2, 2.3 and 2.4 to assess current wind strength, cloud types and present and past weather at some convenient daytime and place.

7. Ideally how high should be the mast of a fully instrumented meteorological ocean buoy?

8. List all the ways of getting information about distributions of temperature, cloud and precipitation throughout the troposphere.

LEVEL 2

9. Suppose that a pitot tube anemometer gives an output of 1 unit in a wind speed of 5 m s^{-1}. What maximum output should be allowed for in its design to enable it to register the presence of a hurricane?

10. Careful work with a water-filled barometer shows that atmospheric pressure falls by about 1.2 cm of water for each 10 m rise near sea level. Use realistic values for g and p in eqn (2.1) to show that the pressure fall is about 78 Pa (1.2 mbar).

11. The rain collected by a 127 mm (5 in) diameter rain gauge on a certain occasion filled the 25.4 mm diameter measuring cylinder to an actual depth of 300 mm. Find the true depth of rainfall, and the consequent rise in water level, if half of this rain falling on a 100 km² catchment is collected by a 500 m square reservoir with vertical sides.

12. The surface of the Dead Sea lies 392 m below MSL. If the air density is assumed to be 1.2 kg m^{-3}, what correction must be applied to the station pressure at Dead Sea level to find the associated MSL value?

13. What must be the time of launch of a radiosonde so that its flight to the 20 km level (from MSL) is completed by 1200 Z? Given that for the first half of this period it is borne by a 20 m s^{-1} westerly wind, and for the second half it is borne by an 80 m s^{-1} southwesterly, find the horizontal distance and direction of the final sonde position from its launching site. Such extreme translations

give the Japanese meteorological service problems when the subtropical jet stream is blowing strongly overhead.

14. The lag of a certain moderately fast response thermometer is 3 s. What is the error of the thermometer 10 s after an instantaneous ambient temperature rise of 3 °C?

LEVEL 3

15. In strong sunlight, sunshine often seems more intense to the human senses inside a greenhouse than in the open air. Actually the reverse must be the case. Explain.

16. Reconsider problem 10 to show that the density of water is 833 times the implied density of air, regardless of the value of g. Consider what would be observed if the same experiment were conducted inside a station on the Moon filled with air at normal terrestrial pressure and temperature. Assume the station to be on the lunar surface where the local g is 0.17 of the terrestrial value.

17. In cool windy weather, large rain drops on the outside of a singly glazed window are often accompanied by mist patches on the warm inside of the window. Explain.

18. In the absence of a dense gauge network, the spatial variation of rainfall round small objects such as hedges and houses can only be guessed. However, typical patterns of snowfall should provide some evidence. Consider factors which support and others which reduce the value of such evidence.

19. In section 2.3 it is suggested that a small thermometer responds rapidly merely because of its small heat capacity. This is too simple, since its small size must tend to slow the response by limiting the heat fluxes between the thermometer and the ambient air. Argue the case more thoroughly for the simple case of a spherical thermometer to show that the speed of response is likely to be inversely proportional to the bulb diameter.

The constitution of the atmosphere

3.1 The well-stirred atmosphere

It is found that up to about 100 km above sea level the atmosphere consists largely of a mixture of gases in remarkably uniform proportions, together with a small and variable quantity of water vapour concentrated almost entirely in the troposphere. We will call the uniform mixture *dry air* to distinguish it from *moist air* which includes water vapour as well. In addition to these gases there are small quantities of water and ice in the form of clouds and precipitation, and a population of even tinier particles known as *aerosol* particles (strictly the term aerosol refers to the air and the particles), all heavily concentrated in the troposphere. And there are traces of many other gases, of interest mainly to atmospheric chemists. Figure 3.1 represents the vertical distributions of several of the most important of the materials just mentioned, together with other distributions and processes to be mentioned in this chapter.

Over 99.9% of the mass of dry air in the atmosphere consists of a mixture of molecular nitrogen, oxygen and argon in the proportions shown in Table 3.1 — proportions which vary only minutely up to the 100 km level. Of course the absolute amount of any constituent (in kilograms per cubic metre for example) diminishes very rapidly as we ascend through this large height range, but so does the density of the air (see section 4.5), with the result that the ratio of the two is nearly uniform. Let us call this ratio the *specific mass* of the constituent, and examine the significance of the uniformity of specific mass of some of the most important constituents.

DIFFUSIVE AND CONVECTIVE EQUILIBRIA

The atmosphere is not a static fluid whose properties are to be explained in terms of some dynamic origin long since past; it is incessantly dynamic and active, and yet continually maintaining a nearly steady state. Therefore its properties, including the uniform distributions now in question, must arise from a balance between continuing competing processes. Let us distinguish between processes promoting

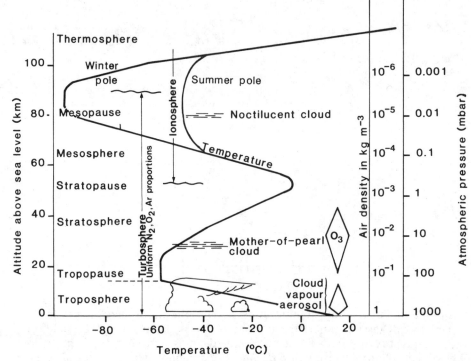

Fig. 3.1 Schematic vertical distributions of atmospheric temperature, pressure, density and atmospheric material. Note the large seasonal variation in the temperature of the high-latitude mesopause, induced by variation of direct input of solar hard UV.

uniformity and those promoting variability. Clearly the former must be dominant in the cases of the uniformly distributed constituents of dry air.

It is fairly obvious that mixing tends to promote uniformity. For example an uneven distribution of ink in a beaker of water is replaced by a uniform distribution almost as soon as stirring begins, whether by a stirring rod or convection from a heated base; and the uniformity is maintained by the stirring for as long as it continues. Uniformity would also result from the intermingling of ink and water by molecular diffusion, even if there were no stirring, but the equilibrium state would take months rather than seconds to achieve, because molecular diffusion is so very much slower than bulk stirring.

In domestic circumstance, diffusion in gases is much faster than diffusion in liquids, because of the much freer molecular motion, but it is still easily swamped by stirring if the dimensions of the gas volume exceed a few millimetres. Atmospheric dimensions are enormously larger than this, with the result that diffusion is relatively ineffective by comparison with stirring on all but the smallest scales. At least this is so in the lower atmosphere; at great heights the air is so tenuous that molecules can move considerable distances between consecutive collisions, and since they move at about the speed of sound, the speed of diffusion can be very high indeed.

In a room-sized volume of air, mixing by either stirring or diffusion tends to even out all distributions, just as in the case of ink in water. However, if the volume of air is at least 1 km tall, gravity will produce a significant lapse of air density from bottom to top, and the two types of mixing tend to produce different vertical distributions.

In the case of diffusion in a tall air column, each molecular species settles into an equilibrium distribution in which the downward drift of its molecules under the pull of gravity is balanced by the upward net diffusion maintained by the upward

Table 3.1 Composition of the turbosphere

Gases Component	Specific mass	Molecular weight	Specific gas constant
Dry air			
N_2	0.755	28.02	297
O_2	0.231	32.00	260
Ar	0.013	39.95	208
CO_2	0.0005	44.01	189
plus hundreds of trace gases			
Moist air			
Dry air	0.997	28.96	287
H_2O vapour	$\simeq 0.003$ overall	18.02	461
	(~ 0.03 maximum)		
Particles and droplets			
Cloud	$\sim 5 \times 10^{-5}$ overall	Water and ice	
	(~ 0.004 maximum)		
Precipitation	$\sim 3 \times 10^{-6}$ overall	Water and ice	
	(~ 0.004 maximum)		
Aerosol	10^{-8} overall	Various solids or concentrated	
	($\sim 10^{-7}$ maximum)	solutions in water	

lapse in numbers of those particular molecules per unit volume (*number density*), regardless of any other molecular species present. Heavier molecules diffuse toward an equilibrium with a steeper lapse of number density than do lighter molecules, and in fact each molecular constituent achieves a distribution defined by its own particular scale height, which is inversely proportional to its molecular weight (section 4.3). In such *diffusive equilibrium* under gravity, therefore, the specific masses of the heavier molecular species are greater at low altitudes, whilst those of the lighter molecular species are greater at high altitudes. This is essentially the state of things in the atmosphere above the 100 km level, where diffusion is rapid in the tenuous air and stirring is relatively weak.

In the case of a tall column of air in which mechanical or convective stirring predominates over diffusion, the different molecular species are being inter-mingled much faster by the random movement and mixing of relatively large eddies than they are being separated by molecular diffusion under gravity. Consider the life cycle of an eddy from the time of its formation from the debris of precursor eddies to the time of its destruction by mixing. In that brief time (~ 10 s in the case of small-scale turbulence) the parcel may move ~ 10 m, conserving its original mixture of molecular species. The combination of conservative motion and sharing-through-mixing, endlessly repeated throughout the stirred volume for as long as stirring persists, leads to the uniformity of specific masses which is charac-teristic of *convective equilibrium* — the term applied to stirred equilibrium maintained by mechanical or convective stirring or both. Since the eddy parcel expands on differential rising through the ambient pressure lapse, and is compressed on sinking, the vertical layering of molecular species in diffusive equilibrium is replaced in convective equilibrium by the vertical layering of the density (section 4.5) of the otherwise uniform mixture.

Clearly the uniformity of the specific masses in Table 3.1 strongly suggests that the great bulk of the atmosphere is in convective rather than diffusive equilibrium. The region in which this dominance prevails is known as the *turbosphere*, and it extends from the Earth's surface to an ill-defined *turbopause* at about the 100 km level, above which diffusive equilibrium prevails. Stirring in the turbosphere is maintained by atmospheric motion of all types and scales, including the great weather systems of the troposphere, as well as the small-scale turbulence cited above.

So far we have considered convective equilibrium distributions of gaseous constituents only. However, much the same balance between gravitational separation and turbulent mixing tends to apply to dust particles etc. whose fall speeds through still air are much smaller than typical vertical speeds of parcels in the stirring air. Particles with larger fall speeds tend to concentrate more at lower levels, the gradients of their specific masses obviously increasing with particle fall speed and decreasing with increasing vigour of atmospheric stirring. In fact the situation rather resembles the distribution of gaseous species in diffusive equilibrium, even though the condition may be one of convective equilibrium with respect to gas molecules. This kind of quasi-diffusive distribution is observed in some of the range of atmospheric particles discussed in sections 3.6 and 3.7.

The distributions discussed so far result from the homogenizing effects of diffusive and convective mixing and the selective effects of gravity acting on the various constituents. There are of course other selective processes at work in the atmosphere, including most obviously a great range of chemical reactions. In many cases, such as most of those in Table 3.1, the selective effects are not strong enough to compete with atmospheric mixing in the short term, but in some cases they are able to compete at least marginally, and in many they have profound effects in the long term. In the rest of this chapter we consider some important examples of the distributions of atmospheric gases and particles (including droplets), and of the ways in which these arise from competition between mixing on the one hand, and gravity and other selective effects on the other.

NEAR AND ABOVE THE TURBOPAUSE

Above about 100 km, diffusive equilibrium produces marked stratification of constituent gases according to molecular weight, with the heaviest components concentrating at the lowest levels. Lighter gases therefore predominate aloft, though the concentrations of hydrogen and helium (the two lightest and therefore the two with the fastest-moving molecules) have been reduced far below their cosmic abundances by persistent leakage to space throughout the Earth's history.

As height increases there is increasing *photodissociation* of molecules into their constituent atoms, and this considerably affects the composition of air above about the 80 km level. For example, above about 120 km the reaction

$$O_2 + \text{photon (solar UV)} \rightarrow 2O$$

maintains more than half of the population of oxygen atoms and molecules in the form of atomic oxygen (O), and in so doing strongly absorbs solar ultraviolet with wavelengths between 0.1 and 0.2 micrometres (μm), thereby contributing to the warming which maintains the *thermosphere* (Fig. 3.1). Incidentally the atomic oxygen acts like a gas with half the molecular weight and therefore half the density of molecular oxygen (O_2), so that this widespread photodissociation significantly reduces the density of air at these levels.

Photoionization also increases with altitude. For example the reaction

$$O + \text{photon (solar UV)} \rightarrow O^+ + \text{electron}$$

maintains a population of ionized oxygen atoms and free electrons, absorbs solar ultraviolet with wavelengths less than about 0.1 μm, and helps warm the ambient air over a wide height range. Actually there are sufficient electrons maintained by photoionization in the upper half of the turbosphere to justify its description as the lowest part of the *ionosphere*. Its electrical conductivity is sufficient to reflect radio waves emitted by radio transmitters on the Earth's surface, which is very important for telecommunications since radio propagation around the spherical Earth would be impossible otherwise. The specific masses of ions of all types increases sharply with height above the turbopause, and at altitudes greater than about 150 km they significantly affect the movement of air in the presence of the Earth's magnetic field. The atmosphere at these and higher levels is a diffuse *plasma* with properties which are very different from those of the electrically nearly neutral lower atmosphere [11].

3.2 Dry air

Because of the rapid fall of air density with increasing height only one millionth of the mass of the atmosphere lies above the turbopause; the rest lies in the well-stirred turbosphere and consists almost entirely of dry air, and in particular of molecular nitrogen (N_2), molecular oxygen (O_2) and Argon (^{40}Ar), as shown in Table 3.1. Let us consider these major gases in order of increasing chemical activity.

Argon is chemically inert and is therefore the simplest of the major gases in its chemical behaviour. Though stirred ceaselessly through the turbosphere it is not otherwise affected in the short term, so that the uniformity of its specific mass throughout the turbosphere, despite its having the heaviest molecules of the major atmospheric gases, is a measure of the predominance of convective over diffusive equilibrium prevailing in the turbosphere.

In the long term the picture is less simple. The present very considerable abundance of argon consists almost entirely of the isotope ^{40}Ar, which has accumulated throughout Earth's history by radioactive decay of potassium (^{40}K) within the solid bulk of the planet and subsequent gradual diffusion to the surface (section 3.8). However, the rate of accumulation of ^{40}Ar is so slow in comparison with the rate of its dispersion throughout the turbosphere by mixing, that any resulting excess of specific mass near the surface would be infinitesimal. The half-life for radioactive decay of ^{40}K is 1.3×10^9 years — nearly one third of the age of the Earth.

Nitrogen in its molecular form accounts for nearly three quarters of the mass of the atmosphere. Although chemically not very active in terrestrial conditions it is by no means inert, and is more heavily involved in the biosphere than its early name Azote (lifeless) suggests. Atmospheric N_2 is *fixed* (chemically combined with other substances) and removed from the atmosphere in three main ways whose magnitudes have been estimated in detailed studies over recent decades [12]. Expressed in billionths (10^{-9}) of the mass of the atmospheric reservoir of N_2 per annum, these removals were as follows in 1968:

1. fixation by biological micro-organisms, mainly on land   14
2. fixation by lightning and other ionizing processes in the atmosphere   2
3. industrial fixation, mainly to make artificial fertilizers   8

Before man's intervention, the annual fixation was presumably balanced by the

Fig. 3.2 A high-atmosphere research balloon soon after launch. The instrument package contains special radiometers to detect water vapour and other trace gases in the stratosphere and mesosphere, and hangs about 15 m below the base of the slack balloon envelope. As the balloon ascends, the bubble of helium expands, fully filling the spherical envelope at the flotation altitude of about 50 km.

annual *denitrification* (return to N_2) by anaerobic bacteria. However, the industrial fixation and the four units of the biological fixation which arise from agricultural cropping of legumes, both of which have developed largely in the present century, are believed to be making current annual denitrification rates fall about 2 units short of the total fixation rate.

Consider the consequences of annual fixation and denitrification of about 24 billionths of the atmospheric reservoir of N_2, ignoring for the moment the probable slight imbalance between the two. On average a particular N_2 molecule must spend about 42 million years in the atmosphere between one denitrification and the next fixation. This is the *residence time* for nitrogen in the atmospheric branch of the nitrogen cycle, found by dividing the mass of the reservoir by the rate of influx or efflux (Appendix 3.1). Although any particular molecule may move through the atmosphere very much more quickly or slowly than this, the residence time is a useful concept in this and many other studies of nearly balanced cyclic systems.

In the present discussion, although fixation and denitrification rates are very impressive when expressed in tonnes per annum, the residence time for atmospheric N_2 is so long that stirring of N_2 throughout the turbosphere is almost instantaneous by comparison. We should therefore expect the specific mass of N_2 to be very uniform throughout the turbosphere, as indeed it is. The distributions of N_2 and ^{40}Ar are therefore similar in this respect.

Note that if the current slight excess of fixation over denitrification were maintained for 250 million years, half the N_2 would be removed from the atmosphere. However, long before this stage could be reached, the relatively small reservoir of nitrogen in organic debris would be grossly inflated, producing oxygen depletion and *blooms* of algae in rivers and seas. In fact the problem is much more immediate than this would suggest. Industrial fixation may well treble in the last thirty years of the twentieth century, and there have already been many localized blooms in rivers and lakes affected by run-off from intensively fertilized land nearby.

Oxygen is chemically the most reactive of the major atmospheric gases. In fact it is so reactive that at first glance it is surprising that molecular oxygen (O_2) should comprise nearly one quarter of the mass of the atmosphere. However, virtually all inorganic terrestrial materials are already fully oxidized, so that the residual reactions are mainly with the continuously forming organic materials.

O_2 is consumed by many types of *respiration* through which living organisms produce energy by highly regulated oxidation of food, and by *decay*, in which bacteria similarly act on the food stored in the corpses of dead organisms. Detailed estimates [13] suggest that the annual consumption is about 200 units, where a unit is about one millionth (10^{-6}) of the mass of O_2 in the atmosphere. This is nearly balanced by the release of O_2 as a by-product of *photosynthesis* in the presence of sunlight by green plants on land and by phytoplankton in water. Photosynthesis is the process by which these organisms build themselves from carbon dioxide and water, using some of the energy in sunlight

$$CO_2 + H_2O + photon \text{ (solar visible)} \rightarrow CH_2O + O_2$$

where CH_2O represents the sugars which are the basic foods for the photosynthesizing organisms.

O_2 is also consumed by combustion of all sorts, from forest and bush fires to the coal and oil-fired furnaces of industry, but the total consumption is currently about 10 units annually, which is only about 1/20 of the consumption by respiration. Clearly the residence time of O_2 in the atmosphere is dominated by photosynthesis, respiration and decay, and the quoted values for reservoir and throughput imply a residence time of about 5000 years. This is much shorter than the residence time for N_2 (proving oxygen's much greater reacivity), but it is still very long compared with the speed of stirring of the atmosphere, so that we should expect the specific mass of O_2 to be very uniform throughout the turbosphere, which is almost always the case. A couple of exceptions prove the rule: in the immediate vicinity of a large bush fire, or a runaway fire storm in a bombed city, the rate of consumption of O_2 by combustion may be so high as to cause temporary local depletion.

On account of its chemical reactivity, the present large amounts of atmospheric O_2 cannot have survived the Earth's convulsive origin; and for the same reason no O_2 is to be found in gases vented by volcanoes. Almost all atmospheric O_2 is believed to have arisen instead by means of a slight excess of photosynthesis over respiration and decay which has persisted throughout much of the long development of life on Earth, and which has been especially marked in the almost explosive development of life in the last few hundred million years (section 3.8). This excess has probably been maintained by the burial of organic debris before completion of decay. Radioactive tracing shows that of the two oxygen atoms produced in each photosynthetic reaction, one comes from water input and the other comes from the carbon dioxide. There are very large reservoirs of each on the Earth, both largely in the oceans (which contain over 50 times more carbon dioxide than the atmosphere, in dissolved form), though of course the small quantities of each in the atmosphere are crucial for the land-based part of the biosphere. Using these reservoirs and the power of sunlight, life maintains the reservoir of free oxygen on which it now depends.

The current phase of man's wholesale combustion of fossil fuels, which cannot last more than a few hundred years even in the case of the most abundant reserves (coal), represents the very sudden completion of decay left incomplete by organisms which began life hundreds of millions of years ago. The oxygen freed very gradually by the action of ancient sunlight is now being suddenly reclaimed, with consequences which are barely understood, except that they will probably be felt more immediately in the reservoir of free CO_2 in the atmosphere and oceans than in the much larger reservoir of O_2 (section 3.4).

3.3 Ozone

A minute fraction of total atmospheric oxygen is maintained in the form of ozone (O_3) by photochemical reactions involving solar ultraviolet. A maximum of specific mass of O_3 is maintained in the stratosphere between about the 20 and 40 km levels by a series of photochemical and chemical reactions:

$$O_2 + \text{photon (solar UV)} \rightarrow 2O$$
$$O_2 + O + M \qquad\quad \rightarrow O_3 + M$$
$$O_3 + \text{photon (solar UV)} \rightarrow O_2 + O$$
$$O \;\; + O_3 \qquad\qquad\quad \rightarrow 2O_2$$

The first is the one already mentioned in section 3.1 as being very important at even higher altitudes. It maintains a proportion of oxygen in the form of atomic oxygen which is very small in the upper stratosphere and almost zero at lower levels. In the second reaction atomic and molecular oxygen combine to form ozone. The third body M of the triple collision can be any other gas molecule, and is needed to take away surplus energy and so prevent the newly formed ozone from flying apart immediately because of its initial excitement. Once formed, the ozone is a powerful absorber of *soft* solar *ultraviolet* (wavelengths between about 0.2 and 0.3 μm) through the photodissociation described in the third reaction. Some of the atomic oxygen produced in this way combines with ozone to reform molecular oxygen in the fourth reaction.

This series of reactions maintains a maximum *number density* of ozone molecules (i.e. number per unit volume) between the 20 and 30 km levels, and maximum specific masses ~ 10 km higher. The maximum is maintained in a

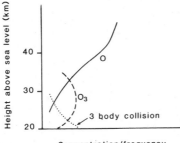

Fig. 3.3 Schematic vertical profile of ozone concentration and mechanisms maintaining its stratospheric maximum.

region where the first and fourth reactions balance to maintain a small but not negligible proportion of oxygen in the form of *odd oxygen* (O and/or O_3) as distinct from molecular oxygen (O_2). The second and third reactions then maintain a dynamic equilibrium between O_3 and O, which favours O_3 in the height region specified because it is high enough to have significant O and yet low enough for the triple collision to be reasonably frequent (Fig. 3.3).

The O_3 maximum maintained in this somewhat tortuous way is only a very small proportion of the atmospheric mass at these levels, maximum specific masses being $\sim 10^{-5}$. And yet this apparently remote and slight proportion is extremely important to most living things, because the soft ultraviolet which it very effectively absorbs is potentially very damaging to exposed living tissue. Although individual photons of the soft ultraviolet are less energetic than those of the hard ultraviolet, they are much more numerous than the latter, being nearer in wavelength to the peak of the solar spectrum (Fig. 8.1). If the apparently insignificant veil of ozone were to be removed, then all living things not shielded by several metres of water or its equivalent would die almost immediately of catastrophic sunburn as their weak molecular bonds were disrupted by the destructive deluge of soft ultraviolet. The development of this literally vital filter is obviously related to the growth of atmospheric O_2, and through this to the development of the biosphere which it largely protects, and an understanding of these complex relationships is central to current attempts to trace the origin and development of life on Earth (section 3.8).

The maintenance of the O_3 maximum in the stratosphere demonstrates the presence of selective reactions which can act quickly enough to compete with the moderately vigorous stirring at those levels. The competition is quite nicely balanced, however, and ozone would be strongly concentrated in the low-latitude stratosphere, especially in winter, if meridional air currents did not carry it poleward from the low-latitude formation zones.

In recent years, observations over most of Antarctica have shown a sharp temporary fall in the amount of stratospheric ozone in late winter and spring (September to December — Fig. 3.4). Although the detailed chemistry is difficult, it is now widely accepted that this ozone 'hole' is being produced by domestic and industrial *CFCs* (chlorofluorocarbons used in refrigerators and some aerosol cans). These gases are vented mainly from the heavily populated northern hemisphere, and are gradually mixed throughout the troposphere and into the stratosphere. Although CFCs are manufactured because they are chemically stable in the situations where they are used, they can break down in the stratosphere to release chlorine, which is potentially very active chemically. The release of chlorine is encouraged in the low Antarctic stratosphere in winter, because a thin haze of ice-cloud particles forms in the unusually cold conditions there. As winter begins,

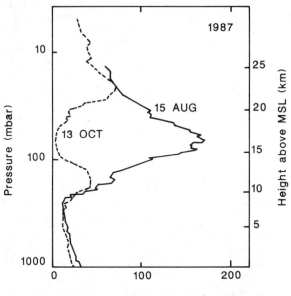

Fig. 3.4 Observed profiles of O_3 content above Halley (75° S, 28° W) Antarctica in 1987, during the heavy springtime depletion (13 October) and after summer recovery (15 August). (After [14])

the low stratosphere settles into a huge closed westerly circumpolar vortex, spinning and cooling in relative isolation from the surrounding atmosphere. Chemical reactions occur on the myriads of tiny ice surfaces and release relatively large quantities of molecular chlorine (Cl_2) which is photodissociated to the very active atomic chlorine (Cl) on the return of weak sunlight in spring. Ozone is then consumed at several percent per day by a series of fast catalytic reactions. The destruction ceases when the cold circumpolar vortex breaks down in early summer, as intrusions of air from lower latitudes bring heat and ozone from lower latitudes.

In view of its importance for life on Earth (section 3.8), there is great concern over any reduction of the Earth's ozone shield. Though Antarctica is virtually uninhabited, and the loss there is only seasonal, it represents a net loss of global stratospheric ozone, and there are some signs of a smaller seasonal loss beginning to appear in the North polar regions, though the winter circumpolar vortex is less isolated there. In principle too, losses could escalate as local solar absorption reduces with the ozone, perpetuating the cold conditions on which the loss depends. It is therefore quite proper that the discovery of the ozone hole has triggered international efforts to stop the manufacture of CFCs, and has played a major role in bringing environmental matters up the political agenda nationally and internationally.

Note that ozone and CFCs are significant greenhouse gases (section 8.5), and that ozone is produced sporadically in the low troposphere in photochemical smogs (the Los Angeles type — section 11.5), where its action as a strong oxidization agent makes it damaging to plant and animal tissues.

3.4 Carbon dioxide [15]

Carbon dioxide (CO_2) interacts with the biosphere in ways which complement the O_2 reactions, being produced by combustion and respiration and consumed by photosynthesis. Indeed the mass fluxes of CO_2 exceed the fluxes of O_2 specified in the previous section in precisely the ratio of their molecular weights (44/32). However, the mass of CO_2 in the atmosphere is so much smaller than the mass of O_2 (Table 3.1) that the residence time of CO_2 in the atmosphere between successive involvements with the biosphere is only about five years. This time is so short that we might expect to observe temporary localized gradients of the specific mass of atmospheric CO_2, as is indeed the case. The situation is complicated somewhat by continual exchange with the relatively very large reservoir of CO_2 dissolved in the oceans, which takes place at a considerably slower rate than the exchange with the biosphere.

Photosynthesis can proceed so rapidly in sunlit vegetation, especially when the vegetation is dense and well-watered, that CO_2 levels in the immediate vicinity may fall 20% below average values. At night, respiration may raise CO_2 levels as far above the average. The large diurnal variation from maximum depletion to maximum production of CO_2 overcomes the mixing capacity of the local air, which is considerably reduced in dense vegetation, to produce substantial but highly localized diurnal variations in CO_2 concentrations.

A similarly cyclic imbalance produces the substantial seasonal variations in CO_2 levels in the troposphere (Fig. 3.5). The burst of photosynthesis by land plants in middle latitudes in spring and early summer consumes more CO_2 than is released by respiration; and the reverse happens in autumn and early winter when respiration and decay exceed photosynthesis. The effect is particularly pronounced in the northern hemisphere because of the concentration of land masses there with large seasonal climatic range.

The weak oscillation observed in Antarctica with a six-month phase shift (Fig. 3.5) gives a clue to the speed of mixing of CO_2 across the globe in the low troposphere — a speed which must apply to all unreactive and moderately reactive gases since mixing rates are independent of the type of gases. On the one hand the speed of mixing is not so great that the dominant northern hemisphere maximum appears even in the Antarctic. But on the other hand the speed is sufficient to carry the southern hemisphere oscillation from its source in middle southern latitudes to the biological desert of Antarctica. This suggests that in the troposphere, atmospheric mixing throughout a hemisphere is largely complete in less than a year, whereas a considerably longer period is needed to complete mixing across the equator — inferences which are consistent with the restricted air flow between the hemispheres apparent in the wind pattern discussed in section 4.7.

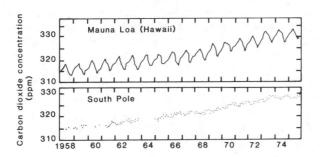

Fig. 3.5 Concentrations of carbon dioxide monitored since 1958 in sites remote from major industrial zones. (After [8] and C.D. Keeling)

57

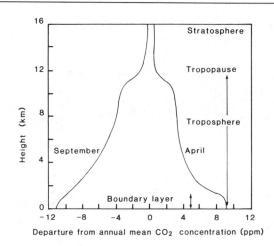

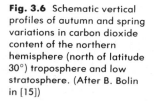

Fig. 3.6 Schematic vertical profiles of autumn and spring variations in carbon dioxide content of the northern hemisphere (north of latitude 30°) troposphere and low stratosphere. (After B. Bolin in [15])

The vertical distribution of the seasonal oscillation of CO_2 levels (Fig. 3.6) confirms and extends this picture. The seasonal oscillation is largest in the planetary boundary layer, whose extremely well-mixed volume extends about 1 km above the surface, where all sources and sinks of CO_2 are localized. The oscillation is smaller, though still quite large, throughout the troposphere, which extends to a little over 10 km in the middle and high latitudes depicted, and is noticeably smaller in the high troposphere than in the low — consistent with the picture outlined in later chapters of tropospheric weather systems pumping air sporadically from the boundary layer up through the troposphere, reaching highest levels least often. Only a small fraction of the seasonal oscillation survives in the stratosphere, where stirring and connections with the troposphere are quite feeble.

The steady rise in the seasonally averaged levels of CO_2 apparent in Fig. 3.5 is a consequence of the rapid combustion of fossil fuels which began on its present scale with the industrial revolution. Rapid though it is, the increase in CO_2 is much slower than its speed of mixing through the atmosphere, with the result that it is observed as clearly in the most remote locations as it is in the most industrial (Fig. 3.5). It is quite another matter in the oceans, where only the surface layers are in even sluggish equilibrium with the atmosphere, and the nearly still, voluminous depths will not be reached for centuries.

Figure 3.7 sets the recent rise in atmospheric CO_2 levels in the context of the large natural variations associated with climate change in the last 160 000 years. Levels have shot up by over 20% since pre-industrial times, and are now well above the values reached in the previous, rather warmer interglacial, 120 000 years ago. In fact CO_2 levels are now probably higher than at any time since the global climate slid into the current series of ice ages several million years ago (section 4.9). On the time scale of Fig. 3.7 the current rate of rise (about 10% per decade) is too fast to be resolved by the thickness of the printed line. The rise in the atmospheric reservoir probably accounts for less than half of the anthropogenic total, the other half being dissolved in the oceans, which will eventually (hundreds of years, as above) take a much larger fraction. All this represents a very large and rapid shift away from the natural CO_2 economy, and there is great concern that it may be inducing significant changes in global climate, because of the important role of CO_2 in the greenhouse effect (sections 4.10 and 8.5). As with the case of ozone

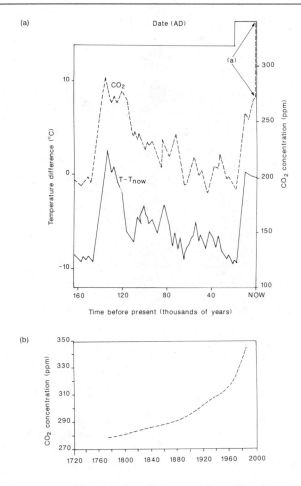

Fig. 3.7 Variations in atmospheric CO_2 in the last 160 000 years.
(a) High correlation between CO_2 concentrations and global mean surface temperatures as deduced from O^{18} abundance in ice cores from Vostock (Antarctica). After the end of the last glaciation CO_2 levels approached values prevalent in the pre-glacial optimum 140 000 years ago.
(b) Recent Antarctic data and direct measurements from Hawaii since 1958 show a large and accelerating rise in CO_2 levels since the onset of industrialization 200 years ago. Current levels are believed to be well above the highest values occurring since the onset of recent glaciations several million years ago. (After [17] and [18])

depletion, this has attracted international concern, but reduction of anthropogenic CO_2 is a much more difficult matter than reduction of CFCs, which play only a marginal role in human development. Global industrialization currently depends on continued and increasing consumption of fossil fuels and consequent increasing production of CO_2. With continuing public concern about the safety of power production by nuclear fission, and in the absence of useful progress in power production by nuclear fusion, the only way out of this impasse is to encourage energy conservation and the sustainable exploitation of alternative energy resources. In the context of this section, there is considerable scope for the burning of fast-growing trees, which simply returns to the atmosphere the CO_2 which was extracted during the growth of the tree over preceding seasons.

3.5 Sulphur dioxide [19]

Sulphur dioxide (SO_2) is one of the most biologically destructive of the common atmospheric trace gases. Background concentrations of $\sim 10^{-4}$ ppm (specific mass

Fig. 3.8 Norway spruce in Bavaria showing 'classic' crown thinning attributed to acid rain and associated air pollution.

$\sim 10^{-10}$) are maintained by natural oxidation of reduced sulphur compounds such as hydrogen sulphide (H_2S) and dimethyl sulphide (($CH_3)_2S$), which are themselves emitted by living and decaying organisms. However, artificial combustion of sulphur-rich coal and some oils, and smelting of sulphide ores, add greatly to local and total production of SO_2.

Irritation and damage to plant and animal tissues are caused when concentrations are raised significantly above the natural background, though complex details of this sensitivity are still being established. If sources are too strong or inadequately ventilated, local concentrations may rise dramatically, and in the past there have been examples of total destruction of vegetation around potent industrial sources. Even after the implementation of stringent air pollution controls in recent decades, concentrations of about 1 ppm are still recorded in some modern cities, with the result that anthropogenic SO_2 is one of the biggest causes of serious lung damage after cigarette smoking. The situation is worse in countries where governments have not been able or willing to meet the considerable cost of implementing pollution controls (e.g. in Eastern Europe and much of the Third World). The most serious pollution episodes occur in calm conditions or light winds, especially when there is a stable 'lid' only a few hundred metres above the surface source region (sections 9.12 and 11.5).

Continuing oxidation of SO_2 tends to produce SO_3 which is hydrated to H_2SO_4 in cloud droplets, producing acidity far in excess of the natural pH limit of about 5.6 maintained by solution of atmospheric carbon dioxide. In this way the vast megalopoli spawned this century in the industrial heartlands of Europe, North America and Japan now maintain zones of seriously acidified rain and snow in broad, persistent, downwind swathes, with consequent biological damage to soils, vegetation (especially trees (Fig. 3.8)), rivers and lakes, over areas which can approach continental scale. The complex mechanisms which underlie the effects of such *acid rain* are still only partly understood, and it seems likely that oxides of nitrogen and other substances and factors are involved, but the fundamental point in respect of all such pollution is that raised concentrations result when local production or deposition outruns the atmosphere's dispersive capacity [19].

3.6 Aerosol [20]

An *aerosol* is a suspension in air of solid and liquid particles. The term is conventionally applied to the very large numbers of solid and liquid particles which are found in the lower atmosphere, and which are too small and slow-falling to be regarded as precipitation. Especially highly concentrated in the low troposphere, they range in size from aggregates of only a few hundred molecules ($\sim 10^{-3}$ μm across) to particles about ten thousand times larger (~ 10 μm). The latter are comparable in size with true cloud droplets but are distinguished from them by a much larger proportion of solid or dissolved material.

Particles with radii less than 0.1 μm are called *Aitken nuclei* because they make up the great majority of particles able to nucleate cloud droplets in a cloud chamber — a method used by the Scottish pioneer Aitken to isolate and count them. They dominate the total aerosol population numerically, but because of their small individual masses comprise only about one fifth of the total aerosol particulate mass. They are produced by natural and man-made combustion, by gas-to-particle conversion of gases produced by the metabolism and decay of living organisms, and by volcanoes. Since most of them therefore enter the atmosphere at or near land surfaces, their concentrations are particularly high in continental air, often exceeding 10^6 per litre. Specific masses fall somewhat above the planetary boundary layer, but are then fairly uniform through the middle and upper troposphere. The fall speeds of such tiny particles are ~ 10 cm per day at most (Fig. 6.10), which is slower than the random Brownian motion imparted by impacts with air molecules and is quite negligible in comparison with even the weakest updraughts. Once carried away from their source regions, the Aitken population reduces quite rapidly by Brownian collision and coagulation of adjacent Aitken nuclei to form larger particles. Aitken nuclei are also removed by being dragged by water vapour diffusing onto the surfaces of growing cloud droplets. For all these reasons maritime air masses have relatively few Aitken nuclei. Above the tropopause there is an even smaller and fairly uniform distribution of specific mass which is consistent with an input of micro-meteorites from space.

Particles with radii between 0.1 and 1.0 μm are called *large nuclei*. They are usually at least ten times less numerous than the Aitken nuclei but comprise nearly half of the total particulate mass. Fall speeds are still much smaller than updraughts, but even so there is a very marked decrease in population above typical cloud base levels in the low troposphere (never more than a few kilometres above sea level). This is associated with the efficient removal of large nuclei during the process of cloud formation above these levels, a process in which they play a very important part on account of their ability to nucleate cloud droplets at very low humidity supersaturations (section 6.9). Large nuclei are produced directly by the same processes which also produce Aitken nuclei, by coagulation of Aitken nuclei, and by the injection of small salt particles when bubbles burst on the breaking sea surface. Though they are more numerous in continental air than they are in maritime air, the difference is less than in the case of Aitken nuclei. Dense populations of large nuclei are effective in scattering sunlight, producing the dry haze which can reduce visibility quite considerably in anticyclones. In fact the haze is largely confined to the convecting boundary layer which in these conditions is confined below a pronounced temperature inversion at a height which rarely exceeds one kilometre (Fig. 12.3).

Particles with radii larger than 1.0 μm are called *giant nuclei*. Though they are less numerous again than large nuclei, their numbers rarely exceeding 10^3 per litre even in the very low troposphere, they comprise nearly half of the total particulate mass. Giant nuclei are produced and lost in the same ways as are the large nuclei, except that they are not formed by Brownian coagulation since this is inhibited by their large masses, but they are produced in significant quantities in the form of fine dust lifted by wind off arid land surfaces. They too are very effective nuclei for cloud-droplet formation, but being too large to swerve round falling raindrops they can be efficiently swept to the underlying surface by rain and showers, causing dramatic reductions in haze beneath cloud base. Fall speeds of giant nuclei are sufficiently large (Fig. 6.10) to ensure that the giant population reduces significantly by gravitational settling when updraughts are low — an effect which increases sharply with particle size. For the same reason the reduction in the giant population above the very low troposphere is even more marked than it is in the case of large nuclei.

The total mass of aerosol particles suspended in a column of air resting on a square metre of the Earth's surface amounts to only a few tenths of a gram — more than ten million times smaller than the mass of the air column itself. But as well as playing a vital role in cloud formation, and considerably affecting visibility, this minute fraction of the atmosphere significantly influences some chemical processes in the troposphere. Indeed it seems probable that periods of enhanced volcanism can influence global climate quite significantly by injecting aerosol particles into the stratosphere in sufficient quantities to interfere with incoming solar radiation (section 4.9). As in the case of ozone (section 3.3), it is remarkable how, in a complex system like the atmosphere, minority constituents can have an importance out of all proportion to their abundance.

The present discussion has emphasized how the behaviour of aerosol exemplifies competition between uniform dispersion by stirring and localization by selective influences — of which the most obvious is differential gravitational settling.

3.7 Water

From previous sections it might seem that the variability of atmospheric constituents is in inverse proportion to their abundance: all the major constituents have essentially uniform specific masses; only minor variations occur in carbon dioxide; and large percentage variations occur only in trace materials such as ozone and aerosol. There is however an extremely important exception to this pattern of behaviour: the atmospheric *water substance* (water in all its states and forms — vapour, cloud and precipitation) is very unevenly distributed through the atmosphere, being almost entirely confined to the troposphere, indeed concentrated heavily in the warm, low troposphere in particular (Fig. 3.1). For example, the overall *specific humidity* (the specific mass of water vapour) of the atmosphere is only about 0.3%, whereas the specific humidity of the warmest parts of the troposphere often exceeds 3%. And it is observed that cloud, and of course therefore precipitation, are almost entirely confined to the troposphere.

About 97% of the *hydrosphere* (all the water substance on and close to the Earth's surface) is currently in the oceans, comprising a mass of water which would

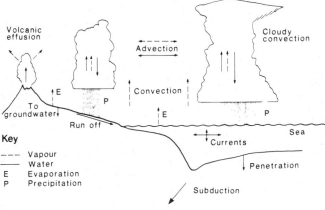

Fig. **3.9** Schematic outline of the hydrologic cycle, emphasizing the atmospheric branch.

cover the Earth's surface to a depth of about 2.8 km if evenly distributed. Of the remainder about 3/4 is in the ice-caps of Greenland and Antarctica, 1/4 is in lakes and groundwater, and only 1/3000 is in the atmosphere, almost all of it in the form of vapour. In fact the total vapour content of the atmosphere would produce a rainfall of only 3 cm if it were suddenly and completely precipitated (this being therefore the *precipitable water content* of the atmosphere (appendix 5.2)), which shows by comparison with the equivalent depth of the oceans that only 10^{-5} of the total hydrosphere is in the atmosphere.

The hydrosphere is believed to have been formed by outgassing of steam from volcanoes during the early life of the Earth, the vapour condensing on the cool exterior to form the oceans (section 3.8). The mass of the current hydrosphere is effectively constant on all meteorologically relevant timescales, except perhaps those related to very slow climatic change. However, within the hydrosphere, a much more rapid exchange takes place continually as the water substance moves through the *hydrologic cycle* (Fig. 3.9) — evaporating from the oceans and moist land surfaces, moving extensively but briefly as the vapour content of moist air, condensing to form cloud, and being precipitated back to the surface to be available for recycling either immediately or after movement through groundwater, rivers and lakes. In the ice ages which have dominated substantial periods of Earth's history, including most of the last few million years, polar regions have been cold enough to support ice caps, fed by falling snow and drained by glaciers. But although these have diverted as much as 5% (over twice the present amount) of the hydrosphere from the oceans, and have affected some parts of the Earth directly by burial under ice and many coastal regions indirectly by alteration of sea level, they represent the diversion of only a small part of the global hydrologic cycle [21].

The flux of water through the hydrologic cycle is very rapid indeed. Globally about 1 metre of rainfall equivalent falls annually, most of it in the form of rain. Since the total mass of the hydrosphere is fixed on meteorological timescales, this precipitation must be balanced on average by evaporation. Indeed the precipitable water content of the atmosphere is so much smaller than the annual throughput of water vapour that the balance between precipitation and evaporation on a global or hemispheric scale must be nearly as close instantaneously as it is on average. Comparing throughputs and capacities, we see that the residence time of water substance in the oceans is about 2800 years, but that it is only 11 days in the atmosphere.

Fig. 3.10 Cloud forming as
moist gases vent from hot
springs in a volcanically
active region. Chilling is
probably by mixing rather
than expansion.

Water is sent on its way along the atmospheric branch of the hydrologic cycle when it is evaporated from the surface of the sea or moist land, or when it sublimes from an ice surface. All of this conversion to vapour is maintained by solar heating, either directly as when moist land surfaces are dried quickly in strong sunlight, or less directly as when a drying wind (air warmed by the sun elsewhere) does the same. As discussed in Chapter 8 this evaporation accounts for a substantial part of the solar input to the Earth's surface, which would otherwise be considerably warmer in sunlight and colder at night.

Water vapour moves rapidly on the winds and updraughts and downdraughts, and is obviously closely involved in some of the most vigorous motion and activity of the troposphere. But since the vapour remains highly concentrated (in terms of specific humidity) in the warm, low troposphere, it is clear that there must be selective processes working quickly to counter the homogenizing effects of mixing. These are the rapid cooling and cloud formation which take place in rising air, and the equally effective precipitation which returns water and ice to the surface from all but the smallest clouds. This behaviour is considered in some detail in Chapters 5 and 6, but a simple outline now will serve to complete our summary of the constitution of the atmosphere and act as a preface to much of the rest of this book.

CLOUD AND PRECIPITATION

As mentioned in the previous section, many aerosol particles and droplets act as condensation nuclei, encouraging cloud formation so readily in supersaturated air that the vapour content of any air parcel is effectively limited by saturation at the parcel temperature. This has two important and related consequences.

1. In the presence of the sharp temperature fall with increasing height imposed by nearly adiabatic cooling of rising air and warming of sinking air (section 5.6), the saturation limit ensures that the great bulk of water vapour is concentrated in the warm, low troposphere. In fact specific humidities at saturation in temperatures typical of the high troposphere ($-40°$ to -60 °C) are so low that this region acts as a very effective barrier to the passage of vapour into the stratosphere and beyond.
2. As moist air rises from the surface layers in convection and other updraughts, it is brought to saturation as the air expands and cools (though just why there should almost always be an unsaturated layer between the surface reservoirs and the lowest clouds layer is surprisingly difficult to explain — section 6.3). Cloud forms and thickens as the cloudy air ascends still further, and does it so rapidly that the air is only marginally supersaturated in even the most vigorous updraughts. When clouds are more than a few hundreds of metres deep some of the cloud droplets or crystals are able to grow large enough to fall through the rising air and reach the surface as precipitation. The efficiency of precipitation varies with the surprisingly complex microscale and cloud-scale mechanisms involved (sections 6.11 and 10.4), but generally it increases with the vertical extent of the precipitating cloud. Deep frontal clouds and shower clouds quite quickly return a very considerable fraction of their cloud water and ice to the surface.

In these ways cloud and precipitation as well as water vapour are largely concentrated in the lower troposphere. There is, however, an important difference between the horizontal distribution of precipitation as compared with cloud. A large fraction of the vapour entering the atmosphere rises little more than a kilometre or two before condensing to form the clouds which usually cover more than half of the Earth's surface at any instant. However, much of this cloud is so shallow that it precipitates little if at all before re-evaporating as the cloudy air sinks or mixes with dryer surroundings. At any instant only the relatively small fraction of the surface which is covered by deeper clouds (extending into the middle and upper troposphere) receives precipitation at the moderate and heavy rates which dominate the total input in almost all locations (Fig. 2.5). The instantaneous distribution of precipitation is therefore relatively much less extensive horizontally than is the distribution of cloud. The thin and sometimes extensive veils of ice cloud often seen in the upper troposphere can give a quite misleading impression of the concentration of cloud there; they are usually very tenuous and are important mainly for their ability to veil sunlight and partly close the atmospheric window to terrestrial radiative emission to space (section 8.3).

There is almost no cloud above the tropopause. Only the strongest updraughts in vigorous showers may overshoot the tropopause a little way before falling back by negative buoyancy and spreading out as anvils just beneath the tropospause (section 10.4). The *nacreous* or mother-of-pearl clouds occasionally seen in the lower middle stratosphere (about 20–30 km above sea level) are believed to be ice cloud forming in the crests of waves of air flowing over hills far below (section 10.8). In the very cold Antarctic winter stratosphere, such clouds are believed to play an important role in ozone depletion (section 3.3). *Noctilucent* clouds (Fig. 3.11) are occasionally observed well after dusk (hence their name) at altitudes of about 80 km, but their constitution is not well established. Each of these is a rarity occurring only in high latitudes and providing very minor exceptions to the rule that clouds are confined to the troposphere.

Though cloud is extremely widespread, its observed transience on many occasions, and the speed with which shower clouds in particular are seen to form

Fig. 3.11 Noctilucent clouds over Scotland, looking north from near Glasgow (56° N) at 2315 Z, 22 June, 1981. At this time and date the nearest daylight at sea level is over 10° of latitude further north.

and precipitate themselves out of existence, suggest that residence times of the atmospheric water substance in the form of cloud and precipitation are very much shorter than the eleven days found for water vapour. It follows that the atmospheric water substance exists in the form of water vapour most of the time, or equivalently that at any instant most of the water substance is in the form of vapour.

Although the water substance is actually only a very minor constituent of the atmosphere in terms of specific mass, it is so heavily involved in several important aspects of atmospheric behaviour that it must be regarded as a major constituent in effect. Many of its roles have been mentioned in this section, and are dealt with in detail later, but it is obvious even at this stage that the confinement of the water substance to a rather shallow layer overlying the Earth's surface must have important consequences, the most obvious of which is the localization of weather as we know it in the same shallow layer.

3.8 The evolution of the atmosphere

ORIGINS

The Earth is believed to have formed in the aftermath of a supernova (the catastrophic explosion of a heavy star) which left a nebula of debris from which the solar system emerged by gravitational agglomeration of gas and dust about 4.6 Gy BP (4600 million years ago). It is likely that the very young Earth had a primordial atmosphere consistent with the make-up of the parent nebula, but that this was swept away by violent activity in the very active young solar system. Certainly the current amount of ^{36}Ar (not the abundant ^{40}Ar mentioned in section 3.2) is about 10^6 times less than the value for the solar system as a whole, and there seems to be no other way of losing this very inert, heavy gas. It is therefore very likely that our present atmosphere is a secondary atmosphere produced by outgassing from the Earth's interior as it warmed by self-compression and radioactive decay. Regardless of its mode of formation it is clear that this *paleoatmosphere* has since been greatly altered by terrestrial processes, including the evolution of life.

Current evidence [22] suggests that by about 4.4 Gy BP, initial fast outgassing of water vapour, hydrogen, carbon monoxide and dioxide, hydrogen chloride and molecular nitrogen (N_2) was largely complete (unlike the slow outgassing of ^{40}Ar). The water vapour probably condensed quite quickly to form deep archaean oceans, and the very light hydrogen molecules escaped from the Earth's gravitational field, leaving the atmosphere dominated by CO_2 and N_2. The large amounts of atmospheric CO_2 (probably hundreds of times present values) may have kept the Earth's surface a little warmer than at present despite the young Sun's heat output being about 30% below current values — the mechanism being a greenhouse effect several times larger than today's (section 8.5).

This paleoatmosphere was very different from today's, in which carbon dioxide is reduced to a trace, and molecular oxygen (O_2) is the most abundant gas after N_2 and these differences must be explained by any account of the subsequent evolution of the Earth's atmosphere. The current range of differing models shows how little we confidently understand of events so removed from the familiar present by the abyss of time, but the following summary represents what seems to be an emerging consensus.

EVOLUTION

It seems likely that much of the CO_2 was removed by weathering of exposed or submarine rock, or by hydrothermal pumping of sea water through fissures in submarine mid-ocean ridges, becoming locked in a variety of carbonates in the Earth's crust and upper mantle. Indeed there is speculation that such removal may have prevented a runaway greenhouse effect like that suffered by our neighbouring planet Venus as the Sun slowly increased its heat output — which would have abruptly ended the prospects of life on Earth. Instead, it seems that by about 3.8 Gy BP, atmospheric CO_2 had been reduced to an important minor component, able to maintain surface temperatures near modern values by a modest greenhouse effect involving water vapour and cloud, much as now (section 8.5).

There is no fossil evidence of life at this early stage, and it is known that in the absence of photosynthetic production, atmospheric O_2 levels must have been limited to $\sim 10^{-9}$ present values, maintained by photolysis of H_2O and CO_2 by solar UV. The subsequent increase in atmospheric O_2 is closely related to the evolution of life on Earth (Fig. 3.12). From about 3.5 Gy BP there is increasingly unambiguous fossil evidence of single-celled bacteria capable of feeble, *anaerobic* (in the absence of oxygen) photosynthesis (section 3.4), and by 2.7 Gy BP large populations of *blue-green algae* (Fig. 3.13) were laying down beds of limestone in shallow seas, achieving net production of atmospheric O_2 as their numbers increased. Such early life was aquatic, since several metres' depth of water were needed to filter the damaging UV from the necessary sunlight. Oxygen levels in the atmosphere probably rose only very slowly, but as they did, so did the much smaller concentrations of ozone, maintained as outlined in section 3.3, which in turn increased the protective UV filter and therefore the size and range of aquatic habitats suitable for primitive life.

When atmospheric O_2 reached about 1% of present values, a new and much more efficient type of *aerobic* photosynthesizing organism developed, raising oxygen levels and widening habitats much more rapidly. There is fossil evidence of such organisms from about 1.2 Gy BP, distinguished by the appearance of a nucleus in each single cell. Even more efficient multicelled organisms followed, so that by about 0.6 Gy (600 My) BP there were marine organisms with shells, which are believed to have required dissolved oxygen in equilibrium with atmospheric O_2 levels of at least 10% of present values. As they died they formed fossil-rich beds of calcium carbonate in the shallow seas, further raising O_2 levels and O_3 levels to the point where organisms no longer needed a shield of water against solar UV. Whether or not it followed immediately, the result was the vast Cambrian explosion of life, and the colonization of land surfaces. Widespread intense photosynthesis raised O_2 levels to present levels and, in the Carboniferous period, laid down the rich beds of hydrocarbon debris which are currently supporting mankind's brief spree of hectic combustion. It is possible that this led to a temporary depletion of CO_2 which reduced the greenhouse effect and encouraged the great Permian ice ages (280 My BP). All this time, ^{40}K was decaying radioactively to ^{40}Ar which slowly diffused through the crust to produce the third major component of the present atmosphere, albeit a biologically and chemically inactive one.

This brief survey shows how far the present atmosphere has evolved from its paleoatmospheric origins; how a variable greenhouse effect has maintained surface temperatures in the narrow range needed to allow the accelerating evolution of life, and how that evolving life has in turn built up the reservoir of free oxygen on which it now depends. Such obvious self-regulation has recently encouraged the view that the terrestrial biosphere should be regarded as a coherent

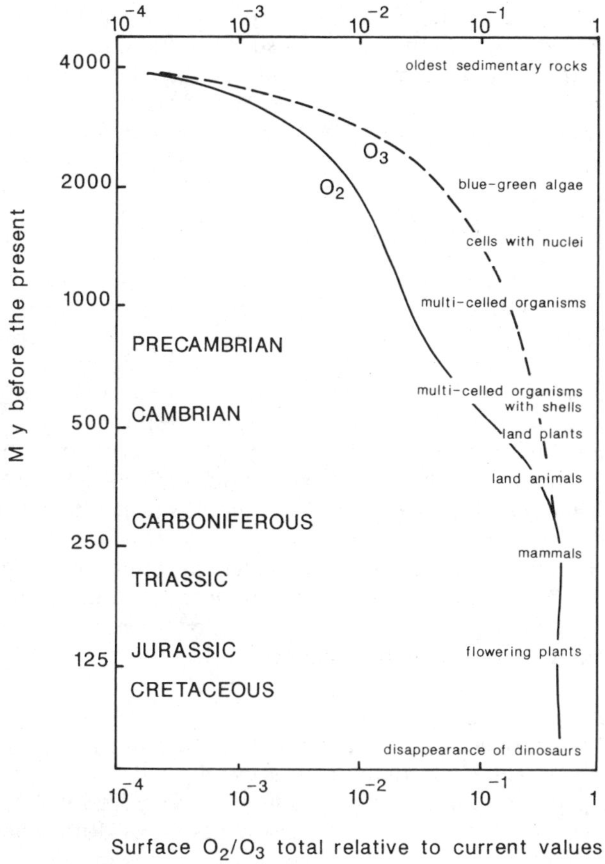

10^{-4}
10^{-3}
10^{-2}
10^{-1}
1
4000
2000
1000
500
250
125
My before the present
oldest sedimentary rocks
O_3
O_2
blue-green algae
cells with nuclei
multi-celled organisms
PRECAMBRIAN
multi-celled organisms with shells
CAMBRIAN
land plants
land animals
CARBONIFEROUS
mammals
TRIASSIC
JURASSIC
flowering plants
CRETACEOUS
disappearance of dinosaurs
Surface O_2/O_3 total relative to current values

Fig. 3.12 Probable co-evolution of atmospheric O_2, O_3 and life on Earth. (After [23])

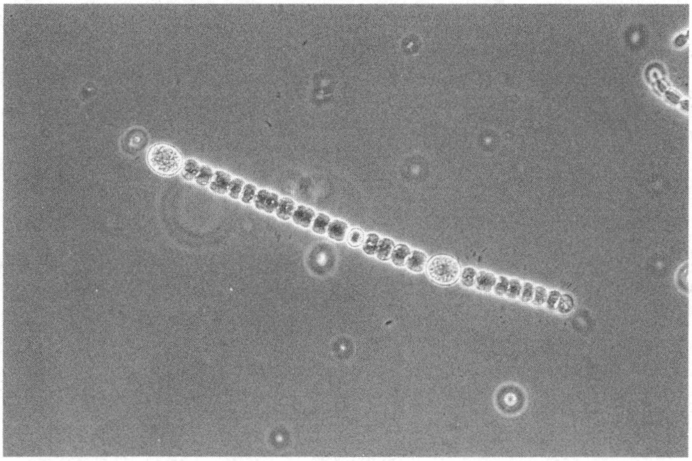

Fig. 3.13 *Anabaena*: a modern blue-green alga, descendent of a group of organisms first apparent in the fossil record of more than 2000 million years ago. The filament is about 200 μm long and contains heterocysts capable of fixing atmospheric N_2.

superorganism, Gaia [24], whose intrinsic purpose is to regulate its environment to its own advantage. Whatever the merits of such a view, it is impossible not to be impressed by (and grateful for!) the continued cooperation of the Earth's biosphere and atmosphere. It would be a sad irony if the only part of the biosphere capable of being consciously aware of such cooperation should choose to ignore its obvious lessons.

Appendix 3.1 Residence time

Consider atmospheric N_2 as an example of any reservoir maintained in dynamic equilibrium by balanced production and loss. Let M be the mass of atmospheric N_2 in kg, and let F be the rate of denitrification (production of atmospheric N_2) in kg per year, which must also be the rate of fixation (loss of atmospheric N_2), since M is constant. If the individual molecules of N_2 entered and left the atmosphere in the same chronological order, each would spend time M/F years in the atmosphere, which is by definition the *residence time T* of N_2 in the atmosphere. Since of course the molecules do not enter and leave in such an orderly manner, some individuals leave very soon after entry and others remain unfixed for very long periods of time. T is therefore an average period of residence.

$$T = M/F$$

The concept of residence time is potentially useful in all systems which are in or near dynamic equilibrium. The quantity M is usually a mass or a measure of mass, as for example in the treatment of the hydrologic cycle in section 3.7, but it may be any conservative quantity, such as in the discussion of atmospheric energy in section 13.3. Quantities in M and F may be expressed in any convenient way provided the same units are used in each. For example in section 3.7 the quantities of water substance are expressed effectively in volumes of water, which is satisfactory because water is so incompressible that volume is a measure of mass. The residence time T will be given in whatever time units are used in F — years in the case of the annual throughput of nitrogen cited above.

If the quantity passes through successive reservoirs in series without gain or loss, the flux F is conserved along the sequence, and the ratio of quantity to residence time must be the same for each reservoir.

$$\frac{M_1}{T_1} = \frac{M_2}{T_2} = \frac{M_3}{T_3}$$

This chain rule applies nicely to the sequence of vapour, cloud and precipitation in the atmospheric branch of the hydrologic cycle (section 3.7).

Problems

LEVEL 1

1. Uniformity of specific mass of a component of a compressible mixture under

gravity indicates which type of equilibrium? Does this follow also in the absence of gravity (e.g. in an orbiting space station)?

2. List the order of component layering from the top down in the diffusive equilibrium of a planetary atmosphere consisting of moist air, with some photodissociation of molecular oxygen and water vapour.

3. Briefly define the following terms: denitrification, residence time, respiration, photosynthesis.

4. Why is the altitude of maximum specific mass of ozone (or any other component with a level of maximum concentration) greater than the altitude of maximum ozone density?

5. List the effects producing variations of the specific mass of CO_2 in space or time, indicating in each case whether the tendency is upward or downward.

6. Why does the number density of large nuclei fall rapidly in the vicinity of cloud-base level?

7. Starting from evaporation, sketch and label a likely sequence of events experienced by a water molecule which is eventually precipitated onto land. Follow the sequence back to the sea. Add branches describing the cases of evaporation from land and precipitation onto the sea.

LEVEL 2

8. Assuming a reasonable value of 4 l for accessible capacity of a pair of human lungs, find the maximum mass of contained oxygen assuming an air density of 1.2 kg m^{-3}. Why is the actual value likely to be a little lower?

9. Suppose that the maximum specific mass of ozone is 10^{-5} at a height of 32 km above sea level. Using a reasonable value for air density there (about 2 decadal scale heights above the surface), find the density and number density of ozone in quantities per litre.

10. The reservoir of carbon in the living and dead terrestrial biosphere (ignoring fossilized reservoirs like coal and oil) is believed to be about 2400 units, and the annual exchange with the atmosphere is about 110 units [14, 21], where a unit is 10^{12} kg. Find the implied residence time for carbon in the terrestrial biosphere. Since the exchange is almost entirely in the form of CO_2, find the annual exchange of CO_2 in the same units.

11. Find the specific mass of an aerosol dominated by giant nuclei of radius 1 μm and number density 10^3 per litre, where the air density is 1 kg m^{-3}, and compare with Table 3.1.

12. Assuming realistically that 97% of the Earth's water substance is in the form of oceans which would cover the planet uniformly to a depth of 2.8 km, find the average ocean depth given that they actually cover only 70% of the total surface. If all of the present ice caps were to melt and run into the oceans, what would be the maximum rise in sea level?

LEVEL 3

13. If convective equilibrium always tends to encourage uniformity of distribution, why does it maintain the observed sharp vertical lapse of air density rather than a uniform profile?

14. Speculate on possible short- and long-term effects on the total mass of

71

atmospheric carbon dioxide of a brief large-scale nuclear war followed by a
very prolonged winter induced by resulting smoke and dust.

15. Discuss critically but sympathetically the possible physical origins of the story
of the universal deluge which is widespread in the early culture of the Middle
East.

16. The answer to problem 12 is the maximum possible rise in sea level. Why
should the actual rise be less than this? And consider the apparent rise in sea
level relative to a coastal landmark in a region initially under the edge of an ice
cap, remembering that the apparently solid land surface is substantially
depressed by an ice burden.

17. Assuming that the typical precipitation particle or droplet falls at about
4 m s^{-1} from about 4 km above the surface, equate the implied fall time with
the residence time of water in the form of precipitation, and hence find the
global fraction of the atmospheric water substance which is in the form of
precipitation at any instant. You can assume a reasonable value for the
residence time of water in the form of vapour.

State and climate

4.1 The equation of state

We have seen that all but a minute fraction of the mass of the atmosphere consists of the mixture of gases called air. Let us look closely at how the condition or state of an air parcel depends on its temperature and pressure. Since a gas has neither rigidity nor shape the only remaining simple property is its volume. So let us look for a relationship between temperature, pressure and volume.

In terrestrial conditions the major atmospheric gases, and many of the minor ones too, behave very nearly as *ideal gases* in that they obey the *gas laws* discovered by Boyle, Charles and others in the early stages of the scientific revolution. It is shown in Appendix 4.1 that these laws can be combined in a single *equation of state* applicable to all ideal gases. For one *mole* of ideal gas (i.e. a quantity whose mass in grams is numerically equal to the molecular weight of the gas) the equation of state is

$$p V = R_* T \tag{4.1}$$

where p, V and T are respectively the pressure, volume and absolute temperature of the gas, and R_* is the *universal gas constant* whose value is 8.314 J K^{-1} mol^{-1}.

It is useful to change the format of eqn (4.1), since the presence of the parcel volume V is inappropriate to the case of an unconfined gas like atmospheric air. It is shown in Appendix 4.1 that it is relatively easy to find an equivalent equation involving the density of the air rather than its volume V. This is the meteorological form of the equation of state

$$p = \rho R T \tag{4.2}$$

where R is the *specific gas constant* for the gas, so called because its value varies from gas to gas, unlike R_*. Since R is given by $10^3 R_*/M$ it follows that the densities of ideal gases at the same pressures and temperatures are inversely proportional to their molecular weights.

Equations (4.1) and (4.2) apply to single ideal gases. To deal with a mixture like air we must combine the equations of state for the constituent gases — nitrogen, oxygen etc. It is shown in Appendix 4.1 that a mixture of gases, each with molecular weight M_i and specific mass x_i, behaves as a single ideal gas with specific gas constant R given by

73

$$R = \sum_i x_i R_i \tag{4.3}$$

and that the equivalent molecular weight of the mixture is

$$M = 1/\sum_i x_i/M_i \tag{4.4}$$

Substituting values for x_i and M_i from Table 3.1, we find that dry air behaves as a single ideal gas with specific gas constant 287 J K^{-1} kg^{-1} and molecular weight 29.0 (which means that 29.0 g of dry air obey eqn (4.1)).

It is apparent from eqn (4.3) that a mixture of gases with component specific masses uniform throughout a certain volume (e.g. dry air in the turbosphere) has a uniform specific gas constant. This means that the mixture can be treated as a single gas for purposes depending on the modified equation of state (eqn (4.2)). Though the value of 287 J K^{-1} kg^{-1} for R is strictly applicable to dry air only, it can be used for moist or even cloudy air in all but the most precise calculations. In principle, however, the presence of water vapour with its relatively low molecular weight reduces the density of moist air below that of dry air at the same temperature and pressure, and this can be significant when considering the very small buoyancies typical of cumulus convection (Appendix 4.1 and section 7.12).

The meteorological form of the equation of state (eqn (4.2)) is used mainly to determine the remaining variable when the other two (out of pressure, temperature and density) are known. This is often done to put equations into their most convenient form, as for example in deriving the pressure–height relationship in section 4.3. There, as in several places throughout this book, the equation of state is used to express air density in terms of pressure and temperature. This matches laboratory practice, where direct measurement of air density is so slow and clumsy (for example by weighing a rigid container of known volume before and after evacuation of air) that instead density is calculated from simultaneous measurement of pressure and temperature. As an exercise in this procedure let us find the approximate value of air density in the low troposphere using typical values of pressure and temperature. Pressure is always close to 1000 mbar (10^5 Pa) near sea level, and let us suppose that the air temperature is 20 °C (293.2 K). Using the value for the specific gas constant for dry air, the air density is found to be 1.19 kg m^{-3}. Since this value is not very sensitive to realistic variations in pressure and temperature in the low troposphere, we can generalize roughly that air densities near sea level are just a little greater than 1 kg m^{-3}. Though this is only one thousandth of the density of water, it is far from being negligible, as related properties such as atmospheric pressure and the momentum of a high wind clearly show.

4.2 The vertical profile of temperature

Figure 3.1 contains a greatly simplified picture of the major features of the vertical profile of temperature observed over most of the globe — only the upper turbosphere in middle and high latitudes in winter being systematically different on account of the prolonged absence of solar radiation there. Let us consider these

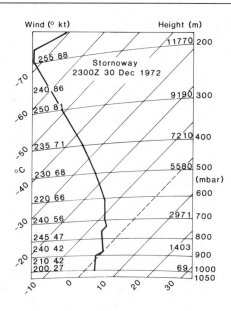

Wind (° kt) Height (m)

255 88
Stornoway
2300Z 30 Dec 1972
240 86
250 81
235 71
230 68
220 66
240 56
245 47
240 42
210 42
200 27

11770 200
9190 300
7210 400
5580 500
(mbar)
600
2971 700
800
1403
900
69 1000
1050

−70
−60
−50
°C
−40
−30
−20
−10
0 10 20 30

Fig. 4.1 A typical winter sounding in the vicinity of an Atlantic depression, obtained by routine radiosonde from the synoptic upper air station in north-west Scotland. It is displayed on a tephigram (section 5.8), with temperature, height and wind azimuth and air speed plotted or listed against pressure. The tropopause is sharply defined a little above the 200 mbar level on this occasion, being surmounted by a temperature inversion in the base of the stratosphere.

features with the aim of identifying physical mechanisms at work in the different height zones.

The first 10–15 km above sea level stand out as being a region in which temperature falls sharply and fairly consistently with increasing height. This is the troposphere, whose characteristically large *temperature lapse rate* is closely related to the widespread occurrence of convection there, both cloudy and cloudless. Values of this lapse rate vary somewhat with time and geographical location when averaged between the surface and the tropopause, but generally lie fairly close to 6 °C km^{-1}. For example the value on the occasion of the radiosonde flight displayed in Fig. 4.1 was 6.5 °C km^{-1}.

Consider how the steep lapse rate of temperature in the troposphere is related to that region's incessant commotion, and in particular to the associated vigorous vertical movements of air. As an air parcel rises into regions of lower ambient air pressure, it expands and cools. The cooling occurs as air molecules in the expanding parcel collide with their retreating neighbours and use up some of their molecular kinetic energy in the process. When there are no complicating gains or losses of heat by any kind of exchange between the parcel and its near or distant environment (by conduction or radiation for example), the process is said to be *adiabatic*, and the degree of cooling is fairly readily given by theory, as discussed in section 5.6 and following sections. Sinking parcels undergo a corresponding warming.

Since vertical motion is so widespread and vigorous in the troposphere we might expect there to be a convective equilibrium temperature profile with corresponding lapse rate(s), just as there is a convective equilibrium distribution of specific masses of the constituents of dry air. This expectation is confirmed by further discussion and observation (sections 5.9 and 5.10), which show that there are different equilibrium profiles depending on whether the air is cloudless or cloudy. In cloudless air there is a single well-defined temperature lapse rate of very nearly 10 °C km^{-1}, whereas in cloudy air the lapse rate ranges continuously from around 5 °C km^{-1} in high temperatures to the cloudless limit at the lowest temperatures.

On the very cloudy occasion of Fig. 4.1, the cloud-free zone extended only about 250 m above the surface, and was therefore too shallow to appear on the rather gross vertical scale of the diagram. However, in the deep cloudy layer filling the rest of the troposphere, there is a clear tendency for the lapse rate to increase with increasing height (i.e. with decreasing temperature), in qualitative agreement with the convective equilibrium model for cloudy air.

To avoid oversimplifying this picture of the relationship between convection and tropospheric temperature profile we must consider two major complicating factors.

1. Although convective air movements in the troposphere are nearly adiabatic, they are not exactly so. Air parcels actually gain or lose heat by direct interaction with solar and terrestrial radiation at rates which produce significant temperature changes over periods of a day or more, as discussed in section 8.10. Though this has relatively little effect on air rising rapidly in a shower cloud or a front, it can considerably influence the great bulk of tropospheric air which rises and falls much more slowly than this.

2. Two apparently insignificant temperature inversions visible in the low troposphere in Fig. 4.1 arise from the second important complication. Although small in appearance, a detailed examination of their effects on the buoyancy of air parcels trying to rise through them would show that they can inhibit all but the most violent convection. In fact there was no showery convection on the occasion of Fig. 4.1. The deep and horizontally extensive layers of precipitating cloud which were present were produced instead by the completely different type of vertical motion associated with a complex system of vigorous fronts then moving north-east across the British Isles. In such fronts the large sheets of cloud are produced and maintained by ascent which is very much slower but more persistent and extensive than that associated with shower clouds (section 12.4). In such slow ascent air parcels usually take several days to rise from the low to the high troposphere, in which time they may well move 1000 or more kilometres horizontally in the relatively strong winds typical of frontal regions (notice the high wind speeds in the upper troposphere in Fig. 4.1). It follows that such air is ascending along horizontally extensive, slightly sloping paths (gradients $\sim$ 1:100), and not along the nearly vertical paths of either radiosondes or air parcels in showery convection. In fact the temperature profile recorded by a radiosonde in or near such fronts, which are quite commonplace in middle and high latitudes, is a nearly vertical section through a complex sandwich of nearly horizontal strata, each layer of which may be related to a geographically distinct surface source region (Fig. 11.14). In such circumstances we cannot expect to be able to interpret details of the temperature profile in terms of processes, such as showery convection, which are confined to the profile itself.

TROPOPAUSE AND LOW STRATOSPHERE

The characteristically steep lapse of temperature ends at the tropopause, often abruptly as in Fig. 4.1. Above this there may or may not be a temperature inversion below the deep, nearly isothermal layer which makes up most of the low stratosphere. Clearly the marked contrast between tropospheric and stratospheric temperature profiles must be associated with a significant difference in atmospheric behaviour. The confinement of clouds below the tropopause suggests that updraughts are confined similarly. In fact it is observed that atmospheric motion in all directions is much more subdued in the stratosphere than it is in the troposphere, though quite high wind speeds are found up to several kilometres above the tropopause in the vicinity of fronts and their associated jet streams (as in Fig. 4.1). In the absence of vertical motion comparable with that found in the troposphere, we must expect the non-adiabatic (or *diabatic*) effects of solar and terrestrial radiation on the temperature profile to be more pronounced. Indeed the largely isothermal stratification of the low stratosphere depends much more on the radiative transfer of heat than it does on convection, which would otherwise naturally tend to maintain the steep lapse rate associated with cloud-free convection.

The upper stratosphere is even more obviously dominated by the diabatic effects of radiation. The steep and deep temperature inversion extending up to the *strato-pause* (Fig. 3.1), where air temperatures approach those of the Earth's surface about 50 km below, is maintained by the absorption of soft ultraviolet radiation by ozone, as discussed in Section 3.3.

Such mild conditions contrast strongly with the abysmal cold which would prevail if thermal equilibrium were maintained solely by convection from the surface to those altitudes. In fact temperatures are maintained by a combination of diabatic heating and cooling, and weak vertical and horizontal movements of air from hotter and cooler regions. For example, in middle and high latitudes, temperatures throughout the stratosphere fall sharply in winter, as solar warming is reduced and cooling by net infra-red emission continues (both through inter-actions with O_3 — section 8.6). However, the fall is smaller than would be expected by these radiative processes alone, because air currents import heat from warmer regions. In some Northern hemisphere winters, there is a dramatic *sudden warming* of the low stratosphere as air moves and sinks rapidly when some dynamic instability of the stratosphere is triggered by patterns of gross airflow in the tropo-sphere. In high latitudes the average temperature of the high stratosphere varies from about $+20$ °C in summer to -40 °C in winter.

Above the stratopause the temperature falls to values at the *mesopause* (Fig. 3.1) which can reach -100 °C in winter in high latitudes — the lowest values in the atmosphere. No doubt some of this lapse is maintained adiabatically by weak vertical air motions, but observed average lapse rates of about 2.8 °C km^{-1} are much smaller than the value of 9.8 to be expected in cloud-free convective equilibrium (sections 5.9 and 5.12), showing that air currents are carrying significant amounts of heat, as in the stratosphere. In summer, continuous local warming by photodissociation and photoionization of air molecules (section 3.1) raises mesopause temperatures to -30 °C.

Above the mesopause the temperature rises sharply with height in the thermo-sphere because increasing absorption of the harder components of solar ultraviolet swamps the opposing convective effect (see section 3.1).

The variations of temperature along a vertical profile through the turbosphere discussed above may seem impressive on the Celsius scale, especially in relation to the narrow temperature tolerances of most living organisms, but it is important to note that they are quite modest on the absolute scale. It is apparent from Fig. 3.1 that temperatures in most of the atmosphere up to the 100 km level lie within 50° of 250 K. In fact to an approximation which is surprisingly good for most purposes which explicitly involve the absolute temperature scale, the whole turbosphere can be considered to be isothermal, with a temperature of 250 K (-23 °C).

4.3 The vertical profile of pressure

By contrast with temperature, atmospheric pressure varies enormously with height, decreasing by a factor of about a million between the surface and the turbo-pause. As mentioned in section 1.1, and as apparent in Fig. 3.1, this huge variation in pressure is very regularly distributed in the vertical, having a nearly uniform scale height throughout the full depth of the layer. In fact the decadal scale height (the height interval in which pressure drops to one tenth of any initial value) approximates quite closely to 16 km throughout. More detailed observations in Fig. 4.2 show that the pressure–height relation is not quite so simple, but the discrepancies are so small that for the moment they serve to emphasize how closely actual soundings approximate to the simplicity of Fig. 3.1. Let us consider in particular the significance of the smoothness and regularity of the vertical distribution of atmospheric pressure.

Smoothness. Atmospheric pressure varies very smoothly with height; there are no signs of any fine structure in the vertical pressure profile corresponding to those in the associated temperature profile (Fig. 4.1). Horizontal pressure variations are very much smaller, as will be seen in section 4.6, and are also very smoothly distributed; and pressure variations with time at a fixed position are very modest and well ordered. In fact pressure is by far the most regularly and smoothly distributed of the readily observable atmospheric variables. To explain such smoothness from the viewpoint of Chapter 3, we must look for a mechanism capable of smoothing the atmospheric pressure field much more quickly than it is disturbed by normal atmospheric commotion. Let us try to clarify the issue by considering what would happen if a substantial pressure variation were suddenly introduced in some small part of the atmosphere.

Suppose that a toy balloon is blown up and burst. A water manometer will show that the excess pressure inside the balloon just before bursting is a few tens of millibars. In the typical low troposphere this is equivalent to the pressure difference over several hundreds of metres of altitude, over hundreds of kilometres in the horizontal, or over a day or more at a fixed position. It is therefore a very substantial pressure difference which on this occasion is concentrated across a fraction of a millimetre of stretched rubber. On bursting, the rubber is almost instantaneously removed, leaving a large, localized pressure excess in the air. This excess spreads outward behind a three-dimensional shock wave expanding at about the speed of sound, so that within a fraction of a second it is distributed over a volume five or six orders of magnitude larger than that of the balloon, and is

Fig. 4.2 Log pressure from Fig. 4.1 plotted to show its near linearity with height. The logarithmic isobar spacing in Fig. 4.1 (and all such diagrams) therefore corresponds to a nearly linear height axis.

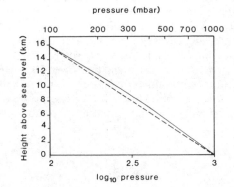

diluted accordingly. The dilution is so rapid that within a second of the burst we should expect no trace of the originally large pressure excess to be detectable by even the most dense and extensive network of sensitive pressure gauges. And since not even the most vigorous atmospheric disturbances contain air moving at more than a small fraction of the speed of sound, the smoothness of the observed pressure fields is explained at least qualitatively.

Regularity. The foregoing does not argue that atmospheric pressure should be uniform. The smoothness which is explained is instead the absence of significant deviation from the pressure distribution to be expected when the air is fully adjusted to the prevailing force, which is gravity. Assuming that such adjustment implies that the air is stationary, let us examine the equilibrium of air at rest under its own weight, looking for an explanation for the marked regularity of vertical distribution apparent in Fig. 3.1. The assumption of an absolutely static atmosphere is obviously unrealistic in principle, but we will see in section 7.12 that deviations from static equilibrium are very slight in the cases of most natural air motions.

In a static atmosphere the upward and downward forces on any air parcel must balance. Gravitational attraction between the parcel and the Earth produces a downward force which by definition is the weight of the parcel. Starting from some imaginary unbalanced state, we can envisage the air settling under gravity and raising the pressure at lower altitudes to the point where the net upward pressure force on any parcel (i.e. the excess of the upward pressure on its base over the downward pressure on its top) exactly balances its weight. This balance is known as *hydrostatic equilibrium*, and it can obviously explain the observed upward lapse of pressure throughout the atmosphere at least qualitatively. To show that it does so quantitatively as well, we need a formal statement of hydrostatic equilibrium — the *hydrostatic equation*.

4.4 Hydrostatic equilibrium

Several forms of the hydrostatic equation are derived in Appendix 4.2, the simplest of which relates the difference in atmospheric pressure between two heights to the weight of the intervening layer of air. Choosing an air column with unit horizontal area, the pressure difference is simply equal to the weight Mg of the air column (mass M), i.e.

$$p_1 - p_2 = Mg \tag{4.5}$$

where p_1 is the pressure at height z_1 and p_2 is the smaller pressure at the greater height z_2.

The appearance of the gravitational acceleration g as a separate factor in eqn (4.5) implies that it has the same value at all heights, whereas it actually diminishes with increasing height according to the inverse square of the distance from the centre of the Earth, and also varies slightly with latitude (section 7.4). However, the depth of the turbosphere, especially the troposphere, is such a small fraction of an Earth radius that g can be treated as a constant in all but the most precise meteorological calculations; and even there (e.g. in measuring the heights of standard isobaric surfaces from radiosondes) the variation in g can be accommodated by replacing actual height by the very slightly different *geopotential height* (Appendix 4.3).

By setting p_2 equal to zero in eqn (4.5), it can be seen that the atmospheric

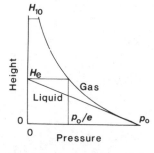

Fig. 4.3 Many geologically young mountains like this reach above the 500 mbar level, and therefore have less than 50% of the atmospheric mass above their peaks. The percentage of atmospheric water substance above peak level is vey much smaller for reasons discussed in section 3.7.

pressure at any level is proportional to the total mass of the atmosphere in a vertical column above that level. For example, since a typical value of atmospheric pressure at sea level is 1010 mbar, you can substitute the familiar value of g into eqn (4.5) to find that about 10.3 tonnes of atmosphere (almost all of it air) rest on each horizontal square metre of the sea surface. And since it is apparent from Fig. 4.2 that the pressure about 5.5 km above sea level is only 500 mbar, it follows that only about half of the atmospheric mass lies above this height.

Regarding the vertical distribution of atmospheric pressure (Figs. 3.1, 4.1 and 4.2) as a measure of the distribution of atmospheric mass, we see that the mass is heavily concentrated at low altitudes and is more and more thinly distributed at greater and greater altitudes. Since the concentration of mass is expressed by density, we can reformulate eqn (4.5) as shown in appendix 4.2 to produce

$$\frac{\partial p}{\partial z} = -g\,\rho \tag{4.6}$$

The term $\partial p/\partial z$ in eqn (4.6) is the instantaneous vertical gradient of pressure, and ρ is the atmospheric density at the same location. The minus sign ensures that pressure decreases with increasing height, as it obviously must.

Equation (4.6) does not only apply to a pressure profile like that in Fig. 4.4. In fact it is the most generally useful form of the hydrostatic equation, describing the hydrostatic equilibrium of any fluid, whether liquid or gas. In an almost incompressible and homogeneous liquid like the sea, the near uniformity of water density is associated with a nearly uniform pressure gradient according to eqn (4.6), and hence with a nearly linear pressure profile (Fig. 4.4). However, in the case of the atmosphere, the compressibility of the air results in a highly nonlinear profile (Fig. 4.4 again). Although apparently very irregular, this profile has an underlying order which we can reveal by replacing air density using the equation of state (eqn 4.2). It follows that

$$\frac{\partial p}{\partial z} = -p/H_e \tag{4.7}$$

Fig. 4.4 A pure exponential decay of pressure with height. The straight-line construction is discussed in Appendix 4.4.

where $H_e = RT/g$ and is termed the *exponential scale height* for reasons which become clear when eqn (4.7) is integrated to relate pressures at two substantially different heights z_1 and z_2. This integration becomes very simple when the air temperature T does not vary with height, giving an exponential relationship between pressure and height.

$$p_2 = p_1 \exp\{-(z_2 - z_1)/H_e\} \tag{4.8}$$

According to eqn (4.8), in an isothermal layer of thickness H_e, the pressure at the upper limit of the layer is $1/e$ times the pressure at the lower limit, where e is the base of natural logarithms (value 2.7183 — see Appendix 4.4). It can then be shown that the height interval in which the pressure falls to one tenth of the value at the lower level (the decadal scale height H_{10}) is given by $2.3026 H_e$.

Equation (4.8) describes an exponential decay of pressure with increasing height which resembles the observed pressure distribution both in its shape (Fig. 4.4) and in the uniformity of its scale heights (Fig. 3.1). To complete the identity of theory and observation we should consider why H_e should be so nearly uniform in the real atmosphere, and check that its numerical value is realistic.

Of the three factors which determine H_e, the gravitational acceleration g is effectively constant in the shallow height zone relevant to meteorology, and the uniformity of the specific gas constant R throughout the turbosphere has been established in section 4.1 and Chapter 3. The remaining factor is the absolute temperature T, whose percentage variability has been shown in section 4.2 to be surprisingly small. Since the temperature of most of the turbosphere lies within 20% of 250 K, it follows that we should expect the scale heights for pressure to be constant to the same degree, as is indeed the case. The observed profiles of pressure against height are nearly but not quite exponential, as shown by the nearly linear height profile of the logarithm of pressure (Fig. 4.2). In comparison with a purely exponential profile, pressure falls slightly less sharply in the warm, low troposphere than it does in the relatively cold high troposphere, where the scale height is somewhat smaller. However, these are relatively minor amendments to the underlying exponential profile.

Lastly, let us insert realistic values to check the realism of the theoretical scale height values. Substituting values for g and R, and assuming a typical mid-tropospheric 250 K for air temperature T, we find a value of 16.8 km for the decadal scale height, in good agreement with observation (Fig. 3.1).

As a general conclusion, it seems that the vertical profile of atmospheric pressure is determined essentially by the material constitution of the atmosphere, its temperature profile, and the prevalence of a very close approximation to hydrostatic equilibrium — the last in spite of the commotion which keeps it so well mixed.

4.5 The vertical profile of air density

The vertical distribution of air density in the atmosphere follows from the distributions of pressure and temperature, using the equation of state (eqn (4.2)). Indeed since pressure varies so strongly in the vertical, whereas temperature variations are quite modest on the absolute scale, the vertical profiles of air density and pressure must be very similar. In fact the density profile shows a nearly exponential decay of density with increasing height and a decadal scale height which is never very far from 16 km throughout the turbosphere.

Since we live most of our lives within a few hundred metres of sea level, the air we breathe is very nearly the most dense in the atmosphere, with values exceeding 1 kg m^{-3} as demonstrated in section 4.1. Only 10 km above sea level the air density is usually a little less than one third of sea-level values and is too low for sustained human breathing. The peak of Mount Everest is at about this level, so that a climber there, inhaling in a normal fashion, obtains less than one third of the mass of air to which he is accustomed at sea level, and the uniformity of specific masses ensures that oxygen intake is reduced in the same proportion. Such a reduced intake cannot sustain the life of even the fittest individual for more than a few weeks, nor allow hard exertion for more than a few hours at a stretch. Air-breathing aircraft engines, such as the common turbojet, are similarly constrained, though they are of course expressly designed to operate at such altitudes to take advantage of the reduced airflow drag on the airframe there.

At greater altitudes air is quite useless for human breathing, and as its density falls away to the extremely low values found in the upper stratosphere and meso-sphere, it even becomes ineffective in conducting or convecting heat to or from an immersed body. For example, high-altitude research balloons floating at an altitude of 50 km above sea level become extremely cold at night despite being immersed in the relatively warm air there (Fig. 3.1), because radiative cooling swamps the feeble conductive and convective warming powers of the very thin air. However, at an altitude of 80 km, the air density is still sufficient to vaporize meteorites and returning artificial satellites by frictional heating of these very fast-moving bodies, even though the density is five orders of magnitude smaller than at sea level.

Many other important consequences flow from the nearly exponential vertical profile of air density. Given the tendency of the vigorous stirring of the turbo-sphere to maintain spatially uniform specific masses of atmospheric constituents, it follows that these too tend to have exponential vertical profiles of absolute density. This in turn governs the vertical distribution of density-related phenomena such as chemical reaction rates and interactions with sunlight. For example, those solar wavelengths which are almost completely absorbed before reaching sea level may be relatively abundant only a few kilometres higher up, because so much absorption is concentrated in the relatively dense surface layers of the atmosphere. This is the case for wavelengths at and just beyond the violet end of the visible spectrum, which explains why people on mountain holidays experience such rapid suntan and sunburn. For similar reasons the familiar blue of the sky, resulting from preferential scattering of the blue end of the visible spectrum by air molecules, deepens rapidly towards black with increasing altitude, and the blue atmosphere seen from space is confined to a relatively shallow layer overlying the Earth's surface (Fig. 1.2).

4.6 Equator to poles

In previous sections we have examined typical vertical distributions of temperature and pressure. When we begin to examine horizontal distributions of these and other meteorological variables, especially annual or seasonal averages, we enter the conventional domain of climatology. In fact the vertical distributions too are climatologically very important, since they are closely related to prevailing weather types, but this is often left implicit in introductory climatological treatments, and is regrettably ignored in the more superficial treatments of climatological *air masses*. We will examine horizontal and vertical distributions of meteorological variables and their relationship with the state and behaviour of the atmosphere at many points throughout the rest of the book, but here we introduce the topic with a review of average distributions between equator and poles.

PRESSURE

Because of the very squashed shape of the atmosphere, horizontal gradients of temperature, pressure, etc., are very much smaller than their respective vertical gradients on all but the very smallest scales, but they are nevertheless very important. The largest and most persistent horizontal gradients are those in *meridional* directions (i.e. along a meridian), and we must expect some of these to be directly or indirectly associated with the strong meridional gradient of solar input to the planet. Significant gradients do occur in other horizontal directions, but they tend to be much less pronounced in annual means since they are mostly associated with seasonal land–sea contrasts and transient weather systems.

Consider first the meridional distributions of monthly mean pressure at the base and in the middle of the troposphere (Fig. 4.5). The data used have in addition been *zonally* averaged (i.e. along lines of latitude) to remove the differences which occur between different longitudes. Data are from the northern hemisphere only, since observations are much more complete there than in the southern hemisphere. However, we should expect the latter to look broadly similar after allowing for the six-month shift in seasons.

In Fig. 4.5 the distribution of pressure at the base of the troposphere is represented by the heights of the 1000 mbar isobar above mean sea level (MSL). On average this is between 100 and 200 m, which corresponds to MSL pressures of between 1008 and 1016 mbar, after allowing for the lapse of about 1.2 mbar per 10 m which is typical of the layer. The global annual average pressure at MSL is 1013.2 mbar. Note that it is standard meteorological practice to represent the horizontal distribution of pressure by means of the elevation of isobaric surfaces in this way. Strictly speaking, the method compounds the effects of vertical and

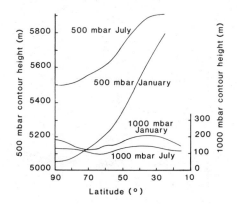

Fig. 4.5 Monthly and zonally averaged height contours of the 1000 and 500 mbar surfaces in the northern hemisphere in January and July. The vertical height scales are chosen to enhance contour gradients rather than absolute heights. (Data from US Department of Commerce, Weather Bureau)

83

horizontal pressure gradients, but the former are so much stronger that isobaric slopes are very small and there is no confusion. Indeed it is shown in appendix 7.8 that the resulting expression for the horizontal pressure gradient force is nicely simplified by the use of isobaric surfaces.

The elevation of the 1000 mbar surface varies little with latitude or season in Fig. 4.5, but there is a slight downward slope toward low latitudes, a broad plateau in the subtropics and an equally broad depression in middle latitudes. In fact these slight pressure differences are associated with quite dramatic differences in weather and climate, for example between the calm, dry subtropics and the disturbed and wet mid-latitudes. However, the mechanisms involved in these associations are quite subtle, depending as they do on the nature of the dominant weather systems (Chapters 11 and 12). For the present we can simply say that the 1000 mbar surface is nearly horizontal from low to high latitudes, i.e. that the average pressure at MSL is nearly uniform.

The pressure distribution in the middle troposphere is represented by the 500 mbar isobar, which lies between 5 and 6 km above MSL and differs considerably from the 1000 mbar isobar in meridional profile. Figure 4.5 shows that the 500 mbar surface slopes downward consistently from low to high latitudes, and that there is a considerable fall in altitude in middle and high latitudes in passing from summer to winter. Actually these variations are large only in a relative sense; a slope of 600 m in nearly 60 degrees of latitude (over 6000 km), and a fall of 400 m in six months, are each very modest, and are only clearly observable because the radiosonde network is operated so meticulously, for example by allowing for variations of g with latitude (Appendix 4.3 and section 7.3).

Combining all parts of Fig. 4.5, we see that the level of the 500 mbar surface falls by about 15% between low and high latitudes in winter, and by about half this amount in summer, whereas the 1000 mbar surface is practically level. It follows that the vertical depth of the lower half of the atmosphere (by mass) varies by the same amounts as the 500 mbar level, as must its volume per unit horizontal area, and hence its average density. Figure 4.6 shows the rather obvious origin of such variations: the mean temperature of this layer falls consistently and significantly from low to high latitudes, the fall being steepest in winter because high latitudes are coolest then.

Fig. 4.6 Monthly and zonally averaged isotherms in January and July. Average tropopause locations are shown by thick lines, and areas above 0 °C are shaded. Note the consistent vertical temperature lapse throughout the troposphere, the concentration of tropospheric baroclinity in mid-latitudes, especially in winter, and the reversal of normal baroclinity about the 200 mbar level in summer. (Redrawn from [27])

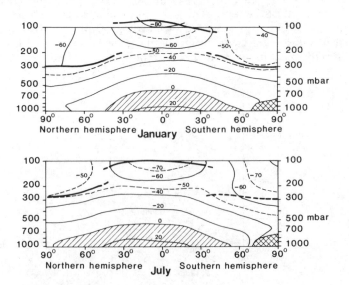

The relationship between layer depth (or *thickness* as it is conventionally called) and layer mean temperature can be quantified by considering the expression for the appropriate scale height. The *binary* scale height H_2 is relevant in the case of Fig. 4.5 since the bounding isobaric surfaces differ in pressure by a factor of two, and according to Appendix 4.4 the expression for H_2 is

$$H_2 = 0.693RT/g.$$

H_2 has a value of nearly 5.5 km when the layer mean temperature is 273.2 K (0 °C). The crucial point about this, and all atmospheric scale heights, is that the layer thickness is proportional to the layer mean temperature on the absolute scale. The meridional gradients of the thickness and layer mean temperatures are therefore in direct proportion, as are their seasonal variations.

In view of the sizeable and extensive horizontal pressure gradients apparent in the middle troposphere in Fig. 4.5, it is perhaps surprising that sea-level pressure is so nearly uniform. Such uniformity clearly shows that the atmosphere tends to distribute its total mass quite evenly despite significant concentrations and depletions localized in some layers.

TEMPERATURE

Figure 4.6 depicts the meridional distribution of air temperature up to the 100 mbar level in summer and winter. Note that this layer contains 90% of the mass of the atmosphere and that its thickness is therefore one decadal scale height.

The distribution of the tropopause in Fig. 4.6 indicates that in low latitudes the layer coincides very closely with the full depth of the deep troposphere there, whereas in middle and high latitudes it includes a substantial part of the low stratosphere above the much lower tropopause. The troposphere is deepest in low latitudes fairly obviously because of the deep and vigorous convection maintained by the strong solar heating there. But although the surface layers in low latitudes are amongst the warmest anywhere on Earth, the great depth of the troposphere there, together with the typically steep tropospheric lapse rate of temperature, produces the rather surprising result that temperatures at the 100 mbar level are lower there than at any other latitude. In fact the poleward lapse of temperature, which dwellers in middle and high latitudes regard as the natural order of things, reverses when we consider levels which lie sufficiently above the mid-latitude tropopause. In summer, when the low stratosphere in middle and high latitudes is relatively warm because of the long hours of daylight, the poleward temperature lapse reverses above the 200 mbar level, and the equatorward lapse exceeds 30 °C at 100 mbar.

Recalling the connection between thickness and layer mean temperature, it is clear that the poleward downslope of isobaric surfaces discussed previously in relation to the 500 mbar surface must persist and increase up to somewhere between the 300 and 200 mbar levels. Above this the increasing depth of relatively warm stratosphere in middle and high latitudes progressively reduces the isobaric slope, reversing it in much of the summer hemisphere.

Comparing Figs. 4.5 and 4.6, it is apparent that the patterns of isotherms and isobars are similar in the troposphere, each tilting downward from the subtropics toward the pole. However, the slopes of the isotherms are much larger than those of the isobars, often by an order of magnitude. It follows that isotherms and isobars must intersect throughout the region, the intersections being most numerous in middle latitudes in winter. An atmosphere with intersecting isotherms and isobars is said to be *baroclinic*, whereas it is called *barotropic* when the isotherms and isobars are parallel. It is apparent that the troposphere as depicted in Figs. 4.5 and 4.6 is nearly barotropic between the tropics but definitely baroclinic in higher latitudes, especially in the winter hemisphere. As will be mentioned at many points in the rest of this book, the distinction between the near barotropy of low latitudes and the strong baroclinity of middle latitudes is basic to the quite different dynamic behaviour in these regions, and the consequent differences in weather and climate. Note that baroclinity is most generally defined by the intersection of isobars and *isopycnals* (lines or surfaces of uniform density) since density is a more fundamental factor dynamically than temperature. However, the behaviour of air as an ideal gas means that the above discussion of baroclinity and barotropy can be repeated with isopycnals replacing isotherms at every stage.

4.7 The general circulation

In many geographical regions even quite casual observation indicates that winds have a preferred direction which either persists throughout the year or varies seasonally. This implies that within these periods the incessant variations of direction associated with weather systems do not cancel out on averaging, but leave a bias towards some particular directions. For example, surface winds in middle latitudes show a pronounced westerly bias, in that winds with azimuths lying between 181 and 359° are more common and stronger than winds with a component from the east. Such observations reveal the presence of mean motion in the troposphere which is amply confirmed by quantitative measurement: when flow at a particular location is averaged over time periods much longer than the longest-lived weather system (which in practice means averaging over at least a month, and preferably a season), a residual speed and direction often remain. The global distribution of these residuals defines the mean flow over the month, season or year, according to the averaging period.

Let us consider the steady component of the *general circulation* in and just above the troposphere. Mean flow is everywhere nearly horizontal. Clearly this is not true close to steeply sloping land surfaces, but only a tiny proportion of the total tropospheric volume is affected by such slopes. Everywhere else mean wind speeds are three or more orders of magnitude larger in the horizontal than they are in the vertical; indeed the latter are so small that they often have to be inferred rather than measured directly. Such asymmetry is of course yet another consequence of the highly squashed configuration of the atmosphere.

Many important features of mean horizontal flow appear on the type of vertical meridional section already used to display pressure and temperature. Consider first the zonal components of mean flow, i.e. components of flow perpendicular to meridional sections (Fig. 4.7).

WESTERLY COMPONENTS

Notice that throughout the troposphere in middle and high latitudes westerly components predominate strongly over easterly components. Surface winds are

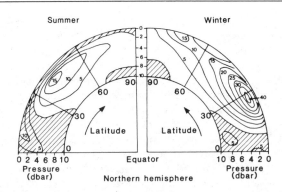

westerly in middle latitudes, in agreement with everyday experience, but even more extensive and intense westerlies are found in the upper troposphere, with maximum wind speeds occurring between the 200 and 300 mbar levels. In winter there is a single very strong maximum in the high troposphere somewhat poleward of the tropics, where seasonally and zonally averaged wind speeds exceed 40 m s^{-1} (nearly 80 knots) in the northern hemisphere, and are even higher in the southern winter hemisphere. In summer, the maximum is considerably weaker and lies substantially further poleward — a seasonal displacement which is much smaller in the southern hemisphere. In both seasons and hemispheres a flange of strong westerly flow extends poleward over middle latitudes in the middle and upper troposphere. The scene being described is really very impressive: in each hemisphere a vast belt of the troposphere is moving from west to east at speeds which even after seasonal and zonal averaging would be equivalent to gale force at the 10 m level in the summer and far more than that in winter. Each belt is entitled a *circumpolar vortex*, a name which nicely points to its continuity around the hemisphere and truly enormous scale.

Note that in middle and high latitudes the maximum winds occur in the upper parts of the baroclinic zones mentioned in the previous section, and that the circumpolar vortices are rotating in the same direction as the Earth itself, though faster. Each of these features is related by dynamics involving the rotation of the Earth in ways which can be quite subtle. We will soon outline the simplest of these mechanisms, but the more subtle must await a fuller discussion of dynamical principles (Chapter 7) and later descriptions of the behaviour and energetics of large-scale weather systems (Chapters 11, 12 and 13).

Above the high troposphere, mean westerly wind speeds fall away quite sharply, at least in the summer hemisphere, where mean winds become easterly everywhere above the lower middle stratosphere (i.e. above the 25 km level). In the winter hemisphere mean winds remain westerly at all levels poleward of the tropics and increase again with height above a minimum between the 20 and 30 km levels. As in the mid-latitude troposphere there is a conspicuous association between baroclinity and zonal flow: westerly flow increases as we rise through a layer with poleward temperature lapse, and decreases when the temperature gradient reverses.

EASTERLY COMPONENTS

In Fig. 4.7 easterly components are apparent at low latitudes in both the low and high troposphere, and at high latitudes in the low troposphere. Though the high latitude easterlies are locally significant, they cover only about 5% of the total surface area of the Earth, as compared with the 40% or more of the surface covered by low-latitude easterlies.

The surface and low troposphere easterlies of low latitudes are the zonal components of the *trade winds*, so called because their day-to-day persistence is so high that they were relied upon in the days of commercial sailing ships. Their weakness and confinement to very low altitudes make them largely irrelevant to modern airborne trade. Unlike the westerlies of high latitudes and levels, the trades are not true zonal winds, having substantial and systematic meridional components to be described shortly. The marked weakening of the zonal component of the trades in summer is confined to the northern hemisphere and is a consequence of the gross distortion of atmospheric flow patterns by the southwesterly summer monsoon of southern Asia (section 12.2). The summer easterlies in the high troposphere in low latitudes are a downward extension of the easterlies of the summer stratosphere, which in turn are associated with the meridional temperature contrast between the cold winter and the warm summer hemispheres at these altitudes.

Fig. 4.8 Schematic plan view of airflow in the low troposphere in the Hadley circulation. The trade winds (TW) blow from the subtropical high-pressure zones (STH) to the intertropical convergence zone (ITCZ), where they feed the cloudy updraughts of tropical weather systems. The very symmetrical flow shown is found when the ITCZ lies near the equator; when it does not, the cross-equatorial flow is deflected to become roughly south-westerly in the northern hemisphere (for example).

Fig. 4.9 Seasonally and zonally averaged meridional wind speeds in the northern hemisphere in winter and summer. Axes and labelling are as in Fig. 4.7, with shaded areas distinguishing northerly (i.e. from the north) from southerly components. Notice that speeds are much lower than in Fig. 4.7, and that patterning is strongest in the region dominated by the Hadley circulation. (Redrawn from [27])

MERIDIONAL COMPONENTS

As depicted in Fig. 4.8, the trade winds converge towards equatorial latitudes, i.e. they have an equatorward meridional component of motion as well as an easterly component. Figure 4.9 shows that seasonal mean equatorward speeds of flow in these latitudes exceed 2 m s^{-1} throughout the year, and 3 m s^{-1} in the winter hemisphere. Another pronounced meridional flow appears in the high troposphere in

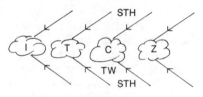

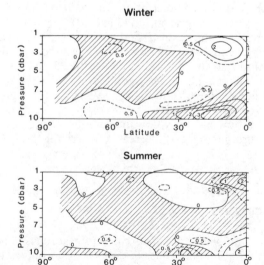

the same low latitudes, directed polewards. These equatorward and poleward flows are the horizontal branches of an enormous pair of meridional circulations (one in each hemisphere, though not quite symmetrical about the Equator) shown schematically in Fig. 4.8 and named after George Hadley whose discussion of 1735 was one of the earliest successful mechanistic descriptions of atmospheric motion. Each *Hadley cell* is completed by a rising branch in the rainy equatorial zone, and a descending branch in the arid subtropical zones (see section 8.13 and Fig. 8.14). The zone in which the opposing trade winds converge is known as the *intertropical convergence zone* or ITCZ, and this convergence clearly feeds air into the rising branches. Similarly the diverging flow near the surface in the subtropics removes the air which is descending there. The descent and divergence are associated with the zones of elevated sea-level pressure often called the *subtropical high-pressure* zones, and these coincide with the great warm and arid zones which virtually girdle the Earth in the subtropics of each hemisphere. Although the consequences of vertical motion are immediately obvious in the association of ascent with the cloud and rain of the ITCZ, and of descent with the arid subtropical highs, the horizontal areas over which the ascent and descent are spread are so large that mean vertical speeds are too small to be observed directly, and have to be deduced by various indirect methods.

In Fig. 4.8 it is apparent that the low-level equatorward flow of the Hadley circulation is twisted westward. Indeed, since the mean meridional and zonal components are comparable in strength, the resultant trade winds must be roughly north-easterly in direction in the northern hemisphere, and south-easterly in the southern hemisphere, directions which are frequently used to name the trades in their respective hemispheres. As in the case of the intertropical convergence zone, the lack of symmetry about the equator is associated with the summer monsoon in southern Asia (section 12.2). In the high troposphere too, the meridional motion is accompanied by zonal motion, though here the eastward flow is so much stronger than the poleward flow, particularly in the poleward half of each Hadley cell, that the result is a poleward spiral in which air parcels encircle the Earth at least once between their equatorial origin and subtropical end.

Throughout the rest of the troposphere, mean meridional components of motion are very small. There is some evidence in Fig. 4.9 of a relatively weak meridional circulation poleward of the Hadley cell but in the opposite sense. This is called the *Ferrel cell*, but though it seems to be quite significant energetically (section 13.3), it is so weak that motion at these and higher latitudes is almost completely dominated by the sporadic, intense air movements associated with middle-latitude weather systems. Even in low latitudes, the relevance of the Hadley circulation to day-to-day patterns of air flow and weather conditions must not be exaggerated. Here too, though less obviously than in higher latitudes, the instantaneous distribution of rain and sunshine, and gentle and strong winds, is explicable largely in terms of the distribution of a range of transient weather systems, each of which may be associated with large deviations of local conditions from the mean (Chapters 11 and 12). The Hadley circulation is very important, but it is so largely because it exerts a controlling influence over the distribution of these weather systems.

ZONAL AND MERIDIONAL COMPONENTS

The association of westerlies with poleward motion and easterlies with equatorward motion is quite evident in the Hadley circulation, and arises because of the rotation of the Earth, and the movement of air across its spherical shape. Air at low latitudes at rest on a weather map is actually moving very quickly eastward because of the Earth's rotation — at over 460 m s^{-1} on the equator (appendix 4.5). At higher latitudes the eastward motion is smaller because points on or near the Earth's surface are closer to its axis of rotation. For example, the eastward speed at latitude 30° is only 402 m s^{-1}. It follows that air moving from equatorial to higher latitudes will tend to preserve its initial eastward momentum (unless it is acted on by a significant zonal force) and hence develop a westerly component of flow relative to the weather map. In fact the effect is even more pronounced than the 58 m s^{-1} implied by the above speeds, since a fuller discussion of rotational dynamics shows that the quantity preserved is the product of the tangential momentum and the length of the normal to the axis of rotation — the *angular momentum*. When this is included (section 7.14), air starting at rest on the equator will arrive at latitude 30° as a westerly wind (relative to the weather map) with speed 134 m s^{-1}. In fact air seldom moves so far poleward on a grand scale, and there is some drag exerted on the air in the upper troposphere by types of friction arising from air flow over mountains, and from deep convection. Nevertheless, the observed subtropical maximum in mean zonal wind speeds in the upper troposphere is maintained essentially by the tendency toward conservation of angular momentum in the poleward-moving branch of the Hadley circulation — as described more fully in section 7.14.

The above argument implies equally that air moving from rest at higher latitudes should develop easterly winds as it moves to lower latitudes. Air descending near the bases of the subtropical high-pressure zones usually has little zonal motion relative to a weather map, having lost its earlier eastward momentum during the long descent in the high-pressure systems. As it flows equatorward it therefore develops an easterly component of flow. However, this is kept in check by intense turbulent friction with the adjacent underlying surface, to the extent that meridional and zonal speeds remain comparable. The modest easterly components of the trade winds are the result.

4.8 Weather and climate

The relation between weather and climate is effectively that between instantaneous and mean conditions of the atmosphere. In considering this relation we must avoid the common tendency to consider average conditions, such as those described in the previous sections, to be the continuous background on which individual disturbances are superimposed. Such a distinction may be useful in the case of disturbances whose associated deviations are small and symmetrical about the mean. But although this is at least partly the case with the small, rapid fluctuations which constitute much of the turbulence endemic in the atmospheric boundary layer, it is of little value when considering the relationship between means and deviations on space and time scales applicable to the synoptic and larger scales. For such purposes we must be aware that mean conditions already contain the effects of a large range of important disturbances. To make this point explicitly and draw attention to some important aspects of tropospheric disturbances, let us re-examine several features of the mean conditions outlined in the previous three sections.

The most striking feature of the temperature distribution in Fig. 4.6 is the strongly baroclinic zone in middle latitudes. However, on any particular meridian and at any instant, this baroclinity is much less evenly distributed than it is on a seasonal, zonal mean. In fact baroclinity is usually concentrated in one or more narrow *fronts* separating broad, nearly horizontally uniform (barotropic) *air masses*. A well-marked front is always associated with an extratropical cyclone, of which it forms a vital part. The behaviour of extratropical cyclones and their associated fronts is quite complex, though some aspects have become familiar to the general public through forecasts in television and newspapers. A fuller description is given in Chapter 11, where it is shown that a sharply defined front persists for several days in the form of an intensely baroclinic zone ~ 2000 km long and 200 km broad, having almost any orientation and moving obliquely at speeds ~ 10 m s^{-1}. The transience, orientation and movement of fronts result in their intense baroclinity being smeared out and greatly weakened by zonal and seasonal averaging. The broad baroclinic zone of Fig. 4.6 is therefore not to be regarded as a persistent feature to be found at all times and meridians in middle and high latitudes, and only incidentally disturbed by the sporadic appearances of extratropical cyclones and their associated fronts; instead it is the mean result of the sporadic appearance of these disturbances over a wide latitude range.

Similar remarks apply to the meridional distributions of mean surface pressure apparent in Fig. 4.5. Inhabitants of middle and high latitudes are familiar with the incessant variations of surface pressure associated with the development and passage of large-scale areas of low and high pressure associated with great weather systems. Any one such low-pressure system would appear on an instantaneous section along a particular meridian as a very sharp and deep dip in the 1000 mbar isobaric profile. In fact, comparing it with the relatively smooth profile on Fig. 4.5, the sharp dip would be ~ 300 m deep (corresponding to about 30 mbar on a horizontal surface) and 10° of latitude in meridional extent. The continual development and decay of several such extratropical cyclones (the low-pressure systems) and *anticyclones* (the high-pressure systems) around each latitude zone is further smoothed over the seasonal averaging period, to give the broad, shallow depression evident in Fig. 4.5.

The situation is somewhat different at lower latitudes, however. The high average pressures apparent in the subtropics in Fig. 4.5 are actually considerably more nearly representative of individual instantaneous meridional profiles than the previous remarks might seem to imply, because they result from the presence of unusually stable and persistent weather systems — the large *subtropical anticyclones* which effectively girdle the Earth at these latitudes.

The pattern of mean westerly winds in Fig. 4.7 is similarly related to patterns on instantaneous meridional sections. The core of maximum wind speeds at about the 200 mbar level in the subtropics is associated with the nearly continual presence of the powerful *subtropical jet stream* there, which represents a concentration of the westerly flow maintained by the conservation of angular momentum as described in the previous section. Though the intensity and latitude of this impressive stream of air vary somewhat with time and longitude, the variations are sufficiently small for the mean pattern to be only moderately weakened and spread compared with the instantaneous local patterns. The subtropical jet stream is at its most intense in the winter months, moving poleward and weakening considerably in the summer months, in obvious association with variations in the Hadley circulation. The subtropical jet was the last of the major features of tropospheric circulation to be established by observation, becoming well known only during the Second World War, when military aircraft flying westward at high altitude in the vicinity of Japan consistently reported speeds over the ground dramatically less than their indicated air speeds. It is now known that the subtropical jet stream is especially intense in that region because of the influence of the Asian continent in general, and the Himalayan and Tibetan massifs in particular, the last two obstructing flow in the middle and high troposphere over an extensive region and producing a pronounced confluence in their wake.

The poleward extension of the westerly maximum across middle latitudes is similarly associated with the presence there of the *polar front jet stream*. Like the fronts with which it is intimately connected, the polar front jet is only partly continuous, and is narrow and highly mobile, with the result that the mean section gives no useful impression of the appearance of individual instantaneous sections. In the latter the polar front jet appears as a relatively narrow stream of fast-moving air centred at about the 300 mbar level and closely parallel to the associated fronts, whose wave-like sinuosity it closely matches (Fig. 11.10). Indeed it will become apparent later (section 7.10) that the westerly wind maximum in mid-latitudes is related dynamically to the baroclinic zone in those regions, and that this relationship applies both to averaged and instantaneous sections.

Obviously it is potentially very significant that the subtropical and polar front jet streams appear in the high troposphere in the height zones used by modern civil aircraft. The minimize the fuel wastage in flying against such fast airflows, routes are planned to avoid their forecast positions either by lateral or vertical displacement. In addition, the strong vertical wind shears in the vicinity of the jet core are sometimes associated with a rather unpredictable kind of turbulence which can be uncomfortable or even dangerous (section 10.8). Only supersonic civil aircraft and some military aircraft normally fly well above the levels of the jet cores.

4.9 Climate change

VARIABILITY

So far we have assumed that atmospheric variability ranges in period from the brief fluctuations associated with small-scale turbulence to the week-long fluctuations associated with synoptic-scale weather systems such as extratropical cyclones, with only the relatively regular seasonal variations occurring on longer time scales. But if this scheme accounted for all the variability, we should expect seasonal averages to be almost identical from year to year in any one location, on the apparently reasonable assumption that a season must contain so many randomly occurring weather systems that averages of temperature, rainfall etc. will be well defined. Observation and experience show that this is not the case. For example, Table 4.1 lists summer rainfall totals for ten consecutive years at a station in north-west England. Clearly there are large year-to-year variations which common experience suggests occur also in seasonally averaged temperature, sunshine, etc.

Such variability indicates the presence of non-seasonal periodicities in atmospheric behaviour which are considerably longer than a week. If you live in middle latitudes you can see at least one way in which such longer periods arise if you follow the daily sequence of weather charts. Occasionally an anticyclone becomes established in your particular region and persists there for several weeks, replacing the normal sequence of eastward-moving weather systems by an uneventful lull which may be warmer or cooler than the usual seasonal average depending

Table 4.1   Summer rainfall totals at a coastal station in northwest England (Hazelrigg 54.0 °N, 2.8 °W, 92 m above MSL)
Data are collected over June, July and August in each year

Year	Total in mm	Year	Total in mm
1975	223	1980	295
1976	99	1981	254
1977	218	1982	276
1978	305	1983	152
1979	229	1984	165
	10-year average	222 mm	
	maximum year to year ⎫	+ 83	i.e. + 37%
	deviations ⎬	− 123	− 55
	Standard deviation σ_n	63	28

on season, the region in question, and the location of the centre of the anticyclone relative to local geography (section 11.5). Such anticyclones are known as *blocking anticyclones* or *blocks* because of their apparent role in blocking the movement of other weather systems — but beware of the spurious implication of cause and effect.

If events such as blocks, lasting for half a season at most, were the longest-lived of the range of transient irregular weather systems found in the atmosphere, we should expect annual averages to be considerably more stable than seasonal averages, and averages over a sufficient number of consecutive similar seasons (for example ten consecutive summers) to be virtually invariant. Averages over ten consecutive full years should be even more stable. Though annual averages are a little more stable than seasonal averages, they are still quite significantly variable (Fig. 4.10). For example, the winter of 1963, which was the coldest for 200 years in the British Isles because of the persistence of a winter block (section 11.5), was followed by a relatively cool summer, with the result that the annual average temperatures for 1963 as a whole were quite unusually low. It seems that the atmosphere must be prone to disturbances more persistent than anything associated with the longest-lived weather systems identifiable on conventional weather maps (the blocks). And since such hidden disturbances can apparently influence conditions over several years, it is natural to ask if there is evidence of others acting

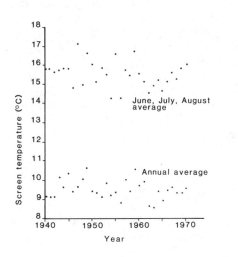

Fig. 4.10 Annual and summer averages of screen level temperatures in central England since 1940. (After [28])

over even longer periods. Examination of climatological records confirm that there is such evidence, and that there is no obvious upper limit to the lengths of associated time periods.

CENTRAL ENGLAND TEMPERATURES

In Fig. 4.11 temperatures in central England are extended backward in time to the first sustained meteorological use of thermometers in England (1659). This record is one of the longest series of direct meteorological measurements anywhere in the world, and of course pre-dates the beginning of the synoptic meteorological networks by nearly two centuries. Early measurements were particularly prone to error in instrumental performance and exposure, and it is fortunate that in recent times people have been prepared to devote painstaking, unglamorous effort to the collation, comparison, quality assessment, and even judicious correction of such data, since without such care the data would be much less reliable [28].

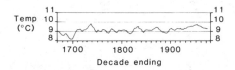

Fig. 4.11 Ten-year running means of annually averaged screen temperature in central England since 1659. Data are most reliable since the institution of the British synoptic observation system about 1860 and are least reliable before 1700, because of poor observation practice. (Redrawn from [28])

To smooth out fluctuations lasting only a year or two, the data have been subjected to a ten-year running mean, and yet there is still very considerable variablity. For example there was a pronounced cold period centred around AD 1700 in the British Isles, and a prolonged warm period peaking around 1950. Despite the obscurity of their origins, and the apparent smallness of their magnitudes, such variations can have pronounced social effects through agricultural production and in other ways. For example, the very cold period at the end of the seventeenth century was the culmination of decades of cooling so widespread and intense that it has become known as the *little ice age*, though it does not begin to compare in scale and intensity with the true ice ages of recent geological epochs (Fig. 4.12). Nevertheless in Scotland the harshness of the climate caused such distress, directly and indirectly by associated famine, that there was large-scale emigration in search of better conditions. A major flow of Scots was directed to north-east Ireland, where conditions were slightly easier and good agricultural land was made available by forcible expulsion of the native Irish, with social and political consequences which are distressingly evident to this day [29]. On a lighter note, the short cool period centred around AD 1820 coincided with the most impressionable phase of the childhood of Charles Dickens, and it is very probable that his early memories of the several harsh winters which occurred then encouraged his subsequent colourful evocations of winter and Christmas, and at another remove gave rise to the Christmas card industry of today.

CLIMATE CHANGE AND PREHISTORY

Direct meteorological measurements hardly existed before the mid-seventeenth century, but current interest in *climate change* (the name given loosely to all

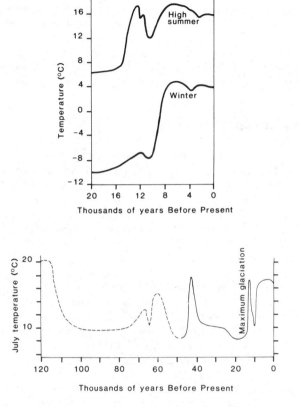

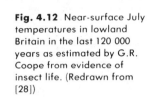

Fig. 4.12 Near-surface July temperatures in lowland Britain in the last 120 000 years as estimated by G.R. Coope from evidence of insect life. (Redrawn from [28])

Fig. 4.13 Near-surface temperatures in central England in the last 20 000 years, estimated largely from pollen analyses. (Redrawn from [29])

climatic variations lasting more than a few years) is so strong that an immense programme of indirect detective work is now under way. Old social records are combed, and a wide variety of physical, chemical and biological clues are sifted for information about past climate. For example, the analysis of pollen grains embedded in buried soil and rock strata shows what flora were growing when those strata were at the surface, and knowledge of the climatic tolerances of the flora then indicates the climate prevailing. Figure 4.12 includes summer and winter temperatures in central England over the last 20 000 years estimated by this method, and shows clearly the rapid rise of temperature associated with the ending of the last glaciation, one of whose ice sheets extended roughly to central England. The plotted temperatures are believed to be the same as would have been logged by thermometers subjected to a 500-year running mean. Figure 4.13 depicts estimates of temperature in lowland Britain derived from analysis of insect fauna over the last 120 000 years, and shows that the present warm spell is one of a very few separated by long periods of extreme cold and associated widespread glaciation. This period of repeated glaciation of high latitudes extends back several million years, before which there was a prolonged warm period in which there was little or no permanent ice even at the poles. Further back still, there is fossil evidence of a long period of ice ages in the Permian (section 3.8 and Fig. 3.12), and again in the Precambrian.

It is apparent from all these data that climatic conditions are variable on almost any timescale we choose to examine. Since only the very shortest of these correspond to the ranges which are the special concern of traditional meteorology,

95

it follows that there is a huge range of longer-lasting events demanding more or less unorthodox explanation. In recent years many people have responded to this challenge, with the result that the field of climate change abounds in speculation and theory. Since much of this is complex and incomplete, and since the subject is in any case too large and specialized to be treated at all thoroughly in the present text, we will consider only a selective outline of some current ideas. Several good specialist introductions are now available [30].

MECHANISMS OF CLIMATE CHANGE

Broadly speaking two types of mechanism for climatic change are envisaged — external and internal. The search for external causes obviously focuses attention on the Sun as the ultimate powerhouse of the atmospheric engine. How variable is the solar output, and do any of its variations correlate clearly with recorded climatic variations? Unfortunately the solar output has been measured with useful accuracy only since the early years of the twentieth century. Variations are small (~ 1%) and seem to correlate with sunspot numbers. These in turn have been recorded for several centuries by astronomers and show a pronounced quasi-regular eleven-year cycle with longer-term fluctuations superimposed. There is some evidence to suggest that sunspot minima (probably corresponding to slightly lower solar output) have been associated with lower global temperatures in that period. In particular the temperature minimum of the little ice age corresponds with a period between 1650 and 1700 AD when there seem to have been very few sunspots indeed.

Variations in the movement of the Earth relative to the sun significantly affect the amount and global distribution of solar input to the Earth (Fig. 4.14). The ellipticity of the Earth's orbit is quite small at present, but nevertheless produces a 6% range in solar input over the yearly cycle. However, the ellipticity itself goes through a cycle lasting about 100 000 years, in which it reaches a maximum corresponding to a 30% annual range of solar input. Secondly, the Earth's axis of rotation moves, varying the season of closest approach to the sun (*perihelion*). For example perihelion now occurs in the northern hemisphere in midwinter, since it coincides with the time when the axis through the northern hemisphere is tilted furthest away from the sun. But just over 10 000 years ago perihelion occurred in the northern hemisphere in midsummer. Following this motion of the Earth's axis back in time through several 21 000-year cycles, there must have been a period when a particularly close perihelion coincided with the northern hemisphere summer (and *aphelion* with winter) to produce much larger seasonal temperature swings than occur now. Thirdly, every 40 000 years the tilt of the Earth's axis of spin relative to the plane of its orbit round the sun completes a small regular cycle of variation about the present value of nearly 23.5°. Clearly the meshing of these three independent cycles must produce a complex variation of solar input involving many rhythms ranging in period from a few thousand to a million years or more, as well as the seasonal cycle itself. These variations are named after Milankovitch,

Fig. 4.14 The precession of the equinoxes. Just as the non-vertical axis of a spinning top precesses (swings round) about its foot, so the Earth's axis of spin precesses about the perpendicular to the plane of its orbit round the sun. As a result, the location of the equinoxes (where the Earth is tilted neither towards nor away from the sun) moves round the elliptical Earth orbit round the sun, completing a circuit in 21 000 years. Other seasonal positions, such as the position of the northern hemisphere midwinter (NMW — shown in its current position) move in step.

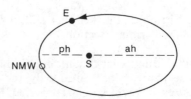

who first discussed their possible climatological significance (in 1930). Evidence is now quite strong that the *Milankovitch* mechanisms play an important role at least in triggering ice ages and intervening interglacials.

Some possible internal causes of climatic variation are easily identified, at least in principle. Since the Earth's atmosphere filters the incoming sunlight before it reaches the surface and so enters the atmospheric engine, it is clear that anything significantly affecting that filtering could be important climatically. Volcanoes do this sporadically, and sometimes very dramatically, by spewing fine, reflective dust into the stratosphere, where it drifts over the globe for months or years, reducing the solar input to the troposphere and surface beneath. For example, a high veil of dust was observed in many parts of the globe for several years after the cataclysmic eruption of Krakatoa (Indonesia) in 1883, and the especially vivid sunsets of the painter Turner come from a period in the early nineteenth century when there was a great deal of volcanic activity in several widely separated areas. There is some evidence that increases in volcanic dust in the atmosphere have been associated with transient coolings of the atmosphere. This connects the atmospheric engine with the independent and much more ponderous engine working in the Earth's interior and producing varying amounts of volcanism as a kind of incidental surface 'spray'. It seems probable that there have been periods in the past when volcanism has been much more vigorous than at present and climatic effects may have been correspondingly large.

More subtle internal mechanisms involve inherent chaotic (section 1.7) instability in the atmosphere/surface/ocean system. These may operate sporadically, just as the baroclinic instability of the troposphere erupts from an inherent instability, though much more sluggishly. For example, it is possible that the opposing extremes between glaciation (very large polar ice caps) and no glaciation at all, may represent two possible conditions of near equilibrium for the atmosphere/ocean/ice cap system, like a dented metal tray which may contain a ball resting in either of two shallow depressions. The system may move between these extremes in response to variations in volcanism and to the Milankovitch mechanism, and moreover may be especially prone to maximum glaciation when a major continent is at one of the poles, as Antarctica has been for the last 30 million years; and of course the continents themselves move significantly in time intervals greater than about ten million years, again in response to the Earth's internal engine. In terms of the ball-and-tray analogy, continental movement has the effect of warping the tray. Another example of inherent instability involves interplay between the atmospheric and oceanic systems. The large heat capacity of the oceans must tend to make any variations much more sluggish than those of the atmosphere alone, and indeed there is evidence that climatic variations lasting a few years are associated with anomalous temperatures over extensive parts of ocean surfaces [31].

4.10 Human society and climate change

GLOBAL WARMING

As mentioned in section 3.4, human industry is probably affecting global climate by increasing the abundance of CO_2 and other gases controlling the temperature of

97

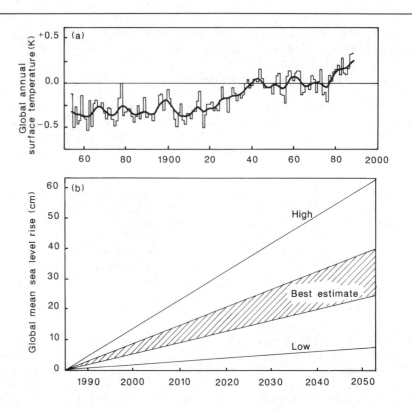

Fig. 4.15 (a) Global annual average surface temperatures since 1854 expressed as a difference from the 1950–79 average. Note how the ten-year running mean smooths the large inter-annual variations and highlights longer periods. (After [32]) (b) Highest, lowest and best estimates of global sea-level rise in the next 60 years. About half comes from melting ice and half from the expansion of warming sea water, both of which are so sluggish that even if global warming stopped at AD 2050, the sea would continue rising for a further ? centuries, doubling or trebling the rises shown. (After [18])

the Earth's surface and atmosphere through the greenhouse effect (section 8.5). So far the effect is too small to be clearly identified amongst the natural variations of climate over fairly short periods (10–100 years). Figure 4.15(b) shows a series of global temperatures which highlights the problem. Is the rising trend apparent from the late nineteenth century something new, or is it a recent stage in the continuing recovery from the little ice age? And note the problem of constructing a single series from a global distribution of temperatures, from areas with different climates and climate changes. Some have argued that a spurious rise in Fig. 4.15 may come from the proximity of many contributing observations to cities which have expanded greatly in this time period, spreading their heat island effects (though this has not been supported by subsequent detailed analysis).

Sophisticated theoretical models, similar to those used for routine five-day weather forecasting, but run for very much longer simulated time periods, now strongly suggest that an anthropogenic increase in the greenhouse effect will soon produce clearly identifiable global warming. These models predict temperature increases in the next century ranging from about 1 to 6 °C, depending on the model [33]. Differences between models arise from uncertainties in modelling the wide range of physical and chemical mechanisms at work, in particular those giving strong feedback (e.g. those involving water vapour and cloud — section 8.5). However, even the smallest predicted warming (ignoring a very few models which predict no change or even a cooling) would be very significant on a global scale, and the largest would take the Earth to its warmest condition since the onset of repeated glaciations of the northern hemisphere over 3 million years ago.

However large the global climate change, some localized effects will be much larger. All models of the changing general circulation during global warming show maximum warming at high latitudes, as poleward heat fluxes in mid-latitude weather systems (section 8.12) increase in the warmer, moister atmosphere. The warming predicted there is so large that melting of polar ice caps will contribute substantially to a global rise of sea levels, already rising through the expansion of the warming seas (Fig. 4.15). Again, the magnitude of the rise varies greatly with the climatological model used, and local rises are very sensitive to local adjustments of land height. Indeed some land masses, like Scandinavia, are still rising so rapidly after losing the burden of their last ice cap, that sea levels along their coasts will continue to fall for some time.

Pronounced meridional variation of climate change arises also from the shifting of the great climate zones (sections 11.8 and 12.5). There is clear evidence that during the most recent glaciation, the relatively slight global cooling and the marked cooling in middle and high latitudes evident in Fig. 4.12 were associated with substantial poleward movement and expansion of the subtropical arid zones associated with the descending branch of the Hadley circulation (Fig. 8.14). For example, ice cores from the Greenland ice cap show that desert dust was carried there in much larger quantities than is now the case. It is very probable that these trends would be reversed by gross global warming, which could be good news for those people currently struggling in marginal conditions in North Africa, for example.

Wind-driven ocean currents like the Gulf Stream are known to have changed substantially in the great climatic events: the north-eastern Atlantic is believed to have been about 10 °C cooler during the last glaciation because the Gulf Stream and North Atlantic Drift lay well to the south of their present locations, which encouraged the persistence of the great ice sheets on Britain and Europe. Such *air–sea* interactions are potentially so important in high latitudes and low (see section 12.5) that climate models have to cover the behaviour of the oceans as well as the atmosphere — a major complication. And evidence of very rapid changes in Greenland's climate during the last glaciation suggest that ocean currents can suddenly shift as some threshold is crossed [34].

SOCIAL CONSEQUENCES

Human sensitivity to climate change can hardly be overemphasized. A general rise in sea level will threaten low-lying coasts with increasing flooding, extending on a smaller scale the huge inundations which drowned the North Sea and Hudson Bay as the great ice caps melted after the last glaciation. Even quite small rises (for example about 10 cm) will force vulnerable populations to improve their coastal defence to avoid increasing disruption during winter storms (Figs 4.16 and 11.13).

The human population is currently growing so rapidly that food production is already inadequate in many areas, and though modern agriculture has helped greatly in some regions, production limits imposed by land availability and social efficiency may well be reached over much of the Earth within the next century. Imagine the possible effects of rapid, large-scale climatic change on this tense situation: established agriculturally productive zones (like the North American 'granary of the world' could be seriously damaged; and in the confusion there might well be serious delay in realizing the increased potential of other zones. And

99

Fig. 4.16 Modern and old-fashioned coastal defences at Morecambe, on the north-east edge of the Irish Sea. A broad, shallow bay encourages 10 m tides and short, steep waves which can threaten the town, especially when combined with a storm surge (section 11.3). (a) Modern defences use 'rock armour' to diffuse and absorb the wave energy which often pounded older sea walls (b) to destruction by coherent impact and reflection (Fig. 11.13).

recent experience in marginal lands like the African Sahel shows how easily climatic deterioration can combine with social change to produce devastation: fifteen years of sporadic drought following a wetter period in which the human and animal populations boomed forced whole societies to mortgage the longer-term future of their agriculture for the sake of short-term survival, leaving them increasingly vulnerable to further climatic anomalies.

Time is the crucial factor in all this. If climate change occurs over many human generations — for example 1000 years for major warming or descent toward another glaciation — then adjustment will hardly be noticed amidst all the other changes we seem to wish on ourselves. If however there is substantial climate change in a period as short as one or two centuries, as suggested by the more pessimistic models of global warming, then adjustment will almost certainly be painful. Faced with our evident ignorance and these possibilities, it is hardly surprising that the study of climate change has recently become very popular after long years in the scientific wilderness.

Appendix 4.1 Ideal gases

Consider a fixed mass of a certain gas at pressure p, absolute temperature T, and filling a known volume V. Then the gas laws established in the seventeenth and eighteenth centuries can be written as follows:

$p \propto 1/V$ at constant T (Boyle's law)
$V \propto T$ at constant p (Charles' law)
$p \propto T$ at constant V

These can be combined in a single equation of state

$$\frac{pV}{T} = \text{constant} \tag{4.9}$$

Through the work of Avogadro and others it became clear at the beginning of the nineteenth century that the constant in eqn (4.9) is proportional to the number of gas molecules in the volume but is independent of the type of gas. In fact if there are n moles of gas (i.e. $n \times M$ grams where M is the molecular weight of the gas), the equation of state takes the form

$$pV = nR_* T \tag{4.10}$$

where R_* is the universal gas constant, so called because the value 8.314 J K^{-1} mol^{-1} applies to nearly all gases in a wide range of conditions. A gas which obeys eqn (4.10) is by definition an *ideal* gas.

By using the definition of the mole we can rewrite eqn (4.10) to involve the mass m of gas filling the volume V.

$$pV = \frac{10^3 \, m \, R_* T}{M} \tag{4.11}$$

The factor 10^3 is needed because the mole is defined in terms of grams whereas the SI unit of mass is of course the kilogram. To remove the volume V, simply note that m/V is by definition the gas density ρ, so that eqn (4.11) can be rewritten in the meteorological form

$$p = \rho RT \tag{4.12}$$

where R is the specific gas constant of the particular gas in question as defined by

$$R = 10^3 \, R_*/M$$

According to Dalton's law of partial pressures, the total pressure produced when a volume is filled by a mixture of ideal gases is equal to the sum of the *partial pressures* which each would produce if they separately occupied the same volume at the same temperature. Since the individual components of the mixture are ideal gases, each must obey a form of eqn (4.12)

$$p_i = \rho_i R_i T$$

where p_i is the partial pressure of the component i, and ρ_i and R_i are respectively its density and specific gas constant. Dalton's law can then be expressed in the form

$$p = \sum_i p_i = \sum_i \rho_i R_i T \qquad (4.13)$$

$$\rho = \sum_i \rho_i$$

where p and ρ are the total pressure and density of the mixture. Rearranging these

$$p = \rho \left(\sum_i \frac{\rho_i R_i}{\rho} \right) T$$

$$= \rho \left(\sum_i x_i R_i \right) T \qquad (4.14)$$

where x_i is the specific mass (ρ_i/ρ) of the component i. By comparing eqns (4.12) and (4.14) we see that the mixture is behaving as a single ideal gas with specific gas constant

$$R = \sum_i x_i R_i$$

With a little rearrangement it is apparent that the equivalent single molecular weight of the mixture is

$$M = 1 \bigg/ \sum_i \frac{x_i}{M_i}$$

The values of R and M are therefore known provided the specific masses and molecular weights are known. The values for dry air (287 J K^{-1} kg^{-1} and 29.0 respectively) follow from the data in Table 3.1. Simple arithmetic shows that the values for moist air with specific humidity 10^{-2} (10 g per kg) are 289 J kg^{-1}K^{-1} and 28.8 respectively, showing that such air is about 0.5% less dense than dry air at the same temperature and pressure.

Appendix 4.2 The hydrostatic equation

Figure 4.17 shows a vertical column of air with unit horizontal cross-section area, which contains mass M of air between heights z_1 and z_2. Because air pressure acts equally in all directions at any point, and because of the choice of cross-section area, the upward force on the base of the segment is simply given by the pressure p_1 there, and the smaller downward force on the top of the segment is p_2. The net upward force on the segment is therefore $p_1 - p_2$, and in hydrostatic equilibrium this must be balanced by the downward force of gravity Mg on the column segment (i.e. its weight).

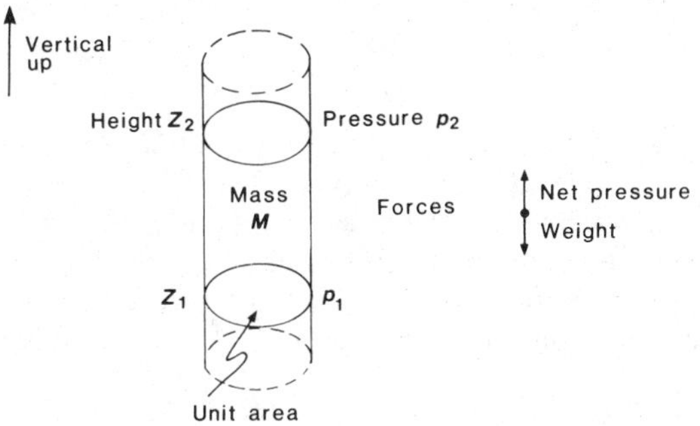

Fig. 4.17 Forces on a vertical column of air in hydrostatic equilibrium.

$$p_1 - p_2 = Mg \tag{4.15}$$

Since the segment volume is $(z_2 - z_1)$, the segment mass M is given by

$$M = \rho \, (z_2 - z_1)$$

if the density ρ is uniform, so that

$$p_1 - p_2 = \rho g \, (z_2 - z_1) \tag{4.16}$$

If the segment is very shallow

$$\frac{p_2 - p_1}{z_2 - z_1} \cong \frac{\partial p}{\partial z}$$

where the right-hand side is by definition the vertical gradient of pressure, and the approximation improves as the segment becomes shallower. In the infinitesimal limit

$$\frac{\partial p}{\partial z} = -g\rho \tag{4.17}$$

which is the most generally useful form of the hydrostatic equation, because density is no longer assumed uniform over a finite range of height. The minus sign ensures that p falls with increasing z, as is always the case with the upward-pointing z coordinate used in meteorology.

Appendix 4.3 Geopotential height

The work done against gravity in raising unit mass through a height interval dz is given by $g \, dz$, and is by definition the increase in its gravitational potential energy (geopotential) $d\phi$. That is

$$d\phi = g \, dz$$

where g is the local apparent gravitational acceleration. Since a truly horizontal surface is one with uniform geopotential rather than uniform height (given the slight horizontal variations in g arising from the rotation and oblate shape of the Earth — section 7.4), it is useful to define a unit of geopotential (the *geopotential metre*) in such a way that geopotentials relative to MSL are numerically very nearly equal to heights in metres. This is done using

$$\Phi = 1/9.8 \int_0^z g\,dz$$

where Φ is called the *geopotential height* of the local surface with height z. Conversions between actual and geopotential heights can be calculated for any geographical location (such as a particular radiosonde station), knowing the local vertical profile of g values. Once this is done, actual heights may be replaced by geopotential heights without changing the form of any equation apart from replacing the slightly variable g by the constant 9.8. For example eqn (4.6) becomes

$$\partial p/\partial \Phi = -9.8\,\rho$$

Note that geopotential height is energy per unit mass, despite its name, and that one geopotential metre is $1/9.8$ m^2 s^{-2}.

Appendix 4.4 Exponential decay

As shown in section 4.4, the hydrostatic equation for an ideal gas under gravity can be written in the form

$$\frac{\partial p}{\partial z} = -p/H_e \tag{4.18}$$

If H_e is a constant, this shows that the pressure gradient is directly proportional to pressure. Equation (4.18) is then a differential equation describing an exponential relationship between p and z, and the minus sign ensures that it is exponential decay rather than growth. Equation (4.18) is identical in form to eqn (2.2), and to many relations which arise in describing the natural world, including radioactive decay of unstable nuclei. By integration and judicious scaling and labelling of axes we can show that each equation has a solution which can be fitted to a common curve which has the same shape as Fig. 4.4, except that in most examples the dependent variable (the equivalent of p) is plotted on the vertical axis rather than on the horizontal one (Fig. 4.18). The equation of the common curve is of the form

$$F = F_0 \exp(-x/L_e) \tag{4.19}$$

where F_0 is the F value at zero x and

Fig. 4.18 Exponential decay in direct and logarithmic presentation.

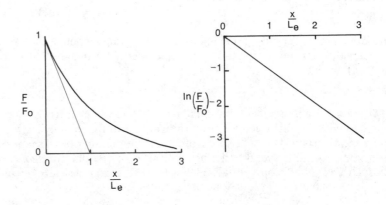

$$\exp(1) = e = 2.7183$$

is the base of natural logarithms. It follows that at $x = L_e$

$$F = F_0 \exp(-1) = F_0/e = 0.3679F_0$$

The constant L_e is the exponential scale length which describes the rate of decay of the curve towards zero. In the case of the vertical profile of atmospheric pressure it is the exponential scale height, and in the case of instrumental response (Appendix 2.1) it is the lag of the instrument in time. It follows that the reduction is $1/e^2$ (0.1353) if the x increase is $2L_e$, and $1/e^n$ if the increase is nL_e.

An exponential relation can always be written in an equivalent logarithmic form. The equivalent form of eqn (4.19) is

$$\ln(F/F_0) = -x/L_e \tag{4.20}$$

which shows that a graph of the logarithm of F against x is a straight line with gradient $-1/L_e$ (Fig. 4.18), which is equivalent to $-H_{10}$ in Fig. 4.2. Figure 4.2 shows that the graph of the logarithm of pressure against height for the occasion of the sounding in Fig. 4.1 is not quite linear, which means that the pressure–height curve is not quite perfectly exponential — a fairly small complication which arises from the vertical lapse of temperature in the troposphere.

We can find an expression for the decadal scale length of any exponential decay by rearranging eqn (4.20), and using the fact that $\ln F = 2.3026 \log_{10} F$ to produce

$$\log_{10}(F/F_0) = -x/(2.3026L_e)$$

Since by definition the decadal scale length L_{10} is the x increase which reduces F to $F_0/10$, setting F/F_0 to $1/10$ shows us that

$$L_{10} = 2.3026L_e$$

In the same way we can show that the binary scale length L_2 is given by

$$L_2 = 0.6931L_e$$

In the case of radioactive decay with time, the equivalent of the binary scale length is the half-life of the decaying species.

Although all scale lengths are equivalent, the exponential scale length has a natural significance not shared by the others. Returning to the case of the vertical profile of atmospheric pressure, we see from eqn (4.18) that the vertical pressure gradient has magnitude p_0/H_e at the level where pressure is p_0. If this gradient were to persist throughout the overlying layer (Figs 4.4 and 4.18), instead of dying away along the actual exponential profile, the pressure would fall to zero in a height interval exactly equal to H_e. Applying this to sea-level observations in the atmosphere, we see that barometric pressure would be as observed if the actual atmosphere were replaced by an incompressible 'sea' of air with depth H_e and uniform density equal to sea-level air density. Realistic values for H_e are about 8.5 km in the atmosphere, which is just over 0.1% of an Earth radius, emphasizing again the extreme shallowness of the atmosphere. In a number of analyses of atmospheric dynamics, it helps and clarifies to replace the actual compressible atmosphere by its incompressible equivalent.

Apart from the unusually pure case of radioactive decay, natural decay profiles in meteorology and elsewhere are seldom perfectly exponential. However, if a profile is only slightly imperfect, the exponential solution is still very useful and can be applied to actual measurements by allowing L_e to vary slightly. This is the case with the vertical profile of atmospheric pressure, where H_e varies with the absolute air temperature (section 4.4).

Appendix 4.5 Motion in a circle

Consider a point describing a circle of radius R in time T, at a steady tangential speed V. Since it travels round the circumference exactly once in time T we have

$$V = \frac{2\,\pi\,R}{T}$$

Now by definition of radian measure the angle subtended by one circumference is 2π radians, so that we can regard $2\pi/T$ as the angular velocity Ω of the orbiting point in radians per second. It follows that

$$V = \Omega R$$

which is often used to find V when Ω is known, or vice versa. As an example of the former, consider the tangential velocity of a point on the Earth's surface arising from the Earth's rotation. The angular velocity of the Earth is

$$2\,\pi/(24 \times 60 \times 60) = 7.27 \times 10^{-5} \text{ rad s}^{-1}$$

and since its radius is nearly 6370 km we find that the tangential speed of a point on the surface at the equator is 463.2 m s^{-1}. At latitude ϕ the radius of the circle described by a point on the surface is $R \cos \phi$ (see Fig. 4.19), so that the tangential speed is $\Omega R \cos \phi$, which is simply $\cos \phi$ times the equatorial value. At latitude 30° therefore the tangential speed is 401 m s^{-1}.

Fig. 4.19 A quadrant of the earth showing how the radius of rotation of a point on its surface varies with latitude.

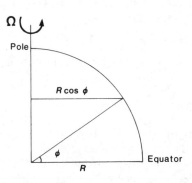

We have just seen that the rotation of the position vector R about O at angular velocity Ω is associated with a velocity V which is instantaneously perpendicular to R on the side favoured by the direction of rotation, and which has magnitude ΩR (Fig. 4.20). By an identical examination of the rotation of the resulting velocity vector V about O, also at angular velocity Ω, it is apparent that the rate of change of velocity is similarly instantaneously perpendicular to V. As shown in Fig. 4.20, this direction is always toward O on a location (rather than velocity) diagram. By direct analogy with the result for tangential velocity derived above, the magnitude of this centripetal acceleration A is ΩV, though either of the equivalent expressions V^2/R or $\Omega^2 R$ is much more widely used. We have the very important result that a body describing a circle at a uniform speed is continuously accelerating toward the centre of the circle at a rate proportional to the square of either the speed or the angular velocity. From previous values we find that a fixed point on the equator has a *centrifugal* acceleration (i.e. toward the Earth's centre) of 0.034 m s^{-2}, which is 0.3% of g.

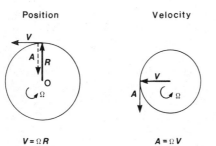

Position   Velocity

$V = \Omega R$ $A = \Omega V$

Fig. 4.20 Position, velocity and acceleration of a body moving steadily around a circle. The position diagram describes path and velocity in terms of a swinging radius, whereas the velocity diagram describes velocity and acceleration in terms of a swinging velocity vector.

Problems

LEVEL 1

1. If the pressure of a rising air parcel halves while temperature remains constant, what variation do you expect in its density? Since in fact parcel temperature always falls during significant ascent, will your first answer be an over- or underestimate?
2. A certain very large, high-altitude research balloon measures 100 m from the top of the envelope to the base of the instrument package. Using Fig. 3.1 as typical, estimate how the temperatures of the base and top should compare as it ascends from the surface to the stratopause.
3. State the conditions for a perfect exponential decay of atmospheric pressure with increasing height.
4. Deep anticyclones have a characteristically warm core throughout all but the highest parts of the troposphere. Qualitatively sketch the expected shape of the 1000, 500 and 300 mbar surfaces on a vertical section across the middle of such a system. (Consider the distribution of thickness.)
5. Sketch isotherms on a vertical section through two atmospheric zones, only one of which is baroclinic. Assume horizontal isobars, and label warm and cold parts of each zone. Name the zone with horizontal isotherms.
6. Using Fig. 4.7 as evidence, brief a party of inexperienced mountaineers about likely wind conditions on Everest.
7. In middle latitudes especially, mean conditions are not to be considered to be a steady background which is disturbed by occasional weather systems. Justify this statement from a typical barograph trace.
8. Currently the northern hemisphere summer coincides with aphelion. Speculate on how the seasons there would compare with the present when aphelion coincides similarly in a time of maximum ellipticity of terrestrial orbit round the sun.

LEVEL 2

9. Use the equation of state for dry air to calculate the air density at the summit of Mt Everest on an occasion where the pressure and temperature there were 313 mbar and -38.5 °C respectively.
10. In a certain isothermal atmosphere in which the exponential scale height is

7 km, find the height intervals between the following standard pressure levels (levels for which radiosonde data is always extracted): 1000, 700, 500, 300 mbar.

11. Use data from Fig. 4.5 to estimate the mean temperatures of the 1000 to 500 mbar layer at latitudes 30 and 70° in winter, assuming an isothermal atmosphere in each case. Use the expression for binary scale height.

12. Inspect Figs 4.7 and 4.9 and estimate the time taken for air to travel (a) around the Earth at constant latitude in the vicinity of the core of the subtropical jet stream, and (b) from latitude 30° to latitude 10° in the winter trades.

13. Estimate the different travel times for aircraft shuttling with an air speed of 250 m s^{-1} between airports A and B which are 1000 km apart, when there is a jet stream flowing from A to B at 75 m s^{-1} at cruising altitude. (Ignore climbs and descents.)

14. Express a typical polar front jet core speed as a fraction of the tangential speed of a point fixed on latitude 50°. Find the centripetal acceleration of a fixed point on the equator and express it as a fraction of g.

LEVEL 3

15. Find the density of moist air with specific humidity 15 g kg^{-1}, temperature 15 °C and pressure 1010 mbar, and compare it with the density of dry air at the same temperature and total pressure. Find the temperature at which dry air would have the same density as the moist air at the same total pressure. Note that this last temperature is known as the *virtual temperature* of the moist air, which is very useful in considering the buoyancy of cumulus etc.

16. Given that the atmosphere of Mars consists very largely of carbon dioxide, and that the gravitational acceleration near the Martian surface is 3.8 m s^{-2}, find the decadal scale height of the Martian atmosphere and compare it with the terrestrial value. (Martian atmospheric temperature ~ 220 K.)

17. Consider possible changes in low latitudes in Fig. 4.6 if deep convection there were to be considerably reduced. And similarly consider how the appearance of Fig. 4.7 might differ in low latitudes if the Earth were not rotating. In fact convection is likely to increase in some parts of low latitudes in the absence of rotation. Where and why?

18. Assuming that the glaciated and unglaciated states of the Earth represent two different equilibrium states between which the surface and atmospheric system of the Earth vacillates in response to random or external factors, consider how a small increase in the total ice-cap extent might lead to a further increase and so to full glaciation. Given that low-latitude conditions alter little, consider how the increased baroclinity implied by the onset of glaciations might on the one hand tend to increase snow and ice cover, and on the other tend to decrease it.

19. Show that the effect of ignoring the meridional variation of g is underestimate the heights of horizontal surfaces at low latitude as compared with high by about 50 m in 10 000.

Atmospheric thermodynamics

5.1 Introduction

Heat enters and leaves the atmosphere in many different ways, and the air is usually warmed or cooled as a result. For example, solar radiation warms the atmosphere, and cooling is usually the net result of the absorption and emission of terrestrial radiation (Chapter 8). In a different type of process, buoyant air rises and expands against the falling pressure of ambient air, and as it expands it cools. Sinking air is warmed by the same process in reverse. And in yet another type of process, large amounts of heat are evolved or absorbed within air parcels as their contained water substance changes state. For example, heat is evolved during cloud formation, the latent heat of vaporization (latent since it was absorbed when the water was last evaporated) being released and shared with the air between the swelling droplets.

In the present chapter we consider in some detail how the temperature changes produced in these ways are related to the physical processes at work. In doing so we make use of some of the techniques and results of the science of thermodynamics, a subject which has grown in the last two centuries from a number of unrelated speculations about the nature of temperature and heat to a powerful and sophisticated part of physical science, indeed one of its cornerstones. During its development, thermodynamics has evolved a subtle, precise and rigorously logical approach which is fully revealed in texts such as [35] but which is not reproduced here. In this book, certain basic thermodynamic concepts are applied to atmospheric behaviour in just sufficient detail to clarify the physical processes at work, and to justify some useful relationships.

5.2 The first law of thermodynamics

To describe the thermal behaviour of a typical parcel of air, we must be able to allow for its response to the injection and removal of heat, as well as to its expansion and compression. We need a relationship between exchanges of heat and changes of temperature and volume, and this is provided by a form of the *first law of thermodynamics* which is particularly suitable for describing the behaviour of unit mass of an ideal gas:

$$dQ = C_v \, dT + p \, dVol \qquad (5.1)$$

where dQ is a very small heat input (from sunlight for example), dT is a very small rise in temperature, and $dVol$ is a very small increase in the parcel volume *Vol* (Fig. 5.1).

The first term on the right-hand side of eqn (5.1) represents the amount of heat input which is used to increase the *internal energy* of the air parcel, i.e. the kinetic energy of its air molecules' random motion, of which temperature is a measure. This term is proportional to the temperature rise, and the constant of proportionality C_v is by definition the *specific heat capacity* of the air at *constant volume*, as can be seen in the special case where $dVol$ is zero. The value of C_v for dry air throughout the turbosphere is close to 717 J K^{-1} kg^{-1}, which means that this number of joules of heat input are required to warm 1 kg of air by 1 °C when the volume of the air is kept constant.

The second term on the right-hand side of eqn (5.1) represents the amount of the heat input which is used in doing work against the surrounding air as the parcel expands. It is proportional to the expansion and to the internal pressure p of the parcel, which in terrestrial conditions is effectively equal to the surrounding pressure, for the reasons discussed in section 4.3.

The layout of eqn (5.1) is a reminder that the heat input to the air parcel is shared between the two processes in question: warming and expansion. The equality expressed is an example of the *principle of conservation of energy*, which is believed to hold under all circumstances. However, any particular example of it, such as eqn (5.1), is valid only in so far as all sources, sinks and transformations of energy are accounted for. Fortunately the very simple scheme represented by eqn (5.1) accounts for several very important processes with useful accuracy.

As in the treatment of the equation of state for air (section 4.1), the parcel volume *Vol* is inappropriate for dealing with an unconfined gas like air, and so it is replaced by the reciprocal of the air density, to which it is numerically equal since we are considering unit mass of air. When this is done, and the resulting density variations $d\rho$ are replaced using the equation of state again (Appendix 5.1), we find

$$dQ = C_p \, dT - \frac{1}{\rho} \, dp \qquad (5.2)$$

Consideration of the special case where the little pressure change dp is zero shows that C_p is the specific heat capacity of the air *at constant pressure*. According to Appendix 5.1, C_p is larger than C_v by the specific gas constant R,

$$C_p = C_v + R$$

and it follows from the values of the terms on the right-hand side that the value of C_p for dry air in the turbosphere is about 1004 J K^{-1} kg^{-1}: a value which can usually be rounded to 1000. Strictly speaking, the different heat capacity of water vapour makes the C_p value for moist air differ slightly from the value for dry air, but the difference is so small that for most purposes the dry value can be applied throughout the turbosphere, regardless of humidity. As will appear in several of the following sections, C_p applies much more widely than might seem proper at first glance, and is much more generally useful than C_v.

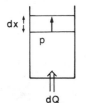

Fig. 5.1 Heat dQ entering air trapped at constant pressure p in a cylinder, warms the trapped air and pushes the light, frictionless piston outward through a distance dx, doing work $p \, A \, dx$ ($= p \, dVol$), where A is the area of the piston face and $dVol$ is the increase in the volume of the trapped air.

5.3 Isobaric heating and cooling

Much heating and cooling of the atmosphere occurs while air pressure is steady or nearly so. For example, the warming of the planetary boundary layer by day and its cooling by night occur at pressures determined by the weight of the overlying atmosphere, and this usually changes very little in percentage terms in the few hours for which the warming or cooling persists. And because of the unconfined nature of the atmosphere the heating or cooling does not in itself change the air pressure in the way that it would for example if the air were trapped in a rigid box. Of course heating may produce convection, with associated significant changes in the pressures of air parcels as they rise or fall through the surrounding air, but even so the individual pressure changes can be ignored if we choose a 'super parcel' large enough to encompass the full depth of the convecting layer.

Equation (5.2) becomes suitable for treating *isobaric* processes when the term involving dp is zero. The first term on the right-hand side ($C_p \, dT$) then completely determines the relationship between heat exchange and temperature change, and can be used to estimate temperature changes from known gains or losses of heat, or vice versa:

$$dQ = C_p \, dT$$

This applies also to a finite slab of air embedded under the rest of the atmosphere, even though there might seem to be a complication as the slab expands and uses some of the energy input to raise its own centre of gravity. A full treatment shows that the complication is illusory and that the simple relation still holds (appendix 13.2).

To consider the isobaric warming or cooling of a mass M of air, rather than unit mass, we must use

$$dQ = MC_p \, dT \tag{5.3}$$

Let us work through an example which shows the meteorological relevance of this very simple relationship. On a sunny morning overland, it is often observed that the air temperature near the ground rises by a couple of degrees per hour for several hours in response to solar heating. Observation in depth shows that the warming layer is often about 300 m deep (Fig. 5.2). What rate of heat input is required to account for this rate of warming of this amount of air?

Assuming the density of air to be 1.2 kg m^{-3} and considering the warming of a

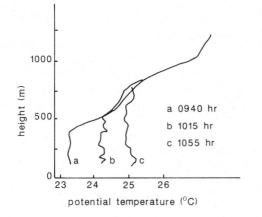

Fig. 5.2 Morning warming and deepening of a convecting layer over extensive grassland in Australia. Some of the temperature wriggles arise because measurements by traversing aircraft cannot be closely synchronized at different heights (potential temperatures should be uniform along vertical profiles in an ideal well-mixed, cloudless layer, according to section 5.7). The situation is similar to that shown in Fig. 5.9, except that potential temperature is used in place of ordinary temperature, and the overlying stable layer is maintained by anticyclonic subsidence (section 12.1). (After [36])

111

column resting on one horizontal square metre of ground, the mass of a 300 m column is 360 kg. Inserting this value for M in eqn (5.3), together with 2 °C for dT and the established value for C_p, we find that the required heat input per hour is 7.2 $\times 10^5$ J m^{-2}, which corresponds to a rate of input of 200 W m^{-2}. In fact the total solar input is likely to be larger than this because some heat is used to evaporate dew and soil moisture, and some heat is absorbed by the warming ground (section 9.5). Values of this approximate magnitude are known to be realistic from direct measurements by radiometer.

The heat in this example is entering the air largely by means of convection from the sun-warmed ground surface: warm, buoyant parcels rise up, while cool ones sink down, and the continual mixing distributes the net heat input throughout the convecting layer. Such heat is called *sensible heat* because it can be sensed directly by thermometers as the warm and cool air parcels pass by. However, in all but the most arid lands, an even larger amount of *latent heat* is carried aloft by the same convection, because air parcels rising from the moist surfaces contain more water vapour than do their sinking neighbours. This latent heat flux has no warming effect until the latent heat is released by condensation of the water vapour, almost always in the course of cloud formation (dew represents a relatively very small sink of atmospheric water vapour). The heat which then becomes sensible is the heat which was absorbed from the surface and adjacent air when the water evaporated in the first place. Since vapour, evaporation and condensation play very important roles in the events discussed in this and succeeding chapters, we must first consider carefully how vapour content is measured and quantified.

5.4 Measures of water vapour

Absolute vapour content is measured in terms of vapour density or pressure. Since vapour behaves as an ideal gas in terrestrial conditions, we can use the equation of state (eqn (4.2)) to relate the vapour pressure e to vapour density ρ_v and absolute temperature T

$$e = \rho_v R_v T, \tag{5.4}$$

where R_v is the specific gas constant for water vapour, which has value 462 J K^{-1} kg^{-1} according to eqn (4.2) and following text. Note that e is the partial pressure of the water vapour (appendix 4.1).

In many circumstances it is more useful to know the proportion of air which is water vapour than it is to know the absolute vapour content, especially since the very well-mixed state of the lower troposphere in particular makes proportional measures much more uniform than absolute measures. We have already used one proportional measure, the specific humidity q, which is the proportion by mass of water vapour in the total moist mass (vapour plus dry air).

$$q = \frac{m_v}{m} = \frac{m_v}{m_v + m_d} \tag{5.5}$$

An alternative and similar measure is the proportion of vapour to the mass of dry air with which it is mixed — the *humidity mixing ratio r*.

$$r = \frac{m_v}{m_d} \tag{5.6}$$

Fig. 5.3 Backlighting highlights dew forming thickly on a spider's web in very slightly supersaturated air. Because the supersaturated zone is confined to within ~ mm of the slightly cool web there is no fog in the background.

In fact the proportions of water vapour in air are so small, even in the warmest and moistest parts of the troposphere, that there is no significant difference between q and r values for any realistic air parcel. A typical value in the low troposphere is 10^{-2}, quoted as 10 g kg^{-1} to avoid exponents, and values in the high troposphere are one or two orders of magnitude smaller. For simplicity we will always use q in this book.

Since the moist air and water vapour occupy the same total volume (the volume of the air parcel in question), q is also the ratio of water vapour density to moist air density.

$$q = \rho_v/\rho \qquad (5.7)$$

Replacing each density using the appropriate equation of state we find

$$q = e\,R/(p\,R_v)$$

where R is strictly speaking the specific gas constant for moist air, but can be assumed to be the dry air value without serious error, and p is the total air pressure. The ratio of the gas constants R/R_v is simply the inverse ratio of their effective molecular weights, which has the value 0.622, often represented by ϵ. Hence

$$q = \epsilon e/p \qquad (5.8)$$

Vapour pressure is normally quoted in millibars, and can be used in eqn (5.8) provided p is expressed similarly, which it usually is. The vapour pressure corresponding to $q = 10$ g kg^{-1} and $p = 1000$ mbar is 16 mbar, which is typical of the low troposphere.

The total amount of vapour in a vertical column extending through the atmosphere can be expressed as the depth of rainfall produced if it were all to be rained out. This *precipitable water content* can be readily calculated from radiosonde profiles of humidity and temperature against pressure (appendix 5.2).

It is found by laboratory study that the pressure and density of a saturated vapour (a vapour in dynamic equilibrium with a plane surface of pure water —

113

section 6.2) depend only on the temperature of the vapour (Fig. 6.2). At any temperature T there is therefore a definite value of the corresponding saturated vapour content, which means that it is possible to express the actual vapour content in any situation as a fraction or percentage of the content needed to produce saturation at that temperature. This is the relative humidity (RH) of the air, and it can be expressed in terms of either vapour pressure or density,

$$RH = 100e/e_s = 100\rho/\rho_s \tag{5.9}$$

where the saturation values are denoted by subscript s. Values of RH range from 100% to about 10% in the troposphere, and are usually in excess of 50% near all but the most arid land surfaces. In growing clouds they may exceed 100% (section 6.9), but the excess is so slight that saturation is the normal limit to vapour content, and therefore 100% is the normal limit to RH.

5.5 Measurements of water vapour

Although direct measurement of relative humidity has long been normal in radiosondes for reasons of instrumental simplicity, the most accurate measurements of vapour content, and the most useful measures of vapour content for some important meteorological purposes, are the wet-bulb and dew-point temperatures (section 2.3).

In the case of a wet-bulb thermometer, water evaporates into the air from a saturated wick enveloping the thermometer bulb (Fig. 5.4), and the excess of this flux over the opposing flux, condensing onto the wick from the surrounding vapour, cools the wick, bulb and surrounding air by net loss of latent heat. Some movement of air over the wick is needed to ensure that the air in contact with it is replaced quickly enough to inhibit the formation of a cocoon of saturated air which would otherwise insulate the thermometer from effective contact with the ambient air. However, the actual rates of net evaporation from the wick are so high in all subsaturated atmospheres that the air in contact with the wick is fully chilled and saturated before it is swept away by the air motion. (The efficiency of net evaporation has another important consequence discussed in section 6.2.) The temperature of the wet bulb is therefore the temperature at which the measured air is saturated by evaporating water into it from the wet bulb. In practice the detailed behaviour of the wet bulb is complex, and the reading is significantly affected by heat conducted from the dry parts of the thermometer (which are warmer in all subsaturated conditions), by absorption of solar and terrestrial radiation, and by differences between the rates of diffusion of water vapour and heat through air. For this reason a simpler but less realistic process is envisaged to define a *thermo-*

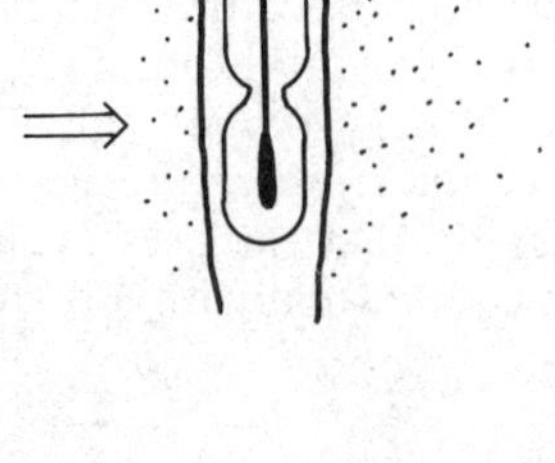

Fig. 5.4 Evaporation from a wet-bulb thermometer into ambient air-flow.

dynamic wet-bulb temperature T_w, which is used as the ideal wet-bulb temperature in all meteorological discussion.

The thermodynamic wet-bulb temperature of an air parcel is by definition the lowest temperature to which the parcel can be cooled by evaporating water into it isobarically and adiabatically (i.e. at constant pressure and in thermal isolation from its surroundings). As water evaporates into the parcel, it is chilled and its vapour content is increased, and the lowest temperature will be reached when the air in the parcel reaches saturation by the combination of these two tendencies. If the evaporation of a mass m_v of water is needed to cool an air parcel of mass m from its original temperature T to its wet-bulb temperature T_w, then the latent heat required for the evaporation (Lm_v) must equal the heat lost in cooling the air, since there is no other source or sink of heat. (L is the *specific latent heat of vaporization* of water — the quantity of heat needed to evaporate unit mass of water. In terrestrial conditions the value of L is always close to 2.5 MJ kg^{-1}, and its great size has many important consequences for meteorological situations involving evaporation or condensation.) By eqn (5.3) we therefore have

$$Lm_v = mC_p (T - T_w)$$

Now according to the process just described, the ratio m_v/m is the difference between the initial specific humidity $q(T)$ and the final saturated specific humidity at T_w, $q_s(T_w)$. (This ignores the change in total mass of moist air in comparison with the change in the proportion of vapour, which is always justifiable in atmospheric conditions.) Hence we find a version of the *psychrometric equation*:

$$q(T) = q_s(T_w) - \frac{C_p}{L} (T - T_w) \tag{5.10}$$

Note that the relatively small heat capacity of the vapour has been ignored throughout. Replacing the q terms by eqn (5.7) or (5.8) allows either the vapour density or the vapour pressure of the air to be calculated from the corresponding values for saturation at the wet-bulb temperature, values which are tabulated, or graphed as in Fig. 6.2.

Similar but slightly different relationships apply to actual wet-bulb thermometers, but by a happy accident of opposing deviations the temperature registered by a suitably aspirated wet-bulb thermometer (ventilation draught 4 m s^{-1} or more) approximates to the thermodynamic wet-bulb temperature so closely that the error can be ignored. Wet-bulb thermometers relying on natural ventilation, in a Stevenson screen for example, may differ more significantly from T_w, and require a modified form of eqn (5.10) for accurate calculation of water content.

In the case of a dew-point meter, the air in contact with the cooling face of the meter is cooled isobarically until it saturates and begins to deposit dew on the face. The temperature of the face and the adjacent air when this happens is by definition the dew-point temperature of the air. Since the proportion of vapour in the air is unaltered by this process, at least until dew begins to form, and since the total air pressure is fixed at its ambient value, eqn (5.8) shows that the vapour pressure of the air before chilling (i.e. at the initial temperature T) is unaltered by the chilling and is therefore the same as the saturation vapour pressure at the dew-point:

$$e(T) = e_s(T_d) \tag{5.11}$$

Measurement of T_d together with tables of saturation vapour pressure therefore gives the vapour pressure of the ambient air, and hence all the other measures of humidity through eqns (5.7) to (5.10). For example, the relative humidity RH is given by

$$RH = 100e_s(T_d)/e_s(T)$$

When it is not zero, the dew-point depression $T - T_d$ is significantly larger than the wet-bulb depression $T - T_w$ (roughly twice as large), because the saturated state at T_d is achieved purely by chilling, whereas saturation at T_w is reached by chilling and net evaporation. Because of the complexity of the thermodynamics involved, no simple relation exists between numerical values of T, T_w and T_d, and thermodynamic tables or diagrams have to be used.

The multiplicity of measures and measurements of atmospheric humidity may seem confusing, but in fact each is particularly useful for dealing with one or more type of meteorological problem. The specific humidity has been used already, and most of the others will be seen in action in this and the following chapter. Although the dew-point meter is occasionally used for slow but accurate humidity measurement, the dew point is used more as a measure of humidity than as a measurable property — one which is very useful in discussions of cloud formation in chilling air (Fig. 6.9), and for displaying humidity on thermodynamic diagrams (Fig. 6.6).

5.6 Adiabatic reference processes

It has already been mentioned (section 4.2) that in a number of important types of tropospheric motion, air rises or falls so rapidly that the cooling caused by expansion, or the warming caused by compression, considerably outweighs temperature changes caused by other factors, such as absorption or emission of radiation. In formal terms the air motion is nearly adiabatic, and it is very useful to define strictly adiabatic *reference processes* with which to compare the inevitably more complex reality of atmospheric behaviour.

The simpler of the two common reference processes is the *dry adiabatic reference process*, in which there is neither condensation nor evaporation of water substance. Strictly speaking it is relevant to the behaviour of all unsaturated air, in addition to the unrealistically completely dry air implied by its title, but the term 'dry' has become standard. The dry adiabatic reference process is defined by eqn (5.2) with the left-hand side (dQ) set to zero, and it follows that the terms on the right-hand side must balance exactly.

In many circumstances of course there is condensation or evaporation of water, whose effects are to supply or remove heat to or from the air (almost always cloudy air), and which corresponds to non-zero dQ in eqn (5.2). Since almost all tropospheric cloud condenses or evaporates in air which is very close to saturation (held there by the surprisingly complex mechanisms considered in the next chapter), it is useful to define a *saturated adiabatic reference process*.

Detailed comparison with observation shows that the dry and saturated reference processes do very well — surprisingly well in view of the deliberately drastic simplifications assumed. In fact a full discussion of why they work so well is beyond the scope of this book and involves matters which are the subject of current research. Using models which work well for reasons which are not clear is of course a symptom of imperfect understanding, but it is a very useful procedure even so, and often tends to highlight particular matters needing further attention. We will now look in some detail at properties of the adiabatic reference processes, knowing that though they are over-simple they nevertheless provide useful keys for understanding the real atmosphere.

5.7 The dry adiabatic reference process and potential temperature

In this process no heat enters or leaves the air parcel, and none is evolved or absorbed within it by condensation or evaporation of water. It follows that dQ is zero in eqn (5.2) and that with rearrangement

$$dT = dp/(\rho C_p) \tag{5.12}$$

Equation (5.12) is clearly consistent with common experience (with bicycle pumps for example), that compression (positive dp) is associated with warming (positive dT), and that decompression is associated with cooling. The relationship is made more obvious when air density ρ is replaced using the equation of state (eqn (4.2)). After rearrangement we find

$$\frac{dT}{T} = \frac{R}{C_p}\frac{dp}{p} \tag{5.13}$$

where T is the absolute temperature of the air parcel, and p is its total pressure. Equation (5.13) shows that small changes in temperature and pressure are in direct proportion, and that the constant of proportionality is R/C_p when each is expressed as a fraction or percentage of its absolute value. The value of R/C_p for dry air is always very close to 0.285 in the turbosphere, and it differs so little in realistic humidities that the same value can be used for moist air in all but the most accurate descriptions. It follows that for example a 3% drop in air pressure (about 30 mbar in the low troposphere) is associated with a fall of just under 1% in air temperature (about 2.6 °C in 300 K).

When percentage changes in p and T are no longer small, eqn (5.13) must be integrated to relate initial and final value (with subscripts 1 and 2 respectively), with useful accuracy. Standard integration then produces *Poisson's equation*,

$$\frac{T_2}{T_1} = \left(\frac{p_2}{p_1}\right)^{R/C_p} \tag{5.14}$$

which can be used to find any of the continuous sequence of intermediate temperatures experienced by an air parcel undergoing compression or decompression. For example, an air parcel in the deep, dry convecting layer over the Sahara desert in the mid-afternoon may rise and expand and cool dry adiabatically from the 1000 mbar level (near sea level) to the 500 mbar level over 5 km higher. If such air starts rising with a temperature of 305.2 K (32 °C) at the lower level, eqn (5.14) shows that it will reach the upper level with a temperature of 250.5 K (-22.7 °C). The initial, final and all intermediate states are said to be on the same *dry adiabat* — the continuous sequence of states, each defined by a pressure and temperature, which is connected by this particular dry adiabatic process. Dry adiabats are specified in thermodynamic tables such as [37], and are plotted on several types of thermodynamic diagram used by meteorologists, one of the most common of which (the *tephigram*) is used in Fig. 5.5. These diagrams are used to produce graphical displays of the nearly vertical profiles of temperature, humidity and pressure measured by radiosondes.

As outlined in the next section, the dry adiabats on a tephigram are straight lines running diagonally from bottom right to top left. Any particular part of the temperature/pressure profile from a radiosonde ascent which corresponds to a dry adiabat will therefore be a straight line on a tephigram, lying parallel to the series of standard dry adiabats printed on it. The Saharan example shows this clearly when plotted on Fig. 5.5, as does a sounding more typical of vigorous convection in middle latitudes, though the dry adiabatic layer is cooler and much shallower in the latter.

It is useful to label each dry adiabat with a numerical value which is the same everywhere along the line, and which is therefore conserved during that particular dry adiabatic process. The convention is to use as a label the temperature at which the dry adiabat crosses the 1000 mbar isobar. This temperature is known as the *potential temperature* of the air at any point along that particular dry adiabat, and it corresponds physically to the temperature which would be reached by an air parcel if it were taken dry adiabatically from its initial temperature and pressure to 1000 mbar. It follows from this definition and eqn (5.14) that the potential temperature θ of air with absolute temperature T and pressure p (expressed in millibars) is given by

$$\theta = T \left(\frac{1000}{p} \right)^{R/C_p} \tag{5.15}$$

Equation (5.15) shows that potential temperature θ is greater than actual temperature T when pressure is less than 1000 mbar, and vice versa, and is equal to T only when pressure is exactly 1000 mbar. It follows that potential temperatures are greater than actual temperatures throughout all but the very lowest parts of the atmosphere, and will be greater even there if the surface pressure is below 1000 mbar, as in a vigorous depression, for example. In the Saharan case in Fig. 5.5, where the potential temperature is 32 °C everywhere along the dry adiabat, the actual temperature at 500 mbar is −22.7 °C. If the dry adiabat in this case extended down to a rather high surface pressure of 1020 mbar (likely only in those parts which are not far above sea level and in the absence of the heat low which is so common in the Sahara in summer — see section 11.7), then the actual temperature at the surface would be 33.7 °C.

5.8 Thermodynamic diagrams

It is very useful to be able to portray thermodynamic processes, such as the meteorological reference processes, on a diagram which will graphically summarize matters otherwise requiring equations or abstract text. A point on such a diagram must correspond to a particular thermodynamic state of an air parcel, i.e. to a definite temperature and pressure, so that a line will represent the series of states corresponding to any particular process being portrayed. It does not follow, however, that we must use temperature and pressure as axes. Indeed there is a long tradition that the most useful axes are those which ensure that equal areas anywhere on such a diagram should represent equal energies — energy gained or lost in going round the cycle of processes represented by the boundary of the area. There are many different diagrams in use which fulfil this criterion, but a detailed discussion of their construction and relative merits involves more than the basic thermodynamics covered in this book. It is sufficient to say that as well as observing the thermodynamic niceties, any particular diagram must provide a clear

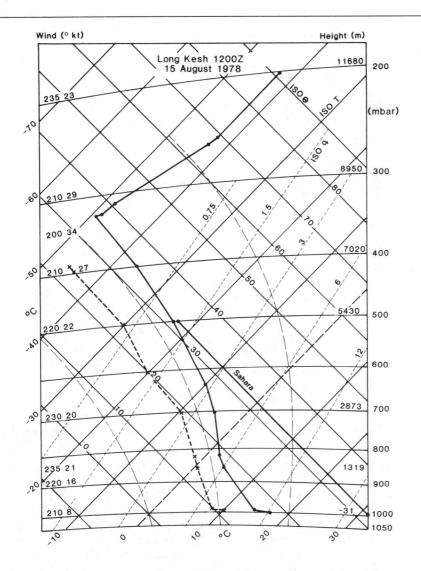

Fig. 5.5 Vertical soundings of pressure, temperature and humidity plotted on a tephigram (sections 5.7 and 5.8), with temperatures and dew points represented by continuous and dashed lines respectively. The main sounding is typical of many situations in middle latitudes, while the layer of much warmer air (further to the right at the same pressure) represents the hypothetical Saharan example quoted in section 5.7. (After Metform 2810, Meteorological Office, Bracknell, Berks. Crown copyright)

display of the processes at work, and if possible distinguish clearly between ones which are felt to be significantly different.

The tephigram is the most widely used thermodynamic diagram in meteorology. It was introduced by the English meteorologist Sir Napier Shaw, who was one of a number of very distinguished meteorologists from a variety of cultures who helped establish meteorology as a quantitative science at the beginning of the present century. The axes are absolute temperature T (usually labelled in degrees Celsius), and the logarithm of the potential temperature θ labelled likewise. In a certain highly idealized type of thermodynamic process (a *reversible* one), the potential temperature axis can be regarded as an entropy axis: hence the name $T\phi$ (for entropy) gram. It can be shown that these axes meet the energy criterion cited above, and this together with the many features of this versatile diagram are discussed in more advanced meteorology texts [8]. The basic construction of the tephigram is probably best appreciated by making a simple outline of one, as described in appendix 5.3. It then emerges that in the limited ranges of T and θ needed to represent conditions in and just above the troposphere, the basic layout

119

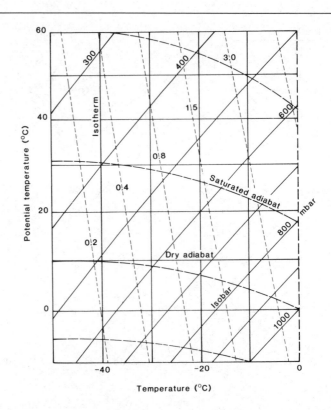

is as sketched in Fig. 5.6. With the temperature axis horizontal and the log θ axis vertical we find the following.

1. By definition, isotherms are vertical straight lines, and isotherms of potential temperature (dry adiabats) are horizontal straight lines.
2. Isobars are slightly curved lines nearly bisecting the right angle between the axes, and are spaced logarithmically, so that a line perpendicular to them is nearly linear in height (because height and log pressure are nearly linear — section 4.4). Highest pressures are to the bottom right.
3. Isopleths of saturation specific humidity are nearly straight lines making fairly small angles with the isotherms.
4. The spacing between dry adiabats decreases slowly as θ increases.
5. Saturated adiabats (section 5.10) make angles of about 45° to the dry adiabats in the high pressures and temperatures typical of the low troposphere, but become more and more nearly asymptotic to them as upper tropospheric conditions are approached.

The diagram is usually displayed as in Fig. 5.5, which is Fig. 5.6 rotated clockwise by about 45°. The isobars now lie nearly horizontally across the diagram, and the undrawn vertical is regarded as a height axis, which can be labelled in actual heights for any particular sounding after calculation of thicknesses. Isotherms now run diagonally upward to the right, with highest temperatures to the bottom right, and dry adiabats run diagonally upward to the left, with highest values to the top right. Saturated adiabats rise nearly vertically in the low troposphere, and turn anticlockwise to join the dry adiabats in the high troposphere.

To illustrate the usefulness of a tephigram, consider the middle latitude sounding plotted in Fig. 5.5. The dry adiabatic layer in the low troposphere shows clearly as a straight line parallel to the dry adiabats. Above this there is a substantial region in which the profile crosses the dry adiabats toward higher potential temperatures, sometimes running nearly parallel to saturated adiabats, and sometimes running across these toward even higher potential temperatures. At the tropopause the profile turns sharply toward the top right, portraying the nearly isothermal profile so typical of the low stratosphere. The sharp graphical distinction between these different segments of the sounding is a very useful feature of the tephigram, since each corresponds to a quite different regime of meteorological behaviour, as will be made clear in subsequent sections. The dashed line on Fig. 5.5 portrays the profile of dew point which is used in conjunction with the temperature profile to describe the humidity structure of the middle latitude sounding, as outlined in section 6.5.

Other diagrams used to display the thermodynamic state of the upper air are the *skew T log p* and *Stuve* diagrams [38]. The former looks very like a tephigram, except that the isobars are straight and the dry adiabats are slightly curved. The Stuve diagram has a long history in meteorology, but is not so convenient in use and does not satisfy the energy: area equivalence.

5.9 The dry adiabatic lapse rate

So far we have considered the dry adiabatic variation of temperature with pressure (eqns (5.12)–(5.14)). In many circumstances, however, pressure variation arises almost entirely because of height variation, so that we can make use of the nearly perfectly hydrostatic relation between pressure and height (section 4.4), to find the equivalent relation between temperature and height for the dry adiabatic process.

By combining eqns (4.6) and (5.12), we find that when an air parcel of density ρ moves dry adiabatically and vertically through ambient air which is in hydrostatic equilibrium and has local air density ρ', small changes in parcel temperature dT and height dz are related by

$$\frac{dT}{dz} = -\frac{g}{C_p}\frac{\rho'}{\rho} \tag{5.16}$$

Note that as usual we assume negligible pressure difference between the air parcel and its surroundings. The minus sign corresponds to the fact that temperature falls (lapses) as height increases.

The appearance of both parcel and ambient air densities in eqn (5.16) threatens to complicate matters. However, the equality of parcel and ambient pressures, together with the behaviour of air as an ideal gas, means that the density ratio ρ'/ρ is equal to the inverse ratio of the absolute air temperature T/T'. In all but the most vigorous types of convection, buoyant air parcels are observed to be only very slightly warmer than their surroundings — about 1 K being a typical upper limit. And even in the most vigorous types of convection (those powerful cumulonimbus associated with severe weather such as torrential precipitation, gale force winds and intense electrical activity), temperature excesses in the rising air seldom exceed 5 K. It follows that the density ratio in eqn (5.16) is almost always very close to unity, the deviation usually being less than 0.5%. To the extent that the ratio is unity, eqn (5.16) simplifies to

$$\frac{\mathrm{d}T}{\mathrm{d}z} = -\frac{g}{C_p} \qquad (5.17)$$

The term on the right-hand side of the equation is effectively constant throughout the lower atmosphere since g variations are $\sim 0.5\%$, and C_p is essentially uniform thanks to very efficient mixing. When values for g and C_p are inserted, the magnitude of the vertical temperature gradient is almost exactly $0.0098 \, °C \, m^{-1}$, or $9.8 \, °C \, km^{-1}$. This is known as the *dry adiabatic lapse rate*, and is usually denoted by the symbol Γ_d. Equation (5.17) now becomes simply

$$\frac{\mathrm{d}T}{\mathrm{d}z} = -\Gamma_d \qquad (5.18)$$

According to eqn (5.18), if an air parcel were to rise dry adiabatically through the full depth of the troposphere, it would cool by about 100° in the shallow troposphere (about 10 km deep) of high latitudes, and by about 150° in the deep tropical troposphere. Such very large temperature lapses are never observed because saturation always intervenes, introducing significant warming which we must now consider.

5.10 The saturated adiabatic reference process

Air in most parts of the low troposphere is so moist that it becomes saturated after rising quite modest distances — often less than 1 km above the local surface (Fig. 1.4). If the air continues rising, the processes of cloud formation and continuing development outlined in the next chapter hold the air very close to saturation and evolve large quantities of latent heat as the excess vapour is condensed. In such conditions the most useful reference process is the saturated adiabatic one, in which we assume that the chilling moist air between the cloud droplets is held precisely at saturation by continual condensation of excess vapour, and that the heat evolved is confined to the air parcel and warms it. The saturated adiabatic process is reversible if we assume that, in descending air, evaporation from available condensed water exactly maintains saturation as the air is compressed and warmed. The condensate is assumed to be water rather than ice, even at temperatures far below the freezing point, and this is reasonably realistic at temperatures down to about $-25 \, °C$ because of the widespread observed tendency for cloud droplets to remain supercooled down to such temperatures (section 6.11). At lower temperatures a more realistic reference process would allow for the liberation of latent heat from the freezing of cloud droplets, but the rather variable efficiency of observed freezing would make the added complication a doubtful asset.

Though the ascending branch of the saturated adiabatic process is more obviously relevant to actual atmospheric behaviour than the descending one, something close to saturated adiabatic descent occurs in cumulonimbus downdraughts, where air often descends in the presence of evaporating precipitation. Somewhat surprisingly, the descending branch also relates to the behaviour of wet-bulb thermometers at fixed altitudes, as outlined in the discussion of Normand's theorem later in this section.

The evolution of latent heat in rising saturated air and its absorption in sinking saturated air reduces the temperature lapse rate below the dry adiabatic value, since in each case the heat exchange partly offsets the temperature change expected according to the dry adiabatic process. To investigate the saturated adiabatic process quantitatively we must include all three of the terms in eqn (5.2).

Suppose that a saturated parcel rises a small distance. It cools slightly by adiabatic expansion and would become supersaturated, because of the sharp decrease of saturation vapour density with decreasing temperature (Fig. 6.2), if the excess vapour were not almost immediately condensed in the form of cloud. Since the excess does condense, and there is no other vapour source or sink, the vapour content of the air between the growing droplets must fall to balance the rise of droplet mass, and this fall must be just what is needed to maintain saturation at the slightly reduced temperature. If the small change in saturation specific humidity is dq_s, then the amount of latent heat evolved in unit mass of moist air is $-L\,dq_s$, where L is the specific latent heat of vaporization. The minus sign ensures that a decrease in vapour content (negative dq_s) corresponds to a liberation of latent heat. The first law of thermodynamics (eqn (5.2)) can now be used to relate the liberation of latent heat to the changes of temperature and pressure:

$$C_p\,dT = -L\,dq_s + \frac{dp}{\rho} \tag{5.19}$$

The opposing signs on the right-hand side confirm the expectation that condensation (or evaporation) offsets the temperature changes induced by adiabatic pressure change.

An important property of the saturated adiabatic process follows from the slope of the graph in Fig. 6.2 relating saturation vapour density (and therefore saturation specific humidity) with temperature. Since the slope increases with increasing temperature, it follows that emissions or absorptions of latent heat are much larger at high temperatures than they are at low, where there is very little vapour left to condense, even at saturation. The contrast between the saturated and dry adiabatic processes is therefore much more marked at high temperatures than it is at low. In fact when the associated saturated adiabatic lapse rates (i.e. values of $-dT/dz$) are calculated, values in the warm, low troposphere may be as low as 5 °C km^{-1}, which is only about half of the dry adiabatic value, whereas values in the cold, high troposphere are very close to the dry adiabatic value.

The saturated adiabatic process is most directly compared with the dry adiabatic one when it is expressed in terms of changes in potential temperature θ, which by definition are zero in a dry adiabatic process. By rearranging the combination of eqns (5.2) and (5.15) as shown in appendix 5.4, we find

$$d\theta = -\frac{L\theta}{C_p T}\,dq_s \tag{5.20}$$

This shows that the effect of condensation (negative dq_s) is to increase potential temperature, whereas evaporation reduces it.

Beginning with any particular state of saturated air (i.e. any particular point on Fig. 5.6 or 5.7) eqn (5.20) could be used step by infinitesimal step to trace out a saturated adiabat. At high temperatures, for the reason mentioned, the saturated adiabats cross the dry adiabats at relatively large angles, whereas at lower and lower temperatures the angles reduce more and more slowly toward zero. In fact saturated adiabats are characteristically curved in conditions corresponding to the low troposphere, but are nearly straight and parallel to dry adiabats in the high troposphere (Fig. 5.7). An associated effect is that adjacent saturated adiabats diverge as they pass from the low to the high troposphere (say 800–300 mbar), with those starting from warmer conditions in the low troposphere cooling less than

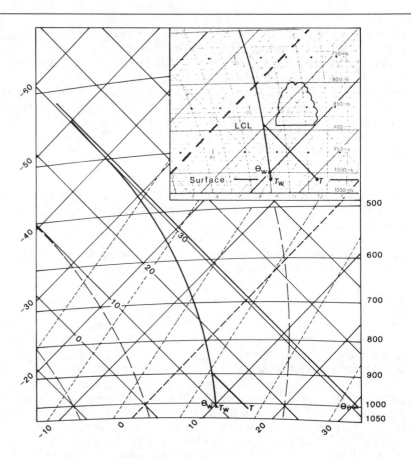

Fig. 5.7 Tephigram depicting
wet-bulb potential
temperature and equivalent
potential temperature. *Inset*:
Enlargement of the base of
the main diagram to depict
lifting condensation level and
Normand's theorem, using
part of Metform 2810. (Crown
copyright)

those starting in cooler conditions, because of the greater vapour content to be condensed. The saturated adiabatic process is therefore much less simple than the dry
adiabatic, in the sense that there is no single saturated adiabatic lapse rate, even at a
particular altitude (i.e. pressure).

Actual soundings through the troposphere often reveal profiles of temperature
(or potential temperature) which are similar in outline to saturated adiabats.
Above the ubiquitous dry adiabatic layer (ignoring nocturnal or seasonal stable
layers near the surface) the potential temperature rises with altitude, usually quite
sharply at first, but then more and more slowly as the cold upper troposphere is
reached (Fig. 5.7a and b), in qualitative agreement with the shapes of saturated
adiabats. However, on close inspection it is usually apparent that the agreement is
imperfect: actual profiles may deviate considerably from saturated adiabats either
over a substantial depth (Fig. 12.3) or over a limited depth in an angular profile
(Fig. 4.1). In many cases such deviations can be associated with details of the
structure or behaviour of weather systems, but this requires specialized discussion
which is best deferred until later (Chapters 11 and 12).

Both the dry and the saturated adiabatic reference processes become completely
irrelevant above the tropopause, where a nearly isothermal layer is often observed
extending many kilometres into the low stratosphere, and the corresponding
potential temperature profile shows a sharp and continuing rise with increasing
height (Fig. 5.5). These profiles point to the virtual absence of vertical or nearly
vertical motion in the low stratosphere, which of course is consistent with the
absence of cloud there.

Finally, consider a point of principle which greatly simplifies the saturated adiabatic model without significantly reducing its utility. In the strictly adiabatic model we must assume that all liquid water is retained in the air parcel after condensation. The cloud burden in any strictly adiabatically moving parcel therefore varies depending on the parcel's previous history, and the resulting slight variations in the total heat capacity of the parcel and contents ensures that there is an infinite number of slightly different saturated adiabats passing through any point on Fig. 5.6. To remove this serious complication it is assumed that cloud water is eliminated from rising parcels as soon as it appears, and that water is made available in descending parcels just in time to maintain saturation by immediate evaporation. This simplified process is called the *saturated pseudo-adiabatic process*, but the differences between it and its strictly adiabatic counterpart are usually unobservably small for realistic concentrations of cloud water and water vapour. Because of its simplicity, the pseudo-adiabatic process is used in the preparation of thermodynamic tables and diagrams. We will use the pseudo-adiabatic process henceforth without further reference to its 'pseudo' nature. To the same degree of approximation we will continue using the specific humidity q to specify vapour content, rather than the humidity-mixing ratio (see section 5.4), even though the latter is apparently more directly relevant to the behaviour of an air parcel having a fixed mass of dry air and a variable mass of vapour.

5.11 Equivalent and wet-bulb potential temperatures

It is useful to have a single temperature with which to label each saturated adiabat, just as each dry adiabat can be labelled by its potential temperature. Figure 5.7 shows two ways in which this can be done.

At high altitudes (i.e. low pressures), temperatures and vapour contents are so low, even at saturation, that any particular saturated adiabat is asymptotic to a particular dry adiabat, whose potential temperature is known as the *equivalent potential temperature* θ_e of the saturated adiabat. In physical terms θ_e can be considered to be the temperature which would be reached by a parcel after saturated adiabatic decompression to nearly zero pressure and dry adiabatic recompression to 1000 mbar.

An alternative procedure is to label each saturated adiabat by the actual temperature at which it intersects the 1000 mbar isobar. Let us call this temperature θ_w and examine the relationship between the two labels for saturated adiabats, θ_e and θ_w.

The equivalent potential temperature θ_e is larger than θ_w (Fig. 5.7) because the condensation of vapour releases latent heat within the air parcel as it rises in a saturated adiabatic process from the 1000 mbar level to the great heights at which the vapour is virtually all condensed (the 200 mbar level, which is about 12 km above sea level, serves in most cases). The size of this temperature excess can be found very accurately from thermodynamic tables or diagrams, but it can also be estimated quite usefully from an integrated and simplified form of eqn (5.20) (appendix 5.4):

$$\theta_e - \theta_w = 2.5q_s \qquad (5.21)$$

Here q_s is the specific humidity of the saturated air at the 1000 mbar level expressed in grams of water vapour per kilogram of moist air, as it always is in practice. A typical value for q_s in temperate latitudes is 10 g kg^{-1}, and the corresponding value for $\theta_e - \theta_w$ is 25 °C according to eqn (5.21). This agrees to within a fraction of a degree with the value found more rigorously from thermodynamic tables and diagrams, and agrees with the value given by Fig. 5.7 to well within the accuracy of reading of even the large-format original used for plotting soundings. Note that the combination of the variation of saturation vapour content with temperature (Fig. 6.2) and eqn (5.21) is consistent with the curvature and divergence of saturated adiabats discussed in section 5.10, both of these being a maximum in the warm, low troposphere of low latitudes.

The physical meaning of θ_w is obvious in the case of saturated air: it is the temperature reached by saturated adiabatic descent to 1000 mbar. However, there is a surprisingly direct relevance to unsaturated conditions which is best appreciated by referring to the inset part of Fig. 5.7. Suppose that an unsaturated air parcel with temperature T has wet-bulb temperature T_w and pressure p. The wet-bulb temperature T_w is less than T by an amount which is a measure of the sub-saturation of the air, and is plotted on the same isobar as T in Fig. 5.7 because it is reached by isobaric cooling by evaporation (section 5.5). According to a thermodynamic theorem named after the meteorologist *Normand*, the dry adiabat through T and the saturated adiabat through T_w intersect at the *lifting condensation level* of the initially unsaturated parcel (LCL on Fig. 5.7), which is the level at which the air would become saturated if lifted dry adiabatically from its original state. Since the theorem applies equally well at all levels (i.e. pressures) below LCL it follows that the saturated adiabat traces out the actual wet-bulb temperature of the adiabatically rising or falling parcel of unsaturated air. If it is traced down to the 1000 mbar level, the temperature reached is known as the *wet-bulb potential temperature* of the air parcel, which is the wet-bulb temperature which the air would have if it were taken adiabatically to the 1000 mbar level. (Normand's theorem is completed in section 6.5.)

Notice that θ_w is conserved in any adiabatic process, whether dry or saturated, and whether isobaric or not. If the process is dry adiabatic, then Normand's theorem applies as already presented. If the process is saturated adiabatic, then θ_w is conserved by definition. If the process involves adiabatic evaporation of water at constant pressure, then it is all or part of the wet-bulb thermometer process outlined in section 5.5, which by definition does not alter the wet-bulb temperature of the air, and therefore does not alter its wet-bulb potential temperature. Since any actual adiabatic process can be constructed out of a sequence of tiny adiabatic changes — dry, saturated, isobaric or not — the conservation of θ_w in a parcel undergoing any adiabatic process has been established. All this makes the wet-bulb potential temperature the most widely conservative thermodynamic marker used in meteorology. To illustrate this, consider two adiabatic processes as they might realistically operate in the vicinity of a cumulonimbus.

Suppose that the cloud whose base is at the lifting condensation level LCL in Fig. 5.7 begins to produce thick, fine rain, consisting of a dense population of small droplets. These have a large total droplet surface area from which there is copious evaporation into the slightly unsaturated air below cloud base. Each droplet chills itself and its surroundings like a wet bulb, so that the net effect is to cool the whole rain-filled parcel to the wet-bulb temperature of the originally unsaturated air at the level in question. The air may then tend to sink with the falling rain, desaturating by descent and resaturating by evaporation so that the air parcel traces out the saturated adiabat extending downward from cloud base. If this continues to the 1000 mbar level, the saturated rain-filled air reaches there with a temperature T equal to θ_w. This sort of process accounts for the downdraughts of cool, moist air observed near the surface just before the arrival of a shower. Evaporation may not keep pace with desaturation by downdraught compression, particularly if the rain drops are large, therefore presenting a much smaller total area of evaporating surface, and in that case the downdraught arrives with the same wet-bulb temperature as before but a higher dry-bulb temperature T.

If, instead, air were to sink completely dry adiabatically below cloud base LCL, as happens in the unsaturated subsidence between adjacent cumulus or cumulonimbus, then it would warm so that its temperature T traced out the dry adiabat through LCL. By the corollary of Normand's theorem, the wet-bulb temperature of the sinking, desaturating air traces out the saturated adiabat through LCL. If such descent should continue to the 1000 mbar surface, the air arrives there with temperature equal to the potential temperature of the subsiding parcel, and the much lower wet-bulb temperature equal to θ_w. The wet-bulb potential temperature is again conserved. Note that the value of θ_w in the air entering the undraught is generally a little higher than the value in the downdraught, the difference indicating the degree of instability which is driving the cumulonimbus convection.

The equivalent potential temperature θ_e is obviously conserved in any process conserving the wet-bulb potential temperature, since it is simply an alternative labelling system for all saturated adiabatic processes. Though clearly relevant to the life cycle of air ascending in very deep convection, with or without subsequent dry adiabatic descent, θ_e lacks the direct identification with conditions near the surface (at the 1000 mbar level, strictly) which is the special virtue of θ_w.

5.12 Convective stability

In previous sections we have considered the sequences of states experienced by individual air parcels undergoing dry or saturated vertical motion. Before beginning to assess the realism of this behaviour, it is important to realize that radiosondes, or the satellite-borne radiometers which are becoming viable alternatives, measure something quite different. They provide snapshots of the vertical structure of the atmosphere — profiles of temperature, humidity, pressure and wind observed at instants of time. It is true that the time taken by a radiosonde to rise to bursting level is about one hour, and that the atmosphere can change somewhat in that time interval, but a great deal of observation shows that such changes are rarely significant, at least in respect of synoptic scale structure. On such scales the radiosonde snapshots are therefore only slightly blurred.

The distinction between the *parcel* viewpoint of adiabatic theory and the snapshot or *environmental* viewpoint of observation is very basic, and might be expected to make theory and observation too different for useful comparison. But although the viewpoints are quite different, it should be clear from observations cited in previous sections that there is often close agreement between parcel theory and environmental observation. For example the dry adiabatic temperature lapse rate (eqn (5.17)) is derived from parcel theory but is observed from an environmental viewpoint throughout extensive layers of the low troposphere. Assuming that the dry adiabatic parcel theory is as relevant as has been claimed, there must be additional factors making its predictions immediately observable. To uncover these we begin by looking more closely at the connection between an individual parcel and its immediate environment.

Suppose that an air parcel is embedded in air with a certain environmental vertical temperature gradient $\partial T/\partial z$. The *partial differential* $\partial/\partial z$ represents rate of change with height alone, ignoring possible variations with horizontal position or time; it corresponds therefore to the vertical gradient (of temperature in this case, but the symbolism is quite general) in an instantaneous picture of the environment of the air parcel. Now suppose that this air parcel moves up or down through its environment, undergoing a parcel process which is dry adiabatic. The parcel's temperature sequence traces out a portion of a temperature profile with vertical gradient given by eqn (5.18):

$$\frac{dT}{dz} = -\Gamma_d$$

where Γ_d is the dry adiabatic lapse rate, value 9.8 °C km^{-1}. The difference in symbolism between $\partial T/\partial z$ and dT/dz corresponds to the difference between the parcel and environmental viewpoints.

The buoyant behaviour of the parcel depends critically on the value of $\partial T/\partial z$. If the environmental lapse rate $-\partial T/\partial z$ is *superadiabatic* (meaning greater than the dry adiabatic value Γ_d), i.e.

$$\frac{\partial T}{\partial z} < -\Gamma_d \qquad \text{or equivalently} \qquad \frac{\partial \theta}{\partial z} < 0 \tag{5.22}$$

we have the state of affairs depicted in Fig. 5.8(a). Upward movement of the air parcel through its environment in the course of convection or turbulence then leads to the parcel becoming warmer and therefore less dense than its surroundings. The resulting buoyancy tends to drive the parcel even further from its original position, generating even more buoyancy, and so on. However, if the parcel moves downwards from its original position, it becomes cooler and therefore more dense than its environment, and the resulting negative buoyancy tends to drive it even further down. (Buoyancy is discussed further in section 7.12.)

It seems that the air parcel can continue in its original position only if it is completely undisturbed. It is therefore in an unstable state in just the same way as is

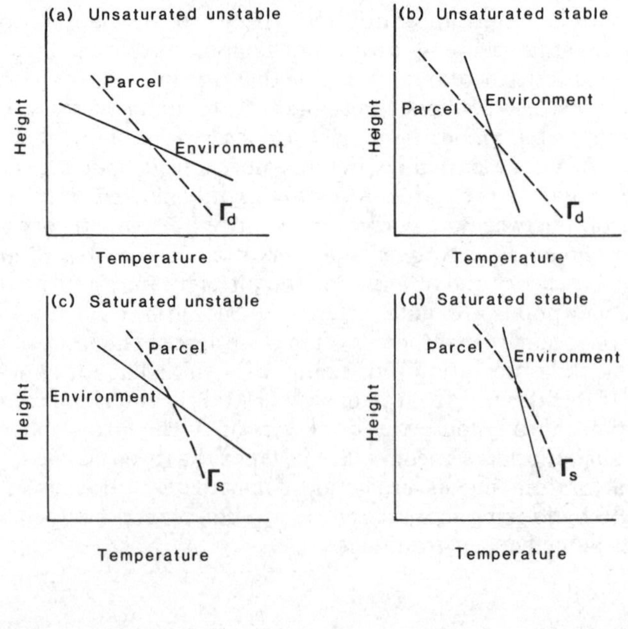

Fig. 5.8 (a, b) Parcel and environmental temperature profiles representing adiabatic convective stability and instability in an unsaturated atmospheric layer; (c, d) the same stability regimes for saturated adiabatic convection in a saturated layer.

a ballbearing perched on the top of a smooth dome, and just as the ballbearing inevitably rolls down off the dome in response to tiny movements in the dome or surrounding air, so the air parcel in Fig. 5.8(a) rises or falls in response to the inevitable stirrings of the surrounding air. In fact every other parcel making up this superadiabatic temperature profile is unstable in just the same way, with the result that the whole layer is said to be *convectively unstable*. The term statically unstable is often used instead, and distinguishes this instability from other types arising dynamically, such as the baroclinic instability which underlies many large-scale weather systems, and the shearing instability discussed in section 10.8.

Now consider the same parcel (obeying a dry adiabatic parcel process) embedded in a layer with a subadiabatic lapse rate. We have

$$\frac{\partial T}{\partial z} > -\Gamma_d \quad \text{or equivalently} \quad \frac{\partial \theta}{\partial z} > 0 \tag{5.23}$$

Figure 5.8(b) and the same type of analysis as before show that this layer of air is convectively stable. For example, when the individual air parcels are displaced upwards through their environment they are driven back to their original positions by negative buoyancy, and when they are displaced downwards they are driven back up by positive buoyancy. The situation is now analogous to a ball-bearing in the hollow of an upturned bowl, which will return to its initial position in the lowest point of the hollow after any displacement.

We can show in just the same way that a layer of air with a precisely dry adiabatic environmental lapse rate, i.e.

$$\frac{\partial T}{\partial z} = -\Gamma_d \quad \text{or equivalently} \quad \frac{\partial \theta}{\partial z} = 0 \tag{5.24}$$

is in a state of neutral convective stability, meaning that any constituent parcel tends to remain in its new position after it has been displaced, rather than move further from or back to its original position, like a ballbearing on a flat horizontal plate.

So far we have considered only the dry adiabatic process. Exactly the same kind of analysis can be used to find the criteria for convective stability when the constituent air parcels are following a saturated adiabatic process. Assuming that there is always enough water to keep descending air saturated we can proceed as before. Figure 5.8(c) looks much the same as Fig. 5.8(a) except that the straight dry adiabats are replaced by the slightly curved saturated adiabats, whose lapse rates Γ_s are always less than the dry adiabatic value, and are quite significantly so in the low troposphere.

The criteria for convective instability, neutral stability and stability are assembled in Table 5.1 for both the dry and saturated adiabatic parcel processes. In each case neutral convective stability occurs when the environmental temperature lapse rate exactly equals the lapse rate defined by the appropriate adiabatic parcel process, with instability or stability resulting when the environmental lapse rate is larger or smaller respectively. Because there are two quite different parcel processes possible (the dry and saturated ones), there is a range of environmental lapse rates which represent convective stability or instability depending respectively on whether the constituent air parcels are following dry or saturated adiabatic processes, which means effectively whether the layer is cloud-free or cloud-filled. Such a layer of air is said to be *conditionally unstable*, the condition being whether or not the layer is saturated. Conditional instability can play an

Table 5.1 Criteria for convective stability

Environmental temperature gradient in T	in θ	Adiabatic parcel process unsaturated	saturated
$\dfrac{\partial T}{\partial z} < -\Gamma_d$	$\dfrac{\partial \theta}{\partial z} < 0$	U	U
$\dfrac{\partial T}{\partial z} = -\Gamma_d$	$\dfrac{\partial \theta}{\partial z} = 0$	N	U
$-\Gamma_d < \dfrac{\partial T}{\partial z} < -\Gamma_s$	$0 < \dfrac{\partial \theta}{\partial z} < \left(\dfrac{d\theta}{dz}\right)_s$	S conditional instability	U
$\dfrac{\partial T}{\partial z} = -\Gamma_s$	$\dfrac{\partial \theta}{\partial z} = \left(\dfrac{d\theta}{dz}\right)_s$	S	N
$\dfrac{\partial T}{\partial z} > -\Gamma_s$	$\dfrac{\partial \theta}{\partial z} > \left(\dfrac{d\theta}{dz}\right)_s$	S	S

Key U unstable
   N neutrally stable
   S stable

especially important role in the middle and low troposphere, since the difference between Γ_d and Γ_s may be very considerable there (section 11.6). A typical sequence is that conditional instability is realized and released in convection when a layer with a supersaturated-adiabatic lapse rate is raised to saturation by slow ascent in a large-scale weather system.

Equivalent diagrams, very like Fig. 5.8, can be drawn on tephigrams, etc., and analysed similarly. For example, the shallow segment at the botton of the Long Kesh sounding in Fig. 5.5 is strongly superadiabatic and therefore convectively unstable (such instability being typical of shallow layers immediately overlying warmed land surfaces — sections 5.13 and 9.9). The next layer, extending nearly to the 800 mbar level, is slightly but definitely subadiabatic and therefore stable if unsaturated (as suggested by the dew point depressions — but see below). The rest of the troposphere has a temperature profile close to the saturated adiabatic, which suggests neutral stability for saturated ascent and descent if the air is saturated. The dew-point profile shows considerable subsaturation, but this may be spurious, since radiosonde humidity sensors are potentially quite inaccurate and especially prone to under-reading [39]. The low stratosphere, above the tropopause at 320 mbar in Fig. 5.5, typically contains a deep inversion which is extremely stable for both saturated and unsaturated convection.

5.13 The maintenance of near-neutral stability

The concept of convective stability associates environmental temperature profiles with thermodynamic processes believed to be at work in the populations of constituent air parcels. But we have yet to trace in practical detail the way in which any particular temperature profile is brought about or maintained. Done thoroughly, this is an elaborate task involving many aspects of atmospheric

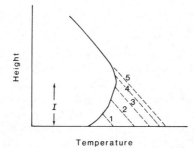

behaviour, some of which are still quite poorly understood, but it is worth trying in outline for the dry adiabatic case, which is important in itself and suggests the kind of mechanisms at work in less simple cases.

A nearly dry adiabatic lapse rate is often observed extending through most of the sub-cloud layer in the low troposphere. However, it is not observed very near a land surface when nocturnal cooling is in progress or has only just ended. The situation then is typically as depicted in Fig. 5.9, which represents conditions close to a land surface toward the end of a cloudless night. The ground and the air close to it have cooled during the night by net emission of infra-red (terrestrial) radiation, producing thereby a convectively stable layer which may extend a hundred or more metres above the surface. The cooling may often be strong enough to produce a temperature inversion in some part of the stable layer, as shown in Fig. 5.9. Now let us consider how such a strongly stable layer is converted in a few hours into a nearly neutrally stable layer which is then maintained throughout the day until the next night's cooling begins to reintroduce stability near the surface.

When the sun comes up, the surface and immediately overlying air are warmed by absorption of solar radiation, the warming being quite rapid initially because of the small heat capacity of the initially very shallow layers of ground and air affected (sections 5.3 and 9.4). In the air close to the warming surface, the environmental lapse rate becomes positive (assuming it was negative initially, i.e. that there was a temperature inversion) and approaches the dry adiabatic value. As soon as this is exceeded the shallow layer begins to convect, since it is slightly super-adiabatic and therefore convectively unstable. The convection has two effects which are worth considering separately in some detail. First, it maintains the lapse rate in the convecting layer at the dry adiabatic value, or more strictly very slightly in excess of it. Second, it very efficiently pumps sensible heat (and water vapour provided the surface is moist) into the convecting layer, thereby warming it (Figs 5.2 and 5.9).

The dry adiabatic lapse rate is maintained because the incessant mixing of the air parcels as they rise and fall in convection holds the vertical temperature profile of the convecting layer very close to the dry adiabatic profile traced out by each moving parcel before it is mixed out of existence. It is true that rising parcels are usually warmer than falling ones (that is why they rise) but the small size and brief lifetimes of the parcels ensure that such temperature differences are small. In essence the environmental lapse rate stays very close to the individual parcel lapse rate because the environment itself consists of an incessant flux of such parcels, each of which moves nearly dry adiabatically in its brief life.

The second major effect of convection, returning to the case of the morning removal of the nocturnal stable layer, is the gradual warming of the convecting layer (Figs 5.2 and 5.9) which comes about because on average the air parcels rising into the layer are warmer than the average temperature of the layer. As these warmer parcels mix out of recognizable existence, their slight temperature excess is given to their surroundings, which therefore gradually warm. The result is that although the instantaneous environmental lapse rate remains close to the dry adiabatic value, or some consistently superadiabatic value if the sun is strong, the whole layer warms gradually but persistently. In Fig. 5.9 the temperature profile moves bodily to the right as time passes, without change of shape or slope in the convecting zone. However, the presence aloft of some remnant of the convectively stable temperature profile developed overnight ensures that the depth of the convecting layer increases as it warms. Since this means that a progressively larger mass of air is being warmed as time goes by, it follows that the rate of warming should tend to decrease. This decrease may be partly offset by an increasing rate of solar input as the sun rises toward the noon maximum, but the net result is that the observed maximum rate of warming occurs earlier rather than later in the morning. In fact the warming tends to become very small indeed if the deepening convecting layer connects up with the upper part of the nearly dry adiabatic layer left over from the previous day's convection, as often happens in the late morning just before cumulus appear.

During the afternoon the solar input declines but usually remains strong enough to maintain convection in the sub-cloud layer, and with it the characteristically dry adiabatic environmental lapse rate. However some little time before sunset the net loss of heat by infra-red radiation from the ground begins to exceed the net input from the sun, and the lapse rate close to the surface becomes subadiabatic. Convection there dies almost at once, and with it all the thermally driven convection through the layers above. Air sufficiently far above the surface may still be stirred by mechanical convection, and even if it is not, may retain a nearly dry adiabatic lapse rate because it is largely disconnected from the cooling surface. However, once formed the ground-based stable layer tends to deepen throughout the night, reproducing by dawn next day another version of the temperature profile with which we began this section.

A complication can arise in the very lowest layers when the sun is fairly strong. It seems that the inevitably very small-scale convection which dominates at such low levels does not transport sensible heat as efficiently as does the larger-scale convection aloft, with the result that it may be unable to cope with the strong solar input. When this happens the surface and overlying air become warmer, and the resulting distinctly superadiabatic lapse rate restores thermal balance by driving the convection more vigorously. The superadiabatic effect may be quite marked in the lowest few metres of air, but diminishes so sharply with increasing height that it is virtually unobservable more than a few tens of metres above all but the most intensely heated surfaces (such as subtropical deserts in summer). This behaviour is examined further in section 9.9. At greater heights, that is throughout most of the convecting sub-cloud layer, temperature lapse rates are observed to lie within a few percent of the dry adiabatic value.

Note that quite vigorous stirring is often observed in the absence of significant surface warming by the sun — for example at night and over substantial depths of water. Provided that a surface is not being strongly cooled by terrestrial radiation, a turbulent tumbling motion is often maintained by moderate or stronger winds near the surface, as shown in Fig. 5.10. This is termed *mechanical convection* to distinguish it from the more obvious thermally driven convection already mentioned. Over many ocean surfaces the sub-cloud layer is more or less continually stirred by a combination of mechanical and thermal convection. In summary it is the vertical movements of stirred air, whether the stirring is driven by wind or surface heating, which together with the dry adiabatic parcel process hold the environmental lapse rates in the sub-cloud layer so close to the dry adiabatic value for so much of the time.

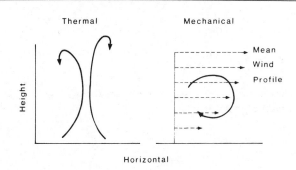

Fig. 5.10 Thermal and mechanical origins of turbulence.

Now that we have examined the particular case of the sub-cloud layer, we can consider the maintenance of neutral convective stability more generally. In a cloudy layer the individual parcels tend to move according to the saturated adiabatic process, and convection therefore tends to maintain a similar temperature profile. The convection may originate in a sub-cloud layer, or in the cloudy layer itself by release of latent heat from some triggering uplift, for example, or it may be mechanical in origin. Though the principle established in the cloud-free case clearly still applies, the behaviour in cloudy air may be considerably complicated in practical detail, especially if the layer is only partly full of cloud, as is often the case.

The maintenance of neutral convective stability may be seen as an example of the operation of an extremely general law of mixing which can be stated as follows: the distribution of each component of a continually stirred mixture tends toward a dynamic equilibrium which is unaffected by further mixing. The mixing of ink and water mentioned in section 3.1 is a simple example; only during the initial stirring is there any visible change in the appearance of the mixture, which quickly settles into a uniform shade apparently unaffected by further stirring. The uniformity of specific masses of gases in the turbosphere (section 3.1) exemplifies the law again, and shows that the result need not be as obvious as in the case of ink and water. And in the present case the result is even less obvious. However, the law is still clearly applicable: if the parcel process is dry adiabatic and the stirring has a component of vertical exchange, the only vertical temperature profile which is unaffected by further stirring is one with a dry adiabatic lapse rate, and this is the one so widely observed.

Appendix 5.1 The first law of thermodynamics (meteorological form)

Since the volume of unit mass of material is inversely proportional to its density, the first law of thermodynamics (eqn (5.1)) can be written

$$dQ = C_v \, dT + p \, d\left(\frac{1}{\rho}\right)$$

$$= C_v \, dT + d\left(\frac{p}{\rho}\right) - \frac{1}{\rho} \, dp$$

$$= C_v \, dT + R \, dT - \frac{1}{\rho} \, dp$$

133

since $\dfrac{p}{\rho} = R\,T$ (equation of state — eqn (4.2))

$$dQ = (C_v + R)\,dT - \frac{1}{\rho}\,dp$$

By considering the case where pressure is kept constant ($dp = 0$) we see that $C_v + R$ is the specific heat capacity at constant pressure C_p.

Hence $dQ = C_p\,dT - \dfrac{1}{\rho}\,dp$

Appendix 5.2 Precipitable water content

Consider a vertical column of moist air, with air density ρ and specific humidity q. The mass m of water vapour in a slice of depth $(z_2 - z_1)$ and unit horizontal area is therefore given by

$$q\,\rho\,(z_2 - z_1)$$

which, according to eqn (4.16) (a form of the hydrostatic equation), can be rewritten as

$$\frac{q}{g}(p_1 - p_2)$$

where p_1 and p_2 are the air pressures at the base and top of the slice respectively.

If this vapour were to be precipitated into a rain-gauge at the foot of the column, it would yield a depth

$$h = m/\rho_w$$

where ρ_w is the density of the precipitated water. It follows that

$$h = [q(p_1 - p_2)]/g\,\rho_w$$

The term in square brackets can be readily calculated for any slice of an upper air sounding in which humidity is reported, and the total *precipitable water* content Σh found by adding the contributions from all moist slices. Note that Σh is given in millimetres of equivalent rainfall by

$$10^{-2} \Sigma\, q_i\,\Delta p_i$$

where q_i is in g kg^{-1} and Δp_i is in mbar. Values can be read directly off a plotted tephigram or its equivalent.

Appendix 5.3 A skeleton tephigram

Let us outline a tephigram in the temperature (T) range $-60\,^{\circ}$C to $+60\,^{\circ}$C and the potential temperature (θ) range $-60\,^{\circ}$C to $+100\,^{\circ}$C. Convert θ values to degrees kelvin and find their logarithms to the base ten. Draw the $\log \theta$ axis in the vertical and the T axis in the horizontal, and draw and label the isotherms and potential isotherms (dry adiabats) at 20 °C intervals including 0 °C.

1. Draw and label the 1000 mbar isobar as the locus of points where $\theta = T$. Prepare to draw and label the 500 and 250 mbar isobars by taking the logarithm of eqn (5.15).

$$\log \theta = \log T + \frac{R}{C_p} \log \left(\frac{1000}{p} \right)$$

Since the first two terms represent the 1000 mbar isobar, any other isobar (p) can be found by sliding the 1000 mbar isobar up the $\log \theta$ axis by a distance given by the numerical value of

$$\frac{R}{C_p} \log \left(\frac{1000}{p} \right)$$

We can therefore find the 500, 250 and 125 mbar isobars by sliding up multiples of the distance 0.0861 (i.e. $(R/C_p) \log 2$).

2. Prepare to draw and label the 10, 5 and 2.5 g kg^{-1} isopleths of saturation specific humidity q_s as follows. Convert the q_s values into straight fractions (the largest being 10^{-2}), and use these in eqn (5.8) to find the corresponding saturation vapour pressures on the 1000 mbar isobar. The values will agree very closely with some of the following table, and the corresponding temperatures (i.e. the temperatures at which these are the saturation vapour pressures) can be used to locate the intersections of the 1000 mbar isobar with the three q isopleths. Repeat for the other three isobars, and connect the intersections to form the q isopleths.

e_s/mbar	16	8	4	2	1
T/°C	14	3.7	-5.6	-14.7	-24

3. After reading section 5.10, sketch and label the saturated adiabat which passes through the intersection of the 10 g kg^{-1} isopleth of q_s with the 1000 mbar isobar, using the following approximate method, which simulates the continuous release of latent heat in a rising air parcel by a three-step process.

(a) Warm air by isobaric release of latent heat by condensing 5 g kg^{-1} of vapour (note that T increases by almost exactly 2.5 °C when q_s falls by 1 g kg^{-1} — eqn (5.19) with dp zero applies). Cool this air dry adiabatically until it saturates again by moving it along a dry adiabat until you reach the 5 g kg^{-1} isopleth for q_s. (The vapour content of the air has now of course fallen to 5 g kg^{-1}.) This is the intersection of the chosen saturated adiabat with the 5 g kg^{-1} q_s isopleth.

(b) Repeat process (a) starting at its end point and condensing half of the remaining vapour to reach the intersection of the saturated adiabat with q_s 2.5 g kg^{-1}.

(c) Find the asymptotic dry adiabat at the cold end of the saturated adiabat (the equivalent potential temperature θ_e) by using eqn (5.21) and noting that θ_w is simply the θ value at the 1000 mbar isobar (the beginning of the saturated adiabat).

(d) Connect up the intersections and the asymptote by a smooth curve which is your representation of this particular saturated adiabat (θ_w 13.7 °C). The simulation could be improved and made much more realistic in execution by using many small alternately isobaric and dry adiabatic steps instead of three very gross ones, but there would be no change in principle or in the essential shape of the resulting saturated adiabat.

Some parts of your completed diagram represent conditions not observed in the atmosphere. Pressures very seldom exceed 1050 mbar; dry adiabats extending up from hot dry surfaces seldom exceed a potential temperature of 45 °C; deep saturated convection (the agent of tropospheric warming in depth) seldom rises from surface layers with specific humidity more than 25 g kg^{-1}. Since this saturates air at a temperature of just over 28 °C at 1000 mbar, it follows that the maximum equivalent potential temperature is just over 90 °C, and this is realized at about the 125 mbar level. A rough saturated adiabat can be sketched to indicate the limiting conditions. Hatch the areas excluded by all of these limits, and compare with Fig. 5.6.

Appendix 5.4 Equivalent potential temperature

Taking the logarithm of the equation defining potential temperature (eqn (5.15) and appendix 5.3) and then differentiating to find the relationship between very small changes in potential temperature (dθ), temperature (dT) and pressure (dp), we finally obtain

$$\frac{d\theta}{\theta} = \frac{dT}{T} - \frac{R\,dp}{C_p\,p}$$

which shows that θ increases with isobaric warming (positive dT and zero dp) and decreases with isothermal pressurization (positive dp and zero dT).

If we now rearrange the meteorological form of the first law of thermodynamics (eqn (5.2)) by replacing the air density using the equation of state (eqn (4.2)), and dividing across by $C_p T$, we find

$$\frac{dQ}{C_p T} = \frac{dT}{T} - \frac{R\,dp}{C_p\,p}$$

Since the two equations now derived have the same right-hand side, we have

$$\frac{d\theta}{\theta} = \frac{dQ}{C_p T} \tag{5.25}$$

which shows that the potential temperature increase dθ is proportional to the heat input dQ divided by the absolute temperature T at which this occurs. This quotient is in fact the increase in *entropy* of the system if the heat input occurs reversibly (i.e. in a manner so close to thermal equilibrium that an infinitesimally small change in conditions could reverse its direction). Entropy occupies a fundamental place in the more advanced analysis of systems close to thermodynamic equilibrium [35], and though its relevance to natural and man-made systems (all of which are quite significantly irreversible) is complex and still unclear, it has a general qualitative significance as a measure of the degree of disorder of complex systems. The dry adiabatic reference process is often termed *isentropic* since by definition it is reversible and assumes zero dQ.

In the saturated adiabatic reference process, the heat input during a small adiabatic ascent is given by $-L\,dq_s$ (eqn (5.19) and preceding discussion), so that eqn (5.25) becomes

$$\frac{d\theta}{\theta} = -\frac{L\,dq_s}{C_p T} \tag{5.26}$$

which becomes eqn (5.20) after trivial rearrangement. In fact $L\,dq_s$ is not a heat input from outside the parcel — it is energy released to one part of the parcel (the air) by a substantial decrease in the molecular disorder in another part (the water substance which has condensed from vapour to liquid). The saturated adiabatic process is therefore also isentropic. The analysis of air motions following dry or saturated adiabatic processes is therefore generally termed isentropic analysis (section 11.4).

To find a relation between the equivalent and wet-bulb potential temperatures (section 5.11) we follow a saturated adiabat from 1000 mbar to a pressure so low that effectively all vapour has been condensed. By definition the initial θ is the wet-bulb potential temperature θ_w and the final θ is the equivalent potential temperature θ_e. We need to integrate eqn (5.20) (eqn (5.26) multiplied by θ) from initial to zero q_s. The simplest approach is to notice that most of the condensation occurs in the warmest parts of the saturated adiabat, where of course θ and T are nearly equal because pressures are close to 1000 mbar. To the extent that they are equal the right-hand side of eqn (5.20) can be integrated immediately to give

$$\theta_e - \theta_w = \frac{Lq_s}{C_p} \tag{5.27}$$

which is eqn (5.21). The ratio L/C_p simplifies to 2.5 when q_s is expressed in grams per kilogram. Equation (5.27) is a special case of a general relationship between rise in potential temperature and condensation of water vapour (fall in q_s). More thorough analysis [8] shows that the degree of approximation is even better than suggested above, as is confirmed by close comparison with thermodynamic tables and diagrams.

Problems

LEVEL 1

1. By applying the first law of thermodynamics qualitatively to a parcel of air first in a rigid container and second in a slack plastic bag, argue why the specific heat at constant pressure (C_p) should be larger than C_v.
2. Examine the following list and decide which processes are likely to be approximately isobaric: daytime warming of the full depth of the atmosphere; air in a shower cloud updraught; air flowing horizontally through the core of a jet stream; air spiralling horizontally into the core of a tornado; the atmospheric boundary layer during nocturnal cooling.
3. The humidity mixing ratio of adjacent air tends to rise when there is evaporation from a land surface. What surface activity tends to make it fall? Can this happen at ocean surfaces? Can it happen at frozen surfaces? Briefly describe what happens in these last two cases.

4. Air in the upper branches of the Hadley circulation may take 10 days to move from the equatorial to the subtropical zones. Consider the applicability of the adiabatic assumption to this process.

5. If meteorology had been pioneered on the Tibetan Plateau (about 4 km above mean sea level), the reference level for potential temperature might well have been chosen to be 600 rather than 1000 mbar. In this case, consider the following pressure levels and decide in each case whether actual temperatures would be greater, equal to or less than Tibetan potential temperatures: 1000, 700, 600, 300, 100 mbar.

6. Which of the following vertical temperature profiles is likely to be spurious? Well clear of the surface: lapse of 1 °C in 200 m, in 80 m; rise of 2° in 300 m. Close to the surface: lapse of 2° in 50 m, rise of 2° in 50 m.

7. On a certain saturated adiabat, place the following in order of ascending numerical value: temperature at 500 mbar, wet-bulb potential temperature, equivalent potential temperature, potential temperature at 500 mbar.

LEVEL 2

8. On a certain cloudless night, an infra-red radiometer showed that the net rate of loss of radiant energy from a ground surface averages 50 W m^{-2} for 8 hours. If the heat lost is drawn from 30 mbar 'depth' of air, find the resulting fall of air temperature, assuming it to be uniformly distributed through the layer.

9. The saturation vapour pressures at 15 and 10 °C are respectively 17.0 and 12.3 mbar. Find the associated saturation specific humidities when the atmospheric pressure is 1000 mbar. Also find the relative humidity of air which has a temperature of 15 °C and a dewpoint of 10 °C.

10. On a certain summer's day in mid-latitudes the air temperature at screen level is observed to be 25 °C. If there is no significantly superadiabatic layer near the surface, what is the minimum height of the 0 °C level above the surface?

11. Given that air at 1000 mbar pressure and 15 °C temperature has a wet-bulb temperature of 10 °C, find its specific humidity and relative humidity, assuming the measured wet-bulb temperature is identical to the thermodynamic wet-bulb temperature.

12. In a mishap which was fortunately not fatal, an airliner depressurized suddenly by losing a cargo door at altitude. If the internal and external pressures were 750 and 400 mbar respectively just before depressurization, and the internal air temperature was 22 °C, estimate the internal temperature just after depressurization, assuming dry adiabatic cooling. No wonder passengers complained of the cold!

13. In middle latitudes, air at the 300 mbar level is often observed to have temperatures as high as −40 °C in the vicinity of jet-stream cores. Find the potential temperature of such air. Use Fig. 5.5 to find its wet-bulb potential temperature and the equivalent potential temperature.

LEVEL 3

14. Recalculate the temperature fall in problem 8 if the heat loss is shared with a 20 cm deep layer of ground surface in addition to the air layer. Assume ground

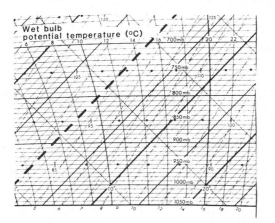

Fig. 5.11 Tephigram section for the low troposphere (problem 5.15). Axes and isopleths are as in Fig. 5.5 (Crown copyright)

density 2000 kg m^{-3} and specific heat capacity 2000 J kg^{-1} K^{-1}. In which direction would the answer be changed if the cooling produced fog?

15. Given that air at 1000 mbar and 20 °C has relative humidity 60%, find values for all possible humidity parameters (including the lifting condensation level), using the low-level section of the tephigram in Fig. 5.11. Estimate the equivalent potential temperature by using Fig. 5.5.

16. In the event described in problem 12 passengers reported a temporarily dense fog in the cabin just after depressurization. Given that the cabin wet-bulb temperature was 17 °C before the event, use Fig. 5.5 to make a more realistic estimate of the cabin temperature just after depressurization. Why was the fog only temporary?

17. Discuss the realism of dry adiabatic ascent from the surface to the jet core in problem 13. If instead there was saturated adiabatic ascent, what global position for the surface source is suggested by the data?

18. Consider a situation in which nocturnal cooling under clear skies establishes a convectively stable layer in the first 300 m of the atmosphere, with a temperature inversion in the lowest 100 m. Sketch an example on the tephigram fragment above, and on an equivalent graph of potential temperature versus height. Now suppose that an approaching depression produces cloudy overcast which stops the cooling and rising winds which encourage mechanical convection through the lowest 200 m. Sketch the altered temperature profile on the same two diagrams.

19. Estimate the precipitable water content of the Long Kesh sounding (Fig. 5.5), assuming the humidity profile to be accurate, and compare it with values from saturated cold and warm atmospheres (0 °C and 20 °C saturated adiabats).

Cloud and precipitation

6.1 Cloud

The Earth's lower atmosphere is typically cloudy. At any instant about half the planet's surface is overlain by cloud varying in thickness from a few tens of metres to the full depth of the troposphere (the site of all but a minute fraction of terrestrial cloud). All this cloud is extremely variable in both time and space, and individual cloud elements are usually quite ephemeral. For example a cumulonimbus may develop so rapidly that within half an hour of its first appearance as a little cloudlet it effectively fills the troposphere in a region ten or more kilometres across; and within a further hour it may vanish again, leaving no trace of its majestic maturity beyond its legacy of precipitation on the underlying surface. Lifetimes of constituent cloud droplets and crystals may be considerably shorter. If you watch the outer parts of a cumulonimbus for a few minutes, you will become aware that everything is in a state of ceaseless commotion, with bumps and filaments emerging from the main cloud mass and evaporating to extinction within a few tens of seconds. Simple calculations based on observed rates of precipitation and estimates of cloud water masses suggest that even in the cloudy interiors, individual droplet and crystal lifetimes are often just a few tens of minutes.

The great swaths of cloud associated with middle-latitude fronts (nimbostratus) are apparently much more persistent, often lasting for a week or more. But careful studies show that air flows through them so quickly that any particular air parcel remains within its parent cloud mass for only a day or two before emerging and evaporating its cloud content.

Because cloud is so prevalent and yet so transient, its formation and destruction must be both widespread and rapid, and the mechanisms involved commonplace. In this chapter we are going to examine these mechanisms in some detail, but first we must consider carefully the state of saturation which is observed to be essential for the formation and maintenance of all significant cloud.

6.2 Saturation

Consider the state of affairs in a very shallow layer of air in contact with an exposed water surface (Fig. 6.1). As well as dry air, it contains a vapour consisting of a gas of water molecules which is continually being fed by evaporation from the water surface and drained by recondensation there. In terrestrial conditions, if we consider a layer extending only a millimetre or so from the water surface, the feeding and draining fluxes are so nearly equal that the layer is effectively in a state of dynamic equilibrium which is termed *saturation*. As the name implies, saturation is a state of maximum vapour content, since any tendency to exceed the saturated state, by migration of more vapour from outside the layer, for example, would be offset by increased recondensation at the surface. The analogy with a saturated sponge is obvious, but it is unfortunate that it has become conventional to speak of the air as being saturated, since it is clearly the vapour rather than the air which is saturated; indeed the state of saturation is quite unaffected by the presence of the dry air. However, the convention will be followed in the rest of this book, with the tacit understanding that saturated air means air whose vapour is saturated.

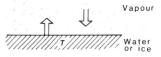

Fig. 6.1 A saturated vapour overlying a plane surface of pure ice or water.

The dynamic equilibrium between a water surface and its adjacent saturated air ensures that both surface and air (including both vapour and dry air) have the same temperature. When the water is pure and its surface is plane, it is found by observation that both the density and pressure of the saturated vapour depend only on their common temperature. (Note that when the vapour is mixed with dry air we must use the partial pressure or the density of the vapour.) Both vapour density and pressure increase with temperature in the nearly exponential manner shown in Fig. 6.2, the rising gradient corresponding to the increasing ease with which water molecules escape from the surface as its temperature and their associated thermal agitation increase. Across the atmospheric temperature range the increases are very large — a fact which proves to be very important for cloud formation and development.

Note that Fig. 6.2 also includes values for saturation close to a plane surface of pure ice. Remember that ice can evaporate (*sublime*) without melting first, as is evidenced by the disappearance of snow patches at sub-freezing temperatures. The saturation vapour pressures and densities for an ice surface are less than those for a supercooled water surface at the same temperature because the stronger inter-molecular bonding in ice inhibits evaporation.

Careful observation shows that saturation occurs very close to exposed surfaces of water or ice even when precise equilibrium between evaporation and recondensation does not apply. The disappearing snow is a case in point, as is the often very rapid disappearance of water puddles in the presence of sun and wind — both observations indicating an excess of evaporation over recondensation. The formation of dew or hoar frost indicates an imbalance in the opposite direction. Such disequilibria are normal in terrestrial conditions, but on detailed enquiry it appears that in almost all such cases the difference between evaporating and condensing fluxes is so much smaller than either of the fluxes separately that the adjacent vapours differ insignificantly from the saturated ideal. We will therefore assume from now on that a saturated vapour always exists within a millimetre or so of an exposed plane surface of pure water or ice. At greater distances saturation may or may not be present, depending on factors such as the mixing of the moist air with drier air from other sources, just as the temperature of more distant air may differ significantly from that of the surface and closely adjacent moist air.

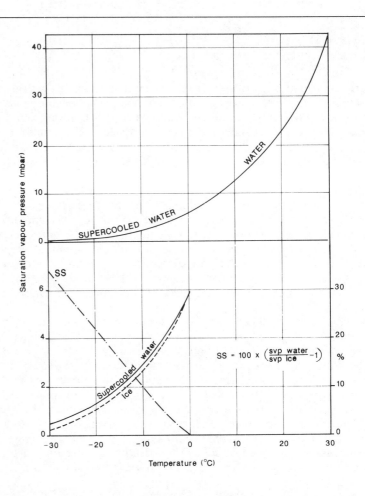

Fig. 6.2 Variation of saturation vapour pressure with temperature. Below 0 °C values for supercooled water exceed those for ice, and the chain line shows the supersaturation (SS in %) relative to ice of a vapour which is saturated relative to supercooled water. Find corresponding vapour densities by using the equation of state for vapour.

Finally note that saturation has been defined in terms of plane surfaces of either pure water or pure ice. In fact dynamic equilibria are as readily set up over curved surfaces and impure water or ice, but they may then differ markedly from those for plane, pure surfaces. Indeed the differences are crucial to the formation of all terrestrial cloud and will be examined in some detail in succeeding sections. To avoid confusion we will reserve the term 'saturation' for the case of plane, pure water or ice.

6.3 The sub-cloud layer

Clouds in the lower troposphere are normally separated from the Earth's surface by a cloud-free layer whose depth may range from less than a hundred metres to several kilometres, depending on weather conditions. Indeed the appearance of clouds, or a cloud layer, above a substantial sub-cloud layer is so normal (Fig. 6.3) that we might assume its explanation to be trivial. In this, as in so many meteorological matters, we would be mistaken. However, the almost universal presence

Fig. 6.3 Panorama of small cumulus over low hills in north-west England, with cloud bases at a fairly consistent level throughout the field of view despite the irregular topography.

of the sub-cloud layer in the low troposphere is so important for understanding cloud formation there that we must try to explain it, if only in outline.

Seventy per cent of the Earth's surface is water, and much of the rest is moist ground or ice. According to the argument of the previous section the air very close to all but the most arid surfaces must therefore be saturated at surface temperatures (the effect of impurities is not significant in this context). In the simplest view this would lead us to expect to find a deep layer of saturated air covering most of the surface and maintained by continual evaporation from it. Given the resulting upward extension of the saturated layer, and the efficiency of the cloud-forming processes to be described presently, we should expect much of the Earth to be shrouded in fog. In fact very little fog is observed, which suggests that there are mechanisms at work which are removing water from the atmosphere relatively efficiently. These are of course the precipitation processes, which ensure that some cloud droplets and crystals grow big enough to fall back to the surface under gravity. Though the details of the precipitation processes are not considered until later, we must now look at them from a very general viewpoint.

The least efficient type of precipitation is observed in deep fog layers, which continually lose part of their cloud water through a very fine drizzle of droplets intermediate in size between the tiny cloud droplets and normal-sized drizzle droplets — barely perceptible except when they touch our faces or car windscreens. This loss is usually insufficient to clear the fog by precipitation, though some schemes for fog clearance envisage its artificial enhancement.

More efficient types of precipitation depend on strong or persistent updraughts and much deeper layers of cloud. When these are present, the rate of precipitation onto any given surface area greatly exceeds the rate of evaporation from that area. The situation is then as shown in Fig. 6.4: only a fraction of the Earth's total

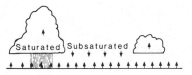

Fig. 6.4 Vertical pathways of vapour and precipitation in the atmosphere.

143

surface is covered by efficiently precipitating cloud, but this cloud and its precipitation are maintained by evaporation from virtually the whole terrestrial surface, the arid regions being relatively small. Between the cloudy updraughts there is sinking air, desaturated by dry adiabatic descent after prior precipitation. As this air approaches the surface it mixes with the saturated air continually produced by evaporation there, thereby maintaining a layer of subsaturated and therefore cloud-free air overlying the thin film of saturated air clinging to the surface. Without going into details it should be clear that the depth of the subsaturated layer, like the fraction of the total surface area overlain by precipitating cloud, is controlled by the efficiency of the precipitating processes at work in the clouds. If these were all as feeble as those in fog, for example, then our initial expectation of a largely fog-bound planet would be fulfilled.

According to Fig. 6.4 the sub-cloud layer is continually fed by subsaturated air from aloft and saturated air from beneath. In and just above the shallow saturated layer, atmospheric mixing is relatively very feeble, with the result that there is a shallow zone overlying the surface in which there is usually a significant vertical lapse of vapour content, as sketched in Fig. 6.5. This zone is seldom more than a few metres deep, and is bounded above by a deep layer, occupying most of the sub-cloud layer, in which the vigorous stirring typical of the atmospheric boundary layer in most conditions maintains a much more uniform distribution of water vapour. This relatively deep and well-mixed layer plays such a basic role in the formation of cloud in the low troposphere that we must consider it in some detail in the following sections.

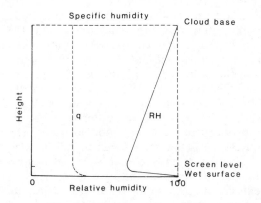

Fig. 6.5 Idealized profiles of specific humidity q and relative humidity RH in a well-mixed sub-cloud layer over a moist surface.

6.4 The well-mixed sub-cloud layer

Consider the vertical distribution of water vapour to be expected in the well-mixed layer extending upward from the shallow transition zone (Fig. 6.5) immediately overlying the surface. Ignoring relatively small volumes dampened by evaporation from falling precipitation, there are no sources or sinks of vapour between the lower and upper limits (the transition zone and cloud base respectively), so that the equilibrium established by continual, efficient mixing must maintain uniformity of specific humidity q, where by definition

q = mass of contained vapour/mass of containing moist air

$$= \rho_v/\rho \qquad (6.1)$$

and ρ_v and ρ are respectively the densities of vapour and moist air. Since air density lapses with increasing height, the vapour density must lapse in proportion, to maintain q. A typical depth of the sub-cloud layer is 1 km, in which both the pressure and density of air fall by about 10%, and it follows that vapour pressure and density must do likewise.

The stirring of the sub-cloud layer also establishes an equilibrium vertical distribution of temperature, which is the dry adiabatic profile as discussed in section 5.13. This temperature profile can be converted into a profile of equivalent saturation vapour density by making use of Fig. 6.2 or thermodynamic tables. The combination of the steepness of the dry adiabatic lapse rate and the rapidity of the variation of saturation vapour density with temperature (i.e. the steepness of the slope of Fig. 6.2) results in very large lapses of saturation vapour density in the well-mixed layer. For example the 9.8 °C temperature lapse expected in a kilometric deep layer is associated with a lapse of saturation vapour density from 12.8 to 6.8 g m^{-3}, assuming that the air at the bottom of the well-mixed layer has temperature 15 °C. In warmer layers the lapse of saturation vapour density is even larger because of the increasing rapidity of variation of saturation vapour density with temperature (i.e. the increasing steepness of the slope of Fig. 6.2).

We have now established that the lapse of saturation vapour density ρ_s in the well-mixed layer far exceeds the lapse of actual vapour density ρ_v. Since the relative humidity (RH) is defined to be the ratio of the two densities expressed as a percentage, that is

$$RH = 100 \, \rho_v/\rho_s \qquad (6.2)$$

it follows that RH must increase with increasing height from a minimum value at the base of the well-mixed layer, as shown in Fig. 6.5. This increase means that at some level, higher or lower depending on the degree of subsaturation of the air at the base of the well-mixed layer, the value of RH approaches 100%. Since this value corresponds to saturation, it follows that any further ascent by an air parcel would produce supersaturation and consequent cloud formation by condensation of excess vapour onto cloud droplets. The level of first saturation therefore corresponds to cloud base and the upper limit of the sub-cloud layer. In fact to the extent that the atmosphere behaves like our simple adiabatic and well-mixed model it coincides with the lifting condensation level defined in section 5.11.

Note that we have now produced an explanation for a fact which is so common-place that we might have supposed no explanation was ever needed in the first place — namely that saturation and cloud develop in rising rather than in sinking air. But if for example the variation of saturation vapour density with temperature had been very much smaller than is apparent in Fig. 6.2, so that the vertical lapse of saturation vapour density in the well-mixed layer was less than the lapse of actual vapour density, then the opposite would have been the case: rising air would have desaturated, giving no possibility of an elevated saturated layer. And without the associated precipitation, the Earth would be as fog-bound as naively assumed at the beginning of this section, with far-reaching consequences for the environment in which life has developed. It is clear that the apparently innocuous balance of factors leading to the production of cloud in rising rather than sinking air is of profound importance to us all.

6.5 The dew-point profile

Humidity profiles in the sub-cloud layer can be depicted most graphically using dew-point temperatures. Consider air at the base of the well-mixed layer (Fig. 6.6). It is significantly unsaturated and so its dew-point T_d is significantly below its temperature T. It follows from the mode of operation of a dew-point meter (section 2.3) that the dew point of this air can be found on Fig. 6.6 by sliding isobarically to the left until saturation is reached, i.e. until the saturation specific humidity q_s (whose isopleths are the faint dashed lines obliquely crossing the isotherms) is equal to the specific humidity q of the air. The temperature at which this happens is the dew-point T_d.

Fig. 6.6 Tephigram of the lower troposphere showing idealized profiles of temperature T, wet-bulb temperature T_w and dew-point T_d in a well-mixed, dry adiabatic sub-cloud layer, with lifting condensation level LCL. This is the completed version of Fig. 5.7(b). Average temperatures and pressures in the sub-cloud layer are such that 1 mbar pressure lapse corresponds to about 9 m elevation.

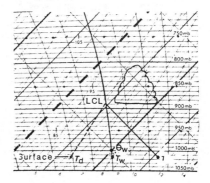

Now according to our simple model of the well-mixed layer, the specific humidity is the same at all levels up to the lifting condensation level. Therefore the dew-point profile in Fig. 6.6 lies along a line of constant q_s. The detailed thermodynamics underlying the tephigram ensure that q_s isopleths are nearly straight lines which are almost perpendicular to the dry adiabats, and so it follows that the temperature and dew-point profiles in the well-mixed layer are effectively straight lines which converge sharply and meet at the lifting condensation level LCL (equality of T and T_d of course implying saturation). In fact we can now complete the full expression of Normand's theorem (section 5.11) by stating that the dry adiabat through the temperature, the saturated adiabat through the wet-bulb temperature and the saturation specific humidity through the dew point all meet at the lifting condensation level.

It can be shown from detailed thermodynamics (appendix 6.1) that the convergence of the T and T_d profiles in typical low troposphere conditions is such that the dew-point depression $T - T_d$ falls by about 0.80 °C for each 100 m ascent through the well-mixed sub-cloud layer — a value which can be roughly checked using the height scale added to Fig. 6.6. It follows that if the dew-point depression at the base of the layer is $\triangle T$ °C, then the height of the lifting condensation level LCL above the base is approximately $125 \times \triangle T$ metres. This rule of thumb, or the slightly more accurate equivalent construction on a tephigram, is often used to calculate cloud-base height from Stevenson screen measurements of T and T_w (converted to T and T_d). However, heights calculated in this way are often observed to undershoot the true cloud base by 10% or more, and a brief discussion of the reasons for such systematic error will usefully qualify the too simple model of the well-stirred layer assumed so far.

The parcel processes assumed in the dry adiabatic and specific humidity conserving model are self-consistent under detailed scrutiny, so that for example Normand's theorem is fully justified; however, they do not exactly match atmospheric behaviour. In particular the actual mixing process falls short of the 100% efficiency assumed in the model, and seems to do this in systematic ways. This has been admitted from the outset in the case of the transition zone between the shallow saturated surface layer and the well-mixed layer, but it also applies to the well-mixed layer itself, though much more subtly. The details are not fully understood, but their upshot seems to be that the actual environmental temperature lapse rate often exceeds the dry adiabatic value by a few per cent in the first few hundred metres above warm ground, and undershoots it to a similar extent in the rest of the layer from there up to cloud base [40]. These effects, together with the observation that the base of the well-mixed layer often lies somewhat above Stevenson screen level (remember that this is only 1.5 m above the ground surface) result in the underestimation of cloud base height mentioned previously. The simple model may be good but it is not perfect.

Finally, note that the simple model discussed in this and preceding sections applies only to a well-stirred layer, i.e. one in which air parcels are continually moving up and down because of thermal or mechanical convection or both. Though such stirring is very widespread in the sub-cloud layer, it is not universal, being either ineffective or absent in the convectively stable layers which often develop in the first 100–200 m over a land surface on a cloudless night. In such conditions there is no simple behaviour observable in the profiles of temperature and humidity in the stable layer, though at higher altitudes the profiles of the previous day's deep convection may still remain. Falling temperatures and rising relative humidities are of course observed near the surface as nocturnal cooling progresses, and saturation and fog may result, but no simple quantitative relations apply to their profiles to compare with those in well-mixed layers.

6.6   Condensation observed

You can watch cloud forming in rising air most easily when weather conditions are encouraging fair weather cumulus (Fig. 6.3). As an updraught begins to reach its lifting condensation level, the air changes its appearance within a few tens of seconds from the slight haziness typical of the upper parts of the sub-cloud layer to a state in which there is an embryonic patch of cloud veined and clotted with thickening elements which quickly merge to form a small cumulus. The nearly horizontal base, representing the lifting condensation level, and the cauliflower-surfaced upper parts, are familiar enough from everyday observation (Fig. 6.7), but the speed and dynamism of the formation and development of such clouds are seldom noticed by the sporadic skyward glances allowed by modern suburban and

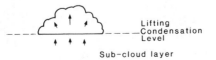

Lifting
Condensation
Level

Sub-cloud layer

Fig. 6.7 Outline of a typical small cumulus forming above the lifting condensation level.

147

urban life. A few minutes spent continuously watching the sky on a bright morning, just as the cumulus are beginning to appear, will offset the quite misleading impression of the static drift of clouds which most people have on the basis of their snapshot observations, and which unfortunately is reinforced by Wordsworth's uncharacteristically inaccurate image:

> I wandered lonely as a cloud
> that floats on high o'er vales and hills.

Careful measurement by sensors mounted on aircraft shows that air throughout the interiors of many clouds is almost exactly saturated with respect to liquid water, having relative humidities of between 100 and 101% (i.e. *supersaturations* of between 0 and 1%). This, and other aspects of behaviour which will prove to play important roles in the cloudy atmosphere, can be observed in a very simple cloud chamber which can either be set up as described in the following, or merely accepted as a descriptive framework for the discussion of cloudy behaviour.

Find a large glass bottle (the five-litre capacity types used for home winemaking are ideal), place a small quantity of water inside and fit the mouth with a bung pierced by a narrow glass tube tipped by a length of rubber tubing soft enough to form an air-tight seal under finger pressure (Fig. 6.8). Illuminate the interior of the bottle strongly with, for example, a slide projector placed close by. A useful extra is a water manometer tall enough to measure an interior pressure excess of 30 mbar (i.e. a 30 cm difference in water level — see appendix 4.2).

After shaking the bottle to ensure that it and its contents are all at the same temperature and that the interior air is saturated, blow into the rubber tube and maintain the excess pressure for about 30 s by pinching. Although in principle the blowing must introduce some warm, moist air, its main effect is to raise the interior pressure and hence raise the interior air temperature by compressional warming. A little further shaking should ensure that the interior air is saturated once again, having been temporarily desaturated by the warming. If the rubber tube is now released the excess interior pressure falls to zero, chilling the saturated air by decompression and producing an apparently tenuous cloud. In fact the cloud seems tenuous only because its volume is small: if a room were to be filled with cloud of this concentration it would be judged to be a very dense fog (Fig. 6.9).

This demonstration shows that cloud can form extremely quickly in air which a little calculation shows is only modestly supersaturated. If for example pressure falls by 30 mbar on depressurization, according to eqn (5.13) the temperature should fall by about 2.5 °C. The actual chilling must be less than this, since the formation of cloud tends to make it a saturated adiabatic rather than a dry adiabatic process, and is probably nearer 1.5 °C. According to Fig. 6.2, air

Fig. 6.8 A simple cloud chamber, showing bottle, inlet and venting tube (hanging right) and water manometer.

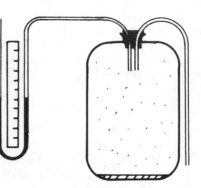

Fig. 6.9 A thin morning fog (visibility ~ 300 m) in winter in England, probably produced by inland advection of warm, maritime air and overnight radiative chilling.

saturated at the initial temperature (i.e. just before depressurization) must be about 5% supersaturated just after depressurization. In fact more sophisticated observation and theory suggest that cloud droplets can grow so quickly at the expense of the supersaturating vapour, that the maximum supersaturation is limited to 1% or less. This certainly seems to be the case in almost all atmospheric cloud, where rates of depressurization are very much smaller than in the cloud chamber.

If you look carefully into the bottle just after decompression, you will see a moving graininess in the strongly illuminated cloud. This is all that can be seen of the individual cloud droplets without magnification, but we can estimate their size very roughly by gauging their fall speed after the initial swirls have died away. The grainy texture is then seen to be sinking at a rate of millimetres per second. From a

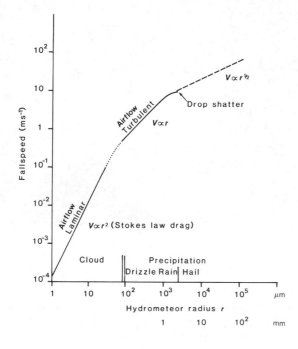

Fig. 6.10 Graph of fall speed versus radius for spherical bodies ranging in size from small cloud droplets to giant hail. Speeds shown apply to the low troposphere. Speeds at higher altitudes are somewhat larger on account of the lower air density there. (Unpublished survey by the author, 1968)

well-established relationship between droplet size and fall speed (Fig. 6.10) it follows that droplets falling at such speeds must have radii of the order of micrometers. This compares reasonably with droplet sizes observed in atmospheric cloud: a variety of techniques indicate average radii of about 5 μm.

A further demonstration using the cloud chamber reveals an unexpected complication. Several minutes after depressurization most cloud droplets have sunk to the bottom of the bottle. If the bottle is again pressurized and depressurized the resulting cloud is noticeably more sparse. Subsequent repetitions separated by similar waiting periods produce less and less cloud even though the degree of pressurization and subsequent cooling is maintained at a constant level. This increasing reluctance to form cloud is dramatically reversed when a small amount of smoke, from a snuffed match for example, is introduced into the bottle. The merest wisp is enough to produce a dense cloud at the next depressurization, whose removal will require many further cycles of depressurization and settling.

It is clear from this demonstration that cloud forms much less readily in clean air than it does in smoky air. In fact as mentioned in section 3.6, smokes are just a part of the considerable range of atmospheric aerosol particles which have been shown by work with cloud chambers to provide effective *condensation nuclei*, i.e. they are able to encourage the formation of cloud droplets at low supersaturations. Aitken and Coulier used cloud chambers in the late nineteenth century to count the particles in atmospheric aerosol by forming cloud on them and then allowing the cloud droplets to settle onto a counting plate. In so doing of course they isolated the condensation nuclei in particular. In industrial regions the connection between smoke and cloud formation is sometimes quite obvious: Fig. 6.11 shows a cumulus developing where smoke from a strong source rises to the lifting condensation level. To examine the mechanism by which condensation nuclei encourage cloud formation we must look closely at the relationship between small droplets and their immediate surroundings.

Fig. 6.11 Small cumulus forming out of an industrial smoke plume over London in Summer.

6.7 Condensation modelled [41]

Saturation has been defined as the state of a volume of water vapour in equilibrium with an exposed plane surface of pure water or ice. However, plane surfaces are not relevant to cloud formation more than a few millimetres above the Earth's surface. Instead there is an aerosol containing roughly spherical particles and solution droplets mostly with radii of less than one micrometre, often much less (section 3.6). Equilibria in such populations are vitally affected by the tight curvatures of the available surfaces, and by the fact that many of the droplets are quite strong solutions. Dissolved impurity is commonplace because many aerosol droplets originate as particles of *hygroscopic* (water-attracting) material, such as sodium chloride, which form droplets of concentrated solution when the relative humidity rises far enough.

Consider first the effect of tight surface curvature on the equilibrium between water (or ice) and the adjacent vapour. A water molecule comprising part of a convex surface of small radius of curvature r is less tightly bound to that surface than it would be to a plane surface in otherwise identical conditions because there are fewer attracting neighbouring molecules within any chosen range (Fig. 6.12). Evaporation therefore proceeds more rapidly, and the equilibrium vapour density ρ_r is higher than the equivalent value for a plane surface in the same conditions (which is by definition the saturation vapour density ρ_s). This is confirmed by detailed observation and theory, which shows that

$$\frac{\rho_r}{\rho_s} = \exp\left(\frac{A}{rT}\right) \tag{6.3}$$

where A is a constant for the droplet liquid (water in the present case) and T is the absolute temperature of the droplet and its immediate surroundings. The effect is named after *Thomson* (Lord Kelvin) who first derived eqn (6.3).

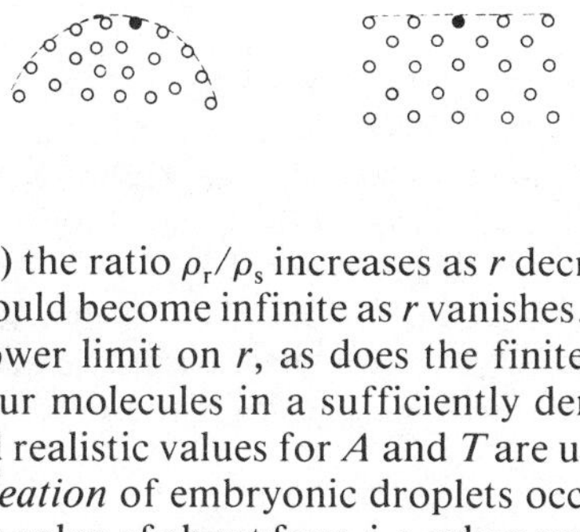

Spherical   Plane

Fig. 6.12 Molecular environment of a molecule at the tightly curved surface of a liquid or solid body, in comparison with a plane surface.

According to eqn (6.3) the ratio ρ_r/ρ_s increases as r decreases (i.e. as the droplet becomes smaller) and would become infinite as r vanishes. However, the finite size of molecules places a lower limit on r, as does the finite probability of multiple collisions between vapour molecules in a sufficiently dense vapour. When these effects are included, and realistic values for A and T are used, it can be shown that such *homogeneous nucleation* of embryonic droplets occurs relatively frequently only when ρ_r/ρ_s exceeds a value of about four, i.e. when ambient relative humidities are about 400%. At smaller values the homogeneous nucleation is ineffective because embryonic droplets form less frequently and evaporate again almost immediately, the ambient vapour density ρ being less than the equilibrium value ρ_r for the tiny droplet. It follows that we should not expect clouds to form in clean, moist air until it has a supersaturation of about 300%. As already mentioned such large values are never observed in the atmosphere, fairly obviously because it is not clean. The role of condensation nuclei now becomes apparent.

Consider a volume of air containing particles much larger than the embryonic droplets formed by homogeneous nucleation. According to eqn (6.3) the ambient vapour density needed to exceed the equilibrium vapour density for such relatively large particles, and thereby initiate condensation onto their surfaces, is much smaller than the value required for effective homogeneous nucleation. In fact the exponential decrease of equilibrium vapour density with increasing particle radius is so sharp that a supersaturation of only 1% is enough to enable a particle of radius

0.15 μm to grow by condensation. (Strictly speaking, to apply eqn (6.3) with the A value for water we must assume that it is already coated with a layer of water molecules. Despite some complicating surface effects if this is not the case, the argument as presented still stands in principle and applies quite well numerically.) Such a low value confirms the importance of condensation nuclei in promoting cloud formation; without them relative humidities would have to rise to about four times the saturation value before cloud could begin to appear. In the simple cloud-chamber experiment, the repeated scavenging of condensation nuclei by cloud formation and settling produced air so artificially clean that further cloud formation was inhibited until smoke was introduced. In the atmosphere the several sources of aerosol mentioned in section 3.6 ensure that there is always an adequate supply of condensation nuclei.

The role of condensation nuclei in encouraging cloud formation is further enhanced when we consider that many of them consist of hygroscopic solids which readily form droplets of quite concentrated solution. According to *Raoult's law* the presence of solute reduces the equilibrium vapour density over a plane surface of solution below the value for a plane surface of pure water (i.e. below the saturation value) by an amount which increases with the solute concentration, and we must expect a similar effect in the case of droplets. If we consider a mass m of solute dissolved in a droplet of radius r, then the Raoult effect is represented by a negative term of form Bm/r^3, where the value of B varies only with the type of solute, for any given solvent (water). If we focus on conditions when equilibrium and saturation vapour densities are nearly equal, then the Thomson effect simplifies to the first two terms of the series expansion of the exponential function, and the resultant of the Thomson and Raoult effects is

$$\frac{\rho_r}{\rho_s} = 1 + \frac{A}{rT} - \frac{Bm}{r^3} \qquad (6.4)$$

The value of B varies only with the type of solute, for any given solvent, and the other symbols are as for eqn (6.3). The first two terms on the right-hand side of eqn (6.4) represent the Thomson effect and the third represents the Raoult effect. Note that the dependence of the latter on solute concentration introduces great sensitivity to the value of the droplet radius.

Figure 6.13 represents the combined Thomson and Raoult effects for several values of m typical of sodium chloride in atmospheric aerosol. Each *Köhler curve* (named after a Swedish pioneer meteorologist) represents the variation of equili-

Fig. 6.13 Kohler curves for droplets containing the indicated mass of NaCl. The scale of the RH axis expands above 100% to clarify the structure there, introducing kinks in the curves which would otherwise be smooth. (After [41])

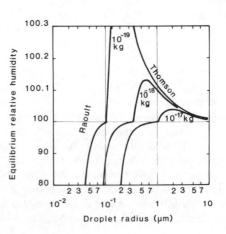

brium relative humidity ($100 \, \rho_r/\rho_s$) with droplet radius r for a fixed mass m of sodium chloride in water. When the salty droplet is small enough the salt concentration is so high that the Raoult effect dominates and the equilibrium RH is reduced below the saturation value (100%) despite the opposing Thomson effect. It is therefore possible for small droplets of sodium chloride solution (and other solutions too — all Köhler curves have the same general shape) to be in equilibrium with a subsaturated vapour, as was observed by Aitken. At much larger radii the Thomson effect predominates, reducing the equilibrium supersaturation towards zero as the radius increases. Between these extremes there is a critical droplet radius r_c for each particular mass of solute at which the equilibrium supersaturation is a maximum. The existence of such a maximum vitally affects the behaviour of growing droplets, as is outlined in the next section.

6.8 Cloud droplet growth

Consider a hygroscopic aerosol particle in air whose relative humidity is rising because of adiabatic expansion and cooling. At some RH value well below 100% the particle becomes a tiny droplet of concentrated solution whose equilibrium relative humidity is equal to the ambient value. As the ambient RH rises further (because the fall of temperature implies a fall in saturated vapour density which is much larger than the fall in actual vapour density — section 6.4) the droplet grows by diffusion and condensation of vapour, its equilibrium RH rising to match the ambient value so rapidly that the difference between the two is negligible in the modest rates of rise of ambient RH associated with even the most vigorous atmospheric updraughts. Hence the Köhler curve (Fig. 6.13) appropriate to the mass and type of solute can be used to describe the growth of the droplet with ambient RH. Strictly speaking we should allow for the effect of falling temperature on the shape of the Köhler curve, but this is relatively small, although the effect on the saturation vapour density is of course very large and important for the reasons discussed in section 6.4. Note that the droplet and ambient vapour are in a state of neutrally stable equilibrium in the sense that any change in the ambient vapour content produces rapid growth (or shrinkage) to a new equilibrium size. In the realistic case of a droplet containing 10^{-18} kg of NaCl, the equilibrium radius increases from 0.1 μm at 87% ambient RH to 0.2 μm at 99%.

Such neutrally stable equilibrium persists until the droplet reaches its critical radius, which occurs when the ambient vapour is supersaturated to some slight degree (0.13% in the case of 10^{-18} kg NaCl). If the ambient RH rises further the droplet enters a phase of runaway growth because the falling equilibrium RH can no longer match the ambient value. The droplet therefore grows at a rate which is limited only by the ability of the diffusive properties of the air to deliver vapour to the swelling droplet and remove the latent heat liberated by condensation there. Despite the typically small excess of ambient over equilibrium vapour densities the droplet radius grows ten or twentyfold beyond its critical radius in a very short time period, accounting for the almost instantaneous appearance of micron-scale cloud droplets in the simple cloud-chamber experiment, and in even faster cloud production (Fig. 6.14). Assuming that ambient vapour pressure remains constant, droplet growth continues indefinitely, but at a slower and slower pace as measured by droplet radius, essentially because the very rapid increase of droplet mass with

Fig. 6.14 Cloud forming quickly in wing-tip vortices. Net rotation of air round the wings (anticlockwise round the near wing) spins off into trailing vortices whose core pressures are low enough to produce saturation and hence cloud. In more humid conditions the cloud may envelope all upper surfaces of the airframe. Assuming an air speed of 150 m s^{-1} and a cloud formation fetch of <1 m, cloud is forming in <7 ms, much faster than in any cloudy weather system.

radius outruns the ability of vapour to diffuse sufficiently rapidly down the gradient of vapour density.

When a growing droplet exceeds its crucial radius, the condensation nucleus is said to be *activated* — a term which recognizes both the crucial change in behaviour as it grows from a state of neutral stability to one of almost explosively unstable growth, and the significance of this change for the conversion of embryonic droplets to fully grown cloud droplets. Note that if the ambient humidity falls below the value associated with the critical radius the droplet deactivates by rapid shrinkage to an equilibrium radius below the critical radius, shooting leftward over the maximum on the Köhler curve (Fig. 6.13).

A non-hygroscopic condensation nucleus has no neutrally stable growth regime corresponding to a hygroscopically nucleated droplet below its critical size. Instead it remains essentially dry until the ambient supersaturation exceeds the value corresponding to its radius of curvature. Vapour then begins to condense and the droplet grows rapidly, like an activated hygroscopic nucleus.

6.9 Cloud development

Previous sections have outlined mechanisms at work on individual droplets during formation and subsequent growth. Let us bring these together to describe the production of a small cloud of droplets just above a well-mixed cloud-free layer. In the first instance we can suppose that this represents the formation of fair-weather cumulus (Fig. 6.3), but the same mechanism applies to air rising into the base of an established cumulus congestus or cumulonimbus. Indeed it applies to all atmospheric cloud formation, from very slow, static cooling in fog, to rapid smooth uplift in lenticular cloud (Fig. 10.30), and from slow persistent ascent in fronts, to rapid horizontal decompression in tornadoes (though size and timescales, and the degree of isolation from other clouds, can enforce very different subsequent development).

We begin with air rising toward the top of the sub-cloud layer, its temperature falling and its relative humidity rising because of dry adiabatic expansion. The population of condensation nuclei in the humidifying air is clearly very important, both in terms of the total numbers and their distribution across the available spectra of sizes and materials. Numbers and sizes in particular vary very widely with position and time, but as a rough guide note that one litre of air from the low troposphere at an inland site may contain 10^7 Aitken nuclei (radii between 5×10^{-3} and $0.1 \ \mu m$), 10^5 large nuclei (radii between 0.1 and 1.0 μm) and 10^3 giant nuclei (radii exceeding 1.0 μm). No nuclei smaller than the largest Aitkens become activated in natural clouds, because as we shall see maximum supersaturations are held sufficiently close to zero by the simultaneous growth of droplets on all larger nuclei. The numbers of the 'large' category can therefore be used as an order of magnitude for the total numbers of nuclei available for activation in the rising air. Over land these are believed to consist mainly of sulphates and sulphuric acid (derived from the combustion product SO_2), fine soil particles and sea salt. Obviously numbers are larger closer to localized sources such as active volcanoes and natural and man-made fires. Over the oceans numbers may be down by a factor of ten or more and the proportion of sea salt much larger.

When the relative humidity of the rising air exceeds about 80% many hygroscopic nuclei begin to swell by condensation, and their growth as RH exceeds 90% may be sufficient to reduce visibility quite noticeably by a wet haze, especially when the line of sight is nearly horizontal. Observers on mountains or aircraft often see a marked haziness extending tens or even hundreds of metres below cumulus cloud base. In heavily industrial regions this elevated haziness may be visible even from the oblique viewpoints of ground-level observers. From Fig. 6.13 and equivalent data it transpires that the radii of haze droplets forming on all but giant nuclei are less than 1 μm. The relatively weak interactions of such small droplets with visible light make them effectively invisible in bulk in small volumes such as cloud chambers.

While the rising air remains unsaturated, all haze droplets remain in equilibrium with the ambient RH, swelling as it rises, but not yet activated. But almost as soon as the air begins to supersaturate, some droplets exceed their critical radius, become activated and so begin to grow very rapidly. At first only the most favoured are activated — those forming on giant hygroscopic nuclei — but as the supersaturation of the rising air increases, increasing numbers of less favoured droplets are activated, as are some of the largest non-hygroscopic nuclei. In a relatively short space of time so many droplets are growing so rapidly that they begin to deplete the ambient vapour quite substantially; the growth of supersaturation checks and then reverses. Once the supersaturation starts to fall no further droplets are activated, and those still struggling towards activation begin to shrink again. However, the supersaturation close to the surfaces of the activated droplets is so nearly zero (imagine the Köhler curves extended far to the right of the critical radius) that they continue their rapid growth for as long as the ambient air remains supersaturated, which means in effect for as long as the rising air remains within the body of the cloud.

The way in which a population of droplets becomes activated has three important consequences.

1. The transition from wet haze to cloud tends to be quite rapid, since most of the droplets are activated at supersaturations of below 1%. The activated droplets grow to micron scale in a few seconds, at which size they become very much

more effective at scattering sunlight than they were as submicron-scale haze droplets. This explains the dramatic visual transformation apparent in the atmosphere and in cloud chambers. Together with the well-mixed nature of the sub-cloud layer this also explains the familiar sharpness and uniformity of cumulus cloud-base layer (Fig. 6.3).

2. The number of cloud droplets formed is equal to the number of droplets and particles activated, which in turn is determined by the aerosol population and the maximum supersaturation reached. Although cloud droplets may coagulate and divide subsequently, to a large extent the cloud population is established at least in respect of its numbers by the conditions in which it is first formed. In the case of cumulus these are the conditions prevailing in the first few tens of metres above cloud base.

3. With the numbers of cloud droplets firmly established, their ultimate size is determined by the amount of vapour available for condensation. For example, consider the case of a litre of air which contains 10^5 activated condensation nuclei and 10 mg of vapour (corresponding to a specific humidity of 10 g per kg — quite normal for the low troposphere). If half of the available vapour is condensed, as might well happen if the air ascends through an appreciable fraction of the troposphere, you can easily check that each cloud droplet must grow to a radius of about 30 μm. In fact growth to such a size by diffusion and condensation is relatively slow — it is rapid only in the initial stages, as droplets activate and race to radii of a few microns — and is usually short-circuited by mechanisms we must consider later to explain the observed ease and speed with which clouds precipitate.

Deactivation of cloudy air occurs most dramatically at the edges of cumulus clouds of all sizes. The cauliflower-like upper parts are in a state of continual and rapid change as parcels of cloudy air erupt from the interior and are turbulently mixed with the clear, subsaturated ambient air. As the cloudy parcels become subsaturated the cloud droplets shrink very rapidly below their critical radii. In fact the

Fig. 6.15 A decaying cumulus, dull compared with the developing cloud behind.

dryness of the ambient air usually takes them so far below their critical size that there is little or no wet haze. On the sunlit side of the cumulus the sharp contrast between the clear ambient air and the strong backscattering of sunlight by the cloud gives rise to its familiar brilliant appearance (Fig. 10.9), and on the shaded side of the strong forward scattering from the edge zone forms the 'silver lining' round the dark body of the cloud (dark because so much light has been backscattered at the far side) (Fig. 1.4). Often deactivation occurs more gradually throughout a larger volume when an updraught weakens and dies. As the supersaturation gradually falls to zero by mixing with somewhat drier air the population of cloud droplets is thinned by selective deactivation, and the cloud loses its brilliant, sharp-edged appearance in favour of a diffuse dullness which finally decays to nothing through a transient haziness (Fig. 6.15).

6.10 Precipitation

So far we have concentrated on the mechanisms at work maintaining the cloudy content of the troposphere. In the long run these are precisely balanced by the precipitation processes which return the condensed water and ice to the surface. Evaporation of cloud and precipitation in the air is widespread but merely represents a redistribution of the water substance within the atmosphere.

Over most of the Earth it is a matter of common observation that water falls to the surface in a variety of forms summarized in Table 6.1, and selectively depicted in Fig. 6.16. Water droplets reach the surface with diameters ranging from about 100 μm to about 3 mm. The smallest of these are felt rather than seen in fog. Normal drizzle (diameters about 200 μm) reaches the surface from low cloud such as stratus, which can be considered to be a fog layer lifted clear of the surface by stirring and slight desaturation of the surface layers. Raindrops cover the rest of the size range and account for the great bulk of precipitation reaching the surface in all but the coldest conditions. Fine rain, verging on drizzle, is observed only when

Table 6.1   Properties of selected precipitation droplets and particles

	Name	Diameter (mm)	Shape	Fall speed (m s^{-1})
Water				
	Drizzle	0.2	Sphere	0.8
	Rain	0.5	Sphere	4.0
	Rain	5.0	Unstable cap shape	10.0
Ice				
	Snow crystals	0.2	{Prism, plate,}	0.3
		5.0	{Star, needle}	0.7
	Snowflakes	1.0	Irregular aggregates of from 2 to 100s	0.5
		20.0	of crystals (often stars)	1.0
	Graupel	0.5	Conical	0.5
	(soft hail)	5.0		2.5
	Hail	3.0	Roughly spherical	8.0
	Giant hail	20.0	Spherical with knobs	20.0

157

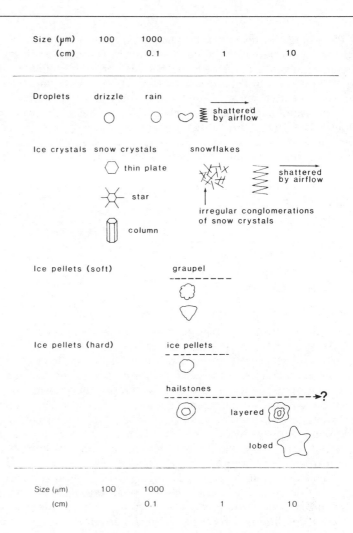

Fig. 6.16 (a) A selection of precipitation particles and drops. Note the hexagonal shape of all the ice crystals, reflecting the basic lattice structure of ice. Snow crystals grow by developing dendrites on the six corners, and snowflakes by the amorphous aggregation of crystals.

cloud bases are low, whereas larger raindrops (of the order of millimetres) require several kilometres depth of cloud above cloud base — such as occurs in many frontal and shower clouds. The largest raindrops fall from the heavy shower clouds which are endemic in equatorial regions and are common in mid-latitudes in summer.

Ice is precipitated in an even greater range of sizes and in literally innumerable shapes (Fig. 6.16). Snow crystals of much the same size as water droplets fall to the surface from stratiform clouds when the low level air is more than a few degrees below 0 °C and the air aloft is colder still, as it normally is. (If unusually a cold low layer is overlain by air above 0 °C, *freezing rain* may result: rain drops, super-cooled by falling through the cold layer, freeze on impact with the surface.) When the low-level air is close to the freezing point, the snow crystals aggregate by collision to form snow flakes which in calm conditions may be several centimetres across, though only millimetres thick.

Snow crystals and flakes are also observed to fall from shower clouds when the low-level air is cold enough and the air aloft is colder. However, the central core of the showers generally contains some hail. In mid-latitudes in winter the hail particles are often millimetre-scale cones of white, low-density ice, known as *graupel* or soft hail, the whiteness and low density arising because air is trapped between the constituent ice grains of the particles. Over land in summer in mid-latitudes larger hail may fall, often including layers of clear high-density ice alternating concentrically with layers of white ice to produce an onion-like appearance. There seems to be no clear upper limit to the sizes of such hailstones, and a few comparable in size with cricket or baseballs fall from the most vigorous showers, often having a knobbly exterior like that in Fig. 6.16.

The minimum sizes of precipitation droplets and particles are quite readily explained. As soon as one falls out of its parent cloud into subsaturated air it begins to evaporate, and it can be shown (appendix 6.4) that the sensitivity of both evaporation rate and fall speed to droplet size combine to make the distance fallen before total evaporation increase very rapidly with initial droplet size. In fact the simplified model in appendix 6.4 suggests that the distance fallen increases with the fourth power of the radius, and that a droplet must have an initial radius of at least 100 μm to survive a fall of a few hundred metres in air subsaturated by only a few per cent. Cloud droplets, being ten times smaller in radius, survive for only a few centimetres, which is consistent with the sharpness of cloud outline already mentioned, whereas raindrops can fall tens of kilometres.

It follows that drizzle-sized droplets and crystals are the smallest which can survive falling through even quite shallow sub-cloud layers. When cloud bases are higher, the drizzle droplets evaporate on the way down with the result that droplets reaching the surface are usually at least millimetric in size, corresponding to normal sizes for rain and snow. Occasionally solar backlighting of precipitation from a solitary cloud shows up the progressive evaporation of the smaller droplets or flakes with increasing distance below cloud base, the tapering veil being called *virga* (Fig. 6.17).

From the quoted sizes of precipitation droplets and particles, it is clear that all are very much larger than typical cloud droplets. In terms of radius, drizzle and raindrops are ten and one hundred times larger respectively, and in terms of volume these ratios become 10^3 and 10^6. The latter values draw attention to the

Fig. 6.17 Virga trailing from a small cumulus congestus, possibly enhanced on this winter occasion by the melting of snow behind cloud base.

enormous growth which has to be explained by any mechanism for precipitation: for example an ordinary raindrop has about one million times the water content of an ordinary cloud droplet.

6.11 Precipitation modelled

In the previous sections we have followed the development of cloud droplets from their hazy origins to the phase of rapid growth by condensation which follows activation. Once the complex Köhler curves around the critical sizes are negotiated, the growth of a droplet by condensation is controlled by the supersaturation of the ambient vapour (the equilibrium vapour density at the droplet surface being almost exactly equal to the saturation value for a plane, pure surface). As outlined in Appendix 6.2 droplet growth by diffusion and condensation in the presence of a constant supersaturation is inversely proportional to droplet radius. Although the model ignores the complications introduced by the warming of the growing droplet by release of latent heat, the predicted lengths of time for growth are known from more elaborate models to be quite realistic [41]. They show that whereas a droplet may grow to a radius of 5 μm in about 20 s in the presence of 0.2% supersaturation, it would take another 20 minutes to grow to a radius of 30 μm and several days to reach the size of a small raindrop. But simple observation shows that a shower of raindrops can be produced by a cloud only half an hour after the cloud's first appearance, and that more generally almost any cloud will precipitate if it lasts more than a few tens of minutes and is at least a few hundred metres from base to top. Clearly this familiar and vitally important behaviour cannot be explained by diffusion and condensation alone.

Note that it is quite impossible for all the activated droplets in a typical cloud to grow to the size of raindrops: this would require a million times more vapour than is available in the cloud-forming air and it would produce an utterly unrealistically catastrophic deluge. It seems that precipitation depends on the favoured growth of a small fraction of the total population of cloud droplets. In looking for precipitation mechanisms we are therefore looking for mechanisms allowing rapid, selective growth.

Let us begin by noting that a cloud must contain a spectrum of droplet sizes. During initial formation giant hygroscopic nuclei will have been activated somewhat earlier than less favoured nuclei, so that there will be a size spectrum in the population of rapidly growing droplets. Now the *terminal velocity* (the equilibrium fall speed at which weight is balanced by drag) of droplets in this size range increases with the square of the droplet radius, as outlined in appendix 6.3 and depicted in Fig. 6.10. It follows that in the developing cloud the larger droplets must be falling a good deal faster than the smaller ones, up to ten times as fast in fact. This must produce collisions, which in turn may lead to the growth of the larger, faster-falling droplets by coalescence with smaller droplets. The process is entitled growth by *collision and coalescence*, to distinguish it from the growth by diffusion and condensation considered earlier. A very important property of growth by collision and coalescence emerges if we consider the volume of air swept out by a falling collector drop of radius r and terminal velocity V (Fig. 6.18). The volume swept per second is proportional to r^2V, which in turn is proportional to r^4, given the dependence of V on r^2 already noted. Assuming that the mass of smaller droplets collected is proportional to the volume swept, then it follows that the rate of growth of the radius of the collecting drop dr/dt increases as r^2 (Appendix 6.5). Such accelerating growth contrasts with the decelerating growth produced by diffusion and condensation.

Observations in the atmosphere and very special cloud chambers, together with computer simulation of the growth of a model cloud population, seem to confirm that collision and coalescence is at work and effective in atmospheric clouds. However, the mechanism is a good deal more complex than has been suggested so far. For example, very small drops tend to flow round the would-be collector drop in the disturbed air flow, and so avoid collision. Even if they do collide, not all colliding droplets coalesce with the collector — significant numbers bounce off in some situations. There are in fact efficiencies of both collision and coalescence which are observed to vary with the sizes of both the collecting and collected droplets in quite complex ways. Observation requires sophisticated and painstaking laboratory work, but its results and theory agree in indicating that collector droplets with radii less than 20 μm are inefficient, and that droplets more than 80% and less than 20% of the radius of the collector are not efficiently collected.

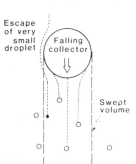

Fig. 6.18 A large cloud or precipitation droplet falling through smaller neighbours and tending to collect those within the swept path (enveloped dashed). Dotted lines are paths of small bodies relative to the falling collector.

From all this it seems that precipitation develops in two stages. First, diffusion and condensation produces a population of cloud droplets which includes some with radii exceeding 20 μm — though it is still a little difficult to envisage how the largest of these are produced in realistically short times. Subsequently these large droplets grow by collision and coalescence with their smaller neighbours, at a rate which increases with size, to produce millimetric-sized drops in a few tens of minutes. This two-stage process is believed to explain precipitation from all relatively warm clouds, i.e. those which do not contain temperatures below about $-10\ °C$.

In fact much of the troposphere is colder than $-10\ °C$, some of it very much colder. Since clouds are at best only marginally warmer than the ambient air, much cloud must be similarly cold. In low latitudes this will apply to the upper parts of tall shower clouds, but in middle and high latitudes it may apply to all but the bases of clouds, or even to the entire clouds, depending on season and geography. We must consider whether the prevalence of cold cloud has any bearing on the development of cloud and especially precipitation.

Everyday experience with water suggests that it freezes as soon as the water temperature falls to 0 $°C$, or a little below this if the water is contaminated by salt or some other impurity. However, it is observed that large amounts of atmospheric cloud persist in the form of supercooled droplets down to $-30\ °C$ or so. It seems that as in the case of equilibrium between vapour and liquid or ice, tiny volumes of water behave very differently from much larger volumes.

Much careful work in the laboratory and atmosphere shows that water freezes by a process of nucleation — the crystal lattice of ice spreading rapidly through the water volume from a very small nucleus. At about $-40\ °C$ the thermal movement of molecules in liquid water is so sluggish that clumps of them settle into the lattice formation by chance and hence freeze the entire volume. This is known as *homogeneous nucleation*, and the identity of name reflects the strong similarity with homogeneous nucleation of cloud droplets in a highly supersaturated vapour. At much higher temperatures homogeneous nucleation is so rare in the small volumes of typical cloud droplets that their freezing requires the intervention of a tiny quantity of a different substance which initiates freezing because its lattice structure is similar to that of ice. These

161

are known as *freezing nuclei* and their action is termed *heterogeneous nucleation*.

A freezing nucleus may be present from the outset within a supercooled droplet, or it may make contact with its outer surface, but in either case they become increasingly likely to be effective (i.e. initiate freezing) in any observation period as the droplet cools, until they are swamped by homogeneous nucleation. A slight practical complication is that most experimental work on ice clouds concentrates on the resultant ice crystals, even though some of these may be produced by the action of *sublimation nuclei* triggering the formation of an ice crystal directly from the vapour. Sublimation nuclei are probably much less important than freezing nuclei, but to avoid being too specific we will refer to all nuclei encouraging the production of ice crystals in clouds as *ice nuclei*.

It is conventional to ascribe the increasing likelihood of heterogeneous nucleation with decreasing temperature to an increasing concentration of ice nuclei, even though it is their effectiveness which is increasing — their numbers presumably being fixed in the absence of a local source. Although there are still discrepancies between observations it is certain that concentrations of ice nuclei are very much smaller than those of condensation nuclei, being about one per litre at $-10\,°C$.

Ice nucleation processes ensure that the ice crystal population in clouds increases with decreasing temperature. Above $-10\,°C$ there is almost no ice; between $-10\,°C$ and $-20\,°C$ there is an increasingly significant minority of ice crystals which may convert to a majority between $-20\,°C$ and $-30\,°C$; and below $-30\,°C$ most clouds are predominantly ice. No supercooled water can persist below $-40\,°C$. It follows that the concentration of ice crystals in clouds increases upwards, that clouds in the middle troposphere in middle latitudes contain both ice and supercooled water (i.e. they are *mixed clouds*), and that ice clouds predominate in the high troposphere (Fig. 6.19). This last statement corresponds to simple observation: ice cloud has a characteristically fibrous appearance (hence the term cirrus, being Latin for hair) which arises because of its slowness to evaporate in the presence of unsaturated air and which can be seen frequently in the high troposphere (Fig. 2.7). The tops of vigorously growing shower clouds maintain their sharp-edged cauliflower appearance until they reach the upper troposphere, but then their edges can be seen to soften and diffuse into a fibrous anvil as the cloud of supercooled water *glaciates* quickly and extensively in the very low temperatures there. The Norwegian Bergeron, observing the development of cumulus in clear westerly air flows reaching Norway from the Atlantic, noted in the 1930s that the clouds began to precipitate very shortly after their tops began to glaciate, and he proposed the following connecting mechanism.

It is apparent in Fig. 6.2 that air which is saturated with respect to water is considerably supersaturated with respect to ice, at least when the temperature is significantly below $0\,°C$. In fact at $-15\,°C$ the supersaturation is over 60%. If an ice crystal is produced in a cloud which consists mainly of supercooled water, it follows that the vapour will be grossly supersaturated relative to the ice crystal, which will therefore grow very much more rapidly by diffusion and sublimation than will its supercooled neighbours by diffusion and condensation — hundreds of times faster according to the ratio of the relevant supersaturations. Indeed as more ice crystals grow more and more rapidly, they may reduce the ambient vapour below saturation relative to super-cooled water and so cause wholesale transfer of water substance from droplets to crystals. This mechanism is therefore powerfully selective and has been named the *Bergeron–Findeisen* mechanism after Bergeron and the German Findeisen who pioneered the use of very large cloud chambers to study this and other cloudy processes. It is believed to be important in the mixed

Fig. 6.19 Shower clouds over the Irish Sea in winter. Since the upper parts show the fibrous appearance characteristic of ice cloud, whereas the low parts are pure water (temperatures ~ 0 °C), the middle parts are mixed ice and water cloud.

clouds present in cumulonimbus and nimbostratus, and therefore to contribute to precipitation from all cold or partly cold clouds.

Crystals growing by diffusion and sublimation preserve the hexagonal shape of the ice crystal lattice, but in different conditions produce different variations on the hexagonal theme (Fig. 6.16). Fall speeds are often considerably less than those of droplets of the same mass, with the result that it is at least possible that some fine crystals falling out of fairly shallow but cold clouds may have had time to grow by this mechanism alone. This could then explain some drizzle from cold strato-cumulus, where the crystals fall into air warmer than 0 °C before reaching the surface.

It is much more likely, however, that the Bergeron–Findeisen mechanism operates in conjunction with the equivalent of the collision and coalescence mechanism, though the latter is complicated in detail by the nature of the collector crystal. An ice crystal falling through smaller supercooled water droplets grows by *accretion*, the droplets freezing on impact with what amounts to a giant freezing nucleus to produce a rapidly growing hailstone. If the droplets are only slightly supercooled they may form a layer of water which freezes subsequently to form clear, dense ice. If the droplets are heavily supercooled they freeze instantaneously and trap the intervening air to form white, spongy, low-density graupel. Hail containing alternate concentric rings of white and clear ice has probably oscillated between higher and lower temperature before finally falling out of the parent cloud: air flow in very vigorous showers is known to be capable of allowing a substantial vertical oscillation of the growing hail (section 10.6).

If a large ice crystal is falling through smaller ice crystals it tends to grow by aggregation, the smaller crystals colliding with and sticking to the collector crystal. Not all collisions lead to aggregation however: it is observed that significant aggregation occurs only when temperatures are above −10 °C, presumably because some of the ice surfaces are coated with a slightly sticky liquid film. Aggre-gation of snow crystals (where diffusion and sublimation has produced pro-nounced hexagonal spikes — Fig. 6.16) is enhanced by interlocking of adjacent spikes or *dendrites*, producing very large snowflakes in calm conditions.

Lastly, note that because so much of the troposphere is so cold, even in equatorial regions, a great deal of precipitation begins in the form of ice crystals and develops into snow crystals and flakes. However, if these fall into air which is warmer than about 2 °C they melt and arrive at the surface as rain. In middle

163

Fig. 6.20 A rainbow segment beneath melted snow. A heavy shower of snow is melting to give rain in the lowest few hundreds of metres. The sun behind the camera produces a rainbow by internal reflection in the nearly spherical drops, but produces unstructured backscatter from the snow.

latitudes almost all rain is melted snow, even in the height of summer (Fig. 6.20). Only the larger hailstones from summer thunder showers reach the surface largely unmelted because of their much higher fall speeds. In winter the melting level for snow often fluctuates between sea level and a kilometre above, as the synoptic scale weather pattern changes, giving the variable snow cover on hills and mountains and the occasional blanketing of lowlands so typical of maritime climates such as Washington State's and Britain's.

6.12 Atmospheric electricity

Showers are often associated with thunder and lightning, and very heavy showers always are. Indeed the very heaviest showers may contain nearly continuous thunder and lightning which, together with their torrential rain, large hail and damaging gusts of wind, earns them the technical title *severe local storms*. Thunder and lightning also occur in some fronts, especially vigorous cold fronts, and in rather special weather systems (called *squall lines* in the USA) intermediate in character between fronts and lines of severe storms. All tropical cyclones produce vigorous thunder and lightning. It seems that at any instant there are several thousand thunderstorms active in the troposphere, concentrated in low latitudes and over daylit land.

The loudness of thunder and the vividness of lightning, often accentuated by the brooding darkness of the clouds in which they occur, have impressed man from his earliest times and teased his powers of explanation. Gods like Zeus and Thor have been invoked, and the latter has left his mark in the name thunder and in the use of the term *anvil* to describe the shape of the glaciated upper parts of a mature cumulonimbus — thunder being presumably the sound of Thor beating on his gigantic anvil. Lucretius in 55 BC supposed that thunder was the noise of great clouds crashing together, which at least recognizes the fact that large clouds are always involved. In AD 1752 Benjamin Franklin and d'Alibard proved inde-

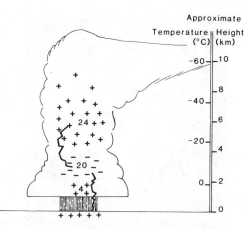

Fig. 6.21 Typical distribution of electrical charge within an active thunderstorm in middle latitudes, as deduced by remote observation of changes in electrical field strength accompanying lightning strokes. Quantities of separated charge are given in coulombs. (See [8] and [41])

pendently that at least part of the activity of thunderstorms was electrical in nature, by the potentially lethal procedure of flying kites nearby on conducting tethers. Franklin in particular was able to determine that the electrical charges involved were sometimes positive and sometimes negative. In the 1920s C. T. R. Wilson established the typical distribution of charge within a thunderstorm (Fig. 6.21) by analysing measurements of electrical field strength made on the ground at a range of distances from the storm. In the next decade this work was confirmed and amplified by G. Simpson and others who flew field strength meters into thunderstorms on balloons. Meanwhile very fast-response cameras had been used to examine the complex structure of lightning — the giant sparks jumping within clouds, and from cloud to ground.

The typical charge distribution shown in Fig. 6.21 implies that positive and negative charges are being separated by the thundercloud. Lightning within the cloud temporarily neutralizes the main separations of charge by shorting them together by a conducting channel, but the observations show that reseparation requires only 20 s or so. In recent years much painstaking work in the laboratory and by instrumented aircraft in and near thunderstorms has been directed at finding out what the charge-separating mechanisms are. Results so far are inconclusive but suggest that several distinct mechanisms may be important. Their common feature is that small particles or droplets tend to become positively charged while larger particles become negatively charged. The greater fall speeds of the larger particles (many of them precipitation elements) enable gravity to drive the required charge separation. For example, according to one proposed mechanism, a small cold ice crystal collides with a hailstone warmed considerably by release of latent heat from the freezing of impacting supercooled droplets. During their brief contact the very mobile positive ions migrate toward the colder end of the crystal (i.e. away from the hailstone) leaving the warmer end and the hailstone negatively charged (Fig. 6.23). On separation the stone carries its negative charge downwards and the crystal its positive charge upwards. The charge separation at each collision is very small, but there are so many collisions in the cubic kilometres of active precipitating cloud that the net result is at least potentially consistent with the observed charging rate.

Lightning is a giant spark produced when the electrical insulation of the air, cloud and precipitation mixture breaks down. This is believed to occur when the electric field strength reaches about 1 MV m^{-1}. From basic electrostatics the field strength at the edge of a 2 km radius sphere containing 24 C of electrical charge is

Fig. 6.22 Lightning in and around a large cumulonimbus at night. Several unrelated strokes have occurred during the opening of the camera aperture. Those within the cloud are seen indistinctly, while the directly visible external strokes are seen to have the familiar jagged forked structure.

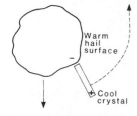

Fig. 6.23 Charge separation during collision between an ice crystal and a warmer hailstone.

166

only about one tenth of this critical value, and it is supposed that the critical value is exceeded because of some local concentration of charge. Once the spark has begun it is able to propagate quickly across regions where the electric field is considerably weaker. For obvious reasons cloud-to-ground lightning has been studied much more than lightning within cloud, and it seems likely that many of these ground strikes begin in the strong field between the small positive charge just below the melting level and the large negative charge higher up. Once begun a so-called *stepped leader* makes its way from the negative charge toward the surface, its jagged steps outlining the familiar forked lightning shape. As it nears the surface the electric fields there, which may already be high enough to support a flickering *corona discharge* around sharp points such as the twigs of trees, cause a spark to jump to meet the stepped leader. The main surge of free electrons toward the earth then occurs in the established ionized channel, but in such a way that the surge propagates upward in a *return stroke* which is the main part of the lightning flash. The current surge in a channel of a few centimetres across may be about 10 000 A, and the resultant heating of the air produces the incandescent flash and the explosive expansion whose shock waves are heard as thunder. As the negatively charged region of the cloud may be only partly discharged by this process, there are often several return strokes at intervals of about one twentieth of a second, each preceded by a downward *dart leader* reactivation of the ionized channel. The succession of several brilliant return strokes is just perceptible to the eye as the flicker of a clearly visible lightning flash. Discharge being now complete, the recharging processes get under way again, preparing for another flash a few tens of seconds later, which occurs in a fresh location, since the ionized channel will have long since dispersed. Storms producing much more frequent flashes to ground must have several separate charging zones at work simultaneously.

In everyday speech it is customary to distinguish between fork lightning and sheet lightning. Sheet lightning may be the loom of a distant fork lightning when seen through an intervening cloud, or it may be a lightning flash of the main type — between the positive and negative charges accumulating within the cloud. The latter appears to differ somewhat from the cloud-to-ground discharges discussed already, for example being longer lasting and duller, although their location makes detailed comparison difficult.

Thunder is the noise of the shock waves of sound rippling out from the incandescent lightning flash. Since this spreads at the speed of sound compared with the effectively instantaneous transmission of the visible flash, a distant flash produces audible thunder after a time delay which is directly proportional to the distance, the delay being about 3 seconds per kilometre. Thunder is seldom audible more than about 10 km away, but lightning is clearly visible for many tens of kilometres at night if there are no intervening clouds. Such silent flashes, indistinct because of great range, are sometimes called sheet lightning, and sometimes *heat-lightning*, because they are often seen on hot summer nights in middle latitudes.

The prolonged booming peals of thunder are rarely caused, as popularly supposed, by echoes from clouds or mountains. Lightning flashes may be several kilometres in length so that a ground strike near the observer may rumble for 10 s or more as the noise from the more distant parts of the flash follows after the hissing, tearing noise of the nearest part of the flash. And the forked path may produce booming concentrations in the long rumble.

Thunderstorms play an important role in the atmospheric electrical cycle. The cloud-to-surface flashes are continually pumping electrons earthward at a rate of several thousands of amperes world-wide, and maintaining the high troposphere and higher atmosphere at an electrical potential of about half a million volts above earth potential. (Individual storms produce potentials ten or more times larger than this in their upper parts, but much of this is wasted by cloud-to-cloud lightning and other leakages.) This maintains the *fair-weather* electric field as depicted in Fig. 6.24, the field strength being a maximum near the surface where it is about 100 V m^{-1}. In this field the small concentration of ions produced by cosmic radiation and natural radioactivity in the surface and atmosphere maintains a gentle, persistent current which world-wide balances the concentrated spasmodic currents driven by thunderstorms.

The dissipations of energy in lightning flashes are very spectacular because they are brief and highly localized. When averaged throughout the volume and lifetime of a thunderstorm they represent only a small fraction of the storm's total energy budget, the largest components of which are the buoyant potential energy of the air which is about to rise and fall, and the kinetic energy of the rising and falling air (Chapter 10). In effect thunder and lightning are spectacular sideshows in the wings of the main drama, in which millions of tons of air, water and ice are in great commotion. The noises which Lucretius thought were cloudy collisions are more

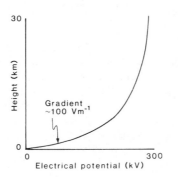

Fig. 6.24 Vertical profile of electrical potential in fair weather. The profile gradient is the strength of the vertical component of the electrical field.

properly the relatively slight sounds of clouds tearing themselves apart under gravity and buoyancy.

6.13 Practical applications

CLOUD SEEDING

Speculation about the importance of the Bergeron–Findeisen process in initiating precipitation led to extensive work in artificial stimulation of precipitation beginning in the 1940s and continuing to the present day. The ability of ice nuclei to trigger precipitation from suitable clouds and the belief that such nuclei might be in short supply in some conditions led people to believe that the artificial introduction of ice nuclei might enhance the production of precipitation. This is clearly potentially valuable to agricultural production in marginally arid areas. Initially some early spectacular claims were made for the effectiveness of such so-called *cloud seeding*, but although it can occasionally change the appearance of clouds quite dramatically, years of exhaustive investigation have shown that the effects are quite slight, probably not increasing precipitation more than 20% above the natural level. Because precipitation is naturally so variable, the detection of such small increases requires prolonged experimental study and careful statistical analysis. Nevertheless such increases could be economically valuable in some circumstances, and work has continued in some countries for several decades now, though sometimes sustained more by hope than objective results. Silver iodide has been used quite widely in this work, because laboratory studies have shown it to be one of the most efficient freezing nuclei, beginning to be effective at $-4\ °C$. Its crystal structure is very similar to that of ice. Myriads of tiny crystals of AgI are produced from small burners which can be flown through clouds or sited on the ground to feed updraughts rising into clouds aloft. Note that seeding must be carefully controlled. If a cloud is swamped by freezing nuclei the effect may well be to produce larger numbers of smaller ice crystals than would have been produced without seeding, which is the opposite of the selective mechanisms which underlie all precipitation.

AIRCRAFT ICING

As soon as aircraft began to fly in supercooled water clouds, the icing-up of exposed surfaces, especially the leading edges of wings, began to be a serious and potentially fatal hazard. Surfaces at ambient sub-freezing temperatures act like giant hailstones and collect masses of ice and freezing rain by the impact-freezing of supercooled cloud droplets (Fig. 6.25). In the worst conditions these can accumulate so quickly that their weight becomes a serious problem and their distortion of the aerofoil shape seriously reduces the lifting capacity of the wing. Many aircraft have been lost in this way and others have survived only by the last-minute melting and loosening of the burden of ice as they fell out of control into the warm low troposphere. Although flying lower or higher (where the collection efficiency of the more numerous ice crystals is so low as to be no problem) does reduce the problem, almost all modern aircraft have systems for dealing with icing, involving heating or even flexing the leading edges of wings and other crucial parts.

Fig. 6.25 Underside port wing of Concorde showing ice (pale on the deliberately blackened wing) accumulating during flight trials to determine heating rates needed to prevent icing in service.

RADAR

Radars emitting wavelengths of about 10 cm receive strong echoes from precipitation particles and droplets, which can be used to measure precipitation over wide areas as mentioned in section 2.6. Radar has also been used to investigate the invisible and fairly inaccessible interiors of cumulonimbus (shower clouds), nimbostratus (precipitating frontal clouds), and the active inner zones of severe tropical cyclones (Fig. 12.12). Indeed some of the major features of cumulonimbus outlined in sections 10.4 and 10.6 have been clarified very considerably with the help of radar, including types which scan vertically to provide a vertical section of echo (Fig. 6.26) rather than the more common plan view. Many large

169

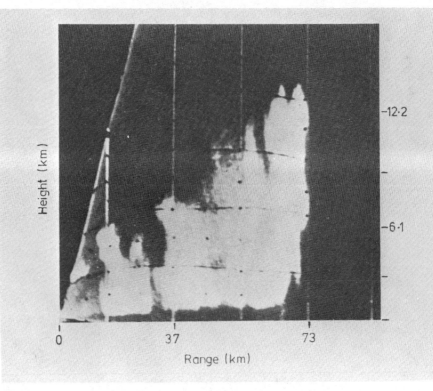

Fig. 6.26 Radar echo from a large cumulonimbus, showing vertical shafts of heavy precipitation. The vertically scanning radar has been programmed to produce a display of echo strength (brightness) against range and height (RHI display).

civil aircraft now carry small radars which enable them to detect and avoid regions of heavy precipitation, mainly because they can be unpleasantly and even dangerously turbulent because of the associated strong updraughts and downdraughts. Nimbostratus often maintains a horizontal band of strongly enhanced echo where snowflakes become coated with water as they melt, reflecting radar waves much more efficiently than either the dry snow and ice above or the rain beneath. Such *melting bands* (Fig. 6.27) confirm that most surface rain in middle and high latitudes is formed by the melting of snow falling from higher levels.

LIGHTNING CONDUCTORS

In what has become a classic example of the connection between pure research and useful application, Benjamin Franklin followed his confirmation of the electrical nature of lightning by proposing that buildings could be protected from lightning damage by connecting a stout electrical conductor from just above the highest part to a grounded metal stake. This he supposed would offer the lightning an easy route to earth — so easy that little damaging heat would be generated *en route*. (Buildings and trees are damaged by the heat generated as the current surges through their substantial natural resistance, explosive boiling of interior moisture or sap being particularly destructive.) His suppositions were amply confirmed by those who followed his advice, and damage to church spires and other very exposed

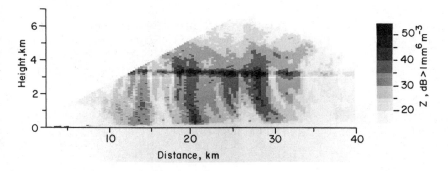

(a) Single-polarization radar data

Fig. 6.27 Radar echo strength from a section of frontal cloud over southern England, showing the horizontal melting or 'bright' band of enhanced echo at altitude 3.2 km separating the snow above from the rain beneath. The upright stripes of stronger echo neural small scale cells of heavier rain which are tilted by the strong wind shear in the lowest 1.5 km.

buildings was dramatically reduced after protection. However, there was resistance from traditionalists and cynics, and many decades were to pass before the lightning conductor became universally accepted, as now.

Sailing ships had long suffered damage and injury to crew as a result of lightning strikes to masts, spars and rigging. Injuries and deaths occurred because crew were often aloft at the time of the lightning strike, adjusting the sails in response to the squalls and shifts of wind associated with thunderstorms. The British Royal Navy then maintained a large number of sailing ships in all seas and weather and suffered from lightning accordingly. However, the Admiralty did not take kindly to advice from an outsider, especially after the American revolution cast Franklin in the role of rebel statesman, and so British sailors continued dying in significant numbers. Eventually as a compromise their Lordships fitted temporary lightning conductors, to be placed aloft when danger threatened, but otherwise to be coiled away below. After a period in which more British sailors were killed hauling lightning conductors aloft, permanent copper strips were attached to the masts and damage and injury was reduced far below their previous levels.

Every year a few thousand people world-wide are killed or injured through inadvertently becoming lightning conductors. Local news reports often note that the unfortunate person was struck on a metal cap badge or something similar, and this is indeed often the case. However, it is probably quite wrong to suppose that the metal object contributed to the accident or that its removal might have prevented it. As the stepped leader reaches down to within about 100 m of the surface the electric fields there increase dramatically, especially in the vicinity of any fairly sharp upward projection from the surface. An upright human form will serve and so localize the upward spark and the subsequent return stroke. If there is a sharp protruding badge or golf club no doubt this will be selected, but the result is the same. The only sensible precaution to take if you are caught in the open by a thunderstorm is to get down on the ground and stay there until the lightning strikes are at least a mile off. Standing under a solitary tree is no protection since a strike on the tree will easily produce lethal fields and currents within many metres of the base of the trunk, and the tree is of course a prime target. Nor is it sensible to console yourself that only a few people are struck by lightning; only a few people are caught in the open by thunderstorms, but a significant fraction of them are struck.

Appendix 6.1 Dew-point lapse rate

According to more advanced texts [8] the variation of saturation vapour pressure e_s with absolute temperature T is given by the *Clausius–Clapeyron* equation

$$\frac{de_s}{dT} = \frac{L}{T(\alpha_2 - \alpha_1)}$$

where α_2 and α_1 are respectively the specific volumes of vapour and water. Since α_2 is very much larger than α_1, the latter can be ignored. Replacing vapour density using the equation of state for water vapour, we find

$$\frac{de_s}{dT} = \frac{L\,e_s}{R_v T^2}$$

where R_v is the specific gas constant for water vapour. A solution of this equation gives a very close approximation to Fig. 6.2. Since the vapour pressure e of an unsaturated parcel is by definition equal to the saturated vapour pressure at the dew-point T_d of the parcel, we can reinterpret this equation as a relationship between vapour pressure and dew point:

$$\frac{de}{dT_d} = \frac{L\,e}{R_v T_d^2}$$

In the particular case where specific humidity is conserved, the vapour pressures and total atmospheric pressures are directly proportional

$$e = \frac{qp}{\epsilon} \tag{eqn (5.8)}$$

so that vapour pressure may be replaced by atmospheric pressure to produce an identical relationship between atmospheric pressure and dew point. Assuming hydrostatic equilibrium (eqn (4.6)), this becomes a relationship between dew point and height, which after inversion and rearrangement using the equation of state for air (eqn (4.2)) simplifies to

$$\frac{dT_d}{dz} = -\frac{g}{L}\frac{R_v}{R}\frac{T_d^2}{T}$$

Since dew point and temperature differ only slightly on the absolute scale, this simplifies further to

$$\frac{dT_d}{dz} = -\frac{g}{L}\frac{R_v}{R}\,T_d$$

At a typical low tropospheric temperature of 283 K the right-hand side has magnitude $1.8 \times 10^{-3}\,\text{K m}^{-1}$, which means that dew point lapses by 0.18 K per 100 m rise through the well-mixed layer, as can be checked from the crossing of isotherms and q_s isopleths on a tephigram. Since ordinary temperature lapses by 0.98 K per 100 m (the dry adiabatic lapse rate), we see that dew point and temperature converge at a rate of about 0.8 K per 100 m in a typical well-mixed layer in the low troposphere — which in other words is the lapse rate of the dew-point depression.

Appendix 6.2 Droplet growth by diffusion and condensation

The mass flux F of vapour diffusing toward a spherical droplet is given by Fick's law:

$$F = 4\pi n^2 D \frac{d\rho}{dn}$$

where D is the diffusion coefficient for water vapour in air and $d\rho/dn$ is the radial density gradient of vapour density ρ at radius n from the centre of the droplet (Fig. 6.28). In a steady state, F must be independent of n and so we can integrate the diffusion equation to find a relation between F and the vapour density difference $\triangle\rho$ between the droplet surface ($n = r$) and the distant environment ($n = \infty$).

$$F \int_r^\infty \frac{dn}{n^2} = 4\pi D \int_\rho^{\rho + \triangle\rho} d\rho$$

which gives

$$F = 4\pi r D \triangle\rho$$

As expected we see that F is positive if $\triangle\rho$ is also (i.e. the flux is toward the droplet if the ambient vapour density exceeds the vapour density at the droplet surface), and also that $F \propto \triangle\rho$.

This vapour flux is condensed at the droplet surface and adds to the mass of the droplet.

$$F = \rho_w \frac{d}{dt}\left(\frac{4}{3}\pi r^3\right) = 4\pi\rho_w r^2 \frac{dr}{dt}$$

where ρ_w is the density of the droplet water (or ice if it is a crystal). Equating the expressions for F we find

$$r \frac{dr}{dt} = \frac{D\triangle\rho}{\rho_w} \tag{6.5}$$

Notice that if the vapour density difference is constant, the rate of growth of droplet radius is inversely proportional to the radius, with the result that the droplet grows quickly at first and then more and more slowly.

We can integrate eqn (6.5) to find the time taken by the droplet to grow from one radius to another. First let us interpret the vapour density difference $\triangle\rho$.

$$\begin{aligned}\triangle\rho &= \rho_a - \rho_s \\ &= \frac{\rho_s}{100}\left(\frac{100\rho_a}{\rho_s} - 100\right) = \frac{\rho_s}{100} SS\end{aligned} \tag{6.6}$$

where SS is the supersaturation ($RH - 100$) of the ambient air. Note that the vapour density at the droplet surface has been assumed to be the saturation value ρ_s at the common temperature of droplet and environment. This ignores both the difference between equilibrium and saturation vapour densities discussed in section 6.7 (which can be important when r is small), and the temperature excess maintained by continual release of latent heat at the droplet surface. However, neither of these

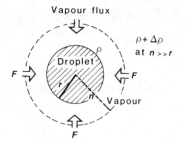

Fig. 6.28 Vapour diffusing radially inwards towards the surface of a spherical droplet in response to greater environmental vapour density.

effects has a substantial effect on the following results. Using eqn (6.6) in (6.5) we find

$$r \frac{dr}{dt} = \frac{SS}{X} \tag{6.7}$$

where $X = 100\rho_w/(D\rho_s)$, which has value 4.5×10^{11} s m^{-2} when saturation vapour density is 10 g m^{-3}, corresponding to a temperature of 11 °C.

Integrate eqn (6.7) to find the time t to grow from radius r_1 to r_2.

$$t = \frac{X}{2SS} (r_2{}^2 - r_1{}^2)$$

Starting from radius 0.5 μm, the times taken to grow to various larger sizes are as follows at 11 °C and SS 0.2%:

r	1	2	4	8	16	32 μm
t	0.9	4.3	18	72	291	1163 s
				(1.2	4.8	19.4 min)

The unreality of assuming saturation over the surface of a small droplet clearly cannot alter the conclusion that it takes a long time for a droplet to grow from a small to a large cloud droplet. These results apply equally to the shrinkage of a droplet by evaporation: SS is then the subsaturation $(100 - RH)$ and the table is read from right to left.

Appendix 6.3 Terminal velocities of falling drops [41]

When a spherical body of radius r and density ρ_w is immersed in air of density ρ, the net downward force on it is its weight less the weight of the displaced air (by Archimedes' principle)

$$N = \frac{4}{3} \pi r^3 g (\rho_w - \rho)$$

Since the density of water or ice is $\sim 10^3$ times the density of air, the latter may be neglected in the bracketed term. As the body accelerates downward under this force, the air resistance to its fall increases until downward weight and upward drag balance at a steady fall speed which is known as its *terminal velocity* or fall speed. Since the weight of the body is proportional to its volume, while the drag is proportional to its surface area, the terminal velocity increases with the size (radius, for example) of the body, though not necessarily in direct proportion.

When the body is very small, the terminal velocity V is so slow that the Reynolds number Vr/ν of the flow round the falling body is less than 0.25. The flow is then completely dominated by the viscosity of the air, and theory and observation agree that the drag X is proportional to the product Vr, being given by

$$X = 6\pi\rho\nu Vr$$

At terminal velocity the values of N and X must be equal, and it follows that

$$V = \left[\frac{2 \, g\rho_w}{9 \, \nu\rho} \right] r^2$$

This is the Stokes regime of fall speeds (Fig. 6.10), and it is a matter of arithmetic to show that it applies to spheres of water or ice up to radii of about 30 μm, which includes all cloud and aerosol, with modification for non-spherical shapes. The proportionality to the square of the radius means that fall speeds increase rapidly with size. In conditions typical of the low troposphere, the bracketed terms have value 1.2×10^8 s^{-1} m^{-1}, which is consistent with Fig. 6.10. Calling this Y we have simply

$$V = Yr^2 \tag{6.8}$$

For droplets with radii between about 30 and 1000 μm (i.e. large cloud to raindrop sizes), the flow around the falling drop is turbulent. The drag increases more sharply with droplet size than in the Stokes regime, being proportional to the product Vr^2. It follows that the fall speed is almost exactly in direct proportion to droplet radius. In the low troposphere in SI units

$$V = 8.0 \times 10^3 r \tag{6.9}$$

which is the middle section of Fig. 6.10.

For larger bodies (large raindrops and virtually all hail) the flow is so strongly turbulent that the drag is proportional to the product V^2r^2, which gives a terminal fall speed proportional to the square root of the radius. In SI units we have

$$V = 250r^{1/2}$$

The turbulence is so great that raindrops shatter soon after reaching radii of about 2 mm. No such limit applies to hailstones, which as a result grow to diameters of as much as 10 cm on rare occasions.

Appendix 6.4 Evaporation of falling drops

Consider a droplet falling at speed V. In a little time interval dt it will fall a short distance dz (inverting our height coordinate for this problem) given by

$$\mathrm{d}z = V \, \mathrm{d}t$$

If the droplet is evaporating, its radius r is decreasing with time at a rate given by dr/dt. In fact the time interval dt can be regarded as the time taken by the droplet to change its radius by dr, where

$$dt = \frac{dr}{(dr/dt)} = \left(\frac{dt}{dr}\right) dr$$

Hence
$$dz = V \frac{dt}{dr} dr \qquad (6.10)$$

and
$$z = \int_r^0 V \frac{dt}{dr} dr = -\int_0^r V \frac{dt}{dr} dr$$

Note that the minus sign ensures that a positive downward distance is fallen when the drop is shrinking (i.e. dr/dt is negative).

In the case of droplets less than about 30 μm in radius, we already have expressions in Appendices 6.2 and 6.3 for V and dr/dt in terms of r, and these can be used to evaluate the integral and find the distance z fallen during the total evaporation of a droplet of any initial radius in a subsaturated environment. We use eqn (6.7), with SS interpreted as the degree of subsaturation $(100 - RH)$ of the environment of the falling, evaporating drop, together with eqn (6.8) to find

$$z = \int_0^r Yr^2 \frac{X r}{SS} dr$$

If SS is uniform along the path of the falling droplet, standard integration gives

$$Z = \frac{X Y r^4}{4 SS}$$

Using the quoted realistic values for X and Y, and assuming an ambient subsaturation of 5 (RH 95%), we can find the distance fallen during the time taken to evaporate completely from chosen initial radii

r	1	10	100	1000 μm
z	2.7 μm	2.7 cm	270 m	2700 km
			{240 m	240 km}

These values exemplify the great sensitivity of z to r which arises from the r^4 dependence, and show clearly that cloud droplets ($r < 30$ μm) cannot survive a fall of more than a few centimetres through even mildly subsaturated air. The z values for larger droplets are too large because these droplets are outside the size range in which momentum and water vapour transfer is caused entirely by the molecular diffusion assumed above, and actual magnitudes of V and dr/dt are respectively reduced and enhanced by turbulence around the droplet. A thorough treatment is complicated, because it involves not only the turbulently enhanced drag and evaporation, but also the wet-bulb chilling of the rapidly evaporating drop below the ambient air temperature. The bracketed z values above follow from combining the $V \propto r$ fall-speed regime, which includes the effects of turbulence on drag (eqn (6.9)), with the above purely diffusive expression for dr/dt. Though the approach is internally inconsistent, and on the simplest view should overestimate z by underestimating dr/dt, it produces values of at least the right order of magnitude in comparison with more thorough studies [42], presumably because evaporation rates are depressed by wet-bulb chilling. At any rate they confirm that drizzle can fall a few hundred metres in moist air, and that rain could fall further than is ever available in the limited depth of the troposphere.

Appendix 6.5 Growth by collision and coalescence

Consider a collector droplet of radius r, falling at speed V and therefore sweeping volume at the rate $\pi r^2 V$ (Fig. 6.18). If there is a population of much smaller droplets with specific mass m, their mass per unit volume is $m\rho$, where ρ is the aggregate density of the air plus cloud. If the small droplets have negligible fall speed and all in the swept volume are collected, then the rate of collection of mass is

$$F = \pi r^2 \, Vm\rho$$

As in appendix 6.2 this produces a rate of increase of collector mass

$$F = \rho_w \frac{d}{dt}\left(\frac{4}{3} \pi r^3 \right)$$

where ρ_w is the density of the collected water or ice. Equating the expressions for F, we find

$$\frac{dr}{dt} = \frac{m \rho}{4\rho_w} V \tag{6.11}$$

Up to 30 μm radius we have (appendix 6.3) $V = Yr^2$, so that

$$\frac{dr}{dt} = \frac{1}{A} r^2 \tag{6.12}$$

where $A = 4\rho_w/(m\rho Y) = 3.3 \times 10^{-2}$ m s if m is 1 g kg^{-1}, and other values are as in appendix 6.3. The time taken for the collector to grow from radius r_1 to r_2 is given by the integral of eqn (6.12)

$$t = A \left(\frac{1}{r_1} - \frac{1}{r_2}\right)$$

With m $= 1$ g kg^{-1}, it takes nearly 14 minutes for the collector to grow from 20 to 40 μm in radius, and this time halves as m doubles and so on. Growth by collision and coalescence can therefore be faster than growth by diffusion and condensation in this important size regime if there is a good supply of droplets suitable for efficient collection (neither too small nor too large).

Collector droplets with radii between 30 and 1000 μm have fall speeds given by eqn (6.9). When substituted in eqn (6.11) it gives

$$\frac{dr}{dt} = \frac{1}{B} r$$

where $B = 4\rho_w/(m\rho 8 \times 10^3) = 500$ s for $m = 1$ g kg^{-1} and previously quoted values. The time taken for the collector to grow from radius r_1 to r_2 is now given by

$$t = B \ln (r_2/r_1)$$

With $m = 1$ g kg^{-1}, growth from 30 to 300 μm (a crucial transition from large cloud to small precipitation) takes 19 minutes, and this time halves as m doubles and so on. Clearly growth by collision and coalescence has the capacity to produce precipitation in the observed short time intervals if conditions are favourable.

Problems

1. Which of the following has no direct effect on the density of a vapour in equilibrium with a water surface: water purity, air pressure, air purity, water temperature, air temperature, wind speed, water surface shape, air?

2. Define a saturated vapour.

3. A friend gazes at rain beating on the window and exclaims in exasperation that if only there could be less rain there would be more sunshine. Gently explain his inconsistency.

4. Rain leaves a hilly road equally wet at two locations separated by 500 m in altitude. If the ensuing sunshine and wind plays equally on each, give two reasons why you would expect the lower level to dry faster.

5. List the deductions about cloud formation which can be drawn from the simple cloud-chamber experiments described in the text.

6. Given the large number densities of aerosol, why are the number densities of cloud droplets so comparatively small?

7. Explain why, when the upper parts of a mixed (ice and water) cloud are dissolving, the last vestiges are always ice cloud.

8. Although rain drops with radii larger than about 2 mm are not observed (appendix 6.3), much larger drops are observed beneath trees in wet weather. Explain.

9. It is a common fallacy to suppose that sheet lightning is not associated with thunder. Explain what actually happens.

10. Figure 6.27 shows a melting band on a range–height radar display. Consider its appearance on the more normal plan–position display (like Fig. 2.6), in which the radar beam continually scans around at a small angle above the horizontal. Remember the curvature of the Earth's surface.

11. Use Fig. 6.2 to estimate vapour densities and pressures at the following surfaces: an ice cap in summer (-10 °C) and the equatorial ocean surface (25 °C). Why might a similar estimate for the Sahara be very misleading?

12. In a certain subtropical desert, dew is observed to form each night when the surface temperature falls to 5 °C. Assuming constant atmospheric pressure and vapour content, estimate the relative humidity when the surface temperature reaches 50 °C during the heat of the day, using Fig. 6.2 and given that the s.v.p. at 50 °C is 123 mbar. Assuming that the air temperature at screen level is then 45 °C and that this is at the base of the well-mixed layer, use the simple rule of thumb for lapse of dew-point depression to find the implied height of cloud base above the surface.

13. Show from the approximate form of the relationship between equilibrium and saturation vapour densities (eqn (6.4)) that on any particular Köhler curve the radius r at which these densities are equal is given by

$$r^2 = \frac{BmT}{A}$$

Given that $A = 3.16 \times 10^{-7}$ m K for water at 0 °C, and that for NaCl $B = 1.47 \times 10^{-4}$ m³ kg⁻¹ K⁻¹, find r values corresponding to salt particle masses of 10^{-18} and 10^{-16} kg.

14. Use the final equation of appendix 6.2 to calculate the time taken for a cloud droplet of radius 5 μm to evaporate completely in an environment with ambient relative humidity 90% when the temperature of droplet and environment is 11 °C. Repeat the process at −20 °C, given that the saturation vapour density is then 1.07×10^{-3} kg m⁻³, and note the relevance of the result to the persistence of ice cloud in the upper troposphere.

15. Find the time taken for a droplet to grow from radius 15 to 500 μm by collision and coalescence in the low troposphere with cloud water of specific mass 3 g kg⁻¹. Allow for the change in terminal velocity regime by working out in two stages: the first to cover growth to 30 μm and the second to cover further growth. How much longer is needed to grow to radius 1 mm?

16. Lightning strikes the surface 200 m away from an observer. Given that the flash extends to a maximum height 2 km vertically above the observer, and has a nearly horizontal section 1 km above the observer, describe the observed peal of thunder, given that the speed of sound in air is 330 m s⁻¹.

LEVEL 3

17. Consider the realistic development of a layer of dew 1 mm thick in 1 hour to find the implied flux of condensing water in molecules per second per unit area. Compare this with the number of impacts (X) of molecules of water vapour calculated from the following expression derived from the kinetic theory of gases [35].

$$X = \frac{10^3 A e}{2\sqrt{3} RTM}$$

where A is Avogadro's number, R is the universal gas constant, e is the vapour pressure, M and T are respectively the molecular weight and absolute temperature of the vapour, and the units of X are again molecules per second per unit area. Notice that the result shows that the deviation from perfect dynamic equilibrium is very small.

18. By analyzing the shape of the Köhler curve described by eqn (6.4), show that radius r_m at which the equilibrium vapour density is a maximum is $\sqrt{3}$ times the radius r at which the equilibrium vapour density is equal to the saturation value (i.e. $\sqrt{3}$ times the radius found in Question 13). Show further that the maximum supersaturation SS_m implied is given by $200A/(3\sqrt{3}rT)$, where terms are as defined in problem 13. Calculate numerical values for r_m and SS_m for the salt particles cited in problem 13.

19. Discuss the role of equilibrium and runaway diffusive growth and evaporation in ensuring that clouds have both sharp bases and edges.

20. Consider the growth by collision and coalescence of a droplet falling through cloud with cloud water specific mass m, and show that droplet growth dr increases in proportion to distance fallen dz (assuming no updraught in the cloud) according to

$$dr = \frac{m \rho}{4\rho_w} dz$$

where ρ and ρ_w are respectively the density of air and water. Hence find the depth of very wet cloud ($m = 4$ g kg^{-1}, typical of warm, low clouds over hills in wet weather) which could produce an increase of 1 mm in droplet radius. Such rapid growth partly accounts for the familiar increase of precipitation over hills.

Atmospheric dynamics

7.1 The equation of motion

When a force F acts on a body of mass M, then according to Newton's second law of motion, which is the law underpinning all dynamics, we have

$$F = \frac{d}{dt}(MV) \tag{7.1}$$

where V is the velocity of the body and MV is called its *momentum*. The right-hand side of eqn (7.1) represents the rate of increase of the body's momentum with time as a result of the action of the force. This rate of change can arise because of changing mass or changing velocity or both, but in the common case of constant mass we have

$$F = M\frac{dV}{dt}$$

or

$$a = F/M \tag{7.2}$$

where $a\,(=dV/dt)$ is by definition the *acceleration* of the body. The form of Newton's second law of motion apparent in eqn (7.2) is used so widely in meteorology that it is known simply as the *equation of motion*, but it is important to remember that it assumes constancy of air parcel mass and is therefore inapplicable to the dynamics of cumulus, for example, where air parcels grow rapidly by entrainment of ambient air (sections 7.15 and 11.2).

The vector notation of eqn (7.2) allows for the fact that both acceleration and force have magnitude as well as direction, and the simple form of the equation means that their directions are the same: acceleration takes place in the direction of the acting force. Being a vector equation it can be broken down into three scalar (i.e. having magnitude only) equations in mutually perpendicular directions or *axes*, and it is these *component* equations which are often most convenient for examination and solution. The convention in meteorology is to use the axes shown in Fig. 7.1, with origin O at a convenient point on the Earth's surface, and with the x, y and z axes pointing horizontally eastward and northward and vertically

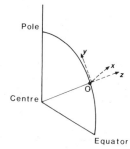

Fig. 7.1 Northern hemisphere quadrant showing conventional meteorological axes with origin O on the surface.

upward respectively. The wind speeds in these respective directions are written as u, v and w by the same convention. Since the axes are at right angles to each other, the component equations can be considered quite independently, although of course all three may be needed for any particular problem. For example, the vertical component of the equation of motion is as follows, regardless of the horizontal components

$$\frac{\mathrm{d}w}{\mathrm{d}t} = F_z/M \tag{7.3}$$

where F_z is the vertical component of the net force or forces acting on the parcel, and $\mathrm{d}w/\mathrm{d}t$ is the parcel's vertical acceleration. Notice that the right-hand side of the equation is the force per unit mass of air parcel, as is always the case in the equation of motion. If the force component is known, the acceleration is determined, and successive integrations give its vertical speed and displacement. If, however, the three-dimensional parcel motion is required, then the horizontal components are needed also.

Although the notation and equations may seem rather abstract and forbidding, it is important to realize that they are used precisely because they provide concise and unambiguous descriptions of events, as well as the hope of manipulation and solution. Solutions are minimal in an introductory treatment such as this, but statements and manipulations involving mathematics are very useful nevertheless. Even in the extensive equations which abound in advanced texts, a vital first stage in any study of dynamics is to identify the physical forces and other factors represented by the various symbols, and to consider the physical reality of the balance represented by the equations combining them. When examined carefully in this way, even the relatively simple equations quoted in this chapter will be seen to summarize relationships and behaviour which would each require many paragraphs of purely verbal description.

7.2 Forces

If the gradient of air pressure at a given location is steepest along a certain axis, then an air parcel there is subjected to a force along that axis toward low pressure, because the pressure on the up-gradient side of the parcel is greater than on the opposite side. In appendix 7.2 it is shown that the force per unit mass of air parcel is $-(1/\rho)/\partial p/\partial n$, where $\partial p/\partial n$ is the instantaneous pressure gradient in the direction of increasing n (distance along axis), ρ is the density of the air in the parcel, and the minus sign indicates that the force acts in the direction of decreasing p. The partial derivative $(\partial/\partial n)$ has already been used in section 4.4 but now that we will be considering variations with time as well as position, it is important to be clear that the partial derivative describes variation (of pressure p in this example) with only one *independent variable* (n in this example) out of several. In dynamics we are usually concerned with four independent variables — x, y, z and t — and they are independent in the sense that we can readily imagine any one of them varying while the others are held constant. For example, $\partial p/\partial x$ represents the eastward gradient of pressure at fixed y, z and t, and $\partial p/\partial t$ represents the rate of variation of pressure with time at fixed position (x, y and z) — the pressure tendency as measured from a fixed barograph.

The pressure gradients along the x, y and z axes are found in the case of any particular axis n by taking components in the usual way (Fig. 7.2). For example, if the angle between the n and z axes is α, then the vertical pressure gradient $\partial p/\partial z$ is given by

$$\frac{\partial p}{\partial z} = \frac{\partial p}{\partial n} \cos \alpha$$

On scales much larger than those of small-scale turbulence and the perturbations of flow around buildings etc., the atmospheric pressure gradient is very nearly vertical, i.e. α is very small in the above and $\partial p/\partial z$ is very much larger than either $\partial p/\partial x$ or $\partial p/\partial y$. Since pressure falls with increasing height, it follows that by far the largest component of the pressure gradient force per unit mass on these and larger scales acts vertically upwards, tornado funnels being rare and highly localized exceptions. For example, on the synoptic scale, horizontal pressure gradients associated with extratropical cyclones rarely exceed 5 mbar per 100 km, in comparison with the vertical pressure gradient of 1 mbar per 10 m which is typical of the low troposphere. However, thanks to the independence of the vertical and horizontal components of the equation of motion, the small horizontal pressure gradient forces are nevertheless very important, as we shall see.

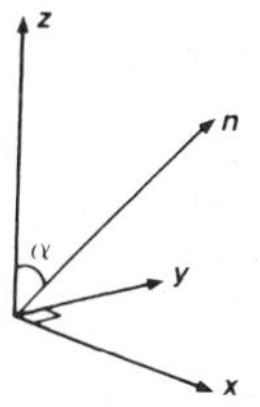

Fig. 7.2 Axes x, y and z related to a generalized axis n.

GRAVITATIONAL FORCE

The mutual gravitational attraction between an air parcel and the Earth produces the downward force on the parcel which we call its weight, and an equal and opposite force on the Earth which ordinarily has no observable effect. Measurements on calibrated weighing machines at rest on the Earth's surface show that the weight of any body of mass M is given by Mg, where g has an average value of 9.81 m s^{-2}, and varies slightly with latitude ϕ and altitude z above sea level as shown in Table 7.1. (We assume that we already have an unambiguous definition and measure of mass, though in fact Newtonian dynamics largely evades this point by circular argument and an appeal to an intuitive sense of the quantity of matter. It is simplest to accept that mass is already defined and measurable.) The factor g is known as the *gravitational acceleration* because it is equal to the downward acceleration which the body would undergo in the absence of any other force (i.e. falling freely in a vacuum). In the meteorological context it is important to notice that in either case (weighing machine or free fall) we make observations from a laboratory fixed to the rotating Earth's surface, so that the measured g is *apparent* g, rather than the absolute value arising purely from gravity which would be measured from a non-rotating laboratory. The distinction will be discussed later (section 7.4). Note that the gravitational force per unit mass is simply g in the negative z direction, and that for many meteorological purposes g can be assumed to have a uniform value of 9.8 m s^{-2} throughout the troposphere. Indeed for rough estimates the value of 10 m s^{-2} is often convenient and accurate enough.

Table 7.1 Apparent gravitational acceleration

Altitude in km	Latitude 0°	45°	90°
0	9.780	9.806	9.832
20	9.719	9.745	9.770
40	9.538	9.684	9.709

FRICTIONAL FORCES

Friction arises when bodies in contact move with different velocities, or tend to do so. Although we may think of it as a property of solids in contact, friction occurs in all normal fluids and plays an important role in the incessant commotion of the atmosphere by tending to reduce all velocity differences induced by other means. Such differences appear in the form of *shears* (gradients) of air flow, and some of the most significant shears, apart from the localized, transient ones associated with turbulence, are shears of horizontal wind. A strong concentration of vertical shear of horizontal wind is almost universal in the planetary boundary layer, particularly in its lowest levels. This is depicted in Fig. 7.3 as an example for visualization and discussion, but the principles outlined are quite general. For the moment, nothing is lost by assuming the shear to be two-dimensional, though in fact it is usually three-dimensional (like most atmospheric shears) in the sense that both wind speed and direction vary with height.

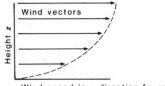

Fig. 7.3 Two-dimensional picture of mean horizontal wind near the Earth's surface, showing vertical shear.

The effect of friction on the sheared flow in Fig. 7.3 is to tend to speed up the slower and slow the faster air, and ultimately to make both move at the same speed as the surface. Over land this would mean the air eventually coming to a halt, as happens occasionally on very quiet nights, and over the sea it would mean the air moving at the very low speeds (~ 1 m s^{-1}) typical of the wind-driven sea surface. Neither of these conditions arises often, because the air flow is usually driven from above by some large-scale weather system, such as an extratropical cyclone or the trade winds. The shear between flowing troposphere and static or nearly static surface is then maintained, and the effect of friction in the sheared layer is to transfer horizontal momentum downwards from the weather system to the surface. The air aloft and the surface beneath experience equal and opposite drags (Fig. 7.4).

Momentum is transferred across wind shears because air itself is transferred, and this occurs in two very different ways in the atmosphere. On the molecular scale, molecules intermingle in the course of their incessant, random thermal motion, and the resulting diffusion of momentum is called *viscosity*. On the scale of turbulent eddies ($\sim$ mm to 100 m or more), individual eddies intermingle in the course of turbulent motion and the resulting very rapid diffusion of momentum is often ascribed for convenience to *eddy viscosity*. Let us consider the molecular scale first because it is much simpler in principle, and because its formalism can then be extended to eddy viscosity.

If τ_{zx} is the viscous drag per unit area in the x direction experienced by the horizontal upper surface of an air parcel because of the relative motion of the air above (so that τ_{zx} is a tangential stress to compare with the perpendicular stress exerted by pressure), then according to eqn (7.1) this drag must tend to increase the momentum of the air parcel in the x direction at just this rate. Noting the presence of the unit area in the definition of stress, it should become clear after a little

Fig. 7.4 A small tree on an exposed part of the South Downs near Brighton, Sussex. The branches have developed systematically on the side sheltered from the prevailing wind giving the whole crown an asymmetrical, streamlined appearance.

thought that τ_{zx} can be regarded as the downward flux of eastward momentum through unit horizontal area. It is called the x component of the horizontal *shearing* stress because of its obvious connection with the shear $\partial u/\partial z$ – the shear depicted in Fig. 7.3.

According to Newton's empirical law of viscosity, the shearing stress and shear are directly proportional, the constant of proportionality being the *dynamic coefficient of viscosity* μ, which has a well-defined value for any given thermodynamic state of the air.

$$\tau_{zx} = \mu \frac{\partial u}{\partial z} \tag{7.4}$$

According to appendix 7.3 the net viscous force per unit mass eastward on a thin horizontal slab of air embedded in the shear $\partial u/\partial z$ is given by

$$\frac{1}{\rho} \frac{\partial}{\partial z} \tau_{zx} \tag{7.5}$$

Note that as in the case of a pressure gradient, the net force depends on the gradient of a stress, though in this case the stress and gradient are perpendicular rather than parallel as they are in the case of pressure. Since in realistic atmospheric conditions gradients of μ are proportionately much smaller than gradients of shear, we can combine eqns (7.4) and (7.5) to obtain

$$\frac{1}{\rho} \frac{\partial}{\partial z} \tau_{zx} = \nu \frac{\partial^2 u}{\partial z^2} \tag{7.6}$$

where ν is the *kinematic coefficient of viscosity* (μ/ρ) of the air. The value of ν depends mainly on temperature in typical atmospheric conditions and is about 14×10^{-6} m^2 s^{-1} in the low troposphere. Note that the net viscous force increases with the curvature $\partial^2 u/\partial z^2$ of the velocity profile (the shear of the shear), which is consistent with the intuitively reasonable tendency for viscosity to remove humps and hollows in the velocity profile and smooth it as much as possible.

Because of basic similarities between molecular and eddy viscosity, we might expect there to be the turbulent equivalent of eqn (7.4) linking eddy stress to the shear of the mean wind $\partial \bar{u}/\partial z$ (where the wind component is averaged to remove the turbulent gustiness)

185

$$\tau_{zx} = K \frac{\partial \bar{u}}{\partial z} \tag{7.7}$$

In fact the statistics of turbulence are observed to be so much more elaborate than those of molecular motion that the *coefficient of eddy viscosity K* defined by eqn (7.7) is not related in any simply predictable way to the thermodynamic state of the air, and indeed is not fully determined by any statistical description of the turbulence discovered so far. This severely limits the usefulness of the concept of eddy viscosity, but its attractive simplicity nevertheless keeps it in widespread use, especially in experimental studies of the connection between eddy stress and mean shear in the enormous range of turbulent conditions encountered in the atmosphere, and in relations which can be derived from these. Values of K derived from field measurements and eqn (7.7) range across many orders of magnitude, but more than a few centimetres above the surface they are almost always at least four orders of magnitude larger than ν. Despite uncertainties of detail the general picture is clear: in almost all of the lower atmosphere eddy viscosity overwhelms molecular viscosity as an agent of momentum diffusion to the extent that we can assume that the motion of an air parcel is affected only by eddy viscosity.

7.3 Relative and absolute accelerations

Newton's second law of motion (eqn (7.1)) relates resultant force and acceleration as measured from an unaccelerated observation platform (an *inertial reference frame*). The need to use an inertial frame should be clear from the example of a man unwise enough to weigh himself on spring scales in an accelerating lift. Though his acceleration relative to the lift is zero if he is standing steadily, it does not follow that the downward force exerted on his feet by the scales (and registered on their display) will be equal to his weight; in fact if the lift is accelerating upward the scales will register more than his true weight, and if it is accelerating downward they will register less than his weight. This illustrates the very fundamental point than an unrecognized acceleration in one direction appears as an inexplicable force in the opposite direction — mysterious weight or buoyancy in this case.

Now atmospheric motions, including accelerations, are measured from a reference frame fixed to the Earth (Fig. 7.1) and therefore rotating with it. The meteorological reference frame is rotating and therefore accelerating continually, and this acceleration must be allowed for when using the equation of motion, otherwise mysterious forces will seem to invalidate Newton's second law of motion and any predictions based on it.

Because the three-dimensional shape of the Earth complicates the issue without altering basic principles, we will consider the effects of rotation first in relation to a flat turntable. Figure 7.5 represents a model train T running at speed V around a circular track of radius R on a turntable which is rotating with angular velocity Ω about the centre of the track O. Let us use the basic kinematic rule that a body whirling at speed v round a circle of radius r experiences a continual *centripetal* acceleration v^2/r toward the centre of the circle (Appendix 4.5). It follows that the centripetal acceleration of the train toward O is

1. V^2/R as measured by an observer rotating with the track and turntable; but that it is

2. $(V+\Omega R)^2/R$ as measured by a non-rotating observer, for example one sitting beside the turntable, where $\Omega R = V_t$ is the tangential speed of any part of the track on account of its rotation (Appendix 4.5 again).

The difference between expressions (2) and (1) represents the centripetal acceleration which is ignored when observations are made from the turntable rather than a non-rotating frame, and which would therefore seem to the observer on the turntable to be an outward (*centrifugal*) force acting on the train. Such forces are often called apparent forces in the sense that they appear only when observing from an accelerating reference frame, but of course in that frame they are quite real. For example in the present case, the toy train would tip outward off its rails if the turntable rotation and train speed were too high for the train's lateral stability, and its toy-town passengers would be justifiably mystified if the derailment were attributed to an unreal force!

Multiplying out the bracket of the *absolute* centripetal acceleration (2) and subtracting the *relative* centripetal acceleration (1), we find that the difference comprises the two terms

$$\Omega^2 R + 2\Omega V \tag{7.8}$$

The first of these is the familiar centripetal acceleration of any fixed point on the track (Appendix 4.5), which obviously must be one of the terms lost in making observations from the turntable itself. The second term combines the motions of the track (Ω) and of the train relative to the track (V), and is known as the *Coriolis* acceleration, after the French engineer who drew attention to its importance in rotational dynamics. It is a less obvious casualty of using a rotating reference frame, but it is important to see that no principle is involved beyond what is implied in the simplest treatment of centripetal acceleration. The Coriolis acceleration is a component of the centripetal acceleration of a moving body (the train in this case) which becomes identifiably distinct because we choose to observe the body from a rotating frame (the turntable in this case) in which the body is not at rest. Note that in the present example the Coriolis acceleration acts perpendicularly to the left of the instantaneous velocity of the body relative to the rotating frame — the left being the side of the train's direction of relative motion which is obviously favoured by the particular direction of rotation of the turntable (Fig. 7.5). It will transpire that this is not just a result of this particularly simple example: like the centripetal accelerations of which they are a part, all Coriolis accelerations are perpendicular to the relative velocity in the direction favoured by frame rotation (section 7.5).

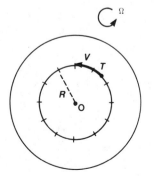

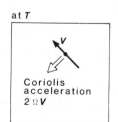

Fig. 7.5 Toy train T travelling round a circular track on a rotating turntable. The inset shows the magnitude and direction of the Coriolis acceleration at T.

7.4 The Earth's rotation and apparent *g*

If the Earth were a perfect sphere with concentric distribution of mass, the true gravitational force on a body on the surface would everywhere act with uniform strength toward the Earth's centre. However, if the gravitational force is measured relative to a fixed point on the rotating Earth's surface, as it normally is of course, true *g* will be offset by a 'mysterious' centrifugal force at all latitudes except the poles, with the result that measured (apparent) *g* will be less than true *g*, and directed slightly away from the Earth's centre unless the point is on the equator (Fig. 7.6). There should by now be no mystery about the centrifugal force: it is a consequence of ignoring the centripetal acceleration of the Earth-bound reference frame, and if we were unwise enough to measure *g* from a frame moving over the Earth's surface we would have a Coriolis effect as well! The discrepancy between true and apparent *g* values will be greatest at the equator because the centrifugal effect is greatest there. As shown in appendix 4.5 $\Omega^2 R$ has value 0.034 m s^{-2}, which is of the same order of magnitude as the variations in apparent *g* in Table 7.1. However, more significantly the angling of apparent *g* away from the local radius would imply a horizontal component of apparent *g* acting toward the equator (i.e. one parallel to the local surface) with a maximum value at latitude 45°.

Consider the consequences of having an equatorward horizontal component of apparent *g*. In the absence of any opposing force, the atmosphere and oceans would move toward the equator and accumulate there, adding to the equatorial girth of the fluid planet by denuding the polar regions. If the oceans were frozen and unable to move, the atmosphere would redistribute its mass until the accumulation of air in low latitudes produced a poleward pressure-gradient force everywhere exactly balancing the equatorward apparent *g* force. A simple estimate in Appendix 7.4 suggests that this would produce permanent horizontal pressure gradients in middle latitudes fully comparable with those found in deep depressions, and isobaric surfaces rising over 10 km from pole to equator. If the oceans were free to move they would flow until mean sea level was similarly tilted, in the process doing very strange things to the relative levels of sea surface, sea-bed and land. No such relative displacement is observed because the 'solid' Earth itself is nearly fluid on geological timescales and has bulged at the equator to adjust to the centrifugal effect of its own rotation in essentially the same way as we have been imagining for the atmosphere and oceans. In fact it has reached the equilibrium distribution in which there is no longer any component of apparent *g* parallel to the Earth's surface, so that the local vertical, as defined by a plumb line, is perpendicular to the local horizontal as defined by mean sea level. The equilibrium is complicated by the onion-like zoning of increasing density toward the Earth's centre, and the distorting effect of the bulged mass on its true gravitational field, but the

Fig. 7.6 True and apparent accelerations (g_t, g_a) on the surface of a spherical Earth. The dashed line represents the actual spherical Earth.

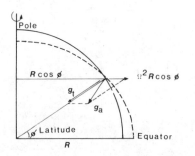

Key:

g_a apparent gravitational acceleration

g_t true gravitational acceleration

upshot is that the Earth's equatorial radius exceeds its polar radius by about 21 km and the strength of apparent g varies with latitudes as shown in Table 7.1. This slight oblateness of the Earth is of no meteorological significance, and the meridional and vertical variations of apparent g have to be considered only in the most accurate representations of large-scale fields of atmospheric pressure. Because of the long-term fluidity of the Earth itself therefore, the static distributions of the atmosphere and oceans over the slightly oblate spinning globe are virtually what would be expected on a non-rotating sphere.

7.5 The Coriolis effect

The static centrifugal bulge of the Earth outlined in the last section accommodates by far the largest effects of the Earth's rotation, and leaves the resting atmosphere and ocean effectively dynamically unaffected by that rotation provided we treat mean sea level as our horizontal datum, which of course we do. But once these fluids begin to move relative to the spinning Earth the smaller but still significant Coriolis effect comes into play, producing terms which are proportional to the wind and current speeds themselves. Because of this proportionality there is no single distribution of fluid mass and flow which will accommodate the Coriolis effects; there are temporary accommodations which change as the atmosphere and oceans ceaselessly shift and vary. We need a simple but comprehensive description of the Coriolis effect to apply to the many different situations in meteorology (and oceanography) in which it is important.

Consider the model train again (section 7.3), but this time moving steadily along a straight track placed at random on the same rotating turntable (Fig. 7.7). As the turntable and track rotate, the train is forced to accelerate laterally, and it is shown in Appendix 7.6 that this lateral acceleration is simply $2\Omega V$, which is identical to the Coriolis term in eqn (7.8), even though this time there has been no simple arrangement of track relative to axis of rotation. The train may be moving straight towards the turntable axis, or away from it, or along any intermediate line, but in every case the Coriolis acceleration is perpendicular to the track (i.e. perpendicular to the train's line of motion relative to the turntable), to the left or right depending on the direction of turntable rotation (Fig. 7.7). Of course there is also an $\Omega^2 R$ type term which does vary with position on the turntable and is always directed toward the axis of rotation, but this is the equivalent of the static centripetal term, which the previous section has shown we can ignore almost entirely.

The Coriolis effect is complicated in detail but not in principle by the fact that the Earth is a rotating sphere rather than a flat turntable. In Fig. 7.8 it is apparent that at any point P at latitude ϕ on the Earth's surface in the northern hemisphere, the Earth's angular velocity vector (magnitude Ω and directed along the axis of rotation in the sense of a right-handed screw) can be resolved into two components — one of magnitude $\Omega \sin \phi$ along the local vertical at P (the z axis), and the other of magnitude $\Omega \cos \phi$ along the poleward-pointing horizontal (the y axis). We can therefore represent the Earth's rotation by means of two turntables, one rotating about the z axis and the other rotating about the y axis, and can apply the turntable form of the Coriolis effect to each to produce all the Coriolis terms for the actually nearly spherical Earth.

The turntables are shown in Fig. 7.9 as seen looking at P down the z and y axes

Fig. 7.7 Coriolis acceleration for relative motion on a counterclockwise rotating turntable.

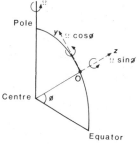

Fig. 7.8 Components of Earth's rotation about the local vertical axis z and about the local northward axis y, at latitude ϕ.

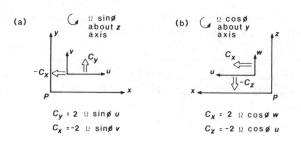

Fig. 7.9 Components of Coriolis acceleration in (a) the local horizontal plane, and (b) the vertical plane perpendicular to the local northward axis.

separately, with the sense of rotation as for the northern hemisphere. Considering the components of wind speed u, v and w in turn and comparing with Fig. 7.7, we find the following components of the Coriolis acceleration in the directions of the conventional x, y and z coordinates axes (Fig. 7.1):

$$\left. \begin{array}{ll} x & -2\Omega v \sin \phi + 2\Omega w \cos \phi \\ y & 2\Omega u \sin \phi \\ z & -2\Omega u \cos \phi \end{array} \right\} \tag{7.9}$$

If for example we are considering a northerly wind (i.e. negative v and zero u and w) we see that the only Coriolis component is a horizontal acceleration toward the east (i.e. in the x direction). Westerly winds and updraughts are each associated with two Coriolis components, and a wind with significant components in all three directions is associated with a similarly three-dimensional Coriolis acceleration. Notice that all Coriolis terms have the same form — a product of relative velocity component, the Earth's angular velocity, and the tell-tale factor 2.

In many realistic meteorological situations not all of these terms are equally important, and some are often negligible, but discrimination is possible only when we have assembled all significant terms of each component of the equation of motion.

7.6 The equations of motion

The various contributions to the equation of motion (eqn (7.2)) may now be summarized symbolically as a balance between accelerations (A terms) and forces (F terms), each of which is a vector quantity.

$$RA + FA = PGF + GF + FF \tag{7.10}$$

The relative acceleration RA is what is observed from the meteorological reference frame. For example, on the large scale, air on a chart of the upper troposphere may be seen accelerating into the core of a jet stream — a linear acceleration in the sense that it is parallel to the wind velocity. Or in the low troposphere, air may be seen moving in a horizontal circular path around a centre of low pressure, which implies a centripetal acceleration toward the centre of the circle. On smaller scales there may be transient intense accelerations as buoyant air rises into vigorous shower clouds, and on even smaller scales air in the turbulent boundary layer is continually accelerating and decelerating in the chaos of transient eddies. Each of these is an example of the acceleration of air parcels relative to the Earth-bound reference frame.

The frame acceleration *FA* contains only the Coriolis acceleration *CA* because the large fixed component of centripetal acceleration has been accommodated by the fluid deformation of the rotating Earth and the use of apparent *g* instead of true *g* (section 7.4). It is customary to deal with the Coriolis effect as if it were a force on the right-hand side of the equation of motion rather than an acceleration on the left-hand side. As discussed in section 7.3 this is a perfectly satisfactory procedure provided the apparent Coriolis force *CF* is equal and opposite to the Coriolis acceleration *CA*. Equation (7.10) then becomes

$$RA = CF + PGF + GF + FF \tag{7.11}$$

The remaining terms on the right-hand side are the pressure gradient force *PGF*, the (apparent) gravitational force *GF* whose magnitude is defined by Table 7.1, and the frictional force *FF*. Of the last two, *GF* always acts vertically downward and *FF* in the low troposphere usually acts to oppose the air motion.

This vector form of the equation of motion is very compact, even when each symbolic term is fully written out, but it is not particularly suitable for detailed analysis or solution. These require that the equation of motion is broken into its *x*, *y* and *z* components, producing in effect three equations of motion which are independent except in so far as they share common terms. We can write down these component equations by collecting the detailed terms discussed in the last few sections.

$$x \text{ component} \quad \frac{du}{dt} = 2\Omega \sin \phi \, v - 2\Omega \cos \phi \, w - \frac{1}{\rho} \frac{\partial p}{\partial x} + F_x \tag{7.12a}$$

$$y \quad \cdot \cdot \quad \frac{dv}{dt} = -2\Omega \sin \phi \, u - \frac{1}{\rho} \frac{\partial p}{\partial y} + F_y \tag{7.12b}$$

$$z \quad \cdot \cdot \quad \frac{dw}{dt} = 2\Omega \cos \phi \, u - \frac{1}{\rho} \frac{\partial p}{\partial z} - g + F_z \tag{7.12c}$$

These equations form an elaborate set whose complete solution, together with any thermodynamic or other constraints which may apply, is far beyond the currently very considerable capabilities of applied mathematicians. Indeed the friction terms are not even reasonably defined by current theory and observation, though they can be written much more specifically for many purposes (e.g. section 9.8).

Fortunately it is possible to gain very useful insights into some of the types of behaviour they describe by means of judicious simplification of the full set, removing terms which are believed for good reasons (ultimately observational) to be relatively unimportant. A first step in this process is to examine the relative magnitudes of the various terms in each component by substitution of observed values, but since the relative sizes of these values often vary enormously depending on the time and space scales of the types of motion being examined, we must first specify these scales. This simple but extremely useful procedure is known as *scale analysis*. Let us use it first to examine the relative importance of the various terms of the equation of motion as they apply to synoptic-scale behaviour.

7.7 Synoptic-scale motion

Typical values observed in the synoptic-scale weather systems which dominate the troposphere in middle latitudes, and in modified form are still important in low latitudes, are as follows.

Horizontal scale length	L	1000 km	10^6 m
Vertical scale length	H	10 km	10^4 m
Timescale	t	1 day	10^5 s
Horizontal pressure changes	Δp	10 mbar	10^3 Pa
Vertical pressure change	p	1000 mbar	10^5 Pa
Air density	ρ		1 kg m^{-3}
Earth's angular velocity	Ω		10^{-4} rad s^{-1}
Apparent g	g		10 m s^{-2}

Although the meaning of any particular scale is deliberately vague, and the values are mostly simple orders of magnitude (such is the variability of atmospheric behaviour), the following indicates the way in which we can interpret them. L is the horizontal distance in which there is usually a substantial change in any observable property of the system, such as pressure or wind field. H and t mean the same thing in relation to the vertical scale and the timescale. Δp is the typical horizontal pressure variation observed in such weather systems, whether spatially from centre to edge, or temporally as observed by a fixed barograph as a system travels past. The variation in pressure in the vertical is dominated by the large background pressure lapse, and its value is comparable with the typical pressure surface pressure p because most large weather systems fill the full depth of the troposphere and hence encompass more than a scale height of pressure. And ρ is a typical average air density.

Values of other measurable properties, and terms in the component equations of motion, can be derived from the above set. For example:

Horizontal wind speed	$U \sim L/t$	10 m s^{-1}
Vertical wind speed	$w \sim H/t$	10^{-1} m s^{-1}
Horizontal acceleration	U/t (or U^2/L)	10^{-4} m s^{-2}
Vertical acceleration	w/t (or w^2/H)	10^{-6} m s^{-2}
Major Coriolis acceleration	ΩU	10^{-3} m s^{-2}
Horizontal pressure gradient	$\Delta p/L$	10^{-3} Pa m^{-1}

Note that the bracketed alternatives remove the timescale in a way which will prove very useful later, and that omission of $\sin \phi$ and $\cos \phi$ from the Coriolis terms is unimportant provided we are not very near the equator.

Ignoring the friction terms for the moment, we can now give an order of magnitude for each term in the eqns (7.12). The units are m s^{-2} (acceleration or force per unit mass) throughout.

$$\frac{du}{dt} = 2\Omega \sin \phi \, v \quad - 2\Omega \cos \phi \, w \quad - \frac{1}{\rho} \frac{\partial p}{\partial x} \qquad (7.13\text{a})$$
$$10^{-4} \qquad 10^{-3} \qquad\qquad 10^{-5} \qquad\qquad 10^{-3}$$

$$\frac{dv}{dt} = -2\Omega \sin \phi \, u \qquad\qquad - \frac{1}{\rho} \frac{\partial p}{\partial y} \qquad (7.13\text{b})$$
$$10^{-4} \qquad 10^{-3} \qquad\qquad\qquad 10^{-3}$$

$$\frac{dw}{dt} = 2\Omega \cos \phi \, u \quad - g \qquad - \frac{1}{\rho} \frac{\partial p}{\partial z} \qquad (7.13\text{c})$$
$$10^{-6} \qquad 10^{-5} \qquad 10 \qquad\qquad 10$$

Several important simplifications now emerge.

1. The second Coriolis term in the x component equation (7.13a) is two orders of magnitude smaller than the first because synoptic-scale updraughts are so much weaker than horizontal winds, and may therefore be ignored, leaving

the two horizontal-component equations (7.13a) and (b) looking much more alike.

2. The relative acceleration terms (du/dt, dv/dt) in the horizontal-component equations are smaller than the major Coriolis and pressure-gradient terms, but by only one order of magnitude. This suggests that though the acceleration terms are often relatively unimportant they are not always so. It will appear in the next section that the relatively very simple situation which arises when the relative accelerations really are negligible corresponds to geostrophic balance, and that much of the lower atmosphere, but not all of it, is in a state of quasi-geostrophic balance on the large scale.

3. In the vertical-component equation (7.13c) only the gravitational and vertical pressure-gradient terms are significant. The balance between these two corresponds to the hydrostatic equilibrium discussed in detail in Chapter 4. The relatively complete unimportance of the other two terms shows how good the hydrostatic approximation really is on the synoptic scale. But despite this, note the absurdity of assuming from this that vertical accelerations are strictly zero: there could then be no updraughts, other than over active volcanoes, no cloud and no precipitation!

Let us now consider the possible importance of the frictional terms which have been deliberately omitted from eqn (7.13). If the friction were due solely to molecular viscosity, then the frictional terms in the equation of motion would include $\nu \, \partial^2 u / \partial z^2$ and similar terms (section 7.2). In fact this particular term and the other terms involving the curvature of the vertical wind profile are always likely to be the largest frictional terms, especially because of the concentration of strong wind shears in the planetary boundary layer. To estimate such terms by scale analysis, note that the second derivative $\partial^2 u / \partial z^2$ is the gradient of a gradient, i.e. a velocity difference divided by a length and divided further by a length. If the length is assumed to be the typical depth h of the planetary boundary layer then

$$\nu \frac{\partial^2 u}{\partial z^2}, \ \nu \frac{\partial^2 v}{\partial z^2} \sim \nu \frac{U}{h^2} \sim 10^{-10} \text{ m s}^{-2} \tag{7.14}$$

since $\nu = 10^{-5}$ m^2 s^{-1} and $h \sim 10^3$ m.

The small size of the largest viscous friction terms compared with other terms in eqn (7.13) confirms the earlier contention (section 7.2) that viscosity is dynamically unimportant even in the strongly sheared boundary layer. In fact it is very much less important even than the very weak synoptic accelerations du/dt and dv/dt.

The comparison between acceleration and viscous terms has been found to be a crucial one across the whole range of fluid dynamics for distinguishing between flows which are significantly controlled by viscosity and flows which are not. It is conveniently expressed by the ratio acceleration/(viscous term), which is called the *Reynolds number* (*Re*) after the pioneer fluid dynamicist Osborne Reynolds. Normally the same scale is assumed in both the acceleration and viscous terms, with the result that

$$Re = \frac{UL}{\nu} \tag{7.15}$$

It has been observed in an enormously wide range of flows, from water in narrow pipes to planetary atmospheres, that when *Re* is less than about 10^3 the flow is always dominated by viscosity, whereas it is usually turbulent when *Re* is larger (Fig. 7.10). In the present case we see that *Re* for synoptic-scale atmospheric flow is about 10^{12}, or 10^6 if we compare synoptic-scale horizontal accelerations with the boundary-layer-scale viscous terms in eqn (7.14). In either case the critical value of

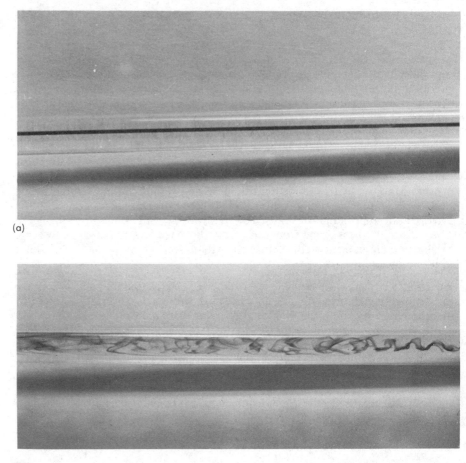

(a)

(b)

Fig. 7.10 Water flow in a glass pipe with dye injected centrally at the right-hand end. In laminar flow (a), the dye plume's downstream motion and divergence are imperceptible. In faster turbulent flow (b), the plume diffuses quickly to the pipe walls while swirling downstream (to the left).

10^3 for the Reynolds number is greatly exceeded and the flow is potentially turbulent.

Synoptic-scale flow is certainly unsteady, with large weather systems developing, maturing and dying like enormous eddies. But there is such a gulf of scale between these huge 'eddies' and the much smaller and unambiguously turbulent eddies which are so common in the the planetary boundary layer that it is difficult to say dogmatically that atmospheric flow is truly turbulent on the synoptic scale. It might be quasi-laminar, as the smoothness of air flow on weather maps seems to suggest, especially in the middle and upper troposphere: compare the smooth jet-stream flow in Fig. 7.13 with the clearly turbulent-matter flow in Fig. 7.10.

There is no doubt, however, that flow on the scale of the planetary boundary layer itself is enormously turbulent. The gustiness of surface winds and the looping, ragged dispersal of smoke plumes are ample evidence (Fig. 9.11), and the value of $Re = U h/\nu \sim 10^9$ confirms the potential for turbulence. It was noted in section 7.2 that turbulence in the boundary layer is similar to molecular viscosity in some important ways, and that it maintains an eddy viscosity which overwhelms molecular viscosity in almost all realistic atmospheric conditions. It follows that we should use the kinematic coefficient of eddy viscosity K instead of ν when estimating the friction terms omitted from eqn (7.13). Values of K are extremely variable but ~ 10 m^2 s^{-1} is a reasonably typical value in the presence of large, efficient eddies. Using this in eqn (7.14) with the original value for $\partial^2 u/\partial z^2$, we find that the eddy viscosity term has magnitude 10^{-4} m s^{-1}, which is just one order of magnitude down on the largest terms in eqns (7.13a) and (b). This result suggests that it is at least possible that the turbulent friction terms can be quite significant in the horizontal components of the equation, as is amply confirmed by observation (section 9.9).

Note that the Reynolds number is one of several *dimensionless numbers* we shall use to describe the importance of terms on the right-hand side of the equation of motion in comparison with the relative acceleration on the left-hand side. These and many other dimensionless numbers have been found to be very useful in the study of fluids, because similar values are found to correspond to dynamically similar behaviour despite other quite dramatic differences of particular situation. For example, the flow of glycerol round a spinning immersed ball, and the flow of air round an identical ball spinning in air, are observed to be similarly laminar when the Reynolds numbers are similarly small (say about 100), which occurs when the ball in the glycerine is spinning about 500 times faster, since that is the ratio of the kinematic viscosities of glycerol and air. These flows are said to be *dynamically similar* in this respect, and the existence of a critical value of *Re* which distinguishes between necessarily laminar and possibly turbulent flow is obviously a special example of the more general use of *Re* as a criterion of dynamic similarity. Other dimensionless numbers act as criteria for other types of flow.

Such numbers are called dimensionless because in each case the physical *dimensions* of mass, length and time (and temperature when it is involved) cancel out when they are inserted into the expression defining the number in question, leaving a pure number whose value in any particular situation is therefore independent of the system of units (SI, CGS, etc.) used. The concept and elementary use of dimensions is outlined in appendix 7.1.

7.8 Geostrophic flow

In the free atmosphere (i.e. above the planetary boundary layer), a great deal of observational evidence suggests that the friction terms in the equation of motion are relatively small for synoptic-scale flow. The scale analysis of the last section therefore suggests that horizontal flow is determined by the following equations:

$$\frac{du}{dt} = fv - \frac{1}{\rho}\frac{\partial p}{\partial x}$$
$$\frac{dv}{dt} = -fu - \frac{1}{\rho}\frac{\partial p}{\partial y} \tag{7.16}$$

where $f = 2\Omega \sin \phi$ is called the *Coriolis parameter*, and represents the Coriolis effect arising from the component of the Earth's rotation about the local vertical. The scale analysis further suggests that the relative accelerations du/dt and dv/dt may be an order of magnitude smaller than the Coriolis and pressure-gradient terms.

We can examine the relative magnitudes of the relative accelerations and Coriolis terms by setting up the ratio

$$\frac{\text{relative acceleration}}{\text{Coriolis acceleration}}$$

which is the second important dimensionless number we consider, and is called the *Rossby number* (*Ro*), after the Swedish meteorologist who pioneered ways of dealing with the meteorological implications of the Earth's rotation.

$$Ro = \frac{\text{relative acceleration}}{\text{Coriolis acceleration}} \sim \frac{U^2/L}{\Omega U} = \frac{U}{\Omega L}$$

When $Ro \ll 1$ the Coriolis terms predominate over the relative acceleration terms, and observations of real and model atmospheres show types of flow characterized by large, flat vortices about the local vertical. Using the characteristic scales for synoptic scale behaviour listed in section 7.7, $Ro \sim 0.1$ for synoptic-scale flow, suggesting that the Coriolis terms tend to predominate but not totally.

Consider what would happen if the Rossby number were zero. Equation 7.16 would then simplify to

$$f v = \frac{1}{\rho}\frac{\partial p}{\partial x} \qquad\qquad f u = -\frac{1}{\rho}\frac{\partial p}{\partial y} \qquad\qquad (7.17)$$

By definition these equations define what is called *geostrophic flow*, some of whose properties we will now consider. The reason for doing this is of course that the actual behaviour of the Earth's atmosphere is approximately geostrophic in important respects. The approximation is not nearly so good as the hydrostatic approximation which dominates the vertical component of the equation of motion (hence the descriptive term *quasi-geostrophic*), but it is good enough to be extremely useful.

In the special case of a purely westerly wind (i.e. positive u and zero v) eqns (7.17) simplify to

$$\frac{1}{\rho}\frac{\partial p}{\partial y} = -f u \qquad\qquad \frac{\partial p}{\partial x} = 0$$

These show that the pressure gradient must be parallel to the y axis, with pressure increasing in the negative y direction (Fig. 7.11). Since isobars are perpendicular to the pressure gradient (just as in maps of terrain height contours are perpendicular to terrain slope) it follows that the isobars in this case must lie east–west, which is parallel to the assumed westerly air flow, as shown in Fig. 7.11. In addition, the fact that the pressure gradient is directly proportional to wind speed means that isobar spacing must increase with decreasing wind speed and vice versa, the constant of proportionality varying with air density and latitude (through f).

The properties of isobaric parallelism and spacing in relation to air flow are in fact quite independent of the direction of air flow. It is readily shown (appendix 7.7) that the horizontal wind speed V and the horizontal pressure gradient $\partial p/\partial n$ are connected in magnitude by

$$\frac{1}{\rho}\frac{\partial p}{\partial n} = f V \qquad\qquad (7.18)$$

where $\qquad \dfrac{\partial p}{\partial n} = \left[\left(\dfrac{\partial p}{\partial x}\right)^2 + \left(\dfrac{\partial p}{\partial y}\right)^2\right]^{\frac{1}{2}}$

and $\qquad\quad V = (u^2 + v^2)^{\frac{1}{2}},$

and that the directional relationship in the northern hemisphere is as shown in Fig. 7.12. The latter is summarized verbally in a law named after the Dutch pioneer meteorologist of the 19th century *Buys Ballot* which is usually stated in a form such

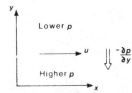

Fig. 7.11 Easterly geostrophic flow in a north–south pressure gradient (northern hemisphere).

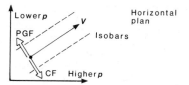

Fig. 7.12 Generalized geostrophic flow (northern hemisphere).

as 'low pressure in the northern hemisphere is on the left hand when facing down wind'. (It is on the right in the southern hemisphere). The law was recognized empirically, particularly from observations of surface winds blowing round extra-tropical and tropical cyclones, before it was formally linked to the Earth's rotation through the equation of motion.

It is customary to rearrange eqn (7.18) in the form

$$V = \frac{1}{\rho f} \frac{\partial p}{\partial n}$$

and to use this to define the *geostrophic wind* V_g as being that which satisfies this balance in magnitude and direction. By definition of V_g therefore

$$V_g = \frac{1}{\rho f} \frac{\partial p}{\partial n} \qquad (7.19)$$

It should be clear from this equation and Fig. 7.12 that geostrophic flow can be regarded as equilibrium between pressure-gradient force and Coriolis force (which acts perpendicularly to the right of the direction of motion in the northern hemisphere).

As mentioned in section 4.6 it is conventional to describe horizontal pressure fields at levels above sea level by plotting contours, in metres above sea level, of a convenient isobaric surface. Since the vertical distortion of such isobaric surfaces from the horizontal is always quite small in the lower atmosphere, no significant confusion of vertical and horizontal gradients arises from this practice. It follows that the contours of an isobaric surface are almost exactly parallel to the isobars on a closely adjacent horizontal surface, and that Buys Ballot's law holds when lower contours are substituted for lower pressures. It is shown in appendix 7.5 that the replacement of $\partial p/\partial n$ and $\partial Z_p/\partial n$ (the contour slope on the p isobaric surface) very conveniently removes the variable ρ from eqn (7.19), giving

$$V_g = \frac{g}{f} \frac{\partial Z_p}{\partial n} \qquad (7.20)$$

The removal of ρ and the near uniformity of g means that at any particular latitude (and hence value of f) the same contour slope corresponds to the same geostrophic wind speed, regardless of whether we are dealing with observations near sea level or in the stratosphere where air density may be 20 times smaller.

Let us use eqns (7.20) and (7.19) to estimate isobaric slopes and horizontal pressure gradients geostrophically associated with typical actual winds in middle latitudes, where $f \cong 10^{-4}$ rad s^{-1}. For a wind speed of 10 m s^{-1} the isobaric slope is about 1 in 10^4, and slopes seldom exceed 8 in 10^4 even in the most powerful jet stream. The pressure gradient corresponding to 10 m s^{-1} in the low troposphere (where air density is about 1 kg m^{-3}) is about 1 mbar per 100 km, but the pressure gradient corresponding to the jet stream is only about double this value because the increased contour gradient is largely offset by the reduced air density in the high troposphere. The greater simplicity of using isobaric contours is obvious.

The accuracy of the geostrophic approximation in any particular real situation is

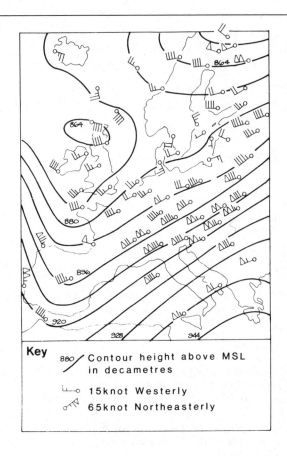

Fig. 7.13 Observed quasi-geostrophic flow at the 300 mbar level over the British Isles and western Europe at 0000 Z, 1 January 1982.

Key

880 / Contour height above MSL in decametres

⌐─○ 15knot Westerly

○⫞⫞ 65knot Northeasterly

apparent from the extent to which actual winds resemble geostrophic winds in both strength (eqn (7.20)) and direction (Fig. 7.12). Typical correspondence between actual and geostrophic winds can be judged from Fig. 7.13, which shows isobaric contours and actual winds in the vicinity of a polar-front jet stream over the British Isles. The data are taken from synoptic radiosonde ascents on the occasion, and the contours sketched by interpolation between the point measurements of the height of the 300 mbar surface. The wind vectors, measured from the horizontal drift of the radiosondes as they rose through the 300 mbar surface, are parallel to the isobaric contours to within 20° everywhere and 10° over most of the area. The actual wind speeds agree with the geostrophic speeds calculated from the contour slopes using eqn (7.20) to within 20% everywhere and 10% in most places. This is typical of the extent to which actual winds are geostrophic provided we avoid the planetary boundary layer, and the equator and its immediate vicinity.

7.9 Why geostrophic?

Before we consider any further aspects of geostrophic equilibrium, note what strange behaviour it embodies. Air, instead of flowing horizontally down the pressure gradients in response to the horizontal pressure gradient force, flows at right angles to this apparently obvious direction. In the vicinity of a pressure minimum like those defining the centres of depressions, the air, instead of flowing directly inwards to fill up the low pressure, flows round the low, forming huge flattened vortices turning anticlockwise on the weather maps of the northern hemisphere, and clockwise in the southern hemisphere. Indeed this majestic wheel-like rotation gave rise to the generic name *cyclone*, from the Greek for wheel.

Note too that we have not in any meaningful sense explained why actual winds should be quasi-geostrophic. The scale analysis of the equations of motion, and the estimation of the Rossby number, are consistent in showing that the geostrophic balance follows from the comparative insignificance of the relative accelerations and frictional terms, but they do not begin to explain why these terms should be so small.

Rewriting the Rossby number in terms of time scale t we find

$$Ro \sim 1/\Omega t$$

from which it follows that Ro is less than 0.1 provided t exceeds about one day, and that this length of time is set by the Earth's rate of rotation. This is indeed the observed timescale of synoptic-scale weather systems, but it is very hard to say why this should be so. It is obviously associated with the relative quietness of the atmosphere on the synoptic scale: depressions take days rather than hours to form and die, and jet streams though strong are very definitely subsonic in speed. Airflow therefore has time to settle into geostrophic equilibrium with pressure fields and vice versa. And it seems intuitively reasonable to link this quietness in turn with the gentleness of the solar input: if this input were ten times larger than it is, then the atmospheric motion would be much more hectic, and the tendency toward geostrophic equilibrium would be continually frustrated by the eruption of tremendous cumulonimbus and the like.

You can consider the onset of geostrophic flow for an individual air parcel by envisaging it starting very unrealistically from rest in the presence of a synoptic-scale horizontal pressure field (Fig. 7.14). After initial motion toward low pressure, it will veer to the right (in the northern hemisphere) in response to the growing Coriolis force, or equivalently in response to the hidden anticlockwise rotation of the weather map. If growing friction critically damps the parcel's tendency to overshoot and undershoot, it will reach geostrophic equilibrium as shown, with the parcel moving parallel to the isobars (or isobatic contours) at the speed required by eqns (7.19) or (7.20). The parcel will obviously take a finite time to reach this state — a time which is related to the magnitude of the Coriolis parameter and hence to the Earth's rate of rotation. If we consider a situation with a much shorter intrinsic timescale — for example air flowing at 10 m s^{-1} in the vicinity of a hill of horizontal scale 10 km ($t \sim 10^3$ s) — then it is clear that there is not enough time to develop geostrophic equilibrium between the flow and the pressure field associated with the obstructing effect of the hill, so that flow round and over the hill cannot be expected to be even quasi-geostrophic, as is confirmed by substituting the time or space scales for the flow in the equivalent versions of the Rossby number. This points to the important fact that the geostrophic tendency

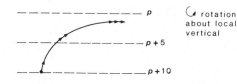

Fig. 7.14 Idealized approach to geostrophic equilibrium from rest (northern hemisphere).

appears only on the large mesoscale and on synoptic and larger scales of atmospheric flow.

Such discussion is suggestive rather than persuasive, since the arguments are in fact largely circular. A sophisticated review of atmospheric geostrophy in a technical journal concluded that 'it seems unlikely that the near geostrophic balance in our atmosphere can be accounted for in any simple manner' [43]. This is an intriguing state of affairs, given the widespread and important consequences of the geostrophic tendency, one of the most important of which we will now consider.

7.10 Thermal winds

We have seen that geostrophic wind speed is proportional to isobaric contour slope, and that geostrophic wind direction is parallel to isobaric contours on a horizontal plan. It follows that if the contour slope varies consistently with altitude, so must the geostrophic wind. Figure 7.15 depicts a vertical section through a part of the troposphere where there is a consistent increase in contour slope with altitude and an associated increase in geostrophic wind speed. For simplicity the situation is two dimensional, in the sense that there is no change of the direction of contour slope and wind with height slope and wind direction with height, but all relationships discussed in the following have quite obvious vector equivalents when directions vary with height.

Applying eqn (7.20) to the isobaric slopes at pressures p_1 and p_2 (the higher pressure p_1 corresponding to the lower altitude), we find the difference in the geostrophic wind speed directly by subtracting the equations for the two surfaces.

$$V_{g2} - V_{g1} = \frac{g}{f}\left[\frac{\partial Z_{p_2}}{\partial n} - \frac{\partial Z_{p_1}}{\partial n}\right]$$

$$= \frac{g}{f}\frac{\partial}{\partial n}[Z_{p_2} - Z_{p_1}] \tag{7.21}$$

The right-hand side of eqn (7.21) contains the thickness gradient

$$\frac{\partial}{\partial n}[Z_{p_2} - Z_{p_1}]$$

which is simply the horizontal gradient of the vertical separation of the p_1 and p_2 isobaric surfaces. In Fig. 7.15 the thickness of the layer between the 1000 and

Fig. 7.15 Vertical cross-section illustrating the thermal wind relation. Tails of geostrophic wind arrows are enclosed by circles whose diameters increase with wind speed (northern hemisphere).

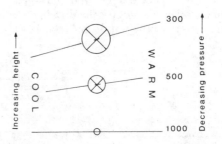

300 mbar surfaces increases sharply to the right, and the magnitude of its gradient could be found by subtracting the gradient of the lower surface from the gradient of the upper.

The thickness of a layer bounded by any two isobaric surfaces is proportional to the mean temperature of the layer (section 4.6) and it follows that the difference in geostrophic wind speed across such a layer is proportional to the horizontal gradient of the layer mean temperature. The presence of a horizontal gradient of layer mean temperature in an isobarically bounded layer means that the layer must be baroclinic, and the strength of the gradient is a measure of the degree of baroclinity. The thermal wind relation is therefore a relation between geostrophic wind shear and baroclinity, and hence highly relevant to the great baroclinic systems of the lower atmosphere, especially the troposphere.

Bearing in mind the smallness of realistic isobaric slopes in comparison with the slopes of isotherms (the ratio can be as large as 100 in a frontal zone, and is usually at least 10 — sections 4.6 and 11.2) the temperature gradient of an isobarically bounded layer is effectively a temperature gradient of a horizontal layer. Equation (7.21) therefore associates a vertical shear of (horizontal) geostrophic wind speed through a layer with a nearly horizontal temperature gradient across the layer. Figure 7.15 shows that the temperature gradient lies along an axis which is perpendicular to the wind shear — perpendicularity being the hallmark of the Coriolis effect. It is apparent in Fig. 7.15 that an extension of Buys Ballot's law connects the directions of vertical shear and nearly horizontal temperature gradient: in the northern hemisphere low mean temperature is on the left when standing with back to the geostrophic wind shear. The very important relationship described by eqn (7.21) and Fig. 7.15 is known as the *thermal wind relation* because it clearly connects the distributions of temperature and geostrophic wind.

It follows from the presence of hydrostatic equilibrium that the connection between layer thickness and layer mean temperature is such (appendix 7.9) that

$$\frac{V_{g_2} - V_{g_1}}{Z_{p_2} - Z_{p_1}} = \frac{g}{f\,\overline{T}}\,\frac{\partial \overline{T}_p}{\partial n} \tag{7.22}$$

which is the finite difference form of the *thermal wind equation*. The differential form follows when we consider a very thin layer of air (appendix 7.9 again)

$$\frac{\partial V_g}{\partial z} = \frac{g}{fT}\,\frac{\partial T_p}{\partial n} \tag{7.23a}$$

Since the thermal wind relation is simply a consequence of geostrophic and hydrostatic equilibria, it will be observed in actual wind and temperature fields to the extent that these equilibria apply. Hydrostatic equilibrium holds so very accurately on the synoptic scale (section 7.7) that we should expect the real atmosphere to obey the thermal-wind relation to the same extent that it is geostrophic, and this is observed to be the case on scales ranging from those of fronts to the hemispheres. Details of this behaviour are given at many points in the book, so only two of the most important examples are very simply outlined in Fig. 7.16.

APPLICATIONS

In Fig. 7.16(a) there is a section of the circumpolar vortex of westerly winds which dominates the troposphere in middle and high latitudes (section 4.7). The thermal-wind relationship appears in the conjunction of the increase of westerly

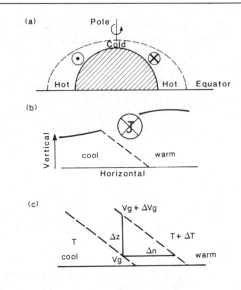

Fig. 7.16 (a) The circumpolar vortex in section, with westerly winds represented by arrow heads and tails, and temperature gradients between low and high latitudes by exaggerated thickness profiles (northern hemisphere).
(b) A polar front in section, with jet-stream arrow tail, zone of strongest horizontal temperature gradient (dashed line), and tropopauses (heavy lines)(northern hemisphere).
(c) A close-up of the foot of the cold front in (b), showing the zone's finite horizontal breadth Δn and vertical thickness Δz.

winds with height and the meridional temperature gradient imposed by the sun's unequal input. The level of maximum westerly winds is associated through the thermal-wind relation with the top of the layer with consistent poleward lapse of temperature. The westerly sense of the shear is consistent with the thermal extension of Buys-Ballot's law, just as the easterly shear (westerlies decreasing with increasing height) above is associated with an equatorward temperature lapse.

In Fig. 7.16(b) there is a section through an active front in middle latitudes — a type of structure described in some detail in Chapter 11. The conjunction of the wind shear up to the high winds of the jet core and the sharp lateral temperature gradient which is the basic feature of a well-defined front is obviously consistent with the thermal-wind relation. The thermal extension of Buys-Ballot's law requires that in the northern hemisphere cold air lies to the left of the jet core, looking down both wind and wind shear, as is always the case. In fronts there is often a significant change of wind direction between the low and high troposphere. According to the vector form of the thermal-wind relation which is derived in more advanced texts, the horizontal temperature gradient is perpendicular to the vector difference between the low- and high-level winds.

Equation 7.23a can be applied in finite-difference form to a simple model of a sloping frontal zone (Fig. 7.16c) with horizontal width Δn

$$\frac{\Delta V_g}{\Delta z} = \frac{g}{fT} \frac{\Delta T}{\Delta n} \tag{7.23b}$$

and rearranged to give an expression for the frontal slope $\Delta z / \Delta n$:

$$\frac{\Delta z}{\Delta n} = \frac{fT}{g} \frac{\Delta V_g}{\Delta T} \tag{7.23c}$$

This shows that a certain increase in geostrophic wind speed across the frontal zone is needed to counterbalance the tendency of the cold air increasingly to undercut the warm air, and hence maintain the frontal slope. Inserting realistic values of 30 m s^{-1} for ΔV_g, 5 °C for ΔT and typical values for f and T, we find the associated frontal slope to be 1:60 in middle latitudes, tending to zero at the Equator, where as usual geostrophic balance is impossible.

It should be clear by now that the importance of the thermal-wind relation for the state of atmosphere can hardly be overestimated. However abstract eqns (7.21)–(7.23) may appear, they describe behaviour which dominates much of the large-scale behaviour of the troposphere. As final reinforcement of this fact, it appears from advanced dynamical analysis [44] that it is the combination of vertical shear of wind and lateral temperature gradient which is baroclinically unstable and which gives rise to the greater weather systems of middle latitudes outlined in Chapter 11.

7.11 Geostrophic departures

Differences between actual and geostrophic winds are often apparently systematic on the synoptic scale. Observations of winds in the entrance region of the core of polar-front jet streams show that there is a systematic tendency for winds to be *supergeostrophic* (slightly faster than geostrophic) and to be angled somewhat across the isobaric contours towards lower levels. The opposite tendency appears in jet exit regions. Such tendencies appear in other cases where synoptic-scale flow is similarly subject to *linear acceleration* (i.e. acceleration effectively along the wind direction). The dynamical sketch in Fig. 7.17 depicts the reason for this behaviour. To provide for the linear acceleration towards the jet maximum, the Coriolis force must exceed the pressure gradient force and be angled forward. From the vector relation between Coriolis force and wind velocity it follows that the wind speed must exceed the geostrophic wind speed (i.e. the speed which will exactly balance the contour gradient through geostrophic equilibrium), and be directed so that there is a component down the contour slope.

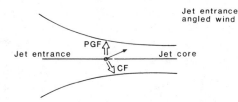

Fig. 7.17 Plan view of jet entrance region with isobaric contours, pressure gradient and Coriolis forces, and typical ageostrophic flow (northern hemisphere).

Significant geostrophic departures also arise from *lateral accelerations* (i.e. accelerations effectively perpendicular to the wind direction) such as those associated with cyclonic and anticyclonic rotation of air. Figure 7.18 contains a dynamic sketch of cyclonic flow of air round a horizontal circular path of radius R

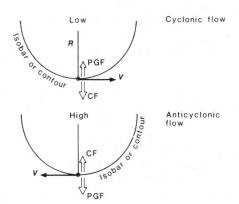

Fig. 7.18 Plan views of cyclonic and anticyclonic flow, with pressure gradient and Coriolis forces (northern hemisphere).

in the northern hemisphere. It is clear that to maintain the centripetal acceleration toward the cyclonic centre, the wind must have speed V such that the pressure gradient force exceeds the Coriolis force by V^2/R. It is apparent that this can be maintained only if the Coriolis force is less than the pressure gradient force, which means that the wind must be subgeostrophic. Examination of the balance of radial forces and acceleration shows

$$\frac{V^2}{R} = \frac{1}{\rho} \frac{\partial p}{\partial R} - f V \qquad (7.24)$$

Since the pressure gradient can always be replaced by the Coriolis term involving the equivalent geostrophic wind, $f V_g$, we have

$$\frac{V^2}{R} = f(V_g - V) \qquad (7.25)$$

confirming that the balancing wind speed is subgeostrophic. Winds satisfying this equilibrium are called *gradient winds*. Expressed as a fraction of the gradient wind speed this geostrophic departure is

$$\frac{V_g - V}{V} = \frac{V}{Rf} \qquad (7.26)$$

which is a Rossby number for the laterally accelerating flow. Observations suggest that deficits of actual winds below geostrophic winds often exceed 20% in vigorous extratropical cyclones.

Figure 7.18 also depicts anticyclonic flow in the northern hemisphere, and it is apparent by the same line of reasoning as above that wind speeds in these cases should be supergeostrophic, as is usually observed. In fact it is shown in appendix 7.10 that if the equivalent of eqn (7.24) is solved to find an expression for the gradient wind speed V corresponding to any particular geostrophic wind speed V_g (and therefore to any particular pressure gradient), then V_g must be no greater than $Rf/4$, which implies that for any latitude and radius of trajectory curvature R there is a maximum pressure gradient which is proportional to R. Although in fact trajectory and isobaric curvatures are not identical (see next paragraph) there clearly must be a strong connection, and the existence of such an upper limit to the pressure gradient therefore suggests that pressure gradients should die away toward the centre of an anticyclone, as is observed. No similar constraint applies in theory or observation to any kind of cyclonic flow.

Although this simple analysis is instructive and beloved of textbooks, it is actually very difficult to check by observation how close actual winds are to gradient equilibrium values. The reason for this is that it is very difficult to observe actual paths (trajectories) traced out by moving air parcels. The *streamlines* of flow which we can produce by careful analysis of observed winds on weather maps may be quite misleading in this respect for at least two important reasons. First, the flow is usually fairly unsteady, so that streamlines are incessantly wriggling. Even so, trajectories could in principle still be established by connecting up a succession of streamline segments from a series of maps at consecutive observation times, if it were not for the intervention of a second reason. Flow almost always has a component of vertical motion which is strong enough to ensure that the actual trajectory curvature of the rising or falling air can differ quite significantly from the curvature deduced by assuming horizontal or isobaric motion. The same uncertainty prevents accurate assessment of the balance of linear accelerations and geostrophic departures in situations such as Fig. 7.13: the actual linear acceleration of a parcel rising through the strong wind shear typical of a jet entrance may be signi-

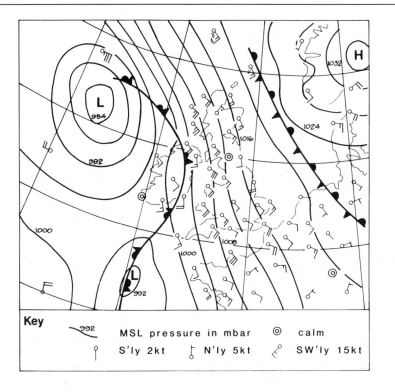

Fig. 7.19 Synoptic surface winds and mean sea-level pressures over the British Isles at 0000 Z, 10 January 1969, showing winds tending to blow obliquely towards low pressure, most obviously in lightest winds. The map shows south-easterly flow east of the occluded front of an old low, with a weak stagnant front between northern Britain and Scandinavia — a typical non-standard situation! (After the Daily Weather Report of the British Meteorological Office for that date)

ficantly different from what appears on a standard isobaric chart. These are part of a very general problem of analysis of air motion discussed in section 11.4.

Another important source of geostrophic departure arises from the enhanced turbulent friction in the planetary boundary layer. Figure 7.19 represents a typical state of affairs near the surface: winds observed at the standard observing height (10 m above the surface) are consistently angled at between 10° and 30° across the isobars, varying with wind speed and local geography, and may have speeds which are less than 30% of geostrophic values. The situation is outlined in some detail in Chapter 9, but for the moment the role of frictional drag can be seen by considering the three-way equilibrium between Coriolis, pressure-gradient and frictional forces (Fig. 7.20). The angular deviation and reduced strength of actual winds near the surface are consistent with the expected opposition of friction to motion over the surface, the rate of working by the pressure gradient on the parcel as it moves toward lower surface pressures being equal to the rate of working against frictional drag.

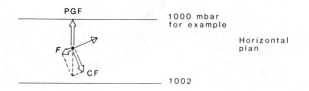

Fig. 7.20 Horizontal balance of pressure gradient, Coriolis and friction forces in the atmospheric boundary layer, showing ageostrophic flow across the isobars (northern hemisphere).

The consistent motion of air across isobars has obvious implications for systems having closed isobars (Fig. 7.21). Friction obviously induces a tendency for low-pressure centres to fill and high-pressure centres to empty — each of which is a

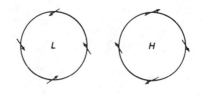

Fig. 7.21 Frictionally induced convergence and divergence in the boundary layers of cyclones and anticyclones (northern hemisphere).

significant aspect in the decay of synoptic scale weather systems (Chapter 11). The effect is quite strong near the surface, but is limited largely to the planetary boundary layer and is seldom significant more than about 500 m above the surface.

Geostrophic equilibrium as defined by eqns (7.16) is clearly impossible at the equator since the Coriolis parameter f is zero there. Very close to the equator, but not at it, the Coriolis terms are observed to be so small at realistic wind speeds that large-scale flow is not even quasi-geostrophic. This has important practical consequences for weather forecasters in the equatorial regions: there is little point in using the horizontal pressure (or isobaric contour) fields to consolidate the wind fields, as is standard over the rest of the globe, since the two are not usefully geostrophically related. Synoptic-scale flow is therefore analysed by drawing streamlines through the observed winds. In fact away from the occasional unusually vigorous weather system the pressure gradients apparent on weather maps in those regions are much weaker than at higher latitudes, and are often dominated by a diurnal tide driven by the Sun's heating. As latitude increases, the local tendency to geostrophy increases, so that for example in the weather systems which thrive on the border between the intertropical convergence zone and the trade winds, there is a fairly obvious relation between pressure and wind fields (section 12.3). At least this is so in the northern hemisphere where this border is some distance from the equator.

(a) (b)

Fig. 7.22 Circumpolar flows of water in a dish pan operating in Rossby-number regimes similar to those in the upper troposphere of (a) middle and (b) low latitudes in the northern hemisphere. (From [45])

More fundamentally, the increasing irrelevance of geostrophic equilibrium as the equator is approached, apparent in the increasing Rossby number U/Lf (expressed in a form which recognizes the special importance of the component of Ω about the local vertical), should lead us to expect atmospheric flow observed in low latitudes to differ in type from that prevailing in middle and high latitudes. Although the detailed dynamical reasoning is far beyond the introductory level, it should seem at least plausible that the increase in this Rossby number with latitude is closely related to the striking difference between the Hadley circulation in low latitudes and the continually contorting circumpolar vortex of middle and high latitudes. This is nicely confirmed by experiments with a laboratory-scale analogue of the atmosphere rather incongruously called the *dish-pan*. An annulus of water is trapped between a cold central pole and a warm equator (Fig. 7.22) in an open-topped pan rotating about an axis through the pole. This represents a flat Earth with only a single value for the Coriolis parameter (which remember is twice the component of rotation about the local vertical). Experiments with relatively slow rates of rotation therefore tend to simulate terrestrial behaviour at low latitudes, provided a special form of the Rossby number is arranged to have a realistic value, and produce Hadley circulations, whereas very realistically waved and slowly eddying circumpolar flow is observed at faster rotation rates.

7.12 Small-scale motion

SCALE ANALYSIS

Consider air flowing in the vicinity of a moderate-sized hill or shower cloud. The horizontal scale L is ~ 10 km, which is effectively the same as the vertical scale set by the depth of the troposphere. This symmetry of scale is very different from the extremely flattened configuration of flow on larger scales. Horizontal wind speeds are by and large set by the synoptic-scale situation, so that 10 m s^{-1} is again typical for U. Vertical wind speeds are about an order of magnitude smaller, so that there is some asymmetry but very much less than on larger scales. Timescales for air being in the influence of such a system range from $\sim 10^3$ s for air passing horizontally through the flow over a hill, to as much as an order of magnitude larger for air rising through a shower cloud (cumulonimbus). Horizontal pressure variations are observed by especially dense networks of sensitive barographs to be ~ 1 mbar.

A list of basic and secondary scales is as follows, others being the same as in the list for synoptic scales (section 7.7):

Horizontal and vertical scale length	L	10 km	10^4 m
Minimum timescale	t		10^3 s
Horizontal pressure change	p	1 mbar	10^2 Pa
Horizontal wind speed	U		10 m s^{-1}
Vertical wind speed	w		5 m s^{-1}
Horizontal acceleration	U/t		10^{-2} m s^{-2}
Vertical acceleration	w/t		10^{-3} m s^{-2}
Horizontal pressure gradient	p/L		10^{-2} Pa m^{-1}

Although the turbulent friction is difficult to scale even roughly, we can get some notion of its maximum magnitude by using KU/L^2 together with a K value comparable with what was used previously for the planetary boundary layer (10 m^2 s^{-1} — see section 7.7), but increased by an order of magnitude to offset the probable error in using L rather than some smaller subscale on which friction is especially effective (like the depth of the turbulent boundary layer). This probably overestimates friction in flow in the free atmosphere around a hill, but may not be too high for the interior of a cumulonimbus, since this is kept very turbulent by convection. The resulting value for the frictional term (see below) shows that friction is relatively unimportant on the chosen scale, but becomes rapidly more important with decreasing scale.

The magnitudes of terms in the components of the equation of motion (eqn(7.12)) are therefore as follows:

$$\frac{\mathrm{d}u}{\mathrm{d}t} = 2\Omega \sin\phi\, v \quad - \quad 2\Omega \cos\phi\, w \quad - \quad \frac{1}{\rho}\frac{\partial p}{\partial x} + F_x \tag{7.26a}$$

$$10^{-2} \qquad\quad 10^{-3} \qquad\quad 5\times 10^{-4} \qquad\quad 10^{-2} \quad\; 10^{-5}$$

$$\frac{\mathrm{d}v}{\mathrm{d}t} = -2\Omega \sin\phi\, u \qquad\qquad\qquad - \frac{1}{\rho}\frac{\partial p}{\partial y} + F_y \tag{7.26b}$$

$$10^{-2} \qquad\quad 10^{-3} \qquad\qquad\qquad\qquad 10^{-2} \quad\; 10^{-5}$$

$$\frac{\mathrm{d}w}{\mathrm{d}t} = 2\Omega \cos\phi\, u \quad - \quad g \qquad\qquad - \frac{1}{\rho}\frac{\partial p}{\partial z} - F_z \tag{7.26c}$$

$$10^{-3} \qquad\quad 10^{-3} \qquad 10 \qquad\qquad\qquad 10 \quad\;\; 10^{-5}$$

HORIZONTAL MOTION

In the horizontal, the balance is mainly between acceleration and pressure gradient, with a marginally significant Coriolis effect. There is therefore little or none of the geostrophic tendency which so dominates flow on larger scales, as is nicely indicated by the much larger value of the Rossby number

$$Ro = \frac{U}{\Omega L} \quad \text{or} \quad \frac{1}{\Omega t} \sim 10$$

which is about two orders of magnitude too large for quasi-geostrophic balance. The time period for air parcels in the influence of such small systems is so short that the Earth's rotation is largely irrelevant.

Local accelerations and pressure gradients are opposed, as shown by the opposing signs of the dominant terms in eqns (7.26a) and (b), so that air undergoing a linear acceleration tends to experience falling pressure, and decelerating air experiences rising pressure. For example, as air impinges on the bluff side of a hill and decelerates into a stagnation zone on the upwind side of the hill (Fig. 7.23), its pressure rises even if it moves absolutely horizontally, with the result that the pressure of the whole zone is raised slightly. Pressure is also raised a little in the lee of the hill (where the pressure gradient force acting away from the hill is associated

Fig. 7.23 Air impinging on a hill and inducing raised pressures in the stagnation zones immediately to windward and leeward.

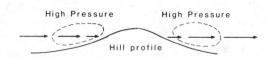

High Pressure   High Pressure

Hill profile

with the acceleration of air out of the stagnation zone there), but usually by less than in the upwind stagnation zone because of the complex influence of friction, with the result that there is a net downwind drag of the wind on the mountain, and of course an equal and opposite drag of the mountain on the wind.

Sometimes, as on larger scales, there are local accelerations which are perpendicular to the local wind, as air flows round tightly curved trajectories (round the sides of an isolated hill for example), and there is then a near balance between lateral pressure gradient force and centripetal acceleration. Such flow is termed *cyclostrophic*, and it applies to a wide range of curved flows centred on our chosen 10 km scale: on a smaller scale it applies to air in the vicinity of tornadoes (section 10.6 and Fig. 7.24) and dust devils, and on a larger scale it applies to air in the active annulus of high winds in a hurricane (section 12.4). The equation for cyclostrophic balance in air moving at tangential speed V around a circular path of radius R is simply

$$\frac{V^2}{R} = \frac{1}{\rho} \frac{\partial p}{\partial R}$$

Air near the funnel of a tornado often blows at speeds well in excess of 50 m s^{-1} around paths with radii of curvature of about 100 m, which indicate very large centripetal accelerations ($\sim$ 25 m s^{-2}). Corresponding pressure gradients are therefore $\sim$ 25 mbar per 100 m at least, and pressures in the funnel cores are depressed by many tens of millibars below ambient pressures. As mentioned in Section 10.6, this heavily distorted pressure field has consequences which are at once visually striking and potentially very hazardous.

Conditions in most systems in the small-scale range are of course much less extreme than in tornadoes, but there are nevertheless tendencies toward linear and lateral accelerations with nearly balancing pressure gradients which account for the dimpling of the small-scale pressure field which is observed by special arrays of sensitive barographs.

Note that the cos ϕ Coriolis term in eqn (7.26a), which was negligible on the

Fig. 7.24 A dust devil — a dusty vortex induced by intense local horizontal convergence at the base of a powerful thermal. They are especially common in dry seasons in tropical regions during the heat of the day.

synoptic scale, is now nearly comparable with the familiar sin ϕ term, because of the near equality of vertical and horizontal wind speeds. It is not clear whether or not the additional term plays a significant role in cumulonimbus, for example, because it is extremely difficult to make detailed observations of winds inside their turbulent, cloudy and precipitation-streaked interiors, but the form of eqn (7.26a) would suggest that there should be an eastward tendency in updraughts and a westward one in downdraughts.

VERTICAL MOTION

Vertical motion on the small scale is still dominated by hydrostatic balance between the vertical pressure gradient and gravity, their terms being four orders of magnitude larger than the next largest terms (vertical acceleration and Coriolis). However, much more rapid vertical accelerations are known to occur at the bottoms and tops of vigorous cumulonimbus, for example, in which the length and timescales of accelerating and decelerating air are locally each an order of magnitude smaller than the listed values. For example cold air pours downwards in shafts of heavy precipitation in cumulonimbus (sections 10.4 and 10.6), and may impinge on the underlying surface so vigorously that the downdraught ($-w$) reduces from 20 m s^{-1} to zero in the kilometre between cloud base and the surface. The corresponding value for dw/dt is $\sim 10^{-2}$ m s^{-2}, which is 0.1% of g and is associated with a corresponding distortion of the hydrostatically expected value of surface pressure (a rise ~ 0.1 mbar). Transient pressure 'footprints' of this and slightly larger orders of magnitude are observed by arrays of microbarographs as big cumulonimbus pass over, though their explanation is complicated by the simultaneous presence of adjacent masses of thermally contrasting air.

This is a rather extreme example of deviation from hydrostatic equilibrium; in convection in particular it is much more common to have an almost immeasurably small and very localized deviation which would be seem to be almost trivial except that it is associated with the very obvious and important production of updraughts and cloud.

The ambient air in which the convecting parcel or parcels are about to move can be considered to be in hydrostatic equilibrium, with only the convecting parcel out of balance. The ambient vertical pressure gradient is therefore related to the ambient air density by

$$\frac{\partial p}{\partial z} = -g\rho'$$

Substitution in the pressure gradient force term of eqn (7.26c) gives $g\rho'/\rho$ for the vertical pressure gradient force acting on the convecting parcel, where ρ is the density of the convecting parcel, which is slightly different from the density ρ' of the environment. The upward pressure gradient force and the downward gravitational force on the convecting parcel can be put together to produce the net upward force.

$$g\frac{\rho'}{\rho} - g = g\frac{(\rho' - \rho)}{\rho} = gB \tag{7.27}$$

This deviation from hydrostatic balance is more familiarly known as the *buoyant force* on the parcel, and in fact eqn (7.27) can be derived very easily from Archimedes' principle (appendix 7.11). The term in brackets contains the crucial density difference between the parcel and its environment which is the source of the net buoyant force, and is known as the *buoyancy B*. When the parcel is less dense than the surrounding air, then the buoyant force is positive and tends to produce an upward acceleration. In the atmosphere the density deficit of a buoyant parcel arises most often because it is warmer than its surroundings, though extra humidity can assist significantly in some circumstances, and extra cloud burden can detract somewhat (section 4.1).

The familiar dynamic effects of buoyancy are as described above, but in any particular situation the direction and magnitude of the associated acceleration depends on what other forces are acting in addition, and eqn (7.26) shows that these can be the Coriolis and frictional forces. To see whether acceleration and buoyancy are closely matched, or whether there are other significant items in the dynamic balance, we can compare vertical acceleration with net buoyant force by examining the dimensionless ratio known as the *Froude number Fr*:

$$Fr = \frac{\text{acceleration}}{\text{buoyant term}} = \frac{\mathrm{d}w}{\mathrm{d}t} / (gB) \sim \frac{w^2}{g\,B\,L}$$

Strictly speaking this is the *internal* Froude number, to distinguish it from the original number used by the pioneer fluid dynamicist Froude in modelling ship wakes and waves. It is another important dimensionless number to rank with the Reynolds and Rossby numbers, and describes the relative importance of gravity in the dynamical balance.

When *Fr* is much less than unity in the present context, the buoyancy is much larger than the vertical acceleration, showing that other factors as well as buoyancy must be significant in determining acceleration. Values for *B* in cumulus are surprisingly small ($\sim 1/300$) on account of the very small temperature excesses in rising air, and those for cumulonimbus are only a little larger (section 10.4). It follows from such values and observed strengths and dimensions of updraughts that *Fr* values in atmospheric convection cover a considerable range centred on about 0.2. It is clear therefore that there are forces at work significantly offsetting the expected effects of buoyancy.

Convection is visibly full of turbulent friction, however difficult it may be to devise a detailed mechanistic framework (Fig. 1.4). Cumulus of all sizes clearly must encounter considerable resistance as they push through their cloudless surroundings. If the buoyant parcels were rigid, the drag they encounter could be modelled in wind tunnels or water flows provided we arranged the details to ensure dynamical similarity with the original, i.e. similar values for the Reynolds and Froude numbers. Strain gauges could be used to measure the resulting drag. But the buoyant masses are not rigid of course; they are barely distinguishable from their surroundings, and they deform and tumble as they rise, in ways which must considerably alter the drag from the equivalent rigid body value and make it impossible to use strain gauges. Some laboratory and theoretical models have been used to show that the drag is very substantial, but the work is incomplete and beyond the scope of an introductory text [46].

Simple observation of cumulus also shows that there is another dynamically significant process at work in convection. Cloudy air expands very noticeably as it rises, showing that the buoyant, cloudy air is incorporating air from its immediate environment at a very considerable rate (Fig. 7.25). Such *entrainment* is a basic property of all vertical convection, from small thermals to large cumulonimbus (section 10.3), and its presence means that we must modify the equation of motion to allow for variation of air-parcel mass with time. This is done by returning to the original, general form of Newton's second law of motion (eqn (7.1)), whose vertical component can be written simply as

Fig. 7.25 A cloudy turret rising out of a parent cumulus congestus, dissipating quickly as a result of rapid entrainment of unsaturated air.

$$\frac{\mathrm{d}}{\mathrm{d}t}(M\,w) = NF_z \qquad (7.28)$$

where NF_z represents the resultant of all vertical components of force on the parcel. The left-hand side of eqn (7.28) can be expanded to make explicit the effects of variations in parcel mass M:

$$M\frac{\mathrm{d}w}{\mathrm{d}t} + w\,\frac{\mathrm{d}M}{\mathrm{d}t} = NF_z \qquad (7.29)$$

The first term is the familiar product of mass and acceleration. The second term represents the rate of increase of momentum which is accounted for in the rate of increase of the parcel mass rather than acceleration. If in any example we know the strength of the net force NF_z and observe only the upward acceleration of the parcel (not its growth by entrainment) we will miss the production of momentum as large masses of the initially static environment are set in motion through the process of entrainment. In fact as usual an effect ignored on the left-hand side of the equation of motion will seem like a mysterious force on the right-hand side, in this case a force logically called *entrainment drag*. Observations and some theory suggest that the obvious frictional drag and the less obvious entrainment drag are of comparable importance in convective dynamics, and that they offset the buoyant forces to such an extent that actual updraught speeds are only a tiny fraction of what they would be in a frictionless and non-entraining thermal. Only large cumulonimbus seem to have the capacity to operate considerably more freely than this (section 10.6).

7.13 Compressing and deforming air

CONTINUITY

Air, being fluid, deforms readily when it impinges on other bodies. Sometimes these are rigid material, such as mountains, or nearly rigid, such as the sea surface, but most often they are other bodies of air. Although the emphasis so far in this chapter has been on the air parcel considered as an isolated, almost solid packet, we must now come to terms with the essential fluidity of the air, and recognize also the compressibility which is so much in evidence in the vertical distribution of its mass.

The expansion of air when rising and its contraction when sinking are large even in the relatively shallow depth of the troposphere, because air is easily compressed. Compressibility may suggest elasticity, but it is important to note that although sound waves and the continual small pressure adjustments which travel at the speed of sound (section 4.3) depend crucially on the elasticity of air (its ability to spring outwards after compression), all other atmospheric dynamics related to weather operate as if the air had no elasticity, even though it is quite obviously highly compressible and deformable. Elasticity is unimportant because the flux of solar energy through the atmosphere is not so high as to encourage near sonic updraughts and jet streams, and this is the same moderation which allows the atmosphere to be so nearly hydrostatic in its vertical distribution of mass. For example air sinking towards the surface in the downdraught of a large cumulonimbus (section 10.4) compresses substantially as its ambient pressure rises, but it nevertheless impinges on the surface so gently that there is no perceptible tendency to compress further and rebound upward by re-expansion. Let us look more closely at the compression and deformation of air involved in synoptic and small-scale weather systems, as defined in sections 7.7 and 7.12.

We can derive an expression which is very useful for this discussion by assuming that the initial mass of any body of air is conserved no matter how it may subsequently deform, expand, combine or divide. The equation expressing the conservation of mass in flowing air is called the *continuity equation* because it is most easily derived by considering the continuity of mass flux as air blows through a volume fixed in space, and its derivation is outlined in Appendix 7.12:

$$\frac{\partial \rho}{\partial t} = - \left[\frac{\partial}{\partial x} (\rho u) + \frac{\partial}{\partial y} (\rho v) + \frac{\partial}{\partial z} (\rho w) \right] \tag{7.30}$$

The left-hand side of this equation is the rate of change of density at a fixed position, known as the *local* rate of change. The terms in brackets on the right-hand side describe the net flux of mass out of the fixed volume in question arising from longitudinal gradients of mass flux. There is a component of this *mass flux divergence* arising from each coordinate axis, as depicted in Fig. 7.26. A component (arising from the x axis for example) is positive if the mass flux in the x direction increases with x, because in that case the flux into the near side of a perpendicular cube fact is less than the flux out of the opposite face. Such divergent mass flux contributes to a fall of air mass in the cube and hence to a decrease in air density, as ensured by the minus sign in eqn (7.30).

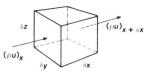

Fig. 7.26 Flows of air mass through opposite faces of a fixed volume. When unequal, they contribute to mass flux convergence or divergence.

HORIZONTAL AND VERTICAL BALANCE

Detailed investigations of atmospheric mass fluxes on scales ranging from the synoptic to the small show that the horizontal mass-flux divergence (the first two terms on the right-hand side of eqn (7.30)) is always very nearly balanced by an opposing vertical mass-flux divergence $\partial(\rho w)/\partial z$, so that the left-hand side is a relatively small residual. In fact the observed magnitudes of the opposing terms are such that if each were not opposed by the other, the local change of air density would often approach 100% per day on synoptic scales and per few minutes on small scales. Nothing remotely like this is observed: local percentage variations in

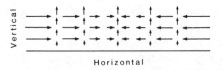

Fig. 7.27 Horizontal convergence and vertical divergence of mass flux on a vertical section through the lower troposphere, with consistent uplift in the middle troposphere.

air density are effectively equal in magnitude to local percentage variations in absolute temperature (local pressure variations being very much smaller than local temperature variations when expressed as fractions) and seldom exceed a very small fraction of the huge figures quoted above for the respective scales. To this extent therefore

$$\frac{\partial}{\partial x}(\rho u) + \frac{\partial}{\partial y}(\rho v) + \frac{\partial}{\partial z}(\rho w) \cong 0 \qquad (7.31)$$

It seems that the atmosphere arranges its flow to minimize local concentrations and depletions of mass: when mass is being gathered horizontally by converging mass flow, as in the lower troposphere in an extratropical cyclone or a cumulonimbus for example, it is stretched (diverged) vertically so that the resulting congestion of atmospheric mass is small. Since the vertical mass flux is zero at the Earth's surface, a vertical divergence of mass flux implies a vertical mass flux which increases with height above the surface throughout the horizontally converging layer, and hence a significant rate of uplift at its upper limit (the middle troposphere in Fig. 7.27), which of course is almost always associated with cloud production. We can investigate the magnitude of this uplift by integrating the terms of eqn (7.31) over the height range from the surface to the level of interest, as sketched in Fig. 7.28. The integration of the first two terms simply gives the net horizontal influx of mass through the sides of the imaginary column from surface to chosen level. The integral of the balancing third term

$$\int_b^t \frac{\partial}{\partial z}(\rho w)\, dz = [\rho w]_b^t = (\rho w)_t$$

simplifies to become the vertical mass flux out of the top of the column (i.e. at the chosen level) because there is of course no mass flux through the bottom surface. It follows from the near balance expressed by eqn (7.31) that

$$w_t = -\frac{1}{\rho_t}\int_b^t \left[\frac{\partial}{\partial x}(\rho u) + \frac{\partial}{\partial y}(\rho v)\right] dz \qquad (7.32)$$

Values for the right-hand side have been measured from careful analysis of synoptic-scale observations, and they show that the horizontal convergence of mass in the lower troposphere is consistent with rates of uplift ~ 10 cm s^{-1} in the middle troposphere, which are known from other evidence (rates of steady rainfall for example — section 11.4) to be quite typical. Mass convergence is difficult to measure accurately on the scale of a cumulonimbus, but the inference from known updraughts in middle levels (~ 5 m s^{-1}) that the convergences are 50 times larger than synoptic scale values is quite reasonable.

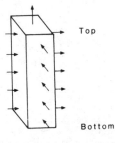

Top

Bottom

Fig. 7.28 Net convergent horizontal mass flux in an atmospheric column, with associated uplift at its top.

COMPENSATION

The reluctance of the atmosphere to concentrate in a confined volume shows itself also in a reluctance to concentrate in a confined column. The horizontal flux of mass into the column shown in Fig. 7.28 adds to the total mass of the column extending up to the top of the atmosphere and hence to the pressure at the bottom of the column (i.e. at the surface). This follows from the hydrostatic assumption, which we can make as usual, and from the fact that the rising air does not of course become weightless just because it rises above the level of our earlier interest. In fact the rate of rise of surface pressure (the pressure tendency as measured by a baro-graph) is simply g times the mass input, as we can see by assuming that the imaginary column has unit horizontal area.

$$\frac{\partial p_s}{\partial t} = g \int_b^t \frac{\partial M}{\partial t} = -g \int_b^t \left[\frac{\partial}{\partial x} (\rho u) + \frac{\partial}{\partial y} (\rho v) \right] dz$$

Since the integral on the right-hand side is given by the compensating rate of rise at middle levels (ρw at the top of the converging layer) we can quickly estimate the pressure tendency associated with uplifts of magnitudes quoted above. In the synoptic case it amounts to an amazing 500 mbar per day (assuming air density to be 0.5 kg m^{-3} at middle levels), and in the cumulonimbus case the same enormous value occurs in about half an hour. Again nothing like this is observed, even though tendencies measured at the bottom of the real atmosphere can obviously be inten-sified in rate (and shortened in time) by the bodily translation of horizontal pressure gradients embedded in moving weather systems. On the synoptic scale, pressure tendencies seldom exceed 20 mbar per day, and on the scale of cumulo-nimbus they seldom exceed a few millibars in the hour-long life of the cloud. The only possible explanation is that almost all of the input of mass to the lower half of the tropospheric column is offset by an output from the upper half, showing that there is horizontal divergence of mass flux there. The simplest picture is presented in Fig. 7.29. The horizontal convergence of mass extends up the middle tropo-sphere and is then replaced by horizontal divergence of mass extending up to about the tropopause. The balancing vertical mass flux increases with height up to the level at which horizontal convergence is replaced by divergence, and then diminishes steadily to about zero at the tropopause. The horizontal convergence and divergence *compensate* (nearly balance), and the slight imbalance accounts for the rate of change of pressure at the bottom of the column. The falling surface minimum pressure in the formative stages of an extratropical cyclone, for example, indicates that divergence aloft slightly but persistently exceeds the convergence beneath. More elaborate schemes with several layers of alternating convergence and divergence are of course possible, and could equally well explain the small pressure changes observed at the surface, but observations suggest that the two-layer scheme accounts for most synoptic-scale and cumulonimbus-scale weather systems. On the synoptic scale, for example, an anticyclone is like an inverted cyclone, with convergence overlying divergence and maximum downward mass flux between. This widespread and persistent subsidence is associated with many of the most striking properties of anticyclones (section 12.1).

We have now seen that the atmosphere arranges its flow on the synoptic and small scales so as to minimize the congestion and depletion of air mass in localized volumes and columns, and that each of these tendencies is associated with

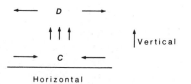

extremely basic and important aspects of weather behaviour. Though the analysis has therefore been useful, it is important to note that we have not made use of any dynamics (forces and accelerations) in the process — the basic physical property assumed is simply the conservation of mass. We have not in any sense explained why the atmosphere arranges its flow in these ways, nor even discussed the mechanisms involved; but the procedure has been valuable even so, and has pointed to quite basic types of behaviour which must be explained by properly dynamic theory. Such theory exists now in respect of synoptic scale motion (the theory of baroclinic instability which is briefly mentioned in section 11.6) but is far beyond the level of an introductory text; it barely exists at all in the case of cumulonimbus.

SYNOPTIC SCALE

Though the basic form of the continuity equation has proved directly applicable, it is useful to develop it a little further for the particular case of synoptic-scale motion, to try to see how compression and decompression can be made explicit in the formal description of deformability begun in eqn (7.31).

A serious snag appears when we consider the horizontal mass convergence in geostrophic flow. We can do this by taking the geostrophic expressions for ρu and ρv from eqn (7.17) and substituting them in the first two terms of eqn (7.31). Ignoring the variation of f with y (i.e. with latitude) which is relatively very small, we get

$$\frac{1}{f}\left[-\frac{\partial^2 p}{\partial x \, \partial y} + \frac{\partial^2 p}{\partial y \, \partial x}\right]$$

which is zero because the terms in the brackets cancel exactly, showing that a pure geostrophic flow has zero horizontal mass divergence. Physically this arises from the inverse relation between wind speed and pressure gradient which is basic to geostrophy. Figure 7.30 depicts a situation in which the geostrophic flow is increasing to the east, which corresponds to positive $\partial(\rho u)/\partial x$. The associated narrowing of the north–south separation of the contours corresponds to a negative $\partial(\rho v)/\partial y$ which exactly balances $\partial(\rho u)/\partial x$, giving zero net horizontal mass divergence. This is disappointing, because it shows that the so far very useful geostrophic approximation completely fails to allow for the convergence and uplift of air which we know is the facet of atmospheric flow which is essential in producing cloud, precipitation, and pressure variations. It appears that geostrophic balance is too well balanced for the real atmosphere. It is still useful as described earlier, but it must not be used in ways which allow it to eliminate vertical motion.

We can make less drastic and more useful simplifications by expanding each of the first two terms of eqn (7.31) into a term in the gradient of wind speed and another one in the gradient of density. Combining the latter from the two horizontal directions we have

$$u\frac{\partial \rho}{\partial x} + v\frac{\partial \rho}{\partial y}$$

which represents the contributions to density variation arising from motion in the presence of instantaneous horizontal density gradients. Since air density tends to move with the air itself, these terms can be regarded as representing the horizontal advection of existing density gradients by the flow. Detailed observations of

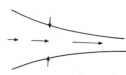

Fig. 7.30 Plan view of balancing longitudinal velocity divergence and lateral convergence in confluent geostrophic flow.

synoptic-scale flow indicate that horizontal advection of density is about an order of magnitude smaller than the other terms produced by the expansion of the original mass flux divergence, which are

$$\rho \left(\frac{\partial u}{\partial x} + \frac{\partial v}{\partial y} \right) \qquad (7.33)$$

representing the contributions arising from longitudinal stretching of flow. The relative insignificance of the advection terms is consistent with the tendency to minimize density concentrations (and therefore density gradients) noted already, and at least in part reflects the tendency of the thermal-wind relationship to align horizontal winds and isopycnals and hence minimize density (and temperature) advection.

DIVERGENCE

The bracketed expression in eqn (7.33) is called the horizontal divergence of velocity, or simply the *divergence*, and is usually denoted by D. The approximate form of the continuity equation (eqn (7.31)) can now be written in the form

$$D + \frac{1}{\rho} \frac{\partial}{\partial z} (\rho w) \simeq 0 \qquad (7.34)$$

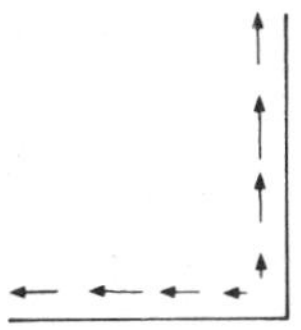

Velocity divergence is similar to the divergence of mass flux discussed above, but it is considerably simpler to interpret, depict and quantify. Figure 7.31 shows an example of flow with positive divergence D; if a patch of this air were marked with smoke at some initial time, the horizontal area A of the patch would increase with time for as long as D remained positive. The patch would also move bodily (northeastward in the example shown), but this is irrelevant to D which describes only the relative motion of the edges of the patch. It is shown in more formal expositions [44] that D is equal to the fractional rate of change of horizontal area of the patch.

Fig. 7.31 Plan view of flow divergent in both horizontal directions.

$$D = \frac{1}{A} \frac{dA}{dt} \qquad (7.35)$$

From a variety of evidence it is known that D values $\sim -10^{-5}$ s^{-1} are typical in the zone of horizontal convergence which virtually fills the lower troposphere in an extratropical cyclone. If divided equally between the two terms of D it follows that each of the stretching terms $\partial u/\partial x$, $\partial v/\partial y$ is similarly $\sim 10^{-5}$ s^{-1}. This amounts to a longitudinal gradient of wind speed of about 1 m s^{-1} in 100 km, which is barely measurable from standard synoptic observations, but corresponds to extremely significant convergence on the synoptic scale. It is a major problem in meteorology that such significant features are only marginally resolvable by direct observation, and one which is compounded by the irrelevance of the geostrophic approximation in this context. Considerable effort has gone into finding alternative ways of estimating divergence from available observations; fortunately there is a useful direct connection with large-scale rotation, as indicated in the next section.

In a region such as the lower troposphere in an extratropical cyclone, according to eqn (7.34) the convergence (negative D) in the lower troposphere is balanced at every level by

$$\frac{1}{\rho} \frac{\partial}{\partial z} (\rho w) \qquad (7.36)$$

If the atmosphere were incompressible and uniform this would simplify to $\partial w/\partial z$, a vertical stretching term, and the continuity equation would become simply

$$D + \partial w/\partial z = 0 \qquad (7.37)$$

showing that at every level vertical stretching balances horizontal convergence, or vertical squeezing balances horizontal divergence. If a D value of -10^{-5} s^{-1} prevails throughout the first 5 km of the troposphere, and is balanced at every level by positive $\partial w/\partial z$, then the cumulative uplift speed w at the 5 km level, which is found by integrating the vertical stretching term up to this level, becomes very simply

$$w = \int_0^z \frac{\partial w}{\partial z}\, \mathrm{d}z = -zD$$

which is 5 cm s^{-1} in this example — a fairly typical value for rates of uplift associated with nimbostratus in extratropical cyclones (see section 11.4). The behaviour described by eqn (7.37) is that of a slack plastic bag containing a fixed volume of water: when squeezed in the horizontal it expands in the vertical and vice versa — an example of simple incompressible deformability.

COMPRESSIBILITY

Because the atmosphere is in fact highly compressible, its behaviour is more complex than is allowed by eqn (7.37). However, the observational studies mentioned above show that it is only the vertical compressibility and non-uniformity which have significant effects, at least on the synoptic scale. If we try to adapt eqn (7.34) to the form of eqn (7.37), we find

$$D + \frac{\partial w}{\partial z} = Xw \qquad (7.38)$$

where appendix 7.13 shows that X is nearly constant with a value of about 10^{-4} m^{-1}. The deviation from simple incompressible behaviour is significant but not dramatic so long as no single layer of consistent convergence is as deep as the scaling thickness $1/X$, which is the case in practice since $1/X$ is about 10 km and most layers of consistent divergence or convergence are no more than half this depth. For example, in the case of the 5 km layer with divergence -10^{-5} s^{-1} quoted above, the balancing rate of uplift at the top of the converging layer is increased from 5 to 6.5 cm s^{-1} by the inclusion of the right-hand side in eqn (7.38) (appendix 7.13). An increase is to be expected since density falls with increasing height and ultimately it is its product with w which provides the balancing mass flux.

In advanced texts it is shown that the complication of the extra term on the right-hand side of eqn (7.38) can be avoided if the vertical coordinate z is replaced by pressure p. In fact all the complicating minor terms in eqn (7.30) vanish and the continuity equation becomes

$$\frac{\partial u}{\partial x} + \frac{\partial v}{\partial y} + \frac{\partial \omega}{\partial p} = 0$$

where ω is $\mathrm{d}p/\mathrm{d}t$, the rate of change of pressure of the moving air parcel, and the horizontal stretching terms are actually evaluated on an isobaric surface. The x, y, p coordinate system is used very widely in theoretical and computational work because of such simplifications, but elegance is achieved at the cost of conceptual subtlety which makes it less helpful for introductory discussion.

FRICTIONAL CONVERGENCE

Let us finish this section by applying the concept of virtually incompressible deformability to the case of frictional convergence in the planetary boundary layer of an extratropical cyclone. Because the wind vectors near the surface are consistently angled somewhat across the isobars, there is an inward component of air flow which may typically average ~ 2 m s^{-1} in the first 500 m. According to Fig. 7.32 this produces a value of about -10^{-5} s^{-1} if it persists across a 1000 km-wide region such as the central zone of a mature extratropical cyclone. This must be balanced by the vertical stretching $\partial w/\partial z$, which is therefore 10^{-5} s^{-1}. Integration through the depth of the boundary layer shows that air at the top of the converging layer (i.e. at the 500 m level) is rising at 0.5 cm s^{-1}. This may seem a very small value, but it is a significant event bearing in mind that it is lifting the overlying atmosphere over an area of 10^6 km^2. This lifting by frictional convergence influences a much greater depth of air than can be reached directly by surface friction, though the effects aloft have been transformed from the simple horizontal drag near the surface.

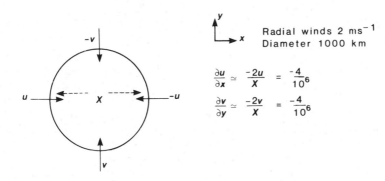

Fig. 7.32 Frictional convergence in the base of an extratropical cyclone, with wind components perpendicular to circular isobars.

7.14 The dynamics of rotation

ANGULAR MOMENTUM

It should be clear by now that the Earth's rotation influences atmospheric behaviour in several important ways, especially on the synoptic and larger scales with their conspicuous quasi-geostrophic balance. We can approach the basis of the dynamics of rotation by examining the extension of Newton's second law of motion to the case of a body whirling about a perpendicular axis of rotation under the influence of a force which exerts a twisting leverage or *torque* about the axis. In the simple case shown in Fig. 7.33 a body of mass M is whirling at a radial distance r about an axis with tangential speed V_t, while acted on by a tangential force F_t. Then the appropriate form of the second law is shown in basic dynamics texts to be

$$rF_t = \frac{\mathrm{d}}{\mathrm{d}t}(r\,M\,V_t) \qquad (7.39)$$

219

Fig. 7.33 Angular
momentum of body P rotating
about axis X.

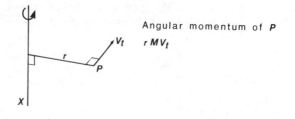

where the term in brackets on the right-hand side is the *angular momentum* (sometimes more logically called the moment of momentum) of the body in its rotation around the axis. It follows from this equation that angular momentum is conserved if there is no torque (no rF_t product) about the axis. In such a case

$$r M V_t = \text{constant} \tag{7.40}$$

Conservation of angular momentum is most usefully applicable to atmospheric motion on the synoptic and larger scales, because the absence of anything corresponding to the entrainment and mixing which so dominates small-scale flow means that the mass M of a large air parcel or mass is usefully constant, and means also that there is relatively little frictional torque outside the planetary boundary layer. Equation (7.40) then becomes simply

$$r V_t = \text{constant} \tag{7.41}$$

A surprising consequence of eqn (7.41) is that if the perpendicular distance r between the rotating body and the axis of rotation is reduced, then the tangential speed V_t must increase and vice versa. In the absence of any torque this may smack of magic, but there are several quite commonplace observations which confirm such a tendency, one of the most dramatic of which involves the spinning manoeuvre beloved of figure skaters. Firstly they generate angular momentum by carrying their initially linear momentum into a spiral trajectory which winds into a spin on an effectively fixed vertical axis. Having done this with whirling arms outstretched, the arms are dropped close to the body, obviously reducing the radial distance of their hands and arms from the axis of spin. In terms of eqn (7.41) r is reduced and the resulting sharp increase in V_t produces the intended visual effect of a spinning blur. To slow down before attempting to set off over the ice again, the arms are stretched out to reduce V_t to manageable proportions by the reverse of the initial process.

ANGULAR MOMENTUM BUDGET

The tendency to conserve angular momentum shows up quite clearly in the Hadley circulation, as already mentioned in section 4.7. Let us reconsider the Hadley circulation with possible conservation of angular momentum particularly in mind. Air rises to the high troposphere in the intertropical convergence zone, usually fairly near the equator, and then moves polewards in the high-altitude branch of the circulation. At these heights there is little friction with the surface; and there can be no overall east–west (zonal) component of pressure gradient force if we consider a zonal ring of air encircling the Earth at any particular latitude, since pressure at any altitude must form a continuous profile round the ring and hence exert as much torque westwards in some longitudes as it does eastwards in others.

To this extent there should be little zonal torque acting on the ring of air, so that as it moves polewards it must tend to conserve the angular momentum it had when it first rose out of effective frictional contact with the surface in the intertropical convergence zone.

If for simplicity the air starts its poleward migration with zero zonal wind speed (relative to the Earth) at the equator, then its initial angular momentum per unit mass is simply ΩR^2, of which ΩR is the eastward tangential speed of the Earth's surface at the equator (Fig. 7.34). If this is to be conserved, then the eastward wind speed U_ϕ (relative to the Earth's local surface of course) at some higher latitude ϕ must be such as to maintain the product of absolute eastward speed and perpendicular distance from the axis of rotation (Fig. 7.34)

$$(U_\phi + \Omega R \cos \phi)R \cos \phi$$

at its initial value. It follows that

$$U_\phi = \Omega R \left(\frac{1}{\cos \phi} - \cos \phi \right) \tag{7.42}$$

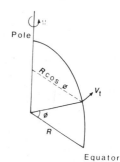

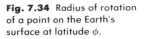

Pole

Equator

Fig. 7.34 Radius of rotation of a point on the Earth's surface at latitude ϕ.

As shown in Fig. 7.35, the implied increase of U_ϕ with latitude is quite rapid, with U_ϕ values exceeding 55 m s^{-1} at latitude 20°, and 130 m s^{-1} at latitude 30°. When averaged zonally, observed speeds seldom exceed 60% of these ideal values, but nevertheless the combination of poleward drift of air in the high troposphere and the sharp and consistent increase with latitude of upper tropospheric westerly wind speeds on the equatorial flank of the subtropical jet stream in each hemisphere strongly suggests that conservation of angular momentum in the poleward drift is a major part of the mechanism maintaining these jet streams.

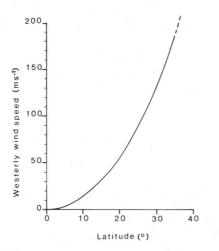

Latitude (°)

Fig. 7.35 Westerly wind speed of air conserving angular momentum from a state of zero wind at the equator.

The lack of complete agreement with predicted values shows that angular momentum is not fully conserved, and it is useful to consider briefly why this should be. Firstly there is still some frictional connection with the surface: cumulonimbus locally connect the bottom and top of the troposphere, exchanging momentum amongst other things. Secondly the argument about continuity of zonal pressure profile breaks down at levels which are pierced by mountains. As sketched in Fig. 7.36, the observed tendency for lee pressures to be a little lower than windward pressures means that there is a net torque opposing the zonal motion of the air. Air flowing over the mountains maintains a pattern of gravity

Fig. 7.36 Large-scale drag arising from slight asymmetry of upwind–downwind pressure fields around hills.

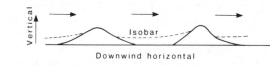

waves up to very great heights (section 10.8), one of whose effects is to extend some of this torque to heights far above the mountain tops. The third effect requires a hemispheric view of atmospheric angular momentum (Fig. 7.37). The atmosphere

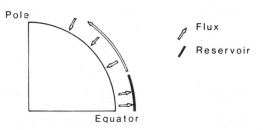

Fig. 7.37 Vertical and meridional fluxes of westerly zonal momentum.

is continually gaining westerly angular momentum in low latitudes and losing it at high latitudes, in each case by momentum exchange with the underlying surface. Large masses of air take up the westerly momentum of the local surface through boundary-layer friction and the mechanisms described above. They may then move away to another region as they become involved in a large weather system, where they meet and conform their flow to other masses from other source regions. In so doing they exchange momentum, even though they do not mix in the way that small parcels do. The general result of such exchanges is that relative westerly momentum is transferred to high latitudes and relative easterly momentum is transferred to low latitudes. The westerlies of mid-latitudes are maintained in this way, largely by the momentum exchanges associated with synoptic and larger-scale weather systems. In fact there is a continual conversion from a large-scale poleward *eddy* flux of angular momentum to the zonal westerlies which is closely related to a similar conversion of kinetic energy discussed in section 13.3. In this context the upper branch of the Hadley circulation is like a flange of equatorial angular momentum moving polewards on nearly frictionless bearings near the equator, but being impeded more and more as it reaches toward the subtropics by momentum exchange with air reaching down from higher latitudes. There is therefore a transition from a smooth poleward flux of zonal angular momentum, in which zonal angular momentum is nearly conserved, to an eddy flux in which no conservation is obvious in the zonal averages (though in principle angular momentum is conserved throughout the eddying process in just the same way as is heat), and the core of the subtropical jet stream meanders in the transition zone.

Conservation of angular momentum lies at the root of the observed tendency of all poleward-moving air to increase its relative westerly motion, or reduce its eastward motion, and of all equatorward-moving air to do the opposite. Such tendencies are widespread in the atmosphere and are often conspicuous in synoptic-scale weather systems. They are apparent in the great cloud swirls of Fig. 1.1, in particular in the eastward curve of the swathes of frontal cloud moving polewards in middle latitudes, known from detailed studies to be associated with huge injections of warm, moist air upwards and polewards from the polar flanks of the subtropical high-pressure systems (section 11.4).

The tendency of air to veer to the right of its current direction of motion (in the northern hemisphere) should remind you of the Coriolis effect. In fact the two approaches are entirely equivalent, the conservation of zonal angular momentum in poleward-moving air being identical to the integrated effect of the eastward Coriolis force associated with poleward motion. The Coriolis formulation is more useful for examining the shape of individual flows, while zonal angular momentum is more useful for looking at large-scale budgets and statistics.

VORTICITY

The large, flattened air masses of synoptic-scale weather systems also tend to conserve angular momentum as they rotate about locally vertical axes in their centres. Indeed the dynamics of this rotational motion is a very important part of the overall dynamics of large-scale motion systems, and is treated at length and in somewhat forbidding detail in advanced texts. However, one of the most fundamental results of such treatments emerges in the following simplified approach.

Consider a narrow horizontal circular ring of air rotating on a weather map as part of a synoptic-scale cyclonic or anticyclonic system (Fig. 7.38). If its angular velocity about a locally vertical axis through the centre of the ring is ω and its radius is r, then conservation of angular momentum about the axis requires (from eqn (7.41)) that

$$r^2\omega = \text{constant}$$

$$\text{or } \frac{d}{dt}(r^2\omega) = 0$$

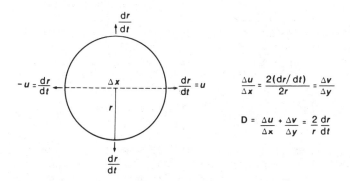

Fig. 7.38 Expansion of a horizontal ring of air in symmetrically divergent flow.

We can expand the latter to find an expression for the rate of change of angular velocity of the ring ($d\omega/dt$) when it is expanding or contracting in response to synoptic scale divergence or convergence (section 7.13).

$$r^2\frac{d\omega}{dt} + \omega\frac{d}{dt}(r^2) = 0$$

Rearranging to isolate the required rate of change of angular velocity, we find

$$\frac{d\omega}{dt} = -\frac{2\omega}{r}\frac{dr}{dt}$$

But it is apparent from Fig. 7.38 that the horizontal divergence D of the circular plate of air bounded by the thin ring is equal to $(2/r)dr/dt$ so that

$$\frac{d\omega}{dt} = -\omega D \qquad (7.43)$$

223

So far we have considered the rotation of the air relative to the map, but since the map is actually rotating with angular velocity $\Omega \sin \phi$ about the local vertical, where Ω and ϕ are the Earth's angular velocity and latitude (Fig. 7.34 again), $\Omega \sin \phi$ should be added to ω in eqn (7.43) to produce

$$\frac{d}{dt}(\omega + \Omega \sin \phi) = -(\omega + \Omega \sin \phi)D \tag{7.44}$$

It is conventional when dealing with rotation in fluids to describe localized rotation by means of *vorticity*, which is defined to be twice the instantaneous angular velocity of the parcel. The usual sign convention means that cyclonic rotation is positive. The component of *relative vorticity* (2ω) about the local vertical is always symbolized by ζ in meteorology. Since the doubled angular velocity of the map about the local vertical $(2\Omega \sin \phi)$ is the familiar Coriolis parameter f, it follows that eqn (7.44) can be rewritten

$$\frac{d}{dt}(\zeta + f) = -(\zeta + f)D \tag{7.45}$$

The Coriolis parameter is sometimes called the *planetary vorticity* on this account, and the total $(\zeta + f)$ is known as the *absolute vorticity* of the air.

A more thorough derivation would have included torque terms. One arises from friction, which fairly obviously exerts a torque tending to reduce relative vorticity. Another very significantly arises from the baroclinity of the atmosphere, since an extended form of the buoyancy force always tends to align intersecting isobars and isopycnals and hence impose rotation. Other terms allow for the twisting of x and y components of vorticity into the local vertical, in particular the substantial component about the local horizontal which is associated with any thermal wind there may be, since shear always implies vorticity about an axis perpendicular to the plane in which the shear is profiled. However, the very simple eqn (7.45) applies quite usefully in many conditions to the evolution of synoptic-scale weather systems, and even more reliably to events on larger scales.

EXAMPLES

As an important example, consider the development of cyclonic rotation in air which is originally at rest on the weather map. Viewed from the weather map the rotation seems to be generated from nothing, but in fact the initially apparently static air has the planetary vorticity (f) of the map itself. In the presence of the crucial ingredient of synoptic-scale convergence (negative D) in the lower troposphere, eqn (7.45) shows that positive relative vorticity will begin to grow there according to

$$\frac{d\zeta}{dt} = -fD$$

If D remains constant and none of the complicating terms becomes significant, the solution of eqn (7.45) (Fig. 7.39) shows that the absolute vorticity $\zeta + f$ increases exponentially with a doubling time of $\ln 2/D$. Assuming a typical value of -10^{-5} s^{-1} for D, we find that the absolute vorticity doubles (which is to say that the relative vorticity increases from zero to f) in about 20 hours, which is observed to be quite typical of a developing extratropical cyclone.

In the same way, eqn (7.45) helps to explain how negative relative vorticity develops in diverging flow. Starting from zero relative vorticity again, divergence

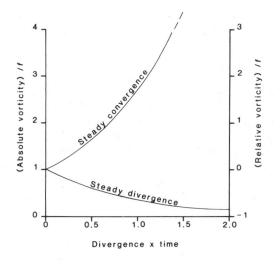

Fig. 7.39 Exponential growth with time of absolute vorticity in a steadily convergent part of a weather system, and exponential decay in a steadily divergent part. The initial relative vorticity is zero in each case, and axes have been normalized by f and $1/D$ for generality.

produces negative relative vorticity, which corresponds to anticyclonic rotation. If the value of D remains constant, and no other terms interfere, the absolute vorticity decays exponentially towards zero, which is to say that relative vorticity decays to $-f$ (Fig. 7.39). Such a limit to the intensity of anticyclonic vorticity is amply confirmed by real synoptic-scale behaviour, and distinguishes anticyclones from cyclonic systems which show no such restraint, and which often produce relative vorticities several times as strong as the planetary vorticity f. To the extent that all synoptic-scale rotation about the local vertical consists of concentrations of the Earth's rotation, rotation in the opposite direction is impossible. This result is consistent with a detailed examination of gradient balance in anticyclonic flow (appendix 7.10).

The association of divergence with anticyclonic vorticity and convergence with cyclonic vorticity is often quite obvious in a single weather system. In terms of the profile of an extratropical cyclone sketched in Fig. 7.29, we should expect to find cyclonic vorticity in the lower troposphere as already discussed, and anticyclonic vorticity in the upper troposphere. The latter is usually much less obvious, but is evident nevertheless after a little scrutiny as outlined in section 11.2.

In the previous section it was shown that synoptic-scale uplift and subsidence are so sensitive to convergence and divergence that routine observations of wind are barely able to detect convergent conditions capable of maintaining widespread, vigorous nimbostratus. We have now seen that synoptic-scale rotation about the local vertical is similarly sensitive. So although it is surprisingly difficult to measure the associated horizontal convergences and divergences of mass, even a casual glance at a satellite picture such as Fig. 1.1 shows striking evidence of both uplift and rotation. In fact the sensitivity of rotation to uplift and subsidence can be used to good effect in theoretical and computational procedures.

By contrast with its inability to cope with convergence or divergence, the geostrophic approximation remains well able to represent synoptic-scale rotation. More advanced texts show that the relative vorticity about the local vertical is given in terms of the usual horizontal wind components by

$$\zeta = \frac{\partial v}{\partial x} - \frac{\partial u}{\partial y}$$

Substituting the geostrophically associated pressure gradients according to eqn

225

(7.17), and ignoring gradients of ρ and f in comparison with gradients of wind speed, we find

$$\zeta_g = \frac{1}{\rho f}\left(\frac{\partial^2 p}{\partial x^2} + \frac{\partial^2 p}{\partial y^2}\right)$$

The terms in the bracket do not cancel, as they did in the case of geostrophic divergence. Each of the pair corresponds to the curvature of the horizontal pressure profile across a horizontal (the actual curvature of the equivalent isobaric surface), and a little thought will show that each is positive if the profile is concave upwards (Fig. 7.40). In other words a pressure minimum is associated geostrophically with positive vorticity, and a pressure maximum with anticyclonic rotation. In fact a synoptic-scale map analysed with the help of the geostrophic approximation provides a usefully accurate representation of vorticity distribution.

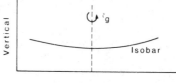

Fig. 7.40 Vertical radial isobaric profile across a depression with cyclonic geostrophic vorticity (northern hemisphere).

If we describe the pressure field in terms of the slopes of isobaric surfaces (eqn (7.20)), the expression for ζ_g is obviously equivalent, and very importantly does not require that we assume uniform air density.

$$\zeta_g = \frac{g}{f}\left(\frac{\partial^2 Z_p}{\partial x^2} + \frac{\partial^2 Z_p}{\partial y^2}\right)$$

Note lastly that the relationship between convergence and rotation also applies to very intense circulations on the small scale, the most extreme being the tornado. Although detailed understanding of the formation of tornadoes is still very incomplete, it is clear that their intense rotation is associated with intense convergence, and is described at least qualitatively by eqn (7.44). As discussed in section 11.6, tornadoes form beneath the bases of very vigorous cumulonimbus, where there is intense convergence feeding the powerful updraughts aloft. This convergence concentrates and intensifies background vorticity to produce the intense and localized vortex which is the tornado. However, in such tumultuous conditions it is not clear which of the many complicating terms missing from eqn (7.44) may also be significant. Most tornadoes are observed to rotate cyclonically, but since conditions in a tornado (section 7.12) imply vorticities $\sim 10^4$ f, it would be very rash to conclude that they are therefore simply extreme concentrations of planetary vorticity. As in the hoary legend of the wash-basin vortex changing direction in ships crossing the equator, there are many local factors which may be much more significant, though these should probably not include the tendency to produce cyclonic vorticity by road traffic driving on the right, as suggested recently by some American meteorologists in playful mood.

Appendix 7.1 Dimensions

All measurable quantities, however complex, can be regarded as a combination of very basic measurements, the most common of which are mass M, length L and time T. For example, every velocity consists of a length divided by a time, as is obvious from the units in which values are expressed. Velocity is therefore said to have *dimensions* of length divided by time — a statement which is often written

$$[\text{velocity}] = [L]\,[T]^{-1}$$

Although the dimensions of a quantity are obviously consistent with the units in which it is expressed, they are quite general and independent of the particular unit system used. Such generality supports the axiom that all equations validly relating observable quantities should be dimensionally consistent, as is justified by experience. For example simple observations and quite elaborate theory show that gravity waves propagate across the surface of a layer of water of depth h at a speed c given by

$$c = \sqrt{g\,h} \tag{7.46}$$

provided the wavelength is considerably larger than h. The dimensions of g are those of an acceleration, which are $[\text{velocity}]\,[T]^{-1}$ and consequently $[L]\,[T]^{-2}$, and the dimensions of h are simply $[L]$. The dimensions of the right-hand side of eqn (7.46) are therefore

$$\{[L]\,[T]^{-2}\,[L]\}^{\frac{1}{2}} = [L]\,[T]^{-1}$$

in agreement with the obvious dimensions of the left-hand side.

Dimensional consistency is a powerful and useful constraint. If incorrect theory or inaccurate measurement had suggested

$$c = g\sqrt{h}$$

its dimensional inconsistency would have shown at once that it could not be valid. Inverting the procedure, if we had guessed that wave speed should depend only on g and h in the form

$$c = g^a\,h^b \tag{7.47}$$

then dimensional consistency requires that a and b must each be $\frac{1}{2}$. This follows at once if we substitute dimensions in eqn (7.47) and require that each must separately balance

$$[L\,T^{-1}] = [L\,T^{-2}]^a\,[L]^b$$

Equating powers we have

$$[L] \quad 1 = a + b$$
$$\text{and } [T] \quad -1 = -2a$$

therefore $a = b = \frac{1}{2}$, and the relation must be of the form

$$c = \sqrt{g\,h}$$

Note that there is no constraint on any pure numbers, such as π, which may be in the full relationship, since they are dimensionless. For example the equivalent of eqn (7.46) for waves whose wavelength λ is smaller than h is

$$c = \{g\lambda/2\pi\}^{\frac{1}{2}}$$

This is dimensionally consistent of course, being like (7.46) with the length λ replacing the length h, but its consistency is quite independent of the numerical factor 2π, which can be justified only by detailed theory or observation.

The dimensional consistency of valid equations enables us to find the dimensions of less simple quantities. For example, from the most familiar form of Newton's second law of motion

force = mass × acceleration

it follows that

$$[\text{force}] = [M] [L] [T]^{-2}.$$

And from the definition of pressure as force per unit area

$$[\text{pressure}] = [M] [L]^{-1} [T]^{-2}.$$

As an important worked example, consider the dimensions of the exponential scale height of the pressure profile of an ideal, isothermal and hydrostatic atmosphere. In so doing we need to recognize another basic dimension — that of temperature $[\theta]$.

$$H = R T/g \tag{7.48}$$

The dimensions of the specific gas constant R can be found from the equation of state for air (eqn (4.2)):

$$[R] = [L]^2 [T]^{-1} [\theta]^{-1}$$

When the dimensions of the quantities remaining on the right-hand side of eqn (7.48) are substituted, we find

$$[\text{RHS}] = [L]$$

which of course is consistent with its interpretation as a height.

Relations are at their most general when they are expressed in the most dimensionless form, the pressure/height relation in an isothermal atmosphere being a case in point.

$$p = p_0 \exp(-z/H_e)$$

If this is rewritten by *normalizing* or non-dimensionalizing pressure by expressing it in units of pressure p_0 at some datum level, and normalizing height by expressing it in units of the exponential scale height:

$$p_* = p/p_0$$
and $\quad z_* = z/H_e$
then $\quad p_* = \exp(-z_*)$

and the curve is the general one of pure exponential decay. A single graph with suitably normalized axes could contain all such pressure profiles, and indeed all other quite unrelated processes involving pure exponential decay (such as radio-active decay of a single species of unstable nucleus). Very sophisticated use is made of such normalization in analysing possible relationships in a field such as turbulence, where observations cannot yet be brought together in a comprehensive theoretical framework, as mentioned in section 9.9 (eqn (10.22)).

Dimensionless numbers such as the Reynolds number Re owe their generality to judicious choice of dynamically significant factors whose ratio is dimensionless, i.e. whose various dimensions cancel out, leaving a parameter which is a pure ratio, independent of whatever system of consistent units is used for expressing the various terms. As an exercise, check that Re is dimensionless.

Appendix 7.2 Pressure gradient force

Consider a block of air in the presence of a component of pressure gradient $\partial p/\partial x$ along the x axis. Because of this gradient, slightly different forces must act on the faces perpendicular to the x axis at slightly different positions x and $x+dx$ (Fig. 7.41a). If the y and z dimensions of the block are dy and dz, then the force on the block face at position x is

$$F_x = p_x\,dy\,dz$$

in the positive x direction, where p_x is the pressure on this face. The corresponding force on the block face at position $x+dx$ is

$$F_{x+dx} = p_{x+dx}\,dy\,dz$$

in the negative x direction.

Now the pressures on the two faces differ slightly because of the pressure gradient, and by the formal definition of the gradient

$$p_{x+dx} = p_x + \frac{\partial p}{\partial x}\,dx$$

for vanishingly small dx (Fig. 7.41b). Hence the net force NF in the positive x direction can be written in terms of the pressure gradient

$$
\begin{aligned}
NF &= F_x - F_{x+dx}\\
&= (p_x - p_{x+dx})\,dy\,dz\\
&= -\frac{\partial p}{\partial x}\,dx\,dy\,dz
\end{aligned}
$$

Since the volume of the block is its mass divided by air density

$$dvol = dx\,dy\,dz = \frac{dm}{\rho}$$

we have

$$NF = -\frac{dm}{\rho}\frac{\partial p}{\partial x}$$

and consequently that the net pressure gradient force per unit mass is

$$-\frac{1}{\rho}\frac{\partial p}{\partial x}$$

(a) (b)

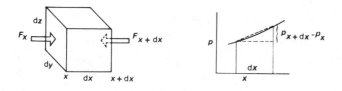

Fig. 7.41 (a) Net force on a pair of opposite sides of an air parcel embedded in a pressure gradient. (b) Pressure difference and pressure gradient in the x direction.

This is the particular case for the component in the x axis. Identical procedure finds $(1/\rho)\,\partial p/\partial y$ and $(1/\rho)\,\partial p/\partial z$ for the y and z components.

Appendix 7.3 Viscous force

Figure 7.3 depicts in profile a flow in the x direction which is sheared in the z direction (i.e. a vertically sheared eastward flow). Consider the frictional forces arising from viscosity and acting on a layer of air of unit horizontal area bounded by heights z and $z + dz$. At level $z + dz$ the faster-moving air above drags the top surface of the layer in the positive x direction with a force numerically equal to the tangential shearing stress $\tau_{z+dz,x}$. At level z the slower-moving air below drags the bottom surface of the layer in the negative x direction with a force which is numerically equal to the tangential shearing stress $\tau_{z,x}$. Just as in appendix 7.2, the small difference in stress across the narrow gap can be written in terms of the gradient of stress there:

$$ND \;=\; \frac{\partial}{\partial z}\,\tau_{z,x}\,dz$$

and the volume of air in the layer (numerically equal to dz because of choosing unit horizontal area) can be replaced by mass/density, with the result that the net force per unit mass is given by

$$\frac{1}{\rho}\,\frac{\partial}{\partial z}\,\tau_{z,x}$$

If the horizontal shearing stress $\tau_{z,x}$ at any level obeys eqn (7.4), then the net viscous force per unit mass is given by

$$\frac{1}{\rho}\left(\mu\,\frac{\partial^2 u}{\partial z^2} + \frac{\partial u}{\partial z}\,\frac{\partial \mu}{\partial z}\right)$$

and if gradients in μ are proportionally much smaller than those in u, we have eqn (7.6).

Appendix 7.4 Centrifugal pressure gradient

As shown in Fig. 7.42, the equatorward component of centrifugal force per unit mass at latitude ϕ is

$$\Omega^2 R \cos\phi \sin\phi \qquad \text{or} \qquad \tfrac{1}{2}\Omega^2 R \sin 2\phi$$

If balanced by an equatorward horizontal pressure gradient

$$-\frac{1}{\rho}\,\frac{\partial p}{\partial y} \;=\; \tfrac{1}{2}\,\Omega^2 R \sin 2\phi \tag{7.49}$$

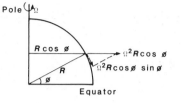

At latitude 45° with ρ 1.2 kg m^{-3} we have

$$\frac{\partial p}{\partial y} = -\tfrac{1}{2}\Omega^2\rho R \cong -2\times 10^{-2} \text{ Pa m}^{-1}$$

which means that near sea-level, pressure is increasing toward the equator at about 20 mbar per 100 km.

To integrate this balancing gradient from equator to pole, notice that $dy = R\, d\phi$ if latitude ϕ is expressed in radians.

$$p_p - p_e = \int \frac{\partial p}{\partial y}\, dy = R\int_0^{\pi/2} \frac{\partial p}{\partial y}\, d\phi$$

Substituting eqn (7.49) for the balancing $\partial p/\partial y$ we find

$$p_p - p_e = -\tfrac{1}{2}\rho\Omega^2 R^2 \int_0^{\pi/2} \sin 2\phi\, d\phi$$
$$p_e - p_p = \tfrac{1}{2}\rho\Omega^2 R^2$$
$$= \tfrac{1}{2}\rho\, V^2$$

where V is the tangential speed of a point on the equator. Using values for V (appendix 4.4) and sea-level air density we find

$$p_e - p_p = 129\ 000 \text{ Pa} = 1290 \text{ mbar}$$

which is nearly 30% larger than mean sea-level pressure.

Appendix 7.5 Curvature terms

The conventional meteorological reference frame (Fig. 7.1) gives rise to hidden accelerations not only because it is rotating, but also because two of its axes (x and y) are curved, and the z axis tilts as we move about over the Earth's surface. Consider a westerly wind of speed u at latitude ϕ. Because the x axis at latitude ϕ is actually a latitude circle of radius $R\cos\phi$, a body moving along it actually has centripetal acceleration $u^2/R\cos\phi$ toward the Earth's axis of the rotation (Fig. 7.43). Similar terms arise from other wind components. Except near the poles, where obviously the conventional meteorological reference frame is inconvenient and should not be used, all curvature terms arising from wind speed V are of order V^2/R. Assuming realistic values for V, the kind of comparison of magnitudes exemplified in this chapter shows that these curvature terms are almost always very much smaller than the dominant terms in each component of the

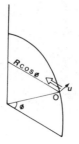

Fig. 7.43 Acceleration implicit in motion along the curved x axis (eastwards).

equation of motion, and can therefore be ignored. In the vicinity of jet streams, where wind speeds occasionally approach 100 m s^{-1}, they are no longer negligible, and they are retained in some of the more elaborate dynamical and computational models of the atmosphere, but we will incur little error by ignoring them in the rest of this book.

Appendix 7.6 The Coriolis effect

Figure 7.44 shows a body moving across a rotating turntable with uniform velocity V relative to the turntable, which is rotating with angular velocity Ω about a perpendicular axis. This means that the body is moving at a steady speed V along a straight line which rotates with the turntable. Since we are interested only in the Coriolis component of the resulting centripetal acceleration, the only rotational effect we need consider is the rotation of AB to A'B' (the obvious centripetal acceleration toward the axis of the turntable is independent of V, and is therefore not a Coriolis component). The body describes an arc AXB' of a circle when viewed from a non-rotating frame, and therefore suffers a lateral Coriolis acceleration toward the centre O of the circle.

Notice that the position of O is quite unrelated to the undefined point about which the turntable is rotating; it simply lies on a perpendicular to the arc on the side favoured by the direction of rotation of the turntable. The oversimplicity of

Fig. 7.44 As the body moves along the rotating straight line AB, the rotation takes B to B' and the body describes the arc AXB' in non-rotating space.

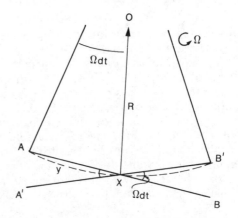

Fig. 7.5 was therefore unnecessary, at least for the purpose of considering Coriolis.

We can find the magnitude of the Coriolis acceleration by finding the radius R of the circular arc in terms of V and Ω. By symmetry the arc must make equal angles $\Omega \, dt/2$ with the chords AX, A′X, XB′ and XB wherever it touches them, where $\Omega \, dt$ is the angle turned by the turntable in the small time dt taken for the body to move from A to B′. By noting that the internal angles of the triangle AOX must total 180°, it appears that the angle AOX is $\Omega \, dt$. However, in radian measure this angle is Y/R, where Y is the length of arc AX. But from the body's motion, $Y = V \, dt/2$, where the approximation improves as $\Omega \, dt$ diminishes. To this approximation therefore

$$\Omega \, dt = \frac{Y}{R} = \frac{V \, dt}{2 \, R}$$

so that

$$R = \frac{V}{2 \, \Omega}$$

The Coriolis acceleration of the body toward O as it moves along the arc at speed V is V^2/R which becomes $2 \, \Omega \, V$, as found in the simple concentric case in section 7.3.

The other components of the total centripetal acceleration arising from the turntable rotation cannot be simply displayed on Fig. 7.44, since they depend on the position of the axis of turntable rotation. However it is easily shown that if L is the perpendicular distance between the body and this axis, then the centripetal components are $\Omega^2 \, L$ and v^2/L, where v is the tangential component of motion relative to the turntable.

The treatment of the kinematics of rotation in two or three dimensions is very elegantly simplified by using vectors, as shown in more advanced texts [44]; however the gain in algebraic simplicity is made at some cost in ease of perception of physical significance.

Appendix 7.7 Geostrophic flow

By definition, pure geostrophic flow is one in which the components of pressure gradient and horizontal flow obey eqn (7.17):

$$f v = \frac{1}{\rho} \frac{\partial p}{\partial x} \qquad\qquad f u = - \frac{1}{\rho} \frac{\partial p}{\partial y}$$

Squaring and adding we find

$$f V = \frac{1}{\rho} \frac{\partial p}{\partial n}$$

where $\qquad V = (u^2 + v^2)^{1/2}$

and $\qquad \dfrac{\partial p}{\partial n} = \left[\left(\dfrac{\partial p}{\partial x} \right)^2 + \left(\dfrac{\partial p}{\partial y} \right)^2 \right]^{1/2}$

As shown in Fig. 7.45, if the resultant wind (speed V) makes an angle α with the

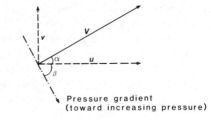

Fig. 7.45 Horizontal wind components and pressure gradient in geostrophic balance (northern hemisphere).

x axis, then $\tan \alpha = v/u$. If the resultant pressure gradient makes an angle β with the x axis, then according to eqn (7.17),

$$\tan \beta = \frac{\partial p}{\partial y} \Big/ \frac{\partial p}{\partial x} = -u/v = -1/\tan \alpha$$

that is

$$\beta = \alpha - 90°$$

This means that the pressure gradient is perpendicular to the geostrophic wind in the sense depicted in Fig. 7.45. Since pressure gradient is positive toward higher pressure, it follows that lower pressure lies perpendicularly to the left when standing with your back to the geostrophic wind in the northern hemisphere, in agreement with Buys-Ballot's law.

Appendix 7.8 Pressure gradient and contour slope

Figure 7.46 shows a vertical section through a region with a horizontal pressure gradient directed to the right, as indicated by the upward slope of the isobars in that direction. Two isobars are shown, with pressures p and $p + dp$. The horizontal pressure difference dp between them is of course identical to the vertical difference, which can be related to the vertical separation dz through the hydrostatic relation $dp = g\rho dz$. It follows that

$$\frac{dp}{dn} = g\rho \frac{dz}{dn}$$

As dn becomes vanishingly small, dp/dn becomes the horizontal pressure gradient $\partial p/\partial n$ and dz/dn becomes the gradient of the p isobar, which we will write as $\partial Z_p/\partial n$. Dividing across by the air density we have

Fig. 7.46 Section through tilted isobaric surfaces, with vertical axis z and horizontal axis n.

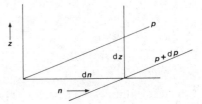

$$\frac{1}{\rho} \frac{\partial p}{\partial n} = g \frac{\partial Z_p}{\partial n}$$

where the left-hand side is the horizontal pressure gradient force per unit mass in the direction of the n coordinate (appendix 7.2). The right-hand side is therefore the equivalent expression in terms of isobaric gradient.

Appendix 7.9 Thermal wind

According to section 7.10 and eqn (7.21) the difference in geostrophic wind speed between two isobaric levels is directly related to the horizontal gradient of the thickness of the bounded layer

$$V_{g_2} - V_{g_1} = \frac{g}{f} \left[\frac{\partial Z_{p_2}}{\partial n} - \frac{\partial Z_{p_1}}{\partial n} \right] \tag{7.51}$$

Since the atmosphere is so nearly hydrostatic we can relate the thickness (the vertical depth) of a column to its mean density and therefore temperature, so that the right-hand side depends on the horizontal gradient of column mean temperature. Let us examine this dependence in detail.

Begin with appendix 4.4 and eqn (4.18), which results from the vertical integration of the hydrostatic equation:

$$Z_{p_2} - Z_{p_1} = \frac{R \, \overline{T}}{g} \ln \left(\frac{p_1}{p_2} \right) \tag{7.52}$$

where the column mean temperature $\overline{T}$ is an average weighted by the logarithm of pressure

$$\overline{T} = \int_{p_2}^{p_1} T \, \frac{d \ln p}{\ln(p_1/p_2)}$$

Since everything on the right-hand side of eqn (7.52) is constant except $\overline{T}$ (remember the isobars are fixed once chosen), substitution in the right-hand side of eqn (7.51) gives

$$V_{g_2} - V_{g_1} = \frac{R}{f} \ln \left(\frac{p_1}{p_2} \right) \frac{\partial \overline{T}}{\partial n}$$

We can use eqn (7.52) again to replace the terms before $\partial \overline{T}/\partial n$ on the right-hand side, and rearrange to find

$$\frac{V_{g_2} - V_{g_1}}{Z_{p_2} - Z_{p_1}} = \frac{g}{f \overline{T}} \frac{\partial \overline{T}}{\partial n} \tag{7.53}$$

If we consider an extremely thin layer then the left-hand side of eqn (7.53) tends to $\partial V/\partial Z$, which is the vertical gradient of shear of the geostrophic wind. On the right-hand side $\overline{T}$ becomes the temperature of the isobaric surface in the middle of the thin layer.

235

Appendix 7.10 Gradient balance in anticyclones

Figure 7.18(b) represents the gradient balance in anticyclonic circular flow. Being entirely radial the balance can be written as a single scalar equation

$$PGF - CF = CA$$

using the symbolism of Fig. 7.18(a). The pressure gradient force PGF can be replaced by the Coriolis force on the equivalent geostrophic wind fV_g, and the Coriolis force CF and the centripetal acceleration CA replaced by their standard expressions to give:

$$fV - fV_g = \frac{V^2}{R}$$

Rearranging in the standard form of a quadratic equation we find

$$V^2 - RfV + RfV_g = 0$$

which has the standard solution

$$V = \frac{Rf \pm \sqrt{(R^2 f^2 - 4RfV_g)}}{2} \tag{7.54}$$

The term under the square root must be positive to avoid physically meaningless imaginary terms, so that

$$4RfV_g < R^2 f^2$$

or

$$V_g < \frac{Rf}{4}$$

which shows that there is a maximum geostrophic wind speed $Rf/4$ which cannot be exceeded, and whose value increases with R and f.

If we substitute this maximum geostrophic wind in eqn (7.54) we find

$$V = \frac{Rf}{2} = 2V_g \tag{7.55}$$

The linear increase of wind speed with radius means that the air is rotating about the centre of the anticyclone like a solid wheel with uniform angular velocity

$$\omega = -\frac{V}{R} = -\frac{f}{2}$$

Since $f/2$ is the component of the Earth's angular velocity about the local vertical, we see that the limiting anticyclone is one which is rotating at a rate which exactly cancels the rotation of the local surface and is consequently non-rotating when viewed from an external inertial reference frame. As discussed in section 7.14 this is expected to be the limiting result of pronounced horizontal divergence of flow with conservation of angular momentum.

Consider the consequences of the restriction on the geostrophic wind speed. Since at any latitude every value of V_g corresponds to a particular isobaric slope through the geostrophic relation, we see that the maximum value of V_g implies a maximum magnitude of isobaric slope

$$\frac{\partial Z_p}{\partial R} = -\frac{R f^2}{4g}$$

where the minus sign means that the isobaric surface slopes downward away from the centre of the anticyclone. When the air density is ρ, the equivalent expression for maximum pressure gradient is

$$\frac{\partial p}{\partial n} = -\frac{\rho R f^2}{4}$$

The radial profile of pressure in such a pressure field can be found by integrating the pressure gradient outward to an arbitrary outermost radius R (1500 km in this case) at which pressure is assumed to be typical of the region outside the anticyclone. The central elevation of pressure Δp is given by

$$\Delta p = \frac{\rho R^2 f^2}{8}$$

and radial profiles of pressure at latitudes 15, 30 and 45° are sketched in Fig. 7.47, from which it is apparent that substantial elevations of surface pressure are hardly possible at latitudes much lower than the subtropics, at least provided the flow is dominated by the gradient balance.

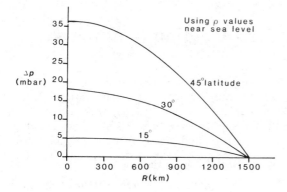

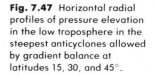

Fig. 7.47 Horizontal radial profiles of pressure elevation in the low troposphere in the steepest anticyclones allowed by gradient balance at latitudes 15, 30, and 45°.

Appendix 7.11 Buoyancy

Consider a parcel of air (or any body) of density ρ totally immersed in air (or any fluid) of density ρ'. According to Archimedes' principle the immersed parcel experiences an upward thrust which is equal to the weight of the air displaced. Since the parcel experiences a downward force equal to its own weight, the net upward force F is given by the difference

F = weight of displaced air − weight of parcel

If the volume of the parcel is Vol, this expression can be rewritten

$F = g(\rho' - \rho)Vol$

Dividing through by the mass of the parcel we find the net upward force per unit mass

$$\frac{F}{M} = \frac{g(\rho' - \rho)}{\rho} = g\,B$$

which is identical to the expression (7.27) derived from the vertical component of the equation of motion.

Note that Archimedes' principle, though still elegant and extremely useful, is no longer regarded as a fundamental principle. The upward force arises because the hydrostatic vertical gradient of pressure in the surrounding fluid, together with the basic property that fluid pressure acts equally in all directions at a point, ensures that the upward pressure force on the lower parts of the body exceeds the downward pressure force on its upper parts. The same thing tends to happen in the presence of any pressure gradient, vertical or horizontal, whatever its cause. In the vortex of a tornado (section 7.12) the horizontal pressure gradient may exceed the vertical by a factor of two or more. The net inward force on an immersed body will maintain a centripetal acceleration exactly matching that of the surrounding air only if the body has the same density as the air. The dense debris picked from the surface and demolished buildings etc. is centrifuged outward (actually tangentially) in a potentially lethal spray which scours the zone at and around the base of the vortex and its cloud funnel.

Appendix 7.12 Continuity of mass

Figure 7.26 shows a very small frame fixed in space with air blowing through the open faces. Considering only air flow in the x direction, the mass flux into the face perpendicular to the x axis at x is given by $(\rho u)_x\,dy\,dz$, where the subscript x implies that the air density and speed vary with x. A similar expression describes the mass flux out of the face at $x + dx$, and the net influx is given by their difference

$$\text{net mass influx} = (\rho u)_x\,dy\,dz - (\rho u)_{x+dx}\,dy\,dz$$

As in appendix 7.2 we can replace the difference in the variable term by the product of its gradient and the small separation dx, to find

$$\text{net mass influx} = -\frac{\partial}{\partial x}(\rho u)\,dx\,dy\,dz$$

Equivalent expressions can be derived for the components of net mass influx in the directions of the y and z coordinates, and the resultant net flux of mass into the volume is the algebraic sum of the three components:

$$\text{net mass influx} =$$
$$-\left[\frac{\partial}{\partial x}(\rho u) + \frac{\partial}{\partial y}(\rho v) + \frac{\partial}{\partial z}(\rho w)\right]dx\,dy\,dz \qquad (7.56)$$

Since mass is conserved, the net rate of mass influx must equal the rate of increase of mass in the volume bounded by the frame:

$$\frac{\partial}{\partial t}(\text{mass}) = \frac{\partial\rho}{\partial t}\,dx\,dy\,dz \qquad (7.57)$$

since the volume bounded by the fixed frame is conserved. Combining eqns (7.56) and (7.57) and simplifying, we find

$$\frac{\partial \rho}{\partial t} = -\left[\frac{\partial}{\partial x}(\rho u) + \frac{\partial}{\partial y}(\rho v) + \frac{\partial}{\partial z}(\rho w)\right]$$

which is one of the standard forms of the continuity equation.

Appendix 7.13 Compressibility and divergence

Although somewhat beyond the general level of this book, the following derivation is included because it does not appear explicitly in texts seen by the author. It is given largely without commentary, but you will see the usual appeals to the equation of state and the hydrostatic equation during manipulation.

The second term of eqn (7.34) can be rewritten

$$\frac{1}{\rho}\frac{\partial}{\partial z}(\rho w) = \frac{\partial w}{\partial z} + \frac{w}{\rho}\frac{\partial \rho}{\partial z}$$

Since $\qquad \rho = \dfrac{p}{RT}$

$$\frac{1}{\rho}\frac{\partial p}{\partial z} = \frac{1}{\rho RT}\frac{\partial p}{\partial z} - \frac{p}{\rho RT^2}\frac{\partial T}{\partial z}$$

$$= \frac{1}{p}\frac{\partial p}{\partial z} - \frac{1}{T}\frac{\partial T}{\partial z}$$

Now $\qquad \dfrac{\partial p}{\partial z} = -g\rho$

and $\qquad \dfrac{\partial T}{\partial z} = -S\dfrac{g}{C_p}$

where S is the environmental lapse rate expressed as a fraction of the dry adiabatic lapse rate.

Hence $\qquad \dfrac{1}{\rho}\dfrac{\partial \rho}{\partial z} = -\dfrac{g\rho}{p} + \dfrac{g\,S}{C_p T}$

$$= -\frac{g}{C_p T}\left(\frac{C_p}{R} - S\right)$$

The value of C_p/R is almost exactly 3.5. Outside relatively shallow inversions or isothermal layers, S values range from about 0.4 (rather more stable than the warmest saturated adiabat) to the limiting value of 1 (a dry adiabat). If for simplicity we assume an S value of 0.5, the error in the bracket is relatively small even in the extreme cases, and less than 5% most of the time.

Thus $\qquad \dfrac{1}{\rho}\dfrac{\partial p}{\partial z} \cong -\dfrac{3\,g}{C_p T}$

$$= -X$$

where X has value $\simeq 10^{-4}\ \text{m}^{-1}$.

Finally

$$\frac{1}{\rho}\frac{\partial}{\partial z}(\rho w) = \frac{\partial w}{\partial z} - X\,w$$

so that eqn (7.34) can be written in the form

$$D + \frac{\partial w}{\partial z} = X w \tag{7.58}$$

Let us solve eqn (7.58) for the simple two-layer case depicted in Fig. 7.48. This includes uniform convergence C from zero z to $z = h$ and uniform divergence $-C$ from there to $z = 2h$; and w is zero at zero z. The solution is simplest if the equation is used additionally in the form $\partial^2 w / \partial z^2 = X \, \partial w / \partial z$. The following solution can be found by standard methods, or simply checked by substitution.

Up to $z = h$ $\qquad\qquad w = \dfrac{C}{X}(e^{Xz} - 1)$

$\quad$ at $z = h$ $\qquad w = w_h = \dfrac{C}{X}(e^{Xh} - 1)$

$\quad$ up to $z = 2h$ $\qquad\qquad w = w_h - \dfrac{C}{X}(e^{Xz'} - 1)$

$\qquad\qquad\qquad$ where $z' = z - h$

$\quad$ at $z = 2h$ $\qquad\qquad w = 0$

A particular case of this solution is plotted in Fig. 7.48, where $C = 10^{-5}$ s^{-1} and $h = 5000$ m, in obvious relevance to an extratropical cyclone. The above solution can be compared with the solution to the equivalent incompressible atmosphere undergoing the same combination of convergence and divergence (the straight dashed lines), and it is clear that the difference is small because h is considerably less than the scaling depth $1/X$, the maximum updraught in the middle troposphere being enhanced by a little over 10% by the inclusion of compressibility.

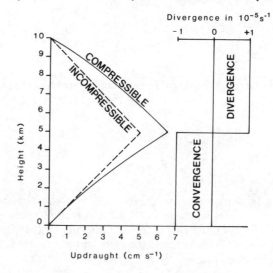

Fig. 7.48 Vertical profiles of updraught associated with a simple two-layer model of typical synoptic-scale convergence and divergence in an incompressible atmosphere (dashed line) and a realistically compressible one (solid line).

Problems

LEVEL 1

1. If pressure increases both northwards and eastwards at 5 mbar per 100 km, show that the maximum rate of increase is 7.1 mbar per 100 km towards the

north-east. What contribution does this make to the pressure gradient in the local vertical?

2. Sketch the sheared planetary boundary layer, including the directions of the shearing stresses, on the top and bottom surfaces of an embedded horizontal layer and the direction of the associated momentum flux.

3. A person is standing on spring weighing scales resting on a rotating turntable. In what way is the reading of the scales affected by the rotation if (a) the scales are horizontal, and (b) they are tilted somewhat toward the axis of rotation?

4. In the case of Fig. 7.5, sketch the directions of the familiar centripetal and Coriolis accelerations when the train and turntable are moving in opposite directions.

5. From the scale analysis in section 7.7, find the basic scales which determine the relative sizes of the horizontal and vertical accelerations on the synoptic scale. Likewise find the basic factors determining the ratio of horizontal to Coriolis accelerations.

6. What would be the effect on the accuracy of the geostrophic approximation near the Earth's surface if all turbulence there were to disappear?

7. In trying to set up a 1:10 scale water model of low tropospheric flow (over buildings for example) at about 10 m s^{-1}, what water speed should be used to ensure that the model is dynamically similar to the atmosphere in respect of turbulence? Note that the kinematic viscosity of water is close to 10^{-6} m^2 s^{-1}.

8. Consider the directions of geostrophic flow (clockwise, anticlockwise) round lows and highs in each hemisphere.

9. Assuming zero geostrophic wind speed in the very low troposphere, use the thermal wind relation to find the wind direction in the high troposphere in the following situations in the northern hemisphere: colder air to the north-west; warmer air to the west; colder air to the north; warmer air to the south-west.

10. Why is there no familiar Coriolis effect at the equator? Consider why this should have no effect on the occurrence of tornadoes there. (In fact they are rare there, but for reasons most apparent in section 10.6.)

11. Consider the advisability of installing a convector heater in the interior of a spacecraft.

12. What is the fundamental significance of the cyclonic direction of rotation? (*Hint:* comparison with clock rotation is not helpful, as reversal between the two hemispheres should suggest.)

13. What directions of rotation should be enhanced in the low troposphere of a cyclone, the low troposphere of an anticyclone and the high troposphere of an anticyclone?

14. If air has a tangential speed of 30 m s^{-1} at a distance of 300 m from the centre of a tornado, what tangential speed should it attain if it spirals in to a radius of 30 m while conserving its angular momentum?

LEVEL 2

15. A toy train is running along a straight track resting on a rotating turntable and passing through its axis of rotation. Given that a lateral force of 1 N per unit train mass is sufficient to tip the train off the track, find the maximum safe speed of the train when the turntable is completing a revolution every 3 s.

16. Carry out a scale analysis of the horizontal equation of motion for air moving through a jet stream, in which individual air parcels pass through the jet core from entrance to exit (1000 km apart) in 6 hours, and other factors are as

specified in section 7.7. Look in particular at the implications for geostrophic balance.

17. In the low troposphere, where air density is 1 kg m^{-3}, what are the geostrophic wind speeds associated with a horizontal pressure gradient of 5 mbar per 100 km at latitudes 60°, 30° and 5°?

18. Repeat the calculations of problem 17 in the high troposphere where air density is only 0.25 kg m^{-3}. Find the equivalent slopes of isobaric surfaces in all these cases (problems 17 and 18).

19. Apply the thermal wind relation to the case of a well-defined front with an isobaric (effectively horizontal) temperature gradient of 5 °C per 100 km throughout a 1 km deep slab in which the mean temperature is 250 K. At a latitude where f is 10^{-4} s^{-1}, and simplifying g to 10 m s^{-2}, find the increase in geostrophic wind speed along the front from bottom to top of the slab.

20. Consider an air parcel with absolute temperature T embedded in a slightly cooler environment (temperature T'). Assuming the parcel to have the same pressure as its surroundings, use the equation of state for air to show that the buoyant force per unit mass is given by $g(T - T')/T'$.

21. Using the expression derived in problem 20, consider the vertical motion of a parcel of constant temperature 300 K through air with temperature 299 K. Find the buoyant force per unit mass and calculate the rate of rise of the parcel after ascending 1 km from rest in the absence of other forces. (Use the standard relation between distance and time for uniform acceleration: $w^2 = 2az$.) Why is this answer unrealistically large?

22. A cumulus cloud is observed to be increasing its volume by 1% per second while rising at 1 m s^{-1}. Find its entrainment drag per unit buoyant mass and compare it with the typical buoyant force per unit mass estimated in problem 21 ($\sim$ $^1/_{30}$ m s^{-2}).

23. A typical value for horizontal convergence in the lower troposphere of an extratropical cyclone is 10^{-5} s^{-1}. Supposing this to remain constant, find the change in horizontal area of a large plate of embedded air. Calculate its area at three consecutive 24-hour intervals after an initial state in which it was 10^4 km^2.

24. In the case of problem 23, find the associated rate of uplift at heights 1, 3 and 5 km above the surface, assuming that the divergence is uniform throughout the height range and that the atmosphere behaves like a simple incompressible fluid.

25. In the case of problem 24, find the absolute and relative vorticities after 1, 2 and 3 days, assuming zero relative vorticity initially and a Coriolis parameter of value 10^{-4} s^{-1}.

LEVEL 3

26. Discuss the roles of drag of all types on synoptic and small-scale dynamics, including cyclonic weather systems and the cumulus family.

27. Sketch the directions of vertical fluxes of easterly momentum in those parts of the trade winds where the easterly flow is surmounted by westerlies in the upper troposphere, and outline the types of mechanism carrying these fluxes through the base and top of the layer of easterlies.

28. Summarize and discuss all points at which atmospheric dynamics is influenced by the flatness of the atmosphere (the gross asymmetry between vertical and horizontal scales). Speculate on how things might differ if the atmosphere were much deeper.

29. Outline the influence of the Earth's rotation on atmospheric motion, discussing reasons for the difference between large- and small-scale motion systems in this respect. Speculate briefly on how things might differ if the Earth rotated (a) much faster than at present (as it did in the distant past before being slowed by marine tidal friction), and (b) much more slowly than at present (as it will do in the distant future for the same reason).

30. There is a widespread popular belief that the direction of water flowing out of handbasins should tend to be anticlockwise in the northern hemisphere and clockwise in the southern hemisphere. Critically examine the likely validity of this belief by identifying the principle which is apparently assumed and assessing the relative importance of other factors involved.

31. The increase of wind speed with height apparent in problem 19 is very sharp. Check this by extrapolating to the case of a layer which fills the mid-latitude troposphere. Such wind speeds are not observed. Consult Fig. 7.16 and a detailed description of the thermal structure of fronts (section 11.3) to explain why it is associated with lesser speeds, and outline the consequences for the shape of associated jet streams as seen in vertical sections across the flow.

32. Assume cyclostrophic balance in a tornado, with air whirling round a radius of 100 m at 100 m s^{-1}. Find the associated slope of an isobaric surface and consider its implications for the shape of the funnel cloud.

Radiation and global climate

8.1 The solar spectrum

The Sun is a powerful emitter of electromagnetic radiation in wavelengths ranging from the ultraviolet to the infra-red (Fig. 8.1). Human vision has evolved to make use of most of the shorter-wavelength half of the total range, apart from the relatively small ultraviolet component, so that the longer-wavelength half is by definition in the infra-red.

Almost all solar radiation comes from the *photosphere*, a relatively shallow shell of incandescent gases in the outer parts of the sun. The photosphere has an outer radius of nearly 700 000 km (over 100 Earth radii) and consists mainly of hydrogen and helium at temperatures of nearly 6000 K and substantial pressures arising from the weight of overlying, less conspicuous gases. In normal laboratory studies, incandescent gases emit line spectra, but at the higher temperatures and very much higher pressures prevailing in the photosphere, the radiating atoms collide with one another so frequently and so violently that the individual lines are smeared out into the relatively smooth hump shown in Fig. 8.1.

The visible spectrum is centred on wavelengths of about 0.5 μm, not far from the wavelength of maximum solar emission per unit wavelength range. On their own these wavelengths would appear green to the human eye, but the substantial range of wavelengths in visible sunlight, together with the variation in human visual sensitivity across it, combine with the effects of the atmosphere on sunlight to produce the familiar golden appearance of the sun when well above the horizon. Although easily seen when the sun is viewed through a spectroscope, the dark Fraunhofer lines produced by selective absorption of certain wavelengths by relatively cool solar gases outside the photosphere would appear as insignificant notches if they were included in Fig. 8.1. The much larger indentations which are apparent there are produced in the Earth's rather than the sun's atmosphere.

The Earth's cloudless atmosphere modifies the solar spectrum quite significantly by selective absorption and scattering, the effect of a standard cloudless atmosphere being included in Fig. 8.1. Clouds are much less wavelength-selective, but can of course reduce the total transmitted power very considerably when thick enough. This is caused almost entirely by back-scattering, and can reduce the solar power reaching the underlying surface by as much as 90% when there are thick cloud layers. Data for Fig. 8.1 have been obtained by careful radiometry at the

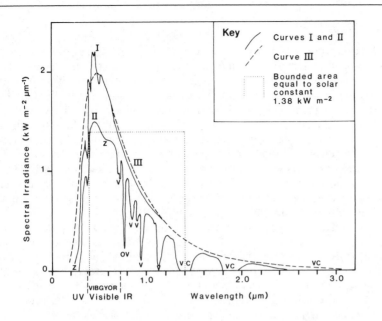

Fig. 8.1 Power spectrum of solar irradiance at the Earth — the distribution of power (vertical axis) across the range of solar wavelengths (horizontal axis). Curve I is irradiance just outside the Earth's atmosphere — the total area under the curve being the solar constant. Curve II is typical of the irradiance at sea level with the Sun in the zenith of a clear sky. Notches marked o, c, v and z indicate atmospheric absorption by oxygen, carbon dioxide, water vapour and ozone respectively. Curve III is the power spectrum for a blackbody at 5900 K reduced to allow for inverse-square attenuation between Sun and Earth. (After P.R. Gast [46])

Earth's surface, and at various altitudes up to tens of kilometres above sea level by special balloons, and more recently of course by artificial satellites above the atmosphere.

Observation and theory agree that the solar photosphere radiates with nearly the maximum efficiency possible at its temperature. Perfectly efficient radiators are known as *blackbodies* because, being also perfectly efficient absorbers, they reflect no electromagnetic radiation; they and the *blackbody radiation* they emit have been studied closely as an important part of thermodynamics. Some properties are briefly outlined in appendix 8.1. The emission spectrum of a blackbody at temperature 5900 K is included in Fig. 8.1 for comparison. After scaling to allow for the attenuating effects of the Earth's distance from the sun, as outlined below, the agreement with the solar spectrum observed outside the Earth's atmosphere is obviously good. It is not perfect, particularly at the peak and on the ultraviolet flank of the spectrum, but it is good enough to allow us to regard the sun as a black-body at the temperature of the photosphere for many purposes. We can therefore use *Stefan's law* (appendix 8.1) to estimate the power radiated from each square metre of the photosphere (its *emittance*) to be about 70 MW m^{-2}. Multiplying by the surface area of the photosphere, we find the total solar output to be about 4.2 $\times$ 10^{20} MW. Though this is a vast figure by terrestrial standards, using Einstein's equation ($E = mc^2$), we find it represents a rate of mass loss which would account for only 0.1% of the present solar mass if it persisted throughout the life of the Sun to date. This output is maintained by thermonuclear reactions in the solar core, and may have changed little during the life of the Earth, though this is obviously difficult to confirm and is subject to debate.

Solar radiation floods outward through the solar system, its *irradiance* (the rate of flow of radiant energy through unit area perpendicular to the solar beam at any location) falling with increasing distance from the Sun. Since only trivial fractions of the total are absorbed by the planets and interplanetary dust and gas, the same total radiant flux must pass through any spherical surface concentric with the Sun. Simple geometry then shows that the irradiance I at any distance R from the sun is related to the photospheric emittance I_s by

$$I = I_s (R_s / R)^2 \tag{8.1}$$

where R_s is the radius of the photosphere. (Note the inverse-square relationship between I and R.) Since the Earth orbits the Sun at a distance of about 200 solar radii, it follows that the solar irradiance at the Earth's orbit is only about $(1/200)^2$ of the photospheric emittance. The Earth's orbit around the Sun is slightly elliptical, the semi-axis increasing from perihelion to aphelion by about 3%. The R dependence produces a corresponding 6% decrease in the irradiation of the Earth, with the maximum occurring in the northern hemisphere winter in the present state of the Earth's orbit. Though the effects of this variation on seasonal climatic variations are currently masked by the very large effects of the 23.5° tilt of the Earth's equatorial plane relative to the plane of its orbit around the Sun (discussed in section 8.7), they may have been large enough at various times in the past to have contributed to major climatic change (section 4.9).

The annual average solar irradiance just outside the Earth's atmosphere is the *solar constant*, whose value is of the greatest importance for the state and behaviour of the atmosphere, since it largely determines the power supply for the atmospheric engine and thereby controls many very important physical conditions prevailing at the Earth's surface, including several such as temperature which are crucial to the evolution and maintenance of life. The solar constant has been measured repeatedly with increasing sophistication throughout the twentieth century, but only those measurements made from artificial satellites in recent decades have been completely free from atmospheric interference and consequent error. As mentioned in section 4.9, there is evidence of small variations linked with sun-spot numbers, and astronomical studies of the lifecycles of stars like our Sun strongly suggest that the output from the young Sun was about 30% less than now. Current values centre on 1.38 kW m^{-2}, with variations of about 1%.

Note that the argument in this section reverses actual observational procedure. Photospheric temperature is actually deduced from measurements of the solar constant, using Stefan's law for blackbody emission and assuming zero interplanetary absorption. However, such estimates agree very well with others made by different means, for example by applying Wien's law (appendix 8.1) to measurements of the wavelength of maximum solar emission, and with theoretical models of the Sun, and it is the consistency of all such results which confirms the validity of the model presented, and with it the value of the photospheric temperature quoted at the beginning of this section.

8.2 The Earth's energy balance

As mentioned earlier (section 4.9), the Earth's composite system of atmosphere, ocean and land surface undergoes irregular vacillations between warm and glacial epochs on timescales ranging from 10^4 to 10^8 years. But despite their dramatic effects on life in middle and high latitudes, such temperature variations are small distortions of the mean temperatures established by planetary equilibrium, and the associated imbalances between energy fluxes to and from the planet Earth are very much smaller than the constituent influxes and effluxes separately. Indeed it is quite likely that the climatic effects of any such imbalances are considerably enhanced by internal instabilities in the composite terrestrial system (section 4.9).

On a much shorter timescale, the big seasonal variations in localized heat fluxes are reproduced year after year with relatively little variation, at least when viewed on the hemispheric scale. To a very good approximation, therefore, the atmosphere, ocean and land surfaces are in a nearly steady thermal state when viewed over one or more full years, and their collective energy budget must be very nearly balanced when its components are averaged over the same periods.

Solar radiation is the obvious energy input to this composite system, and in fact is effectively the only one, since the next largest is the *geothermal flux* from the Earth's hot interior (heat generated *in situ* by early gravitational collapse and subsequent radioactive decay), which is nearly four orders of magnitude smaller than the solar constant, being about 0.05 W m^{-2} when averaged over the whole Earth's surface. Starlight and the gravitational connections with the Moon and Sun through the Earth's ocean tides contribute even less.

Since the Earth's energy budget is closely balanced, there must be an output to balance the continuing solar input, and it too must be in the form of electro-magnetic radiation to cross the near vacuum of interplanetary space. The simplest assumption is that the Earth emits *terrestrial radiation* as a uniform spherical black body whose size and surface temperature completely determine the total power output in the form of terrestrial radiation. Let us consider the balance between solar input and this terrestrial output.

Since the Earth and Sun subtend small angles when viewed from each other, the Earth intercepts solar radiation almost exactly as a disc with radius equal to the Earth's radius R_E. The rate of interception must therefore be $\pi R_E^2 S$, where S is the solar constant (Fig. 8.2). However, not all intercepted radiation is absorbed: significant amounts are scattered back to space and play no part in the Earth's energy balance. The fraction lost in this way is called the *planetary albedo* and is denoted by the symbol a. The rate of absorption of solar energy is therefore given by

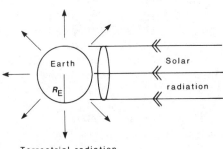

Terrestrial radiation

Fig. 8.2 Nearly parallel solar rays impinging on the Earth, and terrestrial rays diverging from it.

$$(1-a)\,\pi\,R_E^2\,S \tag{8.2}$$

which is the input to be balanced by the terrestrial output. Note that the non-zero albedo of the Earth in solar wavelengths indicates a deviation from pure blackbody behaviour which, according to Kirchhoff's law (appendix 8.1), must have consequences for terrestrial emission of those same wavelengths. However, at terrestrial temperatures, emission of wavelengths as short as solar wavelengths is so extremely small (the Earth does not glow in the dark, after all!) that the shortfall in emission of these wavelengths is quite unimportant. This contrasts sharply with the great importance of the shortfall in absorption of solar radiation which arises from the Earth's non-zero albedo.

Assuming that the Earth emits terrestrial radiation (which we will see shortly is much longer in wavelength than solar radiation) as a spherical blackbody of radius R and absolute temperature T_E, Stefan's law (appendix 8.1 and Fig. 8.2) requires the total radiant power output to be

$$4\pi R_E^2 \sigma T_E^4 \tag{8.3}$$

where σ is the universal Stefan–Boltzman constant, with value 5.67×10^{-8} W m^{-2} K^{-4}.

The terrestrial energy balance can now be stated by equating expressions (8.2) and (8.3). After rearrangement to isolate T_E we have

$$T_E = \left[(1-a)\frac{S}{4\sigma} \right]^{\frac{1}{4}} \tag{8.4}$$

Note that the Earth's radius does not appear in eqn (8.4), showing that the equilibrium equivalent blackbody temperature T_E of the Earth is independent of the Earth's size and would apply as well to an artificial satellite of radius 1 m as to a planet the size of Jupiter. In fact the only planetary factor determining T_E is the albedo a; beyond that we have only the solar constant S, which is part of what we might call the astronomical context of the planet, and the universal constant σ. The albedo is therefore a highly significant property of the planet. It is observed (section 8.6) that the Earth's albedo is determined mainly by clouds, air molecules and dust, as well as by the solid and liquid surface, but that clouds are by far the most important. We might expect this from simple observations of their brightness in sunlight, and indeed it is immediately obvious on all satellite pictures taken in visible wavelengths. Since clouds are produced by the very activity of the atmosphere which we know must be profoundly affected by the balance represented by eqn (8.4), it is clear that the atmosphere is significantly self-regulating at the most fundamental level.

Let us substitute values in the right-hand side of eqn (8.4) to find a value for T_E consistent with radiative equilibrium of the planet. Satellite radiometry shows that the Earth's albedo is about 0.3 in solar wavelengths, meaning that 30% of incident solar energy is returned to space and therefore not absorbed by the planet. If we use 1380 W m^{-2} for S and the established value for σ, we find T_E to be 255 K (-18 °C). This value is well below the observed average surface temperature of the Earth (288 K or 15 °C), so that at first glance it would seem that the model we are using for the irradiated and radiating Earth is too simple to be useful. In fact it will soon appear that the model is unrealistic in only one important respect, whose correction effectively explains the 33 °C anomaly, and is otherwise extremely good. However, to remove the model's single major flaw, we need to know more about the nature of terrestrial radiation. In particular we need to know about the interaction of terrestrial radiation with the Earth's atmosphere and surface.

8.3 Terrestrial radiation and the atmosphere

So far the model suggests that terrestrial radiation is electromagnetic radiation emitted by a blackbody with temperature 255 K. Although not quite true, this is nearly so, and we will obtain a useful first approximation to actual terrestrial radiation if we consider the power spectrum of radiation emitted by a blackbody at this temperature (Fig. 8.3). This shows significant emission spread across wavelengths ranging from about 4 to 100 μm, with maximum emission per unit wavelength centred around 12 μm, which is consistent with Wien's law (appendix 8.1). Virtually all of this terrestrial radiation is beyond the infra-red end of the solar spectrum, and the lack of significant overlap between solar and terrestrial radiation very conveniently allows the two fluxes to be considered separately. Note from the axes of Figs 8.1 and 8.3 that the terrestrial radiation spectrum is very much flatter and more spread out in wavelength than the spectrum of solar irradiance.

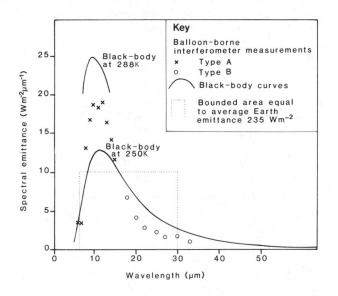

Fig. 8.3 Power spectrum of terrestrial radiation. Measurements A and B were made by balloon-borne spectrometers about 16 km above sea level (above most of the atmosphere active in the far infra-red). Observations on either side of the peak loosely fit the blackbody curve for 250 K (bounded area 221 W m^{-2}). Observations at the peak fall short of the level expected for radiation coming through the window from a 288 K surface (see blackbody peak), consistent with reports of 65% cloud cover at altitude 7 km at the time. Allowing for reasonable variability, the area under the observed spectrum agrees with the 235 W m^{-2} needed for planetary equilibrium. (After J.S. Garing [47])

As implied already, the radiative properties of many terrestrial materials differ quite markedly between solar and terrestrial wavelengths. Virtually all common solids and liquids, including water and all but the most tenuous water and ice clouds, act as almost perfect blackbodies for terrestrial radiation, absorbing these wavelengths completely (and therefore allowing no transmission or reflection), and emitting them as a blackbody would at the same temperature. Further, although the two main atmospheric gases (nitrogen and oxygen) are effectively transparent to terrestrial radiation, and are therefore very poor emitters of it, according to Kirchhoff's law, the minor components carbon dioxide and water vapour absorb it so strongly that the atmosphere is almost completely opaque in substantial parts of the spectrum of terrestrial radiation. The details of this complex behaviour are known from laboratory and theoretical investigations of the radiative behaviour of the gas molecules in question. A greatly simplified picture is sketched in Fig. 8.4, which shows that there are two main features: firstly there is strong absorption by water vapour between wavelengths 5 and 8 μm, and again beyond 14 μm by water vapour and carbon dioxide; and secondly there is near transparency in a *window* centred at about 10 μm. In the window there is some absorption by the much rarer gas ozone, which is unimportant for the atmosphere as a whole, but is quite important for the stratosphere on account of its relatively small heat capacity (sections 3.3 and 8.6). The presence of the window is significant, particularly for conditions at the Earth's surface (Chapter 9), and especially because it nearly coincides with the peak in the spectrum of terrestrial radiation, but it is the opaqueness of the atmosphere to wavelengths on either side of the window which dominates the overall conditions of the atmosphere in the presence of terrestrial radiation.

The atmosphere is really very opaque indeed to the wavelengths strongly absorbed by water vapour and carbon dioxide. The simplified absorption spectrum in Fig. 8.4(b) shows that a layer of air deep enough to contain 300 g of water vapour in a vertical column resting on a square metre of horizontal surface completely absorbs wavelengths between about 5 and 7 μm, and strongly absorbs wavelengths in another micrometre range on either side. In conditions typical of the low troposphere (air density ~ 1 kg m^{-3}, specific humidity ~ 10 g kg^{-1}), such a layer is only about 30 m deep. A similar estimation shows that the same layer contains enough carbon dioxide and water vapour to absorb all terrestrial radiation with wavelengths greater than 14 μm.

Fig. 8.4 (a) Smoothed spectrum of atmospheric absorption of solar and terrestrial radiation. Percentages refer to typical absorptions by the full depth of a cloudless atmosphere. Gaseous absorbers are labelled where they are most effective. (After [48]) (b) Simplified profile of atmospheric absorption in and around the window, used by Simpson in 1928 in the first detailed examination of the Earth's radiation balance — an approach not bettered until ways of accounting for the microstructure of absorption bands and lines were developed. Absorption values refer to a layer containing 300 g of vapour per horizontal square metre.

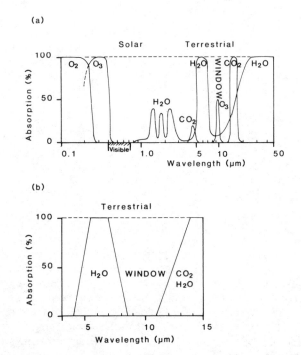

Consider a package of radiant energy emitted from the Earth's surface in these heavily absorbed wavelengths. It will be completely absorbed by the first 30 m of air, warming first the molecules of water vapour and carbon dioxide, and then almost immediately sharing this heat with the surrounding air molecules. But by Kirchhoff's law, water vapour and carbon dioxide must emit these same wavelengths with the same efficiency as they absorb them; in fact a layer deep enough to be opaque to these wavelengths must emit as a blackbody at the layer temperature. (This applies only to the absorbed wavelengths: the same layer will absorb and emit almost nothing in the wavelength range of the window.) If the air and surface were at the same temperature, their emissions would be equally intense in the wavelengths in question. Unlike the surface however, the layer of air emits both upwards and downwards.

Things are becoming complex, so we will pause to present an overall picture of the unfolding situation.

8.4 The terrestrial radiation cascade

The picture emerging is that the Earth's solid and liquid surface radiates as a blackbody and that all of this terrestrial radiation with wavelengths outside the transparent window is completely absorbed by layers of air no more than a few tens of metres deep overlying the surface. By contrast a considerable proportion of all solar radiation passes straight through the atmosphere to the surface, which it warms on absorption, with only a modest proportion being absorbed directly by the atmosphere. Is it possible to imagine the Earth's surface, together with all the overlying atmosphere, being maintained at constant temperatures solely by the absorption of solar radiation and the absorption and emission of terrestrial radiation? In other words can the surface and atmosphere achieve a purely radiative equilibrium? For simplicity assume that the atmosphere consists of a number of discrete layers each of which is just thick enough to be opaque to terrestrial radiation with wavelengths outside the window. Such layers are about 30 m deep in the low troposphere, but are considerably deeper at higher levels where the absorbing gases are much more tenuous. To account for observed totals of CO_2 and water vapour there should be over one hundred of these opaque layers in a deep atmospheric sandwich, but the principle of the argument can be established by considering just a few, in fact only three in Fig. 8.5.

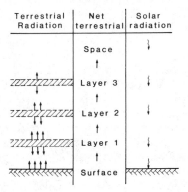

Terrestrial Radiation	Net terrestrial	Solar radiation
	Space	
	↑	
	Layer 3	
	↑	
	Layer 2	
	↑	
	Layer 1	
	↑	
	Surface	

Fig. 8.5 Schematic picture of vertical fluxes of terrestrial radiation in a three-layer model atmosphere in purely radiative equilibrium.

For simplicity we will initially ignore all direct absorption of solar radiation by the atmosphere, leaving the surface as the only absorber, and ignore also the transparent window in terrestrial wavelengths, leaving the atmosphere entirely transparent to solar radiation and opaque to terrestrial radiation.

From Fig. 8.5 we see that the surface receives and absorbs solar radiation, emits terrestrial radiation upwards, and absorbs terrestrial radiation coming downwards from the overlying opaque atmospheric layer. Each opaque atmospheric layer emits terrestrial radiation upwards and downwards, and absorbs terrestrial radiation emitted by the overlying and underlying adjacent layers. The only exception is the highest opaque layer, which absorbs terrestrial radiation coming up from beneath, and emits terrestrial radiation directly to space, but receives no terrestrial radiation from above because it is the highest layer containing significant quantities of carbon dioxide and water vapour and there are no terrestrial bodies close by.

If the surface is to be in purely radiant thermal equilibrium, then it must lose as much energy by net loss of terrestrial radiation (emission less absorption) as it gains

251

by absorption of solar radiation. And if each atmospheric layer is to be in radiant equilibrium, it must lose as much energy by emission of terrestrial radiation as it gains by absorption of the same. In addition, the net upward flux of terrestrial radiation between every adjacent pair of atmospheric layers must be the same, and be equal both to the net upward flux of terrestrial radiation between the surface and the first atmospheric layer, and to the upward flux to space from the highest opaque layer. The end result of all this would be to keep all surface and atmospheric components separately in radiant equilibrium, as well as the planet as a whole, while maintaining a continual conversion from solar to terrestrial radiation at the Earth's surface.

To sustain the net upward flux of terrestrial radiation through the opaque atmosphere, the emissions of terrestrial radiation by the opaque atmospheric layers must decrease with increasing altitude as depicted in Fig. 8.5, and since decreasing emissions require decreasing temperature (Stefan's law again), it follows that there must be a continuing lapse of temperature from a maximum value at the surface to a minimum value at the highest opaque layer.

We can now modify the unrealistically simple initial assumptions by reinstating some direct atmospheric absorption of solar radiation, and allowing for the presence of an atmospheric window for terrestrial radiation.

Since there is in fact significant absorption of solar radiation in the lower layers of the atmosphere (see section 8.6), the net upward flux of terrestrial radiation must decrease with increasing altitude, reaching the value found in the initial model only at levels above the highest absorbing layer. This means that the temperature lapse between adjacent layers must increase upwards when compared with the initial model, beginning at a value smaller than the corresponding value in the initial model and increasing upwards throughout the layer with significant solar absorption. The effect is therefore to reduce the temperature lapse required for thermal equilibrium, at least in the lower part of the radiatively active atmosphere, holding it a little below what would be required if all solar absorption were concentrated at the surface.

The inclusion of the transparent window centred on terrestrial radiation wavelengths simply allows some of the net upward flux of terrestrial radiation from the surface to be carried out to space independently of the cascade of atmospheric absorption and emission. Again the result is to reduce the temperature lapse required for thermal equilibrium by comparison with the initial model, especially between the surface and atmosphere. Since detailed study (section 8.6) shows that the window flux accounts for less than 10% of the Earth's radiative output, the effect is not large.

We can summarize the overall picture by saying that the surface and atmosphere could maintain thermal equilibrium by absorption of solar radiation and an inverted cascade of absorption and emission of terrestrial radiation, if there were an adequate temperature lapse between each atmospheric layer and between the surface and the lowest layer. However, to estimate values for these temperature lapses, thereby detailing the vertical profile of temperature needed for purely radiative equilibrium, we must have a thoroughly quantitative version of the model depicted so simply in Fig. 8.5. This was first attempted by Emden in 1913 and has been repeated since with increasing refinement. All such studies indicate that the temperature profile required for purely radiative thermal equilibrium of the surface and stratified atmosphere is so steep as to be convectively unstable in the lower atmosphere (section 5.9). Since the air is much too fluid to sustain convective instability without breaking into vigorous convection, with consequent significant vertical transport of heat by that convection, it follows that the answer to the question posed at the beginning of this section is 'no': it is impossible for the atmosphere with its observed distribution of radiatively active and inactive gases to maintain thermal equilibrium purely by absorption of solar radiation and the emission and absorption of terrestrial radiation. Thermal equilibrium requires convection as well as radiation, and it follows that the Earth's atmosphere is necessarily rather than incidentally dynamic.

8.5   The greenhouse effect

We are now in a position to explain the apparent anomaly highlighted at the end of section 8.2 — the difference of over 30 °C between the average surface temperature of the Earth and the effective blackbody temperature of the planet as a whole. Contrary to the original assumption, the planet does not act as a blackbody with a single emitting surface situated at the land and sea surfaces: the terrestrial radiation emitted to space comes mainly from the atmosphere and only in small part from the solid and liquid surfaces. In fact the latter is significant only when the atmosphere is cloudless; from the half of the globe on average covered by cloud, even the window radiation comes from the top of the highest cloud layer.

The details summarized in the next section show that about 90% of the terrestrial radiation emitted to space comes from the atmosphere (10% by window radiation from cloud tops and 80% by radiation outside the window emitted by water vapour and carbon dioxide in the highest opaque layer of the atmosphere), with only 10% coming through the window from the underlying land and sea. Detailed analysis also shows that the highest opaque layer is in the upper troposphere, where the temperature is usually 40° or more below the temperature of the underlying surface. Cloud tops are often in the same height and temperature range. The output of terrestrial radiation is therefore dominated by the output from the upper troposphere, and this, together with the temperature lapse imposed by the inverted cascade of absorption and emission of terrestrial radiation through the troposphere (constrained by the associated convection), means that the effective temperature T_E of the terrestrial blackbody is well below the temperatures familiar to us at the bottom of the atmosphere. In fact detailed analysis confirms that T_E is very close to the 255 K estimated in section 8.2.

The effect of the atmosphere's near transparency to solar radiation and near opacity to terrestrial radiation is therefore to raise the temperatures of land and sea surfaces well above those which would prevail in the absence of the atmosphere. This elevation of surface temperatures is called the *greenhouse effect* because the glass of a greenhouse is similarly transparent to solar and opaque to terrestrial radiation. Actually the name is quite misleading because the interior of a greenhouse stays warm primarily because the glass inhibits convective heat loss to the surrounding air — a fact which was established at a fairly early stage by studying the behaviour of a very special quartz greenhouse, and is now confirmed by the growing popularity of polythene greenhouse (quartz and polythene being transparent to both solar and terrestrial radiation). However, since in spite of this the term 'greenhouse effect' has become conventional, it will be used conventionally in this book, with the understanding that it refers to the behaviour of atmospheres rather than greenhouses.

VENUS

The planet Venus has an atmosphere about 100 times the mass of the Earth's, consisting almost entirely of CO_2 and a thick shroud of sulphuric acid cloud, both of which contribute to a very large greenhouse effect. The highly reflective cloud blanket has an albedo of about 0.7 (making Venus such a brilliant object in the night sky), which results in an effective blackbody temperature of only 245 K, compared with the Earth's 255 K with less than half the solar constant (section 8.2 and problems 1 and 11). Strongly protected landing probes have measured surface temperatures of about 730 K, implying a greenhouse effect of nearly 400 K, over ten times the terrestrial value. Estimates of total amounts of carbon in the surface layers (atmosphere, ocean and crust) of Earth and Venus show that they are similar, but on Earth it is mostly chemically locked up in carbonate rocks, while on Venus it largely degassed into the atmosphere as CO_2 during an early runaway greenhouse effect (section 3.8).

RADIATIVE FORCING

As noted in sections 3.8 and 4.10, the Earth's greenhouse effect is believed to have changed considerably over geological time, and to be increasing now in response to wholesale injection of carbon dioxide into the atmosphere by artificial combustion. Other anthropogenic gases are known to add to the greenhouse effect by absorbing terrestrial radiation and detailed estimates have been made of their present and future effectiveness, assuming realistic future production rates. The simplest way to assess this greenhouse effectiveness is to calculate the *radiative forcing* of the terrestrial energy balance — the resulting increase in the trapping of outgoing terrestrial radiation.

Table 8.1 presents recent results for two overlapping periods, each starting from the dawn of the industrial revolution, and the second projecting into the middle of the twenty-first century, by which time the atmospheric CO_2 level is expected to have doubled. Values represent the increases in the rate of absorption of terrestrial radiation by gases in the periods, and are calculated from detailed knowledge of absorption spectra and likely distributions through the troposphere. The results show significant contributions from increases in CO_2 and a number of other *greenhouse gases*, differing sharply in efficiency per molecule — the CFCs being effective out of all proportion to their very small concentrations. The calculations suggest that the increase in total forcing since pre-industrial times is likely to be over 8 W m^{-2} by the middle of the twenty-first century, if human society continues growing and industrializing.

Table 8.1 Changes in radiative forcing from 1765 assuming the following IPCC scenarios for industrial and population growth [33]:(a) business as usual; (b) accelerated control policies (CO_2 emission half of 1985 value, stringent controls in developed countries, moderate controls elsewhere)

Period (AD)		CO_2	CH_4	N_2O	CFCs	Total	(all W m^{-2})
1765–2000	(a)	1.85	0.69	0.12	0.29	2.95	—
	(b)	1.75	0.59	0.11	0.29	2.74	—
1765–2065	(a)	5.49	1.37	0.40	1.03	8.28	—
	(b)	2.77	0.52	0.24	0.68	4.22	—

By agreed definition, radiative forcing is the reduction in net upward flux of terrestrial radiation at tropopause levels produced by increasing absorption by greenhouse gases. Assuming this flux to be about 240 W m^{-2} initially (equivalent to an effective terrestrial blackbody temperature of 255 K — section 8.2 and Fig. 8.6), this would be reduced to 232 W m^{-2} by the above forcing. The Earth would then warm by excess of solar input over reduced terrestrial output, until the flux had been restored to its original value. The global distribution of such warming depends crucially on details of how the atmosphere and surface layers react to forcing, which is probably very complex (see below), so the simplest approach is to find what rise in the equivalent blackbody temperature of the Earth would recover the balancing outward flux of terrestrial radiation. According to appendix 8.5, a forcing of 8 W m^{-2} requires a rise of just over 2 K at around 255 K. This is a very substantial global warming which, depending on its distribution at the Earth's surface, could make the Earth warmer than at any time since the onset of the present ice ages, and fulfil the worst fears expressed in section 4.10. The distribution of the warming depends strongly on the complicating role of the water vapour and cloud.

Table 8.1 omits the most effective greenhouse gas of all (water vapour) because it is not directly anthropogenic. It has an important indirect effect however, because any warming of the low troposphere will increase the vapour content there by increasing evaporation, and the resulting increased cloud will blanket more of the atmospheric window. Both these effects would lead to even greater warming, but the increase in global cloud cover will also have a cooling effect, being very effective in cutting down the solar input to the Earth (see next section). Details of this complex of positive and negative feedbacks are still poorly understood, because we need to be able to predict the vertical and horizontal distributions of increased cloud cover, and indeed changes in many other features of the whole climate system. Most current estimates of average surface warming range from about 2 to 5 °C for a doubling of atmospheric carbon dioxide (with further additions for other greenhouse gases), but the confidence limits on precise values are not high [33]. It is very unfortunate, to say the least, that our understanding of such global consequences of our modern way of life remains so poor, but the problem is exceptionally severe and is likely to tax our pooled scientific resources for decades before the situation becomes clear, for good or ill.

8.6 Average radiant-energy budget

Simplified models of interactions between the atmosphere and solar and terrestrial radiation typified by Fig. 8.4 are very useful for indicating basic aspects of behaviour, but they are inadequate for more detailed studies. In the last 50 years, people have developed ways of dealing with the complex fine structure of lines and bands underlying the gross absorption spectrum sketched in Fig. 8.4(a), with the result that interactions between the real atmosphere and the radiation streams crossing it are now understood in some detail. (Methods are discussed in more advanced texts [8].) It has therefore been possible for some time to use observed distributions of all the radiatively active materials to calculate all significant items in budgets of radiant energy for the Earth's atmosphere and surface. An immense amount of data is required, including appropriately averaged distributions of temperature, pressure, water vapour, carbon dioxide, ozone, cloud and surface albedo. Since the early 1960s, data from meteorological satellites have been used to improve some of the data and to check some of the predicted fluxes.

A classic study of the radiant energy budget on the global scale was made by London in 1957 and has been refined since [49]; results are summarized in Fig. 8.6. Input data were averaged over many years from a large number of geographical locations. The data were further averaged over ten degree zones of latitude before being used in the extensive calculations needed to estimate radiant fluxes. The results reveal the extremely important variation with latitude which is discussed in the next section. For the global picture, the data were averaged further to produce what is in effect an annual global average of the budget of radiant energy (Fig. 8.6). Since all horizontal variations have been removed by averaging, the picture represents the vertical distribution of radiant fluxes. For arithmetical simplicity all values in Fig. 8.6 are quoted as percentages of the annual global average rate of input of solar energy per unit horizontal area at the top of the atmosphere. Note that the latter is one quarter of the solar constant (i.e. 345 W m^{-2}) since the surface area of the spherical Earth is four times its cross-sectional area (Fig. 8.2). Each unit in the budget in Fig. 8.6 is therefore very nearly 3.5 W m^{-2}.

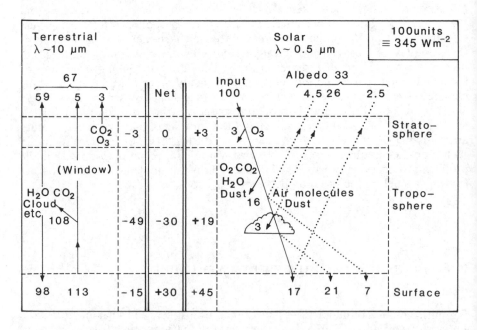

Fig. 8.6 Global annual average flux densities of solar, terrestrial and net radiation in the stratosphere, troposphere and at the Earth's surface. On the solar radiation side, dotted lines represent reflection and scattering, and solid lines represent transmission and absorption. On the terrestrial radiation side, solid lines represent emission, transmission and absorption. The central column contains the net radiative input to each part of the system. (Values are from [49])

THE ENERGY BUDGET OF THE EARTH'S SURFACE

After traversing the atmosphere, solar radiation reaches the surface as both *direct* and *diffuse* radiation. The major diffuser is cloud, though air molecules and haze are also important. The scattering of light by air molecules was first described theoretically in the nineteenth century by Rayleigh, who showed that shorter wavelengths are much more efficiently scattered than longer wavelengths when the scattering bodies are much smaller than the scattered wavelengths (molecules ~ 10^{-4} μm as compared with wavelengths ~ 0.5 μm). This accounts for the blue appearance of cloud-free sky away from the Sun's disc, since the scattered light we receive is enriched in the short (blue) wavelengths by the scattering from other initial directions. It also accounts for the reddening of the solar disc when observed directly through the atmosphere, since the preferred scattering of the blue out of the line of sight leaves the direct sunlight reddened. The reddening is particularly noticeable when the sun is near the horizon because the long path length through the atmosphere enhances the blue-biased molecular scattering, and the unbiased scattering by much larger droplets and particles (see below) makes the solar disc dim enough to view in comfort.

By contrast clouds, mists and most hazes do not alter the colour of sunlight scattered by their droplets and particles, and scattering theory shows that this is because the size of the scatterers is comparable with or larger than the scattered wavelengths. Clouds therefore appear nearly white in normal sunlight, and are red near dawn and dusk only because they are illuminated by reddened direct sunlight. The chromatically uniform scattering of all available wavelengths is called *diffuse reflection*, and Fig. 8.6 shows that this accounts for 21 out of 47.5 units of solar radiation reaching the surface, compared with 7 units scattered by air molecules and dust. Collectively labelling these as diffuse radiation we see that considerably more than one half of the solar radiation reaching the surface is diffuse rather than direct.

Although the distinction between the largely directionless diffuse component of solar radiation and the highly directional direct component is visually obvious, and can be important for the local climatology of hilly terrain (section 9.2) for example, it is largely irrelevant to the energy budget of the Earth's surface as a whole, where the total input is all that matters. Over the whole Earth almost all solar radiation incident on the surface is absorbed there, if not at once then ultimately (there is multiple scattering between the surface and the low atmosphere which is not depicted in Fig. 8.6), only 2.5 units being scattered back to space. The effective albedo of the Earth's surface is therefore very small, as is apparent in satellite panoramas such as Fig. 1.1, where the surface is very dark in comparison with the highly reflective clouds. Of course in localized areas surface albedos may be much larger, and can exceed 70% in snow-covered areas, but the global average value is low because of the preponderance of dark surfaces. Some values for surface albedoes are given in Table 9.1.

Just above the surface, terrestrial radiation consists of large and opposing fluxes which have a relatively small upward resultant. The upward flux in Fig. 8.6 is equivalent to the output from a blackbody with temperature 288 K (15 °C) — essentially the average temperature of the Earth's surface. Cloud, carbon dioxide and water vapour radiate strongly downwards, but do not quite match the upward radiation from the surface because of the absence of any downward flux in the window when there is no cloud, and because the lowest opaque atmospheric layers are somewhat cooler than the surface almost everywhere. Although each of the opposing fluxes is larger than the solar input to the planet (another aspect of the greenhouse effect), the net upward flux is much smaller, and is significantly smaller even than the net solar input to the surface.

Considering solar and terrestrial radiation together, it is apparent from Fig. 8.6 that there is a net input of radiant energy to the surface which proceeds at an average rate of 30 units — very nearly 100 W m^{-2}.

THE ATMOSPHERIC ENERGY BUDGET

Consider the stratosphere first. Three units of solar input are absorbed, mainly in the form of selective absorption of soft ultraviolet between altitudes of 25 and 45 km as described in section 3.3. This may seem a relatively small input, but the air at those levels is so insubstantial that it would warm very rapidly if the input were not balanced by the small net output of terrestrial radiation from the carbon dioxide and ozone there. (There are no significant amounts of water vapour at these levels.) The stratosphere as a whole is therefore in radiative equilibrium. This does not mean that there are no convective heat fluxes either within the strato-sphere or between stratosphere and troposphere, but it does indicate that convection is not essential to the stratosphere in the way that we have seen it to be in the troposphere.

Of the 97 units of solar radiation entering the top of the troposphere (Fig. 8.6), 16 units are absorbed by aerosol particles and water vapour, 30.5 units are scattered back out to space, mainly by cloud, 3 units are absorbed by cloud, and 47.5 units pass through to the surface either directly or after the scattering discussed above. Note the curious behaviour of cloud: on the one hand its diffuse reflection both upwards and downwards dominates the solar radiation received at the surface and returned to space, but on the other hand it contributes only slightly to tropospheric absorption of solar radiation. Total tropospheric absorption is far from trivial, accounting for 19 units, or about one third of the total solar absorption by the planet. We are misled by our visual impression of the near-transparency of the cloudless atmosphere, because the absorption is concentrated in the infra-red half of the solar spectrum (Fig. 8.4).

Because of the distributions of temperature, water vapour, carbon dioxide and cloud, the output of terrestrial radiation from the troposphere to space is 59 units, much of it from the high troposphere. (By definition the flux in the atmospheric window plays no role in the tropospheric budget unless it is intercepted by cloud.). This large loss is only partly offset by the net gain of 10 units by terrestrial exchange with the surface, so that the troposphere suffers a net loss of 49 units by terrestrial radiation.

Including both solar and terrestrial radiation, it is apparent that the troposphere suffers a net loss of 30 units of radiant energy, which is equivalent to 100 W m^{-2}, and is exactly equal to the net radiative gain by the surface.

THE SURFACE AND TROPOSPHERE TOGETHER

These therefore comprise a system which is in radiative equilibrium overall, like the stratosphere and like the Earth as a whole, but which has a large radiant energy imbalance between its two components, amounting to 100 W m^{-2} on a global annual average. Since there is no appreciable warming of the surface, or cooling of the troposphere, from year to year, it is clear that this imbalance must be almost exactly offset by a non-radiant heat flux of 100 W m^{-2} from the surface to the troposphere. This is the convective heat flux associated with the whole range of weather systems from the smallest buoyant thermals to the mighty Hadley circulation. The radiative balance of the stratosphere strongly suggests that this convective flux is confined to the troposphere, and this is confirmed by a variety of observational evidence, including the virtual confinement of cloud to the tropo-sphere, and the widespread appearance of convectively very stable layers just above the tropopause.

8.7 Meridional distribution of radiative fluxes

The global energy budget introduced in section 8.6 includes as one of its most important parts a calculation of the distribution of radiative fluxes with latitude. Annually averaged values are displayed in Fig. 8.7(a) and (b) and agree well with direct measurements by satellite-borne radiometers. Values should be compared with one quarter of the solar constant as mentioned at the beginning of the last section. The latitude scale has been deliberately chosen so that equal lengths represent zonal rings of equal surface area, and it follows that equal areas under the curves in Fig. 8.7 represent equal energy flows to or from the Earth. This would not be so if the latitude scale were linear, since surface areas and fluxes would be exaggerated at high latitudes, just as in the familiar Mercator map projection.

Figure 8.7(a) depicts the meridional distribution of absorbed solar input S to the surface and atmosphere and the terrestrial output T to space. The solar input is concentrated in low latitudes because the sun passes near the zenith there each day, whereas at higher latitudes it misses the zenith by an angle which increases with latitude, and is equal to the angle of latitude at the equinoxes (Fig. 8.8). The increase of obliqueness of illumination with increasing latitude has the effect of spreading the solar beam over greater surface areas at higher latitudes, and of increasing its path length through the atmosphere (Appendix 8.2). Both these factors enhance the loss from the direct solar beam by scattering and diffuse reflection, and increase the relative proportion of solar energy which is absorbed by the atmosphere rather than the surface. This proportion is enlarged further by the widespread tendency for surface albedo to increase with zenith angle, as graphically evidenced by the brilliance of surface glitter beneath the rising or setting sun.

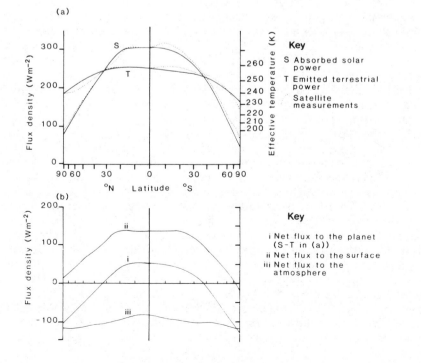

Fig. 8.7 (a) Meridional profiles of zonally averaged fluxes of solar and terrestrial radiation absorbed and emitted by the Earth. The vertical axis shows the radiant flux density per unit area of Earth surface. Appendix 8.3 justifies the cosine weighting of the horizontal axis. (b) Fluxes in (a) analysed into: (i) the net radiant flux to the planet; (ii) the net radiant flux to the surface; and (iii) the net radiant flux to the atmosphere. (After [49])

Key

S Absorbed solar power
T Emitted terrestrial power
Satellite measurements

Key

i Net flux to the planet (S−T in (a))
ii Net flux to the surface
iii Net flux to the atmosphere

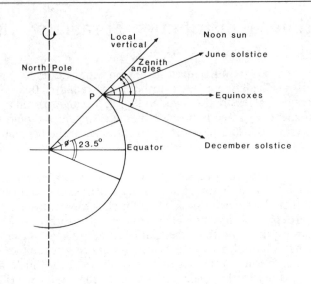

Fig. 8.8 Angles subtended by the Sun at noon at P at equinoxes and solstices, using the Earth's equatorial plane as reference plane (so that the Sun seems to rise and fall by 23.5° seasonally).

In annual averages, the resulting concentration of solar input S in low latitudes would be even more marked than it is, if it were not for the 23.5° angle between the earth's solar orbital and equatorial planes (Fig. 8.8), because of which the midwinter noon zenith angle is increased by 23.5° in middle and high latitudes, and the midsummer value is similarly decreased. The seasonal meridional migration of maximum solar input has the effect of blunting and spreading the low-latitude maximum and the high-latitude minima in the annual averages. This is enhanced by the persistently non-uniform distribution of cloud, with concentrations in equatorial areas corresponding to the intertropical convergence zone, and relative scarcity corresponding to the semi-permanent anticyclones concentrated in the subtropics. The slight bias of S away from the Antarctic is due to the presence of the permanent ice cap there, which scatters more sunlight to space than does the seasonally broken sea ice of the Arctic ocean. In principle there must also be a slight bias of S toward the southern hemisphere because perihelion occurs in the southern summer, but the expected 6% effect (section 8.1) is swamped by those already mentioned.

The meridional variation of absorbed solar input S is very large and is obviously related to the familiar zoning of climate. By comparison, the meridional variation of the emission of terrestrial radiation T is very much smaller. This emission comes mainly from the upper troposphere, as discussed in section 8.5, much of it from the water vapour there. Since continuing convection keeps most parts of the upper troposphere close to saturation, the vapour density there is effectively determined by air temperature alone (Fig. 6.2). And since the highest opaque layer must have a well-defined vapour density, to be only just opaque, it follows that its temperature and its emittance of terrestrial radiation must be similarly well defined. We should therefore expect a relatively uniform distribution of terrestrial output T over the globe. In fact the presence of carbon dioxide, cloud and the atmospheric window ensure that there is some falling away of T in high latitudes, but not very much.

Looking at the very different meridional distributions of S and T in Fig. 8.7(a), it is clear that there is a gross imbalance between low latitudes, which gain much more radiant energy than they lose, and high latitudes which lose much more than they gain. (There is of course no net gain or loss by the planet as a whole, as careful checking of the areas between the S and T curves will confirm.) If these unbalanced fluxes of radiant energy were not offset by other types of energy exchange, then high latitudes would cool and low latitudes would warm, until presumably their terrestrial output achieved the same steep meridional gradient as their solar input. Some of the offsetting poleward fluxes must consist of an exchange of warm and cool air, called *advection* rather than convection because the exchange is horizontal rather than vertical. Clearly the familiar experience in middle latitudes of relatively warm air moving polewards and relatively cool air moving equatorwards (cool northerlies and warm southerlies in the northern hemisphere) represents just such poleward advection of heat, as does the great poleward flux of relatively warm sea water in the Gulf Stream of the North Atlantic, though the contrasting cool equatorward drift is confined to the opposite side of the ocean. The role of the atmosphere and oceans in advecting heat polewards is examined further in the next chapter.

The solar and terrestrial fluxes in Fig. 8.7(a) are regrouped in Fig. 8.7(b) to show the meridional profiles of flux densities of net radiative input to the planet (i), to the surface (ii), and to the atmosphere (iii).

The first of these (i) is $S - T$ in Fig. 8.7(a) and clearly portrays the meridional imbalance already discussed. The net radiative flux to the Earth's surface (ii) has a similar shape to (i), but is everywhere about 100 W m^{-2} larger. It exceeds 100 W m^{-2} (the global average net gain by the surface — see Fig. 8.6 and section 8.6) over the half of the Earth's surface area lying between latitudes $30°$, and falls below 50 W m^{-2} only at latitudes greater than about $60°$. By contrast the net radiative input to the atmosphere (iii) is negative and remarkably uniform, remaining close to -100 W m^{-2} at all latitudes, which obviously therefore agrees closely with the global average tropospheric net loss apparent in Fig. 8.6.

The combination of the steep meridional profile of the net radiative input to the surface ((ii) in Fig. 8.7b) and the near uniformity of the net radiative loss by the atmosphere (iii) means that in low latitudes the net radiative gain by the surface exceeds the net radiative loss by the atmosphere, whereas the position is reversed in high latitudes. To accommodate this strong meridional variation in the radiative imbalance between surface and atmosphere requires more than the simple vertical convection of heat envisaged in the previous section and the poleward advection outlined above: there must be a consistent transport of heat both upward and poleward, taking heat from the radiatively warming surface of low latitudes to the radiatively cooling upper troposphere of high latitudes. Such transport could in principle be accomplished by a combination of convection and advection, but poleward-moving air is generally warmer than the surface it overruns, which tends to suppress convection from the surface. In fact a significant fraction of the required transport is effected by large-scale, slightly tilted flow or *slope convection*, which is particularly associated with the development of extratropical cyclones. Some of these tilted flows become convectively unstable through the interleaving of flows from different surface sources (section 11.4), and through destabilization by vertical stretching associated with large-scale horizontal convergence (section 10.3), encouraging additional vertical transport. In addition, the heat capacities of land and sea surface, especially the latter, allow heat carried polewards in one flow to be pumped aloft by convection in the following equatorward flow.

8.8 Seasonal variations of radiative fluxes

In middle latitudes the seasonal rhythm of solar input is one of the most prominent features of weather and climate, even in the cloudy and relatively temperate

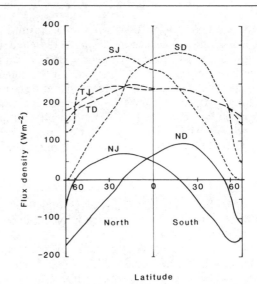

Fig. 8.9 Meridional profiles of zonally averaged solar, terrestrial and net radiant input (S, T and N respectively) to the Earth, averaged through the seasons beginning in June and December (J and D respectively). For example the curve NJ is net radiation in the season June, July and August. (Data from [50])

western margins of continents. Figure 8.9 shows that the difference between the winter and summer values of the solar input to the Earth's surface and atmosphere exceeds 100 W m^{-2} poleward of about latitude 25°, and is about 150 W m^{-2} over a considerable part of the middle latitudes. This represents a variation of 100% at the lower-latitude limit and much more at higher latitudes, and is a direct consequence of the march of the Sun into first one hemisphere and then the other, modified somewhat by the uneven distribution of land and cloud. There is obvious asymmetry between the Arctic and Antarctic summers, the solar absorption in the latter being considerably reduced by the highly reflective ice cap there, as mentioned in the previous section.

The seasonal variation in the output of terrestrial radiation is by contrast extremely small, being significant only in the northern hemisphere because of the concentration of land masses there. The relatively clear skies over the continental interiors in middle and high latitudes inhibit the greenhouse effect somewhat, enhancing the seasonal swing in surface temperatures and encouraging a modest seasonal variation in the output of terrestrial radiation through the relatively large atmospheric window.

The combination of the relatively small seasonal variation of terrestrial output and the very large variations in solar input means that there is a very substantial shift in the net radiative balance between summer and winter, the latitude zone of maximum net input shifting from about 15 °N in the northern hemisphere summer to 25 °S six months later. Note that the southern summer maximum of net radiative input is significantly higher than the northern one, and is centred further from the equator. Figure 8.9 shows that this results partly from the enhancement of solar input in the southern summer and partly from the slight reduction in terrestrial output. The concentration of land in the northern hemisphere, the December maximum of insolation because of the perihelion then, and the complex response of the atmosphere, especially its cloud distribution, all play their parts no doubt, but there is no obvious way of estimating their relative importance.

Apart from the latitudinal position and magnitude of the maximum of net radiative input, the other major feature of Fig. 8.9 is the extensive meridional gradient of net radiative input stretching from the maximum in the summer hemisphere to the minimum at or near the winter pole. The algebraic difference between the positive maximum and the strongly negative minimum is nearly 500 W m^{-2}, and the steep and extensive gradient between them is a major factor in maintaining the relatively intense weather activity typical of middle and high latitudes, especially in the winter. We can trace the connection as follows: the gradient of net radiation extends and steepens the meridional temperature gradient of the troposphere in the winter hemisphere, especially in middle and high latitudes (as noted in Fig. 4.6), and with it the baroclinity which drives much of the large-scale weather activity there, including of course the extratropical cyclones. In the summer hemisphere the difference between maximum and minimum net radiation is smaller and much less extensive, being largely confined to latitudes greater than 30°. To judge from observed behaviour, this smaller and less strongly baroclinic zone is considerably less effective in maintaining synoptic and larger-scale weather activity, with the result that weather systems such as extratropical cyclones are less numerous and vigorous than in the winter hemisphere.

8.9 Diurnal variations of radiative fluxes

The average surface budget in Fig. 8.6 conceals large diurnal variations. The solar input, averaging 45 units overall, is obviously zero at night, and must therefore average 90 units in daylight, assuming a 12-hour day and no variation with latitude. Further assuming a semi-sinusoidal profile of solar influx from dawn to dusk (appendix 9.2), the noon maximum is 141 units (486 W m^{-2}), which is clearly consistent with values estimated from solar elevation and atmospheric transmissivity (appendix 8.2). Maximum noon values (under a zenith sun and a clear sky), can reach about 320 units (1100 W m^{-2}), and may be reduced to about one tenth of this under thick overcast. Maximum values fall with decreasing solar elevation (increasing latitude at a fixed time of year) as outlined in appendix 8.2.

Figure 8.6 also shows that on average the Earth's surface loses 15 units by net output of terrestrial radiation. Since five of these are lost direct to space through the atmospheric window, which is closed about 50% of the time by cloudy overcast, the net loss by the surface in cloudless conditions must be about 10 units (nearly 35 W m^{-2}), with somewhat larger values in areas with less atmospheric water vapour and therefore a larger atmospheric window, and smaller values elsewhere. (These values are consistent with values derived differently in section 9.3, and with radiometer measurements.) Corresponding values are roughly halved under cloud. There is little clear diurnal variation because the tendency to greater emission by the hot sunlit surfaces is offset by the greater vapour content and cloud cover maintained by convection.

It follows from all these values that there can be very large diurnal variations about the average value of 100 W m^{-2} quoted in Fig. 8.6 for the net solar and terrestrial radiative input to the Earth's surface. On cloudless days at low latitudes a maximum net input of about 1000 W m^{-2} at noon may give way to a net output of about 70 W m^{-2} during the ensuing cloudless nights. In thick overcast conditions, the maximum net daytime input could be reduced to about 100 W m^{-2}, with zero nocturnal loss under a warm, low cloud base. At higher latitudes the effect of the reduction in noon solar input is modulated by the increasing seasonal variability of daylength (section 11.8 and [51]), and net losses of about 70 W m^{-2} continue throughout the polar nights.

The impact of these diurnal variations on surface climate is outlined in sections 9.3 and 9.4, where it is shown that the small effective heat capacities of land surfaces allow the large swings of temperature from early afternoon maxima to dawn minima which are such a familiar feature of cloudless conditions. Temperature variations at water surfaces, however, are much smaller on account of their much greater heat capacities. In addition to their relevance to surface climates, these diurnal variations are obviously consistent with the observed tendency for overland convection to peak in the afternoon and die away in the evening.

8.10 The necessity of air motion

We have seen that interactions between solar and terrestrial radiation and the Earth's surface and atmosphere maintain apparent energy imbalances between the surface and troposphere, and between low and high latitudes. When these imbalances were mentioned earlier, it was simply assumed that they must be balanced by convective and advective heat fluxes because the surface is observed to be in a thermally nearly steady state in yearly averages. Actually, before this assumption can be safely made, we need to consider the magnitudes of the imbalances, to assess their significance and to see if they really do require balancing convection or advection of heat, rather than conduction.

We can demonstrate the significance of the imbalances most directly by estimating the rates of change of temperature which would occur if there were no balancing non-radiative fluxes. Consider for example the vertical imbalance between surface and troposphere, by which the troposphere is losing radiant energy at an average rate of 100 W m^{-2} (Fig. 8.6). We can repeat the calculation in section 5.3 to find the associated rate of cooling by estimating the mass and hence the heat capacity of the air in a column of the troposphere resting on unit area of horizontal surface. Since the troposphere is in hydrostatic equilibrium, we can use eqn (4.6) to calculate its mass per unit area from the pressure difference between the bottom and top of a typical tropospheric column. Such columns are consistently taller in terms of pressure as well as height in lower latitudes (Fig. 4.6), so we will choose an intermediate example, bounded below and above by the 1000 and 250 mbar isobaric surfaces. This involves a pressure difference of 750 mbar or 7.5×10^4 Pa, and it follows from section 5.3 that a vertical tropospheric column resting on a horizontal square metre contains about 7.5 tonnes of air and has a heat capacity of about 7.5 MJ °C^{-1}.

In response to a net rate of loss of heat of 100 W, this column of air would cool at about 1.3×10^{-5} °C per second, which corresponds to about 1.1 °C per day. This is obviously a very significant rate of cooling, being much faster than even the most rapid seasonal cooling. Indeed, if it persisted for only a few weeks it would produce catastrophic climatic change. Comparable and even faster rates of cooling are observed in middle and high latitudes, but these do not arise primarily by radiative processes, being associated with the passage of contrasting air flows embedded in extratropical cyclones; as such they are localized and last for only a few days before being as quickly reversed.

A similar calculation related to seasonal variations in net radiant energy budget of the surface and atmosphere in middle and high latitudes (Fig. 8.9) would seem to indicate tropospheric cooling of about 100 °C in a three-month winter season at latitude 45° (where the net rate of loss is again about 100 W m⁻²). However, this cooling ignores the heat capacity of the Earth's surface layers, which are relatively large in oceanic zones. For example it is easy to show that a 30 m deep column of water beneath 1 m² of water surface has a heat capacity of about 126 MJ °C⁻¹, which is over 16 times larger than the tropospheric heat capacity estimated above. Though this additional capacity must help to smooth out what would otherwise be much larger seasonal temperature variations in middle and high latitudes, and play a major role in climatic amelioration over oceans and western continental margins there, it would merely serve to slow the response to any persistent radiative loss of heat in these latitudes, such as the net winter cooling apparent in Fig. 8.9. In fact increasing the heat capacity twentyfold to allow for the oceanic surface layers as well as the troposphere, and dropping the rate of heat loss to 50 W m⁻² to be consistent with the longer-term averages apparent in Fig. 8.7(b), would mean that the 100 °C chilling would then take ten years rather than three months. On the scale of the Earth's life span this is still an extremely fast rate of cooling, and is hundreds of times larger than even the most rapid temperature changes associated with short-term variations in climate.

Now that we have confirmed that the imbalances in the radiant energy budget are very significant, we should check to see if they can be significantly offset by heat conduction. Such conduction is a statistical consequence of the random thermal motion of the myriads of molecules which make up even millimetric-sized quantities of matter at terrestrial densities. If this thermal motion is unevenly distributed at some initial time (i.e. if there are temperature gradients to begin with), then the systematically more energetic molecular jostling in the warmer zones spreads their excess temperature to the cooler zones. Although there is individual

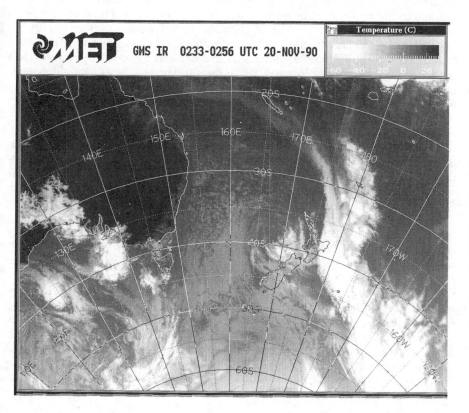

Fig. 8.10 A panorama of southern-hemisphere subtropics and middle latitudes as observed in the infra-red by the Japanese geostationary satellite GMS at about 0240 Z (1400 local sun time), 20 November 1990. A pronounced wave on the cold front east of New Zealand is associated with cyclonic circulation (clockwise in the southern hemisphere) around a surface low pressure centred at 38 °S 169 °E. The satellite radiometer grey scale shows that the brightest clouds are below − 50 °C (the high troposphere), whereas the darker clouds are at about − 20 °C (the lower middle troposphere), and the Australian interior is at about + 30 °C in the early afternoon sun.

molecular movement, heat conduction is effected without gross motion of the conducting material, and occurs in solids, as well as in liquids and gases. Thermal conduction is therefore a diffusion process, with many points in common with molecular diffusion (for example of water vapour through air) and momentum diffusion (viscosity). If conduction can account for some or all of the Earth's imbalances of radiant energy, then to that extent gross motion such as convection or advection of the atmosphere and oceans is not required.

From a wealth of laboratory work it is known that heat is always conducted down the prevailing temperature gradient and at a rate (measured by the heat flux density H — the rate of flow of heat through unit area perpendicular to the direction of conduction) which is proportional to the temperature gradient $\delta T/\delta n$. Symbolically

$$H = -k\,\frac{\partial T}{\partial n}$$

where k is the *thermal conductivity* of the conducting material. The minus sign is necessary because of the convention that the distance n is measured towards higher temperatures, which is up the temperature gradient and in opposition to the conducted heat flow. There is an obvious similarity with the comparable expression for diffusion of momentum (eqn (7.4)).

Values for k are well established for a wide range of materials and conditions. Gases in terrestrial conditions are less efficient heat conductors than are liquids and solids, largely because of their lower densities. In air in conditions typical of the low troposphere, values are about 25×10^{-3} W m^{-2} °C^{-1}, which means that in the presence of a temperature gradient of 1 °C m^{-1}, the conducted heat flux is 25 mW m^{-2}. To account for an upward flux of 100 W m^{-2}, such as is implied by Fig. 8.6 in the very low troposphere, there would have to be a temperature lapse rate of about 4000 °C m^{-1}. This is thousands of times larger than the largest temperature gradients ever observed in the atmosphere on any human scale (those found just above dry, brightly sunlit land surfaces), and is 4×10^5 times larger than the dry adiabatic lapse rate, observed to be the limiting lapse rate for all substantial layers of the free atmosphere. It is therefore clear that thermal conduction can play no significant role in offsetting the gross imbalances of the Earth's radiant-energy budget. Conduction is significant in the laminar boundary layer, the very shallow layer of air which coats all parts and details of the Earth's surface (discussed in section 9.6), and also serves to smooth away the extremely sharp, localized temperature gradients which would otherwise arise from the turbulent juxtaposition of contrasting air parcels, but on more than millimetric space scales it is almost entirely ineffective, as can be justified by a scale analysis similar to that used in eqn (7.14) to demonstrate the ineffectiveness of viscosity.

We have now tried to justify the conclusions drawn rather prematurely earlier in the chapter: the gross imbalances in the Earth's budget of radiant energy are highly significant and must be offset by convection and advection of heat; gross atmospheric (and oceanic) motion is therefore required to account for the maintenance of the present thermal state of the Earth's surface and troposphere. Unfortunately it is tempting to overestimate the extent to which we have explained such a fundamental aspect of the behaviour of our atmosphere, because the discussion began by assuming the present distribution of the radiatively active materials such as water vapour and carbon dioxide. In fact their distributions, particularly in the vertical, are maintained largely by the convection and advection now in question, as discussed in section 3.7 for example. A full discussion of the role of convection in the economy of the atmosphere should therefore include the evolution of atmospheric structure and behaviour from some initial state (and what should we chose for that?) — an awesome problem which is becoming approachable only now that meteorologists and climatologists are getting access to very large computers. What we have done is simply to show that convection and advection play an essential rather than an incidental role in the atmosphere as it is — a much less basic study, which comes close to confusing chicken and egg in the manner forewarned in Chapter 1, but which nevertheless highlights the fundamental importance of the winds and up and down currents which continually trouble our atmosphere.

8.11 Convective heat fluxes

Convection carries the excess heat from the Earth's surface and distributes it through the depth of the troposphere, offsetting the imbalance apparent in the budget of radiant energy fluxes. On a global annual average the convective flux must be 100 W m^{-2} at the Earth's surface and zero at and above the tropopause to be consistent with Fig. 8.6, and it is reasonable to suppose that it decreases upward throughout most of the intervening troposphere as indicated in Fig. 8.11. Such a decrease is consistent with observations of the cumulus family of clouds (an important and obvious type of convection) which show that their prevalence is greatest in the low troposphere and diminishes upwards, with the upper troposphere being reached by a relatively few vigorous cumulonimbus.

If the Earth were arid, all this convective flux would have to be borne as sensible heat, warmer air parcels rising and cooler ones sinking to effect a net upward transport of heat. But as mentioned in the outline of the hydrological cycle (sections 3.7 and 6.3), the very widespread presence of water, ice or moist ground at the surface means that the air there is often nearly saturated, and that rising air is consistently moister than sinking air. Convection is therefore usually associated

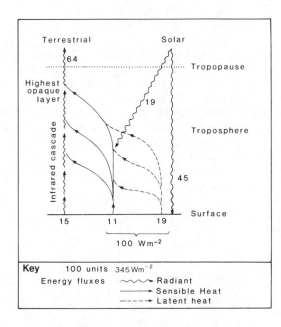

Fig. 8.11 Global annual averages of convective and radiative energy fluxes in the atmosphere. Values are from Fig. 8.6 and other parts of [49].

267

with an upward transport of water vapour, which can be regarded as a transport of latent heat, since there is inevitably an extraction of heat by evaporative cooling at the location of the evaporation (the surface), and a delivery of this heat to the location where the vapour condenses to form cloud (usually in the lower troposphere). Indeed we can convert a vapour mass flux into its associated flux of latent heat simply by multiplying by the coefficient of latent heat L.

The apportionment of the convective heat flux between sensible and latent forms has long been recognized as a very important measure of climatic type, and the ratio of sensible to latent heat fluxes is known as the *Bowen ratio* β. In arid zones such as the subtropical deserts, β values are much larger than unity (approaching 10), while in warm, humid zones they are much smaller than unity (approaching 0.1). The global annual average value of β consistent with the surface fluxes in Fig. 8.11 is nearly 0.6.

The division between sensible and latent heat fluxes at the surface (in Fig. 8.11) is taken from London's study [49]. It is very difficult to make direct measurements of these fluxes on a large scale, though it has been done in a few detailed studies in various locations, and in fact London's estimates are based in part on worldwide measurements of precipitation. These show that the global annual average rainfall equivalent is about 800 mm. From the density of water, this implies a mass flux of precipitated water of 800 kg m^{-2} y^{-1}. Since the global hydrologic cycle is nearly steady over time periods of a couple of months or more (over five times the residence time for water in the form of vapour in the atmosphere — see section 3.7), and since dew and frost are relatively unimportant sinks of atmospheric water vapour, this rate of precipitation to the surface must be very nearly equal to the evaporation rate from the surface. Multiplying the mass flux density by the value for the latent heat of vaporization, we therefore find the latent heat flux density to be 2000 MJ m^{-2} per year, which is 63 W m^{-2} or 19 budget units to the nearest on Fig. 8.6. Since the total convective flux is defined by the radient energy budget, the sensible heat flux is found by deficit to be 37 W m^{-2} or 11 budget units. The accuracy of the estimation is not high, since it depends on assessment of global precipitation and hence on the unreliable measurements over the oceans (section 2.6), but it is probably not in error by more than 20%.

Figure 8.11 depicts a progressive handing over from latent to sensible heat flux, which begins in the low troposphere and is largely complete in the middle and upper troposphere. This corresponds to the widespread appearance of cloud, and associated conversion from latent to sensible heat, in the first few kilometres above the cloud-free layer. Of course there are many tall cumulonimbus, especially in the showery outbreaks of middle latitudes and the intertropical zone of low latitudes, but even there much of the cloud in the upper troposphere is initially formed at or below the middle troposphere before being carried aloft by embedded buoyant updraughts. Latent heat fluxes play a considerable role in increasing the efficiency of convective transport of heat, and are especially effective in the lower troposphere in all but the most arid regions.

Lastly and most fundamentally, Fig. 8.11 depicts the progressive handing over from the sensible heat flux to the cascade of terrestrial radiation, which is effectively complete at levels corresponding to the high troposphere. These are in the vicinity of the highest layer which is largely opaque to terrestrial radiation, and above these levels the thermal equilibrium of the atmosphere is dominated by radiation, with comparatively little scope for convection or advection.

8.12 Advective heat fluxes

It is obvious in Fig. 8.7(b) that the annual average fluxes of solar and terrestrial radiation produce a net heat gain for the planet between latitudes about 32 °N and 38 °S, and a net loss at higher latitudes. The observed nearly steady thermal state of the planet therefore requires a balancing advection of heat from low to high latitudes which must be the net result of meridional exchange of air and water — warmer material moving polewards and cooler material moving equatorwards.

It will appear in Chapter 13 that the thermal expansion of air allows energy to be stored and transported in the form of the gravitational potential energy of vertically expanded air columns, in addition to the more obvious forms of warmer and cooler air at comparable levels. In fact both these forms of energy, together with the flows of latent heat implicit in net fluxes of vapour, are needed in any comprehensive study of the heat/energy budget of the atmosphere and oceans. However, for the present it will be much simpler, and still quite rewarding, to concentrate on the obvious fluxes of sensible and latent heat.

The observed steadiness of thermal state means that the annual average advection of heat across any latitude must equal the net loss of radiant energy from all higher latitudes in that hemisphere, which can be calculated from curve (i) in Fig. 8.7b. We can make successive calculations to produce the meridional profile of poleward advection of heat depicted in Fig. 8.12. Beginning at a pole and choosing successively lower latitudes, the calculated poleward advection of heat increases progressively until we reach the latitude at which the net radiative flux changes from negative to positive, beyond which the advected flux begins to fall away as the net input of radiant energy in each latitude zone lessens the need for heat advection from lower latitudes. In fact the calculated poleward advection reaches zero effectively at the equator, showing that the annual radiant energy budget for the hemisphere is nearly balanced and does not depend on significant net advection to or from the other hemisphere. Calculation of the meridional profiles of poleward advection in Fig. 8.12 is quite simple in principle, and is detailed in appendix 8.3.

For easier appreciation of the numerical values, the advective fluxes in Fig. 8.12 have been divided by the lengths of the local latitude circles to produce average fluxes per unit length of latitude circle. Such values represent the average advection of heat between gigantic imaginary goal-posts set one metre apart on a latitude circle, and reaching up to the top of the effective atmospheric layer and down to the bottom of the effective oceanic layer. In fact this would mean posts extending to the tropopause and down to a depth of about one kilometre below sea level. Of course 'average' is the operative word: at any particular time and place the instantaneous advection between the goal-posts might be poleward or equatorward

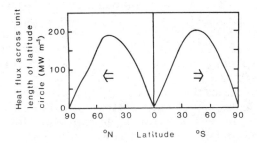

Fig. 8.12 Meridional profiles of annually averaged poleward heat fluxes advected by the atmosphere and oceans, expressed per unit length of latitude circle. (After [49])

269

depending on the local air flow. Only after averaging around a latitude circle and for a substantial time period is there a systematic poleward flow with a reasonably definite value.

One effect of displaying fluxes per unit length of latitude circle is to displace the maxima poleward from the latitudes of reversing radiant balance. The maximum values apparent in Fig. 8.12 are very large indeed, approaching 200 MW m^{-1}, which means that the average advection between goal-posts 10 m apart is equal to the generating capacity of a large power station.

There is significant seasonal variation in the advected fluxes, but it is much smaller than we might expect from the very large seasonal variations in net radiative balance (Fig. 8.9), because of the smoothing influence of the large heat capacities of the oceanic surface layers. For example, the middle latitudes of summer hemispheres would seem to require no poleward heat advection according to Fig. 8.9, but in fact in midsummer the ocean surfaces there are still warming toward their temperature maxima (to be reached in late summer or early autumn) and in the process are significantly offsetting the radiative gain. Despite such smoothing, winter advection in middle latitudes is noticeably stronger than summer advection, as could be guessed for example from the greater frequency and intensity of extratropical cyclones in winter.

8.13 Weather and ocean systems convecting and advecting heat

On the simplest view there are two main types of weather system at work in the troposphere: the nearly vertical *cumuliform* convection associated with the whole cumulus family, from the smallest fair-weather cumulus to the largest cumulonimbus; and the nearly horizontal but systematically slightly tilted *slope* convection associated in particular with extratropical cyclones. Each of these types contributes to vertical convection of heat, and the slope convection in addition plays an important role in heat advection. Although cumuliform convection in particular is extremely widespread, it is especially concentrated in parts of large-scale systems, such as equatorward flows associated with extratropical cyclones and the intertropical convergence zone of the Hadley circulation. Let us consider convection and advection separately, outlining the properties of cumuliform and slope convection which relates to their ability to transport heat.

Updraughts in cumuliform convection are strong (~ 1 m s^{-1} in small clouds and ~ 10 m s^{-1} in large cumulonimbus) and highly localized. They occur typically in concentrated populations (Fig. 6.3), in which, however, only a small fraction of the total horizontal area is covered by updraught. Individual updraughts have horizontal areas ranging from $\sim 10^3$ m^2 to ~ 10 km^2 and are separated by much larger areas of gently sinking, cloud-free air. There is therefore a marked asymmetry between strong localized ascent and weak diffuse descent, which of course is directly responsible for the intermittent nature of the showers produced by clouds big enough to precipitate. The inverse relation between strength and area of the up and downdraughts means that the net vertical mass flux of air in any particular population of cumulus is very much smaller than the constituent upward and downward fluxes. In fact it is often insignificant. However, because the updraughts are consistently warmer and moister than the downdraughts, there are almost always significant net upward fluxes of sensible and latent heat. Temperature excesses in updraughts are usually quite small, seldom exceeding 0.5 °C in small cumulus, and exceeding 5 °C only in large cumulonimbus. Localized downward fluxes of sensible heat sometimes occur when updraughts overshoot into a convectively stable layer, such as the subsidence inversion of an anticyclone (section 11.5), or the base of the stratosphere (section 10.4), but these always surmount layers in which sensible heat is directed in the normal upward direction.

Updraughts in slope convection are much more extensive and much weaker than those in cumuliform convection, and give rise to the large areas of stratiform cloud and persistent precipitation which are so typical of the fronts associated with extra-tropical cyclones. Updraught speeds are too small to be measured directly, but from estimates based on precipitation rates or large-scale mass balances for air (sections 7.13 and 11.4) it is known that on many occasions they are ~ 10 cm s^{-1}. Updraughts often exceed 10^5 km^2 in horizontal area. Corresponding down-draughts are generally even weaker and more extensive, and are displaced from the updraughts by long horizontal distances. The large areas of subsiding, cloud-free air in anticyclones and ridges of high pressure are obvious examples. Though there is some asymmetry between updraughts and downdraughts, it is much less pronounced than in the case of cumuliform convection.

In middle latitudes both cumuliform and slope convection are widespread, contributing significantly to the convection of heat from the surface to most levels of the troposphere. As discussed in sections 11.2 and 11.3, the two types often cooperate: cumuliform convection is often quite vigorous in cold fronts and in the extensive equatorward flows of air which follow them; in other areas small cumuli-form convection pumps heat into the lower troposphere, preparing the air there for subsequent slope convection to the high troposphere in a front.

In low latitudes, cumuliform convection predominates. The intertropical convergence zone (ITCZ) can be regarded as a ragged belt of very tall cumulonimbus pumping heat into the upper troposphere, and the enormous areas covered by the trade winds are populated by smaller cumulus and cumulonimbus which are pumping water vapour (i.e. latent heat) into the dry air spreading out from the

Fig. 8.13 Aerial view of typical trade-wind cumulus and small cumulonimbus. Clouds lean backwards relative to their direction of drift because easterly components weaken and even reverse aloft (section 10.2).

271

Fig. 8.14 A vertical meridional section of the Hadley cell in one hemisphere, with Earth curvature ignored for simplicity. In anticlockwise order from the equator, the main labelled features represent the intertropical convergence zone, the subtropical jet stream, the subtropical high-pressure systems, and the trade winds.

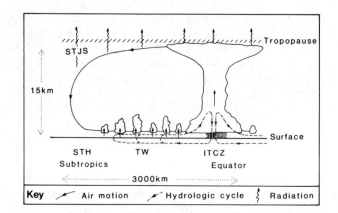

flanks of subtropical high-pressure zones (Figs 8.13 and 8.14). It is true that some of these populations often seem to be organized on larger scales, producing synoptic-scale tropical weather systems which can occasionally be as large as extra-tropical cyclones, but it seems that the cumulonimbus is the essential building block. For example, the severe tropical cyclone (hurricane, typhoon or cyclone depending on geographical zone — section 12.4) can be regarded as being made up of powerful cumulonimbus arranged in a dense circle round a relatively quiet eye, with arms of aligned cumulonimbus spiralling out for several hundreds of kilometres. Such organized populations are often much more efficient at pumping latent heat into the upper troposphere than are more homogeneous distributions, as their heavier precipitation shows.

Vertical transport of heat in the oceans is effected by wind-induced stirring of the surface layers, which carries summer heat downwards for seasonal storage. During the next cooling season, the warmed layer will feed heat back to the surface. In low latitudes the heat is probably stored for shorter periods, smoothing the feeble but nevertheless significant variations in air temperature associated with day and night and passing weather systems. However, all such upward and downward heat transport in the oceans virtually cancels out over time periods of a year at most.

In high latitudes there is persistent downward convection of water which has been chilled at the sea surface, which profoundly affects the temperature distribution in the ocean deeps by maintaining temperatures close to freezing at all latitudes including the equator. This downward flux of cold water in high latitudes (which is of course an upward flux of heat) is balanced by a downward flux of heat which is especially concentrated in the subtropics (Fig. 8.15). There the great semi-permanent atmospheric anticyclones are able, by a subtle mechanism involving horizontal wind drag and the Coriolis effect [52], to maintain large-scale convergence of the warmed surface layers, which pushes the warm water slowly down to greater depths than could be reached by simple stirring. The sinking warm water and the cold water continually filling the deeps from high latitudes maintain a dynamic equilibrium in a region of strong vertical temperature gradient, known as the *permanent thermocline*, which is found at depths ranging from 300 to 700 m below sea level there. But although these processes are vast and continuous, the associated motion is so gentle that it transports heat at a rate which is quite negligible in comparison with atmospheric convection or advection.

Advection of heat occurs in both the atmosphere and the surface layers of the oceans, and usually has a poleward component. The ocean surface layers, with their huge heat capacity, would seem to have the potential to play the dominant role, but despite the sound and fury of an angry sea, the oceans are intrinsically

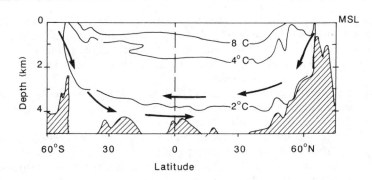

Fig. 8.15 An idealized vertical meridional section through the Atlantic, showing cold water sinking into the deeps at high latitudes, and the permanent thermocline (centred on the 8 °C isotherm) separating the warm and cold water spheres.

much less active than the atmosphere; indeed both their waves and their great surface currents are essentially spin-offs from the much more vigorous activity of the atmosphere. Regarded as heat engines, the oceans are relatively inactive because both their low-latitude heat sources and their high-latitude heat sinks are concentrated at the same level, the sea surface. This is in complete contrast to the atmosphere, where the source is at the surface and the sink is in the high troposphere (section 13.4). It is intuitively reasonable from observation of domestic saucepans of water convecting as their bases are heated, and it can be confirmed theoretically and experimentally, that much more vigorous fluid motion ensues when the heat source lies well below the heat sink, as in the atmosphere. Some of the wind-driven ocean currents do advect heat quite effectively (the Gulf Stream is an example, though the mechanism is curiously indirect) but only in limited regions; and since such currents are maintained by atmospheric motion in the first place, we should expect their contribution to the total heat advection to be considerably less than that of the atmosphere. Detailed analysis of all significant ocean currents suggests that together they carry only about one quarter of the total poleward flux at the mid-latitude maximum, leaving about 150 MW m^{-1} to be carried by the atmosphere, effectively the troposphere. In lower latitudes, ocean currents carry a larger proportion of the smaller total advected flux, averaging about 40% at 30° latitude.

In middle and high latitudes the poleward advection of heat in the troposphere is effected largely by extratropical cyclones and anticyclones. Indeed the sequence of conditions experienced at a particular surface location, as a depression passes by a little to poleward, is obviously consistent with a poleward heat flux: warm moist air in the warm sector moves with a poleward component and is followed after the passage of the cold front by cool, less humid air moving with an equatorward component (section 11.3). In the middle and upper troposphere the flow pattern is different and less familiar, but observation and analysis show that an essential feature of any extratropical cyclone is the broad swath of warm, cloudy air which flows eastward and poleward in the vicinity of the cold front, rising gently from near the surface at lower latitudes to the middle and upper troposphere in higher latitudes (section 11.4). On the western sides of extratropical cyclones cool air flows toward the equator, sinking gently despite containing populations of localized updraughts in cumulonimbus.

In low latitudes the meridional advection of heat is directly related to the Hadley circulation (Fig. 8.14). Air warmed in the great cumulonimbus populations of the ITCZ drifts poleward in the high troposphere. The potential temperature of this air is very high indeed, being little short of the equivalent potential temperature of the saturated adiabats in many cases, because the cloud tops are so high (section 5.11). The air cools somewhat by net radiation as it drifts poleward, and completes its

273

cooling and descent in the subtropical high-pressure zone. As it flows equatorward again at low levels in the trade winds, it has a much lower potential temperature than it had when flowing poleward in the high troposphere, and it is clear that the net result of the whole cycle of ascent, descent and advection has been a strong poleward transport of sensible heat.

Much of the heat evolved in the ITCZ and spread poleward in this way has actually been gathered in the form of latent heat, by evaporation from the huge surface area covered by the trade winds. The smallish cumulus and cumulonimbus characteristic of the trades (Fig. 8.13) pump vapour into the great air streams converging on the ITCZ, so that there is a powerful equatorward flux of latent heat which nearly balances the net flux of sensible heat over a substantial latitude range. However, poleward heat advection predominates where the dry poleward flow aloft extends beyond the moist equatorward low beneath, as happens in the subtropics. Some of this advected heat is leaked to space as terrestrial radiation from the upper parts of the high-pressure systems there, but part descends with the subsiding air and spills out on their poleward flanks, to be drawn into the very different flow regime advecting heat to middle and high latitudes.

Sometimes direct heat advection from low to middle latitudes occurs when weather systems develop in gaps in the chain of subtropical high-pressure systems girdling the hemispheres, and the seasonal disruptions associated with the monsoons of the northern hemisphere land masses in particular (section 12.2) are clearly potentially very significant. Further study shows that symmetry of the advected heat flows about the equator is far from perfect: the concentration of land in the northern hemisphere apparently maintains the mean position of the ITCZ several degrees of latitude north of the equator, after averaging out the seasonal migrations associated with the monsoons. The flow patterns of Fig. 8.14 then ensure a substantial northward flux of latent heat across the equator which is largely balanced by a southward net flux of sensible heat. The precise situation in low latitudes is still obscured by the shortage of reliable data there, and a proper discussion of the relative importance of the deviations from the simple model depicted in Fig. 8.12 and 8.14 is beyond the scope of this book.

8.14 Wind speeds and heat fluxes

We have assumed that heat is carried upwards and polewards in the atmosphere by cumuliform and slope convection, i.e. by weather systems associated with cumulus, cumulonimbus and frontal stratiform clouds. Now such cloud systems have typical properties which are reasonably well defined by observation, and it is interesting to try to make an extremely simplified check that these properties are consistent with the magnitudes of the heat fluxes which we know have to be maintained.

Consider the situation at latitude 45° where the poleward heat flux per unit length of latitude circle is about 200 MW m^{-1} and the vertical heat flux density is 100 W m^{-2}, as it is nearly everywhere over the globe. Assume that one quarter of the poleward heat flux is carried by ocean currents, leaving 150 MW m^{-1} to the atmosphere, and that one half of the vertical and poleward heat fluxes in the atmosphere are in the form of sensible heat. Then the flux densities of sensible heat which must be maintained by atmospheric weather systems are 75 MW m^{-1} polewards and 50 W m^{-2} upwards. Such equal partitioning between latent and sensible heat is apparently inconsistent with the surface data in Fig. 8.11, where latent heat predominates, but we are interested at present in transport in the middle and lower troposphere, where vertical and slope convection are most readily identified with visual observation of cloud, and where some of the latent heat flux at the surface has been transformed to sensible heat.

Now we will make further rather sweeping simplifications of the properties of vertical and slope convection and estimate the vertical and horizontal wind speeds required to maintain the required fluxes of sensible heat. The realism of these speeds will then be a measure of the realism, or at least the consistency, of the whole approach.

Consider first the net poleward advection of sensible heat. In our model this must be effected entirely by slope convection. Assume that on average the extra-tropical cyclones and anticyclones which embody the slope convection effectively occupy one quarter of the 45 ° latitude circle. And assume that these quadrants are entirely covered by equal and opposite meridional air currents which differ in temperature by 10 °C and have wind speed V. Then according to appendix 8.4, the meridional wind speed V must have value 8 m s^{-1} to provide the flux density of 300 MW m^{-1} required in the active quadrant. This is an encouragingly realistic value, bearing in mind that horizontal wind speeds in the more active parts of the troposphere are $\sim$ 10 m s^{-1}.

In considering the vertical transport of sensible heat, we assume that this is borne equally by cumuliform and slope convection, and that cumuliform convection occupies another quadrant of the latitude circle. Unlike the case of slope convection, the opposing air currents in cumuliform convection are highly asymmetric, so we assume that updraughts occupy only one tenth of the active horizontal area, the remainder being occupied by the much weaker downdraughts. In line with aircraft observation of clouds, we assume that updraughts are only 0.5 °C warmer than downdraughts. Then according to the calculations outlined in appendix 8.4, cumuliform and slope convection will each produce vertical heat fluxes with density 100 W m^{-2} in their active quadrants (giving the required 50 W m^{-2} for the zonal average of their sum) if the updraughts have speeds $\sim$ 2 m s^{-1} and $\sim$ 2 cm s^{-1} respectively. Again these values are reasonably realistic. Updraughts in cumuliform systems are known to range from about 1 m s^{-1} in small cumulus to over 10 m s^{-1} in large cumulonimbus. And updraughts in fronts are known from indirect evidence to range from a few centimetres per second in weak systems to several times this in vigorous ones. Notice that the ratio of vertical to horizontal wind speeds in slope convection inferred from required heat fluxes is consistent with air currents having slopes of about 1 in 400, which is known to be realistic from the type of analysis described in section 11.4.

Such oversimplified calculations are not meant to be conclusive; the realistic results for wind speeds and updraughts clearly depend on judicious selection of the various values assumed at several points in the procedure. Nevertheless these values are in themselves quite plausible (for example the fraction of a latitude circle occupied by cumulus systems, and the fraction of such areas covered by updraught), and the consistency of the whole analysis suggests that cumuliform and slope convection can account for the heat fluxes known to be advected and convected through the middle latitude troposphere.

Appendix 8.1 Blackbody radiation

The emission of electromagnetic radiation from the molecules of a body as they jostle in thermal agitation is analogous to evaporation from a wet surface, and just as in the case of a saturated vapour, it is useful to consider dynamic equilibrium between the 'vapour' of *photons* (packets of radiant energy) and the emitting surface. The laws of thermodynamics can be applied to this equilibrium and a fundamental law named after Kirchhoff (1859) can be derived [35] and expressed as follows:

$$E_\lambda = a_\lambda f(\lambda, T)$$

where

1. E_λ is the *spectral radiant emittance* of the surface at wavelength (the power radiated by unit area of surface per unit wavelength range centred on λ).
2. a_λ is the *spectral absorptivity* of the surface (the fraction of incident radiation of wavelength λ absorbed by the surface). It is also known as the *spectral emissivity* ϵ_λ of the emitter because it describes the efficiency of emission as outlined below. When collisions no longer greatly outnumber emissions and absorptions of photons, Kirchhoff's law fails and absorptivity and emissivity are no longer identical. Kirchhoff's law holds good throughout the terrestrial surface and lower atmosphere.
3. $f(\lambda, T)$ is a universal function which varies only with wavelength and surface temperature, and is quite independent of the surface material.

In essence Kirchhoff's law states that the efficiency of absorption of radiation of a certain wavelength by a surface is exactly matched by its efficiency of emission of that same wavelength, where perfectly efficient emission is defined by the universal function $f(\lambda, T)$. It follows that the most efficient emitter of electromagnetic radiation at a given temperature is also a perfect absorber of all wavelengths and therefore a reflector of none. Such an ideal surface is called a *blackbody*, and its emission spectrum is given by the universal function $f(\lambda, T)$, named after *Planck* who first explained its shape in 1900, and in so doing began the quantum revolution in physics. For a blackbody

$$E_\lambda = f(\lambda, T)$$

Planck functions are sketched in Fig. 8.16 for three absolute temperatures T_3, T_2 and T_1 in the ratio 6:5:4.

Real materials approach the blackbody ideal if their molecules are sufficiently energetic and close together to interfere fairly continually with each others' radiatively active atomic energy levels as they jostle and collide, so that their individual line spectra are smoothed into a Planck function hump. In terrestrial conditions solids and liquids are nearly all effectively 'black' in the near and far infra-red, whereas gases are not. Water vapour and carbon dioxide have very complex lines and bands of lines in their far infra-red emission and absorption spectra, which arise from quantized states of molecular rotation and vibration, whereas they are almost completely ineffective as emitters and absorbers in the visible and near infra-red. However, Kirchhoff's law applies throughout. In the visible, a_λ is very small and so is E_λ because $E_\lambda = a_\lambda f(\lambda, T)$. In a thick cluster of lines in the far infra-red, if there is enough vapour or carbon dioxide to make the lines merge in effect, a_λ is nearly 1 and E_λ as nearly approaches the blackbody ideal $f(\lambda, T)$ in those wavelengths.

It is shown by observation, and can be confirmed by manipulating the Planck functions, that the total emittance of a blackbody across all wavelengths is a simple function of absolute temperature T:

$$E = \int_0^\infty f(\lambda, T) \, \mathrm{d}\lambda = \sigma T^4$$

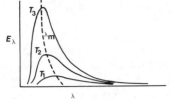

Fig. 8.16 Blackbody emission from bodies with absolute temperatures in the ratios 6:5:4.

This is *Stefan's law*, and the universal constant σ is the *Stefan–Boltzmann constant*, with value 5.67×10^{-8} W m^{-2} K^{-4}. Stefan's law gives the emittance of the blackbody surface, which corresponds to the total area under the Planck curve for that temperature (Fig. 8.16). It clearly shows that blackbody emission increases very sharply with absolute temperature.

Observation and manipulation of the Planck function also show that the wavelength λ_m of maximum spectral emittance of a black body (the peak of the relevant Planck curve) decreases with increasing absolute temperature according to *Wien's displacement law*

$$T\lambda_m = Y$$

where Y is a universal constant which has value 2897 μm K when λ is expressed in micrometres. The reddening glow of a metal bar as it cools from white heat is a classic example of Wien's law in action.

Absorption by a *greybody* is uniformly spread across all wavelengths so that absorptivity (and emissivity) are independent of λ. The wavelength of maximum spectral emittance is unaltered by comparison with a blackbody at the same temperature, and the emittance is given simply by

$$E = a\sigma T^4 = \epsilon\sigma T^4$$

Real solids and liquids are often greybodies across substantial wavelength ranges. For example water and ice are very pale grey (small a_λ) in the visible range, and are very dark grey (a_λ only slightly less than 1) in the far infra-red.

Consider two simple but important examples which involve such radiative behaviour.

1. For reasons mentioned in section 8.1 and the beginning of this section, the solar photosphere behaves as a blackbody even though it is a gas. From Wien's law a blackbody with temperature 5900 K has maximum spectral emittance at wavelength 0.49 μm, near the centre of the visible spectrum, which is consistent with spectroscopic observations of the sun. From Stefan's law the emittance of the solar photosphere is very close to 69 MW m^{-2}, which is consistent with measurements of the solar constant after allowing for inverse-square attenuation by distance.

2. The solid and liquid surfaces of the Earth behave as blackbodies at the local temperature. At temperature 288 K (the mean surface temperature) the wavelength of maximum spectral emittance is almost exactly 10 μm, which lies well into the far infra-red, as confirmed by observation. At the same temperature, Stefan's law gives blackbody emittance 390 W m^{-2}, which is exactly consistent with the 113 units on Fig. 8.6. Using Stefan's law in reverse, the terrestrial output of 67 units to space (231 W m^{-2}) corresponds to the emittance of a blackbody at 253 K, which is fully consistent with values quoted in Sections 8.1 and 8.2.

Appendix 8.2 Surface irradiance

Consider a beam of direct sunlight impinging on a horizontal surface at an elevation angle of β (a zenith angle of $90 - \beta$). According to the geometry of Fig. 8.17, the radiant flux density F_h on a horizontal surface (the horizontal irradiance) is less than on a surface normal to the beam (F_n) as follows:

$$F_h = F_n \sin \beta \tag{8.5}$$

This confirms that the irradiance of a horizontal surface is a maximum when the sun is in the zenith, and tends to zero as the sun approaches the horizon.

Another important factor which varies with β is the path length of the solar beam through the atmosphere (Fig. 8.17), and with it varies the associated attenuation of the beam by scattering and absorption. The details of the interaction between solar radiation and the atmosphere are complex and are dealt with in more advanced texts [8], but an important result emerges from quite simple examination. The small reduction dF in irradiance arising from the presence of small mass dm' of attenuating material in a section of sloping beam is proportional to both F and dm'

$$dF = -k \, dm' \, F$$

where the minus sign ensures that dF is a reduction and k is an attenuation coefficient. (In fact k normally varies strongly with wavelength (Fig. 8.1), introducing considerable complication which we will ignore.) In many circumstances attenuating material is horizontally stratified, like much of the atmosphere, and the simple geometry of Fig. 8.17 shows that dm' is simply related to the element of mass dm in a vertical column bounded by the same horizontals.

$$dm' = dm \, \text{cosec} \, \beta$$

Hence

$$\frac{dF}{F} = -k \, dm \, \text{cosec} \, \beta$$

Integrating along the sloping path from the top of the atmosphere, where F_0 is effectively the solar constant, to the surface where F_n is the surface flux density normal to the beam

$$\frac{F_n}{F_0} = \exp(-km \, \text{cosec} \, \beta)$$

where m is the total mass of the active material in a vertical column through the atmosphere. This is a form of *Beer's law*, and the product km is known as the

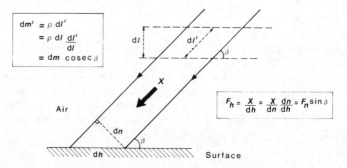

Fig. 8.17 Oblique solar rays passing through an absorbing atmosphere and illuminating a horizontal surface.

$dm' = \rho \, dl'$
$= \rho \, dl \, \dfrac{dl'}{dl}$
$= dm \, \text{cosec} \, \beta$

$F_h = \dfrac{x}{dh} = \dfrac{x}{dn} \dfrac{dn}{dh} = F_n \sin \beta$

optical depth of the atmosphere. The left-hand side of eqn (8.5) is called the *transmissivity* τ_β of the angled beam — the fraction of radiation incident on the top of the atmosphere which penetrates to the surface (in the direct beam). By definition the zenith transmissivity τ_{90} is given by

$$\tau_{90} = \exp(-km)$$

so that we have

$$\tau_\beta = \tau_{90}^{\mathrm{cosec}\beta} \tag{8.6}$$

It is usual to describe the transparency of an atmosphere by its zenith transmissivity even if it is at a latitude at which the sun can never reach the zenith. Thus a clear sky in the British Isles may have a zenith transmissivity of as much as 0.8. It follows that when the sun is 15° above the horizon on the same occasion, the transmissivity of the oblique beam is only 0.42.

The horizontal irradiance at the foot of an oblique sunbeam is found by combining eqns (8.5) and (8.6):

$$F_h = F_0 \tau_\beta \sin \beta \tag{8.7}$$

Figure 8.18 is a graph of horizontal surface irradiance against solar elevation angle for a clear summer's day in Britain.

Obviously this treatment does not try to cover atmospheric transmission when attenuating material is not horizontally stratified (for example when the sky is dotted with cumulus), nor does it deal with diffuse radiation, where photons scattered out of the direct beam may nevertheless reach the surface, sometimes after further scattering.

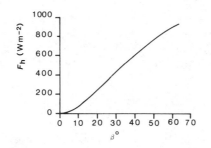

Fig. 8.18 Irradiance of a horizontal surface near sea level in central England on a clear summer's day (zenith transmissivity 0.8). The horizontal axis is solar elevation, and maximum elevation corresponds to noon at the summer solstice.

Appendix 8.3 Meridional advection profiles

According to the global geometry outlined in Fig. 8.19, the surface area of a narrow zone bounded by latitudes ϕ and $\phi + d\phi$ is given by $2\pi R \cos \phi \, R \, d\phi$. The surface area $_1A_2$ of a wider zone bounded by distinctly different latitudes ϕ_1 and ϕ_2 is found by integrating the narrow zone expression between these limits.

$$_1A_2 = 2\pi R^2 (\sin \phi_1 - \sin \phi_2)$$

Substituting values for ϕ_1 and ϕ_2 quickly confirms that the area of a hemisphere is $2\pi R^2$, and shows that half the surface area of a hemisphere lies between the equator

279

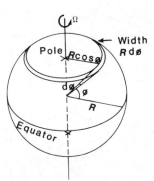

Fig. 8.19 Oblique view of Earth, with zonal ring at latitude ϕ.

and latitude 30°, while only 13.4% lies poleward of latitude 60°. This exposes the grossly misleading nature of the Mercator map projection and justifies the cos ϕ weighting used in Figs 8.7 and 8.9.

Suppose that we divide the hemisphere into latitude zones each 10° wide (obviously a more uniform division could be made as above), and that within each zone we can define a zonal average net influx F of radiant energy per unit horizontal area from the data summarized in section 8.7. Then the total influx into a zone is the product of the zonal area and the flux density. To be in a steady thermal state this heat influx must be exactly balanced by the net efflux of heat, which can only be borne by meridional advection. If advection is described by flux densities M (average flux across unit length of latitude circle), then the advected flux across any latitude circle is $M\,2\pi R \cos \phi$, and the expression for the thermal equilibrium of a zonal strip bounded by latitudes ϕ_1 and ϕ_2 is easily shown to be

$$M_2 \cos \phi_2 - M_1 \cos \phi_1 = FR (\sin \phi_2 - \sin \phi_1)$$

Since M at the poles is necessarily zero (there are no higher latitudes to flow to!), we can work zone by zone to lower latitudes using observed values for F to calculate M values.

Appendix 8.4 Convective and advective heat fluxes

ADVECTION

Consider a vertical plane erected on a line of latitude which is crossed along a fraction N of its length L by a poleward flow of air with speed V, temperature T and surface pressure p. If the remainder of the line is crossed by an equatorward flow with speed V', temperature T' and surface pressure p', then the net poleward flux of sensible heat is

$$N L\, C_p\, \frac{p}{g} V T - (1 - N)\, L\, C_p\, \frac{p'}{g} V'\, T'$$

which is justified by noting that p/g is a mass of air per unit surface area, that $(p/g)V$ is therefore a poleward mass flux per unit length of latitude line, and that this is converted into a flux of sensible heat by multiplying by temperature, specific heat capacity and relevant line length ($N\,L$). The specific heat at constant pressure C_p is used for the reasons mentioned in section 5.3, and actually allows for the presence of heat in the form of thermally extended air columns as well as directly sensible heat (section 13.2).

We assume there is zero net poleward mass flux of air, so that

$$N L\, \frac{p}{g} V = (1 - N)\, L\, \frac{p'}{g} V'$$

Substituting this in the expression for the net heat flux and dividing by L to find the net flux per unit length of latitude line, we obtain

$$N C_p\, \frac{p}{g} V (T - T') \tag{8.8}$$

which shows that the crucial factors are the temperature excess and speed of the poleward flow, together with the fraction of the latitude line it occupies.

In section 8.14 it is assumed that the advective flux density in the active zone is 300 MW m^{-1}, that N is ½ and that $T - T'$ is 10 °C. To allow for the fact that only the troposphere (i.e. the layer from about 1000 to 250 mbar) is effective in advecting heat, we assume p to be 750 mbar. Together with $g = 10$ m s^{-2}, we have p/g 7500 kg m^{-2}, and with C_p 1000 J kg^{-1} K^{-1} we deduce that V must be 8 m s^{-1} to carry the required flux.

CONVECTION

We can treat vertical fluxes of sensible heat in an almost identical way. Suppose that either cumuliform or slope convection is active in a certain region with horizontal area A, so that its associated updraughts fill a fraction N of the horizontal area with air of density ρ, temperature T and speed w, the remainder of the area being filled with air of density ρ' and temperature T', sinking at speed w'. Then a similar procedure to before shows that the net upward flux of sensible heat is given by

$$N A \rho w C_p T - (1 - N) A \rho' w' C_p T'$$

Again assuming that there is no net upward mass flux, we have

$$N A \rho w = (1 - N) A \rho' w'$$

Using this and dividing by A we find the net upward sensible heat flux per unit area to be

$$N \rho w C_p (T - T') \qquad (8.9)$$

which is clearly equivalent to expression (8.8) in all important respects.

In section 8.14 we assume that both slope and cumuliform convection maintain sensible heat flux densities of 100 W m^{-2} in their respective active areas. A realistic value for ρ in the low troposphere is 1 kg m^{-3}. In slope convection N and $T - T'$ are assumed to have the same values assumed above in the advection estimate, because it is the same tilted flows which effect both advection and convection. It follows that an updraught of 2 cm s^{-1} is needed to maintain an upward flux of 100 W m^{-2}. In cumuliform convection, N is $1/10$ and $T - T'$ is 0.5 °C, and it requires an updraught of 2 m s^{-1} to maintain 100 W m^{-2}. Note that this asymmetry of area implies a compensating asymmetry of updraught and down-draught speeds to maintain zero net mass flux (density differences being relatively unimportant).

Appendix 8.5 Radiative forcing

The sensitivity of blackbody emission to changes in temperature is found by differentiating Stefan's Law (appendix 8.1).

$$\frac{dE}{dT} = 4 \sigma T^3$$

This can be rewritten in finite-difference form and rearranged to find the blackbody temperature rise ΔT corresponding to any rise in blackbody emittance ΔE.

$$\Delta T = \Delta E/(4 \, \sigma \, T^3)$$

Applied to radiative forcing of the Earth's atmospheric greenhouse (section 8.5), this expression gives the first-order temperature rise ΔT corresponding to a forcing of magnitude ΔE. Assuming $T = 255$ K, the temperature rise per unit forcing is 0.27 K per W m^{-2}.

Problems

LEVEL 1

1. Some stars are much redder in appearance than others. What does this tell you about their photospheric temperatures?
2. The orbital radius of Venus round the sun is about two thirds of the orbital radius of the Earth round the sun. Find the ratio of the solar constants of the two planets.
3. A black-and-white cat is lying in strong sunlight. Describe its appearance when viewed by a radiometer sensitive to the far infra-red, whose display renders strong radiators white and weak radiators black.
4. How would the shape of the terrestrial radiative spectrum observed in Fig. 8.3 have changed if at the time of observation there had been a complete overcast of clouds with tops at temperature 250 K?
5. Which of the following alterations would tend to enhance or reduce the atmospheric greenhouse effect: widening the atmospheric window; making the flanks of the window more opaque to terrestrial radiation; having a much hotter but smaller sun with the same solar constant; increasing the direct absorption of solar radiation by the atmosphere?
6. Describe everyday observations which suggest that limitation of convection is crucial to the performance of greenhouses?
7. Assuming that the fluxes of terrestrial radiation in Fig. 8.6 are otherwise unchanged, what rate of heat loss do you expect at the surface at night (a) when the sky is clear, and (b) when the sky is overcast by low cloud?
8. Check by squared paper or geometry that curve (i) of Fig. 8.7(b) indicates global radiative balance. Redraw with a linear latitude scale to see the apparent exaggeration of high-latitude losses.
9. Find the angle of solar elevation at noon at latitude 60° at the winter solstice.
10. In Fig. 8.11 the net upward flux of terrestrial radiant energy increases with height through the troposphere, so that every parcel of finite depth loses more through its top than it gains through its base. How is this sustained without steady cooling of the air?
11. Consider the surface wind directions and air temperatures on the western and eastern flanks of an anticyclone which straddles a western continental margin in winter, and find the likely direction of resulting net advection of sensible heat in the low troposphere.
12. The same weather system in summer would effect an equatorward flow of

sensible heat. Consider the directions of net fluxes of latent heat in both situations.

In Fig. 8.14:

13. what happens to the latent heat released by condensation in the intertropical convergence zone;
14. what are the main differences in the middle troposphere between air ascending in the ITCZ and descending in the STH;
15. and what is the advantage for the Hadley circulation that air diverging from the base of the subtropical high pressure systems is unusually dry?

LEVEL 2

16. Venus, Earth and Mars orbit the Sun at average distances of 108, 150 and 228 million kilometres respectively. Given that the sun is a spherical black body of radius 700 000 km and temperature 5800 K, find the solar constants for each planet.
17. Show that the equilibrium radiative temperature T_E for the Earth (radius R_E) is related to the radiative temperature T_S of the solar photosphere (radius R_s) by

$$T_E = \frac{(1-a)^{1/4}}{\sqrt{2}} \left(\frac{R_s}{R}\right)^{1/2} T_s$$

where a is the Earth's planetary albedo. Hence find the equilibrium temperature for Venus, assuming its planetary albedo to be 0.7 (higher than Earth's on account of its thick cloud blanket).
18. Find the radiative emittance of the following surfaces, all except one of which are blackbodies: water at 25 °C, sand at 40 °C (emissivity 0.85), fog at 5 °C, cloud at -20 °C and at -40 °C.
19. Meteorological satellites usually carry two radiometers, one using visible radiation and the other one or more bands of infra-red radiation. The latter have displays in which whiteness of image increases with decreasing radiant intensity. To see that this produces infra-red images which resemble visible ones (in daylight), sketch the infra-red images you would expect to find in a view containing stratocumulus, large cumulonimbus with tops in the middle and high troposphere, and a front with cloud decks at various levels in the middle and high troposphere.
20. Assuming that in Fig. 8.6 the direct solar inputs to the stratosphere and troposphere are spread though the layers 16–48 km above sea level and 0–16 km respectively, find the ratio of the heat inputs per unit mass and hence the relative speeds of implied temperature response.
21. In a certain clear sky the zenith transmissivity for solar radiation is 0.75. Find the irradiance at sea level of normal and horizontal surfaces when the sun is 25° above the horizon. Assume the normal solar irradiance above the atmosphere is equal to the solar constant.

LEVEL 2′

22. Consider the energy loss represented by 100 W m^{-2} output over a three-month season, and express it in terms of the cooling of a 10 m layer of water, the

cooling of the atmosphere, and the depth of ice formed by freezing of the
ocean surface at 0 °C.

23. If the maximum poleward heat advection across the 45° line of latitude were to
be converted with 1% efficiency into useful electrical power, find the power
available per head of world population.
24. Use the approach outlined in appendix 8.3 to show that if the net radiative
input per unit surface area is F when averaged over all latitudes from the
equator to 45°, and the average meridional heat flux to higher latitudes is M
per unit length of latitude line, then

$$M = R F$$

where R is the radius of the Earth. Use this expression to estimate a value for F
from Fig. 8.12 and compare your result with Fig. 8.7(a).

LEVEL 3

25. In another 5×10^9 years or so, our Sun will probably become a red giant, with
its photospheric temperature dropping to 4000 K and radius swelling to 25
times the present value. Find out all you can about the radiative environment
of the Earth which will then apply.
26. Discuss the significance for atmospheric conditions and behaviour of the
atmosphere's transparency to solar radiation and opaqueness to terrestrial
radiation. Speculate on how things might differ if the atmosphere were opaque
to solar radiation and transparent to terrestrial radiation.
27. Consider the following very simple model of the atmosphere in purely
radiative equilibrium. It consists of a single thin sheet of air at absolute
temperature T_a which is transparent to solar radiation but acts as a grey body
with absorptivity and emissivity a in the far infra-red. The underlying surface
acts as a blackbody with temperature T_g. Find the relation between T_g and the
solar input and between T_a and T_g, and calculate numerical values for T_a and
T_g using a realistic value for solar input and $a = 0.8$.
28. Despite using the display mode for infra-red imagery as described in problem
13, some differences remain between visible and infra-red images. Describe
and explain the differences to be expected in the following cases: an area of sea
fog; snow-covered low ground showing through high cloud (not in Antarctica
or Siberia); strongly heated pale desert; a plume of diffuse anvils (section 10.4)
from a cluster of shower clouds.
29. Check on the equatorial value for the solar input in Fig. 8.7(a) by considering
the input there at an equinox. (*Hint:* in this situation solar radiation is
collected continually on a diameter of the Earth's disc, but is spread around
the Earth's circumference by the earth's rotation.)
30. Discuss the climatic significance of meridional and seasonal variations in solar
input and the relative invariance of terrestrial output. In most latitudes,
diurnal variations in solar input are even larger; why are their consequences
relatively small? Note that this is examined in some detail in sections 9.3 to 9.5.

LEVEL 3′

31. Consider how temperatures on the surface and in the atmosphere might alter if
all convection and advection were to cease.

32. The heat capacity of the Arctic ocean obviously tends to limit the seasonal swings of surface temperature there. How is this tendency affected by the seasonal presence, freezing and melting of sea ice there?

33. In a popular science fiction story by a distinguished cosmologist [53], the Earth is supposed to face catastrophe as an opaque cloud of interplanetary dust threatens to cut off solar input and produce 'over 250 degrees of frost within a month'. Persuade the writer that he is being unduly pessimistic.

34. Modify the method for calculating heat advection in appendix 8.4 to analyse the following simplified picture of narrow poleward ocean currents such as the Gulf Stream and Kuroshio: fraction of latitude circle at 40° occupied by currents 1/200; poleward speed 1 m s^{-1}; depth of current 500 m; temperature excess of warmer flow 3 °C. Compare your result with Fig. 8.12.

Surface and boundary layer

9.1 Introduction

The lower boundary of the atmosphere is the solid or liquid surface of the Earth, which is often either complex or dynamic or both. The sea surface is relatively simple when there is little broken water, but it is highly mobile on account of the presence of surface waves and currents. Land surfaces are often more static, though the wavy motion of tall grasses, forests and (on a longer timescale) sand dunes, are conspicuous exceptions, but they are usually highly irregular on space scales ranging from those of sand grains to mountains. Dense vegetation is particularly complex and can significantly blur the boundary between surface and atmosphere by spreading the sites of heat and momentum exchange significantly in the vertical. The importance of the large heat capacities of surface layers of water has already been stressed (section 8.10). By contrast, land surfaces have a very different thermal behaviour which varies widely with surface type and condition. Changes in ground cover by vegetation or snow can produce large changes in thermal and other behaviour of the resultant surface — changes which can significantly affect the overlying air.

In all sorts of natural and artificial fluid flows, it is observed that there is one or more zone, adjacent to the boundary, in which the flow is significantly affected by the nearby presence of the boundary, to the extent that the affected zones are called *boundary layers* to distinguish them from the relatively free fluid beyond. There are several distinguishable boundary layers at the base of the atmosphere, where the presence of the nearby surface has a profound effect on atmospheric behaviour. Figure 9.1 represents the most important of these diagrammatically — layers which together make up the *atmospheric* or planetary *boundary layer*. Dimensions are only notional since the various layers vary tremendously in depth.

Within a few millimetres of any surface, no matter how irregular, air motion is so strongly restrained by friction with the surface, and by the intermolecular friction (viscosity) of the air itself, that gross motion is nearly non-existent. Momentum, heat and matter such as water vapour are transported mainly by molecular diffusion. This is the *laminar boundary layer* — so-called because, apart from random molecular motion, the only significant movement of air molecules resembles the sliding of a sheaf of thin plates (laminae) each of which is able to

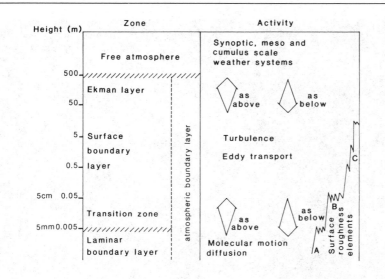

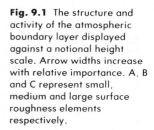

Fig. 9.1 The structure and activity of the atmospheric boundary layer displayed against a notional height scale. Arrow widths increase with relative importance. A, B and C represent small, medium and large surface roughness elements respectively.

move only very slowly between its neighbours (Fig. 7.3). Although occupying only a minute fraction of the total volume of the troposphere, the laminar boundary layer significantly affects large-scale exchanges between the surface and the bulk of the troposphere, by acting as a kind of separating membrane which is only sluggishly permeable by fluxes of heat and matter.

As depicted in Fig. 9.1, another boundary layer known variously as the *turbulent* or *surface* boundary layer extends upwards from the laminar boundary layer for a highly variable and rather poorly defined distance. Just as the laminar boundary layer is the region strongly influenced by the surface through molecular diffusion, so the surface boundary layer is the region which the surface influences strongly by eddy diffusion (see section 7.2). In fact it is the most conspicuously turbulent part of the atmosphere, and is the only part in which the turbulence often seems to tend towards a relatively well-ordered dynamic equilibrium. Even so, the turbulent transport of momentum, heat and matter through it is poorly understood in detail, despite its significance for large-scale exchanges between the surface and troposphere, and despite the fact that the surface boundary layer is the atmospheric environment of Man.

Above the surface boundary layer there is a relatively extensive transition zone in which the direct influence of the surface weakens with increasing height, to the level where it becomes the rather indirect, though still quite strong, influence which the surface exerts on the rest of the troposphere. At this altitude the geostrophic departure induced by turbulent friction (section 7.11) has become unobservably small, having decayed in a curious fashion first explained by Ekman and often named after him. The depth of the whole nest of boundary layers (laminar, surface and Ekman) conventionally comprising the atmospheric boundary layer, is very variable but is often ~ 500 m. Above this the atmosphere is considered to be so free of direct surface influence that it is called the free atmosphere. In fact this is not a very rational title, and much of this book contains strong evidence that the troposphere itself is yet another boundary layer — one influenced by the surface through the action of the common weather systems. Nevertheless we will follow convention and concentrate in this chapter on the atmospheric boundary layer as conventionally defined and as depicted in Fig. 9.1.

Fig. 9.2 Dendrites of hoar frost growing from the edges of small leaves.

9.2 Surface shape and radiation

The surface of the Earth can be regarded as a horizontal plane with projections and indentations ranging in size from the microscopic to the mountainous. These affect the overlying boundary layers in a number of ways which are mentioned at various points in this chapter, the most important of which are the redistribution of heat sources and sinks arising from radiation fluxes, and the local disturbance of air flow. In this section we concentrate on the effects of radiation. Notice that there are many other effects of surface shape which are important or interesting in some contexts; for example the environment of a mountain peak is greatly influenced by the prevalent lapses of temperature and density in the ambient air, while if projections are as small as blades of grass, with correspondingly sharp edges, the diffusion of water vapour is concentrated at the sharp edges, as is often apparent in the distribution of hoar frost (Fig. 9.2).

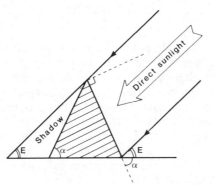

Fig. 9.3 A beam of direct sunlight falling on a ridge with triangular cross-section.

Consider first the effects of surface shape on direct solar input. A simple model of a solitary projection (Fig. 9.3) establishes some important effects. First, solar radiation is concentrated on the sun-facing side at the expense of the other, and the concentration is total if the other side is in shadow. In middle latitudes the typical slopes of steep hills and mountains are such that sunlight often falls nearly normally on sun-facing slopes and very obliquely on others. Since surface irradiance is proportional to the sine of the angle between the plane of the illuminated surface and the line of the incident sunbeam (Appendix 8.2), all normally illuminated slopes are receiving the maximum flux density possible in any given sunlight. It is shown in appendix 9.1 that the ratio of sloping surface irradiance to horizontal irradiance is given by $\sin(\alpha + E)/\sin E$, where α is the angle of slope and E is the angle of solar elevation above the horizon (Fig. 9.3). Values for this ratio are plotted in Fig. 9.4 for a slight slope, a steep slope, and a nearly vertical slope typical of a cliff or tree. This shows that steep slopes are at a great relative advantage when illuminated by the sun at low angles of elevation. Although absolute values of irradiance are then quite small because of strong attenuation of the oblique sunbeams in their long paths through the atmosphere (Fig. 9.4 and appendix 9.1), the advantage can have quite marked effects on the distribution of cumulus convection over hilly country. Since such convection is driven by very small temperature excesses, it can be concentrated similarly over sun-facing slopes, producing lines or *streets* of cumulus drifting downwind from the nursery slopes (Fig. 9.5). In the examples plotted in Fig. 9.4, the absolute value

289

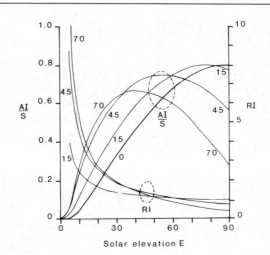

Fig. 9.4 Direct-beam solar irradiance on surfaces making angles of 0, 15, 45, and 70° with the horizontal. Curves RI represent the ratio of slope irradiance to horizontal irradiance. Curves AI/S represent absolute slope irradiance under a clear sky with zenith transmissivity 0.8 (appendix 9.1), normalized by the solar constant S.

of insolation is greatest for all but the slightest slopes when the sun is moderately high in the sky — near the midday maximum elevation in summer in mid-latitudes. The very considerable preferential warming of sun-facing slopes is then quite obvious to anyone walking in hilly terrain. In the case of small projections such as buildings or vegetation, the concentrations of input are too localized to have any significant effects on larger scales, but they can be very important for conditions on the scale of the projections themselves, or smaller, as is often apparent in the distribution of flora and small fauna. The effects of window location in relation to the sun on conditions inside buildings are very familiar, although still hardly acknowledged by much building design.

Another obvious effect of surface shape, already implicit above, is simply that if the projection is steep enough in relation to the solar elevation angle, the sunlit face concentrates insolation which would have been distributed over a much larger area (i.e. larger than the base area of the projection) if the projection had not been there. This is simply to say that a significant area is shadowed by the projection. In the case of the simple projection in Fig. 9.3, shadowing extends beyond the base when the solar elevation is less than the angle of slope. It is easily shown that a steep mountain illuminated by a fairly low sun can intercept sunlight which would otherwise have been spread over a horizontal area twice as large as the base of the mountain, and a very steep solitary projection such as a tree can do several times better than this.

When there are several steep projections close together, each one may shade its neighbour to some extent, so that the solar input is concentrated on the upper parts of each. This can occur in hilly terrain when the sun is low, but is at its most obvious in dense stands of vegetation such as a field of cereal, or a forest. Indeed the infiltration of sunlight through vegetation is more akin to sunlight entering water and being increasingly attenuated with depth below the surface (Fig. 9.6), the leaves playing the same role as the absorbing and reflecting material in the water. This is an example of how a complicated surface shape blurs the normally sharp interface between atmosphere and surface, the radiatively active layer being concentrated in this case in the upper parts of the surface projections.

In the simplest picture the presence of projections redistributes the insolation but does not alter the total input. However, if the projections are so steep and close together that light suffers repeated scattering from adjacent facing surfaces,

Fig. 9.5 A street of cumulus
drifting downwind from a
sunlit slope.

repeated absorption ensures that the total capture by the surface is enhanced. This
happens markedly when sunlight is steeply incident on a deeply indented surface
such as a forest, in which light passing below the treetops is repeatedly reflected
between adjacent trees and leaves, with further absorption at each reflection. As
aerial photographs often show, coniferous forests have a very low albedo as a
result, being amongst the darkest of all common land surfaces.

All of the above has assumed direct sunlight with geometrically edged zones of
illumination and shadow. In many regions, more than half of all surface irradiance
is diffuse, having been scattered from the incident direct beam by cloud, haze and
air molecules. All of the above effects becomes much less marked in diffuse
sunlight because, although not necessarily fully isotropic, diffuse sunlight is very
much less directional than direct sunlight.

291

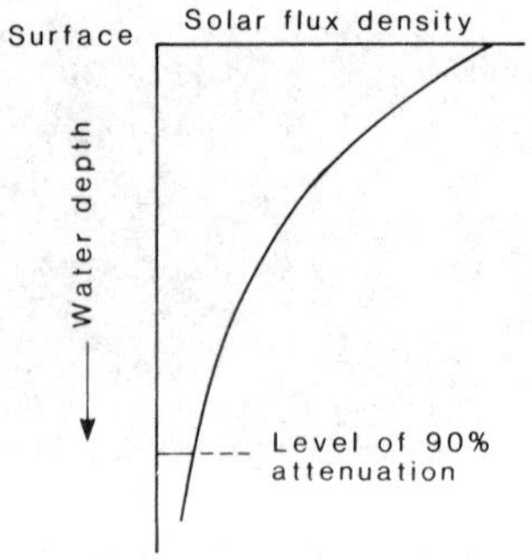

Fig. 9.6 Exponential decay of solar irradiance with depth in water.

Output of terrestrial radiation is complicated by surface shape in ways which are rather similar to those applying to the reception of solar radiation, although there is no equivalent of the direct solar beam and the ways in which it is shaded, since terrestrial radiation is always diffuse. In essence the net rate of loss by emission and absorption of terrestrial radiation is governed by how much sky can be seen by a fish-eye lens at the surface in question. A plane horizontal surface sees the maximum amount of sky (a hemisphere), whereas at the bottom of a narrow cleft between two projections, only a small amount of sky is 'seen' between the projection walls. Since the effective blackbody temperatures of the walls are usually much higher than those of the sky, the net loss is greatly reduced in the cleft. Hoar frost distribution on roads and short grass often demonstrates the radiative sheltering effects of walls and trees very clearly. As with solar input, the terrestrial output from a dense stand of vegetation is concentrated in the upper parts of the stand, though not necessarily in precisely the same height zone as that in which the solar input is concentrated.

9.3 Surface heat input and output

Let us consider the radiative and convective processes injecting and removing heat from representative parts of the Earth's surface, so that we can define typical rates of gain or loss of heat.

Consider first the absorption of solar radiation. The albedos of terrestrial surfaces range very widely from a few % for damp, dark soils to over 90% for fresh snow (Table 9.1). Snow values are localized in high altitudes and latitudes, and the relatively pale land deserts are localized in the subtropics. Over the rest of the globe values are dominated and held low by the presence of water and vegetation. Over land albedos are mostly between 5 and 20%; and over the sea they are somewhat lower, except where the direct beam of sunlight approaches grazing incidence. Rough surfaces generally have lower albedos than equivalent smooth surfaces because of the multiple absorption effect mentioned in the previous section. Albedos in diffuse sunlight are marginally larger than in fairly high elevation direct sunlight because of the inevitable presence of some grazing input in the diffuse light.

Albedo values quoted in Table 9.1 are averaged over the whole solar spectrum,

Table 9.1 Radiative properties of surface materials

Surface material	Solar albedo	Terrestrial emissivity
Soils	0.05–0.40	0.90–0.98
Grasses	0.16–0.26	0.90–0.95
Forests	0.05–0.20	0.97–0.99
Desert	0.20–0.45	0.84–0.91
Snow	0.40–0.95	0.82–0.99
Water		
low solar elevation	0.10–1.0	0.92–0.97
high solar elevation	0.03–0.10	0.92–0.97

and may be applied directly to total solar fluxes to estimate inputs to irradiated surfaces. The strong coloration of many terrestrial surfaces of course shows that reflectivity varies considerably with wavelength in many cases, but though this may sometimes be important (as in the photosynthesis process in plants), it is averaged out in dealing with the heat economies of all surfaces on the human and larger scales.

To establish typical values for solar input, consider a horizontal surface with 15% albedo irradiated by the sun from an angle of elevation of 50° through a relatively clear sky which is transmitting 70% of the solar beam. Then using the value for the solar constant, and the expression for oblique surface illumination (appendix 8.2), we find that solar energy is being absorbed by the surface at a rate of nearly 630 W m^{-2}, a value which is typical of daytime values under clear skies in middle latitudes. Towards midwinter, or at dawn or dusk, values are of course very much smaller, as they are too under thick overcast (which may reduce atmospheric transmission to less than 10% of its clear sky value). In the course of a day the total amount of solar energy absorbed by an element of horizontal surface in middle latitudes may range from as much as 30 MJ m^{-2} in summer to less than 1 MJ m^{-2} under a winter overcast. Daily totals can be measured directly by recording solarimeter, or estimated from day length and midday maxima using the approximation in appendix 9.2.

As mentioned earlier (section 8.3) terrestrial surfaces behave very nearly as blackbodies in the wavelengths which make up terrestrial radiation. The closeness to blackbody behaviour is apparent in the nearness of *emissivity* (Appendix 8.1) to the value 1. Table 9.1 shows that most surfaces have emissivity values of between 0.9 and 0.98. Apart from old snow, only deserts have emissivities less than 0.9, and that is because of the presence of quartz sand which is partly transparent in terrestrial wavelengths. Note that water is almost completely opaque to terrestrial radiation (which means that its emissivity is close to 1.0) although it is nearly transparent to solar radiation — a striking example of how radiative properties can differ in different wavelength ranges. Terrestrial radiation from the sea therefore comes from the topmost millimetres of water.

For present purposes we will simply assume that terrestrial surfaces emit and absorb terrestrial radiation as if they were blackbodies. We can therefore use Stefan's law (Appendix 8.1) together with typical values of surface temperature to show that radiant output from most terrestrial surface lies within 90 W m^{-2} of 390 W m^{-2} — the value corresponding to the global annual mean surface temperature (388 K — see Fig. 8.6). Much of this output is offset by absorption of terrestrial radiation coming downwards from the sky, which has an average value of just over 340 W m^{-2} according to Fig. 8.6. This too varies considerably, exceeding

390 W m^{-2} under warm, low clouds, and falling below 300 W m^{-2} under cold, clear skies. Such variations can have important effects on the net radiative balance, but for the moment let us assume that as a net result of the emission and absorption of terrestrial radiation, a typical portion of the Earth's surface loses energy at a rate of 50 W m^{-2} or about 4.3 MJ m^{-2} per day.

Combining solar gains with terrestrial losses we see that the net result is that on a typical summer day in middle latitudes a net input of about 20 MJ m^{-2} in daylight is followed by a net output of about 2 MJ m^{-2} during the night. In winter there may be a continuous net output since the weak daylight is more than offset by the continuing net terrestrial output, and the daily loss total may be as much as 10 MJ m^{-2}. In summer in particular, convection tends to offset the daylight input very significantly by carrying sensible and latent heat into the atmosphere, and on all but the stillest nights mechanical stirring of the atmospheric boundary layer feeds some sensible heat from the air back to the chilling ground. (Dew and frost formation represent the return of some latent heat, but the amounts are usually relatively small.) The net result of convective and radiative exchanges is that many surfaces suffer a daily alternation between heat gains and losses, each of which is $\sim$ 5 MJ m^{-2}. In middle and high latitudes there is also a large annual alternation between persistent daily gains in summer, when daytime gains outweigh nighttime losses, and persistent daily losses in winter, when night-time losses exceed daytime gains. Allowing for convective and advective offsets, the summer gains and winter losses may each be $\sim$ 500 MJ m^{-2} per season.

9.4 Surface thermal response

We have seen that there are daily and annual rhythms of heat gain and loss by any part of the Earth's surface, and have estimated their orders of magnitude. We know that surfaces respond by heating and cooling, and know also that their responses vary widely. In this section we look at the mechanisms involved in such responses.

WATER

Since water is fairly transparent to solar wavelengths, especially in the visible range, sunlight is attenuated progressively as it travels downwards below the surface — a small fraction of the incident energy usually reaching depths of tens of metres. Most of the attenuation ultimately leads to absorption and heating of the water, but of course there is some backscatter to the surface, contributing a very little to the oceanic albedo and allowing us to admire submarine sunbeams. The simple fact of the considerable transparency of water to sunlight, compared with the almost total opaqueness of even the shallowest layer of ground, accounts for much of the enormous difference in the thermal responses of land and sea, the heat capacity of a typical layer of illuminated ocean being very much larger than that of a land surface, and its temperature response correspondingly very much smaller.

Despite the fact that sea water absorbs different solar wavelengths with different efficiencies, extinguishing the near infra-red more rapidly than the visible, the depth profile of overall solar flux density often follows the ideal exponential shape expected in a homogeneous absorber (Fig. 9.6 and appendix 9.3), with depths for 90% attenuation varying from only a few metres in *turbid* water, usually found near coasts, to as much as 50 m in the relatively clear waters often found in the open oceans (though plankton may contribute to turbidity there). Absorption in turbid waters is principally by suspended sediment and organic matter, and some associated dissolved substances, whereas in clear water it is principally by the water itself.

As mentioned in section 8.13 turbulent mixing in ocean surface layers increases the depth of water effectively influenced by sunlight, enhancing their effective heat capacity. Observations suggest that depths subject to daily and annual cycles of warming and cooling are $\sim$ 10 and $\sim$ 50 m respectively, with corresponding heat capacities $\sim$ 40 and $\sim$ 200 MJ $°C^{-1}$ m^{-2}. If the warming and cooling were distributed uniformly throughout these depths, then daily gains and losses of 5 MJ m^{-2} would produce temperature rises and falls of 0.125 °C, and seasonal gains and losses of 500 MJ m^{-2} would produce temperature changes of 2.5 °C (Fig. 9.9). Because temperature changes are largest at the surface, despite the effects of mixing, actual temperature changes at the surface are often larger than these estimates, but not very much. Marine temperature cycles are therefore very much smaller than those familiar to us on land.

Some of the mixing in the surface layers of the oceans is effected by mechanical turbulence extending down from the troubled surface, but some is convective, and in fact convective stability exerts an overall control on vertical exchange just as it does in the atmosphere. For example the depth of the layer affected by summer warming is limited by that warming itself: warmed surface water is less dense than the cooler water beneath (the intervening layer with the vertical temperature gradient being called a thermocline), and the convective stability of the set-up inhibits vertical exchange of water just as a stable layer does in the atmosphere. The difference is that in the oceans surface warming enhances stability while in the atmosphere it diminishes it.

Cooling in the oceans is localized at the surface in the first instance, since terrestrial radiation, evaporation and wind chill are concentrated there. By increasing the density of water at the surface, cooling encourages convective instability and hence the downward extension of the cooling by vertical mixing. In very sheltered waters such as lakes, the destruction of the summer thermocline in the autumn can be quite sudden, playing an important role in the annual biological cycle by redistribution of nutrients etc., but in the oceans it is masked by constant commotion. However, after the removal of the summer thermocline, further cooling is distributed through a much deeper layer of water than before, which may even extend to the sea bed kilometres below.

LAND

The response of a complex, vegetated land surface to surface heating and cooling is very difficult to analyse, but some important features can be seen by considering the relatively simple case of a homogeneous layer of rock or soil as it responds to a regular cycle of warming and cooling at the surface. This is a standard problem in the theory of heat conduction [53], with a well-known and tested solution which we will now examine and compare with observation (Fig. 9.8).

We should expect that at any depth below the surface, the rate of rise of temperature after the onset of surface heating will increase with the thermal conductivity k

of the intervening layer and decrease with its heat capacity. The full theory shows
that the net effect is determined by the *thermal diffusivity*

$$\kappa = k/\rho\, C \qquad (9.1)$$

where ρ and C are respectively the density and specific heat capacity of the layer
material. The larger the value of κ, the more freely is a change in surface temperature conducted downwards, and the deeper is the layer affected. Values of κ for
typical surface materials are listed in Table 9.2, and range from 0.1 for dry peat to
0.75 for wet sand (units being 10^{-6} m^2 s^{-1}). Trapped air in dry soils is such a poor
heat conductor that it ensures low κ values despite smaller values of ρ and C.

The depth of material affected also depends on the time period of the surface
warming and cooling cycle. If this period is short, then heat which has not penetrated far enough during the surface-warming cycle is conducted back to the
surface and lost during the succeeding cooling cycle. If the period is long, then a
greater depth of soil is significantly warmed and cooled.

The detailed solution shows that if the surface heating and cooling cycle has
period P, then there is an important depth parameter D which is known as the
damping depth and is given by

$$D = (\kappa P/\pi)^{1/2} \qquad (9.2)$$

The damping depth defines the depth of the soil which is significantly affected by
the surface heating and cooling in several different but related ways.

1. If the amplitude of the surface temperature wave is $\triangle T_0$ (Fig. 9.8), then the
    amplitude $\triangle T_z$ at depth z below the surface is given by

    $$\triangle T_z = \triangle T_0 \exp(-z/D) \qquad (9.3)$$

    The temperature wave therefore dies away exponentially with increasing depth,
    its amplitude falling to 37% of the surface value at depth D, and to 10% at $2.3D$
    (Appendix 4.3).
2. As the temperature wave is conducted down into the layer, it lags further and
    further behind the surface wave and is exactly in antiphase (minima coinciding
    with maxima and vice versa) at depth πD (Fig. 9.8).

Table 9.2 Thermal properties of surface materials

Material	Dry peat	Dry sand	Wet sand	Water	Air	units
Density	300	1600	2000	1000	1.2	kg m^{-3}
Specific heat capacity	1920	800	1480	4180	1000	J kg^{-1} K^{-1}
Thermal conductivity	0.06	0.3	2.2	0.57 (4000	0.025 6000	W m^{-1} K^{-1})
$10^6 \times$ thermal diffusivity	0.10	0.23	0.74	$(\sim 10^3$	$\sim 5 \times 10^6$)	m^2 s^{-1}
Damping depth diurnal	**5.2 cm**	**7.9 cm**	**14 cm**	**5 m**	**300 m** (	m
annual	**1.0**	**1.5**	**2.7**	100	7×10^3)	m
Heat capacity diurnal	0.04	0.14	0.60	31	0.63 (	MJ K^{-1} m^{-2})
annual	0.81	2.75	11.4	590 eddy	~ 12 values	( )

3. The amplitude $\triangle T_0$ of the temperature wave at the surface is related to the heat H entering unit area of surface during the heating phase by

$$\triangle T_0 = H/(\sqrt{2}\,\rho CD) \tag{9.4}$$

from which it appears that the effective heat capacity of the affected layer (per unit surface area) is $\sqrt{2}\,\rho CD$, and that the depth of the layer which heat input H would warm uniformly by $\triangle T_0$ is $\sqrt{2}\,D$.

It is clear from the values of D included in Table 9.2 that even the relatively penetrating annual temperature wave affects a soil layer to a depth of no more than a few metres. Even in a responsive wet soil, a worm living at a depth of about 6 m would experience only 10% of the annual temperature range experienced by its cousin living at the surface, and their temperature calendars would differ by 6 months. The daily temperature wave penetrates only tens of centimetres. The near constancy of temperature at quite shallow depths has long been exploited in the construction and use of cellars.

Values for the effective heat capacity of typical ground layers are included in Table 9.2 and show that even relatively responsive wet soils have heat capacities of no more than a few per cent of the values quoted earlier for the sea surface, all of which underlines the fact that heat conduction through terrestrial surfaces is so poor that only very shallow layers are significantly affected by daily and yearly surface heating and cooling. The simple argument presented at the beginning of the subsection on water still stands.

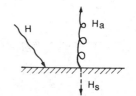

Fig. 9.7 Net radiative heat input H to a land surface being partitioned between convected heat H_a to the atmosphere and conducted heat H_s to the subsurface.

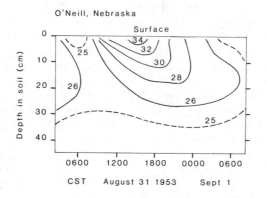

Fig. 9.8 Depth–time section with isotherms labelled in °C, showing downward progress of the diurnal temperature wave under a prairie. (After [54] and [55])

9.5 Partitioning between surface and air

Equation (9.4) represents an incomplete view of the situation at the surface because it ignores the finite effective heat capacity of the air. Using subscripts s and a for surface and air respectively, the total heat input H during the warming phase of the surface temperature cycle must be partitioned according to

$$\frac{H_a}{H_s} = \frac{(\rho CD)_a}{(\rho CD)_s} \tag{9.5}$$

where the symbols are as defined in the previous section for heat conduction. In fact heat transfer in the atmospheric boundary layer is by convection (and some terrestrial radiation) rather than by conduction, and in the surface layers of the seas

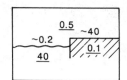

Fig. 9.9 Typical heat capacities (underlined, in MJ °C^{-1} m^{-2}) and associated diurnal temperature ranges ($\sim$ in °C) for subtropical desert and ocean.

it is by solar radiation and turbulent mixing, but for present purposes all such transport is assumed to be effectively equivalent to conduction — very rapid conduction in the cases of turbulent and radiative transport. Note that the relevant H is the excess of net radiant warming over evaporative cooling, since this is the sensible heat to be redistributed by conduction, turbulence and radiation.

By an obvious extension of eqn (9.4) we can now relate the amplitude of the surface temperature wave to the heat input H by

$$\triangle T_0 = \frac{H}{\sqrt{2}[(\rho CD)_s + (\rho CD)_a]} \tag{9.6}$$

Consider now the extremely different cases of heat input to desert and ocean surfaces, to assess the different partitioning of input between surface and atmosphere.

As the sun's rays begin to strike the dry soil of the desert in the early morning, the surface warms rapidly because the heat conducts downward very inefficiently through the air-filled soil. According to Table 9.2, the damping depth for the daily heating cycle is only a few centimetres, and the effective heat capacity is ~ 0.1 MJ °C^{-1} m^{-2}. The gradual elimination of the nocturnal inversion in the overlying atmospheric boundary layer (section 5.13) means that the heat transfer there is complex and changeable, but observations indicate that turbulent transfer often operates as if the eddy thermal diffusivity (the turbulent equivalent of κ) was within an order of magnitude of 5 m^2 s^{-1}. This is equivalent to an effective heat capacity of about 0.6 MJ °C^{-1} m^{-2} for the warming layer of air. It follows that the great bulk of the heat input to the surface passes into the air rather than into the soil beneath, and that the total effective heat capacity of the ground and the air together is ~ 1 MJ °C^{-1} m^{-2}. In a subtropical desert, despite the relatively high albedo and net loss by terrestrial radiation (clouds being scarce), the heat input H to the surface during the warming phase can easily exceed 20 MJ m^{-2}, giving a temperature cycle at the surface with amplitude exceeding 20°, which means a range from maximum to minimum of double this value (Fig. 9.9). In fact these estimates are quite conservative, and surface conditions may be even more extreme. Such violent swings of temperature between blistering daytime heat and freezing in the hours before dawn, shatter rocks by repeated violent expansions and contractions, and impose severe conditions on any organisms living there. However, the excellent thermal insulation of the dry soil, which is one of the factors contributing to the extreme conditions at the surface, offers a refuge which is taken up by many burrowing animals and dormant plants.

Similar estimates for the annual heating cycle in deserts show similar partitioning between air and ground, but at 15 MJ °C^{-1} m^{-2} the total effective heat capacity is much larger because of the greater penetration of heating and cooling into both air and ground. However, the annual cycle of heat input and output in the subtropics is quite muted because the sun is still quite high in the sky even in mid-winter, with the result that the seasonal temperature cycle is often smaller than the diurnal cycle, though still quite large. In middle- and high-latitude deserts, many of which lie in the rain shadows of mountain ranges, the much larger annual range of H values, together with the small effective heat capacities typical of all deserts, give rise to very large seasonal swings of temperature from searing summer heat to sustained sub-zero temperatures in winter. These too make great demands of land-scape and life, and for example severely tested the pioneers opening up the North American Midwest in the nineteenth century.

At the other extreme we have the case of the ocean–air interface. The estimates of ocean heat capacity made in the previous section will do well enough for the effective surface-heat capacity in eqns (9.5) and (9.6), i.e. 40 MJ °C⁻¹ m⁻² for the daily heating cycle. The effective heat capacity of the atmospheric boundary layer will be assumed to be the same as in the case of the desert (0.6 MJ °C⁻¹ m⁻²), although reduction in convective mixing may not be completely offset by increased stirring by the freer ocean winds. Regardless of detail, the heat input to the surface is obviously overwhelmingly taken into the sea rather than into the air, and the total effective heat capacity of the sea and air together is several tens of times larger than in the case of the desert. Since in addition the sensible heat input to the surface is significantly smaller than in the case of the desert (the large evaporative cooling more than offsets the effect of the much lower albedo), the result confirms the explanation for the observed very small oceanic surface temperature range given in the previous section.

9.6 The laminar boundary layer

Air molecules in contact with a liquid or solid surface are so effectively stuck to it by molecular attraction that there is no relative motion. At a very little distance from the surface, molecules can move, but only very smoothly and sluggishly on account of viscous friction, and this laminar flow is always parallel to the surface. With increasing distance from the surface, the speed of flow increases, as do the accelerations and decelerations of air in response to disturbing factors such as surface shape, the influence of turbulent jostling of air beyond the laminar boundary layer etc. At some small critical distance δ from the surface, such unsteadiness overcomes the smoothing and damping effects of viscosity, and the air flow ceases to be laminar. The value of the thickness δ of the laminar boundary layer in any particular set-up is very important, since resistance to the diffusion of momentum, heat and water vapour through the laminar boundary layer increases with its thickness.

The outer limit of the laminar boundary layer is associated with a critical value of the appropriate Reynolds number. As outlined in section 7.7, a Reynolds number consists of UL/ν, where U and L are respectively the characteristic speeds and dimensions of the air flow in the laminar boundary layer, and ν is the kinematic viscosity of the air. The dimension in question is obviously δ, but the choice of U is less obvious since there is a large shear in the flow in the laminar boundary layer. As will become clearer in the next two sections, there is a very important velocity parameter in the turbulent air beyond the laminar boundary layer which is called the shear velocity or *friction velocity u_** and which is defined by

$$u_* = (\tau/\rho)^{1/2} \tag{9.7}$$

where τ is the flux of horizontal momentum toward the surface, which is the tangential stress on an embedded imaginary horizontal surface. This flux is borne by turbulence outside the laminar boundary layer and by viscosity within it, though the latter statement has to be qualified when the surface is rough (see later in this section). Checking by the method of dimensions (Appendix 7.1) confirms that the right-hand side of eqn (9.7) has indeed the dimensions of velocity. Observations in the turbulent air beyond the laminar boundary layer show that u_* is comparable in magnitude with the root mean square of the wind-speed fluctuations there.

Observations in all parts of the atmospheric boundary layer indicate the importance of the friction velocity. In the case of the laminar boundary layer, observations in many different situations, including wind tunnels and water channels as well as in the atmosphere, show that its depth δ corresponds to a critical value ~ 10 for the Reynolds number

$$u_\bullet \, \delta / \nu \qquad (9.8)$$

Values of τ can be measured directly by attaching delicate strain gauges tangentially to horizontal plates, and indirectly as mentioned in section 9.7. Values vary very widely with atmospheric conditions, and when combined with reasonable values of air density ρ in eqn (9.7) form a wide range of values of $u_\bullet$ centred on ~ 0.3 m s^{-1}. It follows from typical values for ν that the associated thicknesses of laminar boundary layers at the base of the troposphere are ~ 0.5 mm, being thicker when the overlying air is less turbulent and vice versa. The layer in which transport is dominated by molecular rather than turbulent diffusion is therefore extremely thin, being in effect no more than a viscous membrane of air adhering to all solid and liquid surfaces.

Although very thin, the laminar boundary layer covers all exposed surfaces, and is only significantly bypassed by momentum (section 9.9). All the other fluxes passing to or from the Earth's surface (heat, water vapour, carbon dioxide etc.) are forced to pass through it, and as a result of the considerable resistances encountered there are often quite sharp gradients of concentrations and associated meteorological quantities. For example, sensible heat diffusing from the surface obeys the equation for heat conduction (eqn (8.5)), which can be written in simple difference form

$$H = k \, \triangle T / \delta \qquad (9.9)$$

where H is the sensible heat flux per unit area and $\triangle T$ is the temperature drop across the laminar boundary, depth δ. A value of 100 W m^{-2} for H is quite typical when fairly dry land is being heated strongly by the sun, and it follows from eqn (9.9) that a fall of over 2 °C is needed to drive sensible heat at this rate through a 1 mm thick laminar boundary layer (using a realistic value for the thermal conductivity k). Notice that this temperature gradient is five orders of magnitude greater than the dry adiabatic lapse rate, which of course applies only when air parcels are able to exchange freely in the vertical — a condition not met in the laminar boundary layer, and not met either, as we shall see, in the lower parts of the overlying turbulent layer. Though such highly localized temperature gradients are not usually perceptible to animals as tall as ourselves, they probably do explain some of the extra warmth we feel on lying down to sunbathe in a sheltered position. They must however be very important to small insects, and contribute to the substantial overestimates of air temperature by normal-sized thermometers exposed to direct sunlight.

The laminar boundary layer diffuses tangential (most usually horizontal) momentum from the moving atmosphere to the surface, and it would seem to follow from the appropriate form of eqn (7.4) that

$$\tau = \mu \, \triangle V / \delta \qquad (9.10)$$

where $\triangle V$ is the difference (i.e. the shear) in tangential wind speed across the shallow layer. Assuming a value of $\sim 10^{-2}$ N m^{-2} for τ (consistent with the value assumed earlier for $u_\bullet$) it follows that $\triangle V \sim 1$ m s^{-1} — a very substantial shear across such a shallow zone. But although shears of this intensity must occur above some surfaces, it is quite misleading to imagine that a jump to 1 m s^{-1} in horizontal wind speed could be widespread only 1 mm above land and sea surfaces. The unreality of such a situation is apparent when we remember that almost all natural surfaces, including all but the calmest water, have roughness elements which are much larger than the depth of the laminar boundary layer. For example each blade of grass in a grassy surface is enclosed in its laminar boundary layer as expected, but the air flowing over and through the grass is impeded by the physical obstruction of the protruding grass stems rather than by the viscous layers, with the result that the vertical profile of horizontal wind speed amongst the grass is determined principally by the distribution, size and shape of the stems. Such a surface is said to be *aerodynamically rough*, and its roughness is influential to levels well above the tallest roughness element, as we shall see in the next section. Virtually all natural surfaces are aerodynamically rough, the only exception being water in almost zero wind and with no waves coming from elsewhere.

9.7 Turbulence

Above the very shallow laminar boundary layer, the atmospheric boundary layer is chronically turbulent. The effects of turbulence are many and varied: the gustiness of all but the lightest winds is familiar (Fig. 9.10), as is the unsteady dispersion of smoke plumes (Fig. 9.11), but other properties are less obvious, though no less significant. Though turbulence in the atmosphere and other fluids has been studied for over a century, and particularly intensively in recent decades, it is surprisingly difficult to define precisely what we mean by turbulence. Sophisticated experimental and theoretical work has served mainly to confirm the subtlety of the topic, so that even an elementary treatment is beyond the scope of this book [54]. The following is therefore a selective and mainly descriptive outline of some important properties.

1. Turbulent flow is irregular in both space and time. For example the wind speed in Fig. 9.10 varies continually and widely but shows no trace of regular oscillations such as might be associated with waves. (In fact recent studies of the nocturnal atmospheric boundary layer in particular show that waves are often present at some height above the surface. However, these are quite distinct from turbulence, though they sometimes give rise to turbulence by breaking.) Virtually every meteorological measurable varies in this way in the presence of turbulence, though wind shows by far the largest percentage range. Variations in temperature and humidity content are relatively small but are especially important because of their contribution to heat and vapour fluxes, as we shall see.
2. The unsteadiness and irregularity typical of turbulent flow represents a large-amplitude, highly chaotic response to inherent instability, which prevents

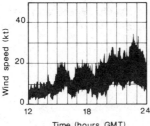

Fig. 9.10 Anemograph for part of a typically breezy day in north-west England. (P. Lursson and M. Beadle, Department of Environmental Science, University of Lancaster).

301

Fig. 9.11 A smoke plume undulating and spreading in the turbulent daytime boundary layer of air flowing along a steep-sided valley.

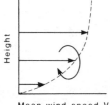

Fig. 9.12 Air tumbling in strong wind shear near the surface.

useful prediction of the precise pattern of flow at any future time. Quantitative description and prediction therefore involve statistics ranging from the simple arithmetic mean and root mean square deviation to the sophisticated power spectrum (which shows the contribution of different periodicities to the total amount of variability) and beyond. In qualitative description we can still make use of the concept of temporarily distinguishable eddies, though with the tacit assumption that they are too numerous and transient to be usefully observable in practice.

3. Turbulence is intrinsically three-dimensional and cannot be adequately described in fewer dimensions. This handicaps pictorial description and mental conception (which is strongly two dimensional), though of course flat pictures like Fig. 9.12 have to be used for limited description. If there is a dominant describable type of motion it is a continual stretching and folding which seems to be common to topological properties of chaos [2]. And although a considerable range of atmospheric turbulence is isotropic (without preferred direction or axis), there is often a difference between longitudinal structure (horizontally across the flow) and lateral structure (across the flow).

4. Turbulence is hierarchical. Inherent instability produces large eddies which in turn produce smaller eddies, and so on down a cascade of diminishing scales until the eddies are small enough to be literally rubbed out by viscosity. According to a famous jingle by a pioneer in the study of atmospheric turbulence, L.F. Richardson:

Big whirls have little whirls which feed on their velocity,
And little whirls have smaller whirls, and so on to viscosity.

The largest eddies have scales comparable with that of the atmospheric boundary layer itself (i.e. ~ 100 m or somewhat larger), and the scale of the smallest is ~ 1 mm for the same reason that this is the depth scale of the laminar boundary layer. At the large end of the scale it is sometimes difficult to distinguish between large eddies and small structured systems, the 'smoke ring' vortex believed to be common in buoyant convection being a case in point.

The fact that the largest recognizable turbulent eddies influencing conditions near the surface are ~ 100 m carries the implication that most turbulent variations (considering a variation to be a cycle of overshoot and undershoot about a notional mean) last less than ~ 100 s at a fixed position, assuming a typical horizontal wind speed. If averages are taken over time periods several times this length, the resultant values are reasonably stable. The meteorological synoptic convention is to take ten-minute averages, and it is these values which are recorded in the standard hourly observations, variability being very asymmetrically assessed by recording the highest wind speed (the strongest gust) in the observation period. There are of course variations on scales longer than a few minutes, and in fact there is a continuous range or spectrum up to time periods corresponding to the passage of major weather systems such as depressions, but the ten-minute average is especially useful because there is something of a gap in the spectrum separating turbulence proper from longer period variations. In the following sections a ten-minute averaging period is assumed for all averages unless otherwise stated.

9.8 The origins and role of turbulence

Turbulence persists because air flow is continually disturbed on scales which are much too large to be smoothed away by viscosity. If disturbances of scale 100 m are being introduced into an air flow of average speed 1 m s^{-1} (a very modest value in the atmospheric boundary layer), the very large value (~ 10^6) of the associated Reynolds number indicates that viscosity is quite incapable of maintaining laminar flow. The disturbed air flow therefore continues jostling in an irregular fashion, so that turbulence continues even if the disturbing factor operates only intermittently.

The single most important source of disturbance is the wind shear which is endemic near the surface in all but absolutely calm conditions (section 7.2). Air parcels embedded in the wind shear tend to roll forward about a lateral horizontal axis, temporarily becoming the rotating parcel which is the archetypal eddy. Rotation is not confined to the longitudinal section portrayed in Fig. 9.12, presumably because as a parcel begins to roll about a lateral axis, it tends to generate by friction opposing rolls about longitudinal axes on either side (Fig. 9.13).

All rolling parcels have descending air on one side and ascending air on the other,

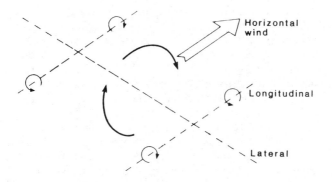

Fig. 9.13 An air parcel rolling about a lateral axis with secondary vortices rotating in both senses about longitudinal axes.

but if the air is convectively stable, as is often the case near the ground at night, vertical motion is inhibited by the effects of reversing buoyancy (section 5.12). It seems that the maintenance of turbulence by shear in these circumstances must depend on a balance between the strengths of the shear (given by $\partial V/\partial z$, where V is the average horizontal wind speed) and of the buoyancy, given by $(g/\theta)\,\partial\theta/\partial z$ (derived from eqn (7.27), where θ is the average potential temperature. Note that $\partial\theta/\partial z$ is positive when the air is convectively stable. This balance was analysed by Richardson who showed theoretically that turbulence should tend to break out when the magnitude of the dimensionless number (subsequently called the *Richardson number* and denoted by Ri)

$$Ri = \frac{g}{\theta}\,\frac{\partial\theta}{\partial z}\bigg/\left(\frac{\partial V}{\partial z}\right)^2 \qquad (9.11)$$

falls below 0.25. Many observations of the atmosphere confirm that critical values of Ri are indeed of this order. Substitution of realistic values for g and θ (in degrees Kelvin) show that very considerable stability is needed to prevent generation of turbulence by typical observed values of wind shear ($\partial V/\partial z$) near the surface. This is consistent with the observed persistence of turbulence there even on nights on which there is pronounced cooling, provided the wind speed is kept up by the large-scale weather situation. If, however, surface cooling is sufficiently strong to raise the value of $\partial\theta/\partial z$ to the point where Ri exceeds the critical value, turbulence is damped out by the static stability. Then turbulence no longer brings down V momentum from air at higher levels, and friction with the ground slows the air near the ground to a virtual standstill. Since surface cooling (assuming this to be a fairly clear night) is no longer partly offset by the flux of sensible heat brought down by turbulence from the warmer air aloft, the cooling and stability increase, and the air remains still until the next day's sun or a large-scale change in the weather intervenes.

Thermally-driven convection also generates turbulence, provided heating from beneath exceeds the ability of the air to transfer heat by molecular conduction, and provided the motion generated is too vigorous for viscosity to maintain laminar flow. In a layer of air of depth $\triangle z$ and potential temperature lapse $\triangle\theta$, this balance is described by the dimensionless *Rayleigh number* (Ra)

$$Ra = \frac{g\triangle\theta}{k\nu\theta}(\triangle z)^3 \qquad (9.12)$$

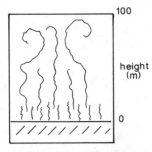

Fig. 9.14 Outline of probable transition from metric-scale 'shimmer' just above a strongly heated land surface to hectometric-scale thermals aloft.

In all terrestrial conditions in which a relatively warm surface maintains a lapse $\triangle\theta$, Ra far exceeds the known critical value for the onset of turbulent convection ($\sim 50\,000$). Although very common, the resulting convection is not well understood. Very near the surface the convection is quite small-scale, and the turmoil of centimetric and metric-scale eddies rising and falling produces the shimmering effects so noticeable when looking at distant objects just above a strongly heated surface. At greater heights it seems that these small movements cooperate to produce larger-scale thermals which rise up through the atmospheric boundary layer, leaving a turbulent wake as they go. The latter stage corresponds well to Richardson's jingle, but the former suggests a reverse sequence of building from smaller to larger scales (Fig. 9.14). Regardless of detail the effect is to maintain turbulence on a wide range of scales in the convecting layer.

As mentioned already, turbulence transports momentum, heat and material such as water vapour and aerosol, and this transport occurs whenever air parcels (eddies) moving in different directions carry systematically different quantities of momentum, heat, etc. Like the molecular diffusion which it overwhelms, turbulence always carries momentum etc. from regions of excess to regions of deficit, i.e. down-gradient. In many cases the gradients are effectively vertical and so therefore are the transports. As a result momentum is always transported down to the surface, and water vapour is almost always transported up. Over land sensible heat is often transported up by day and down by night, though the nocturnal fluxes are usually much smaller.

Consider for example the downward transport by turbulence of V momentum (i.e. horizontal downwind momentum) through horizontal unit area, bounded by a wire frame for ease of visualization. Since most of the effective eddies are larger than the frame itself, even in the SI system where the unit frame is 1 m square, an average downward flux arises because in the sequence of up- and downdraughts successively passing through the frame, the downdraughts systematically carry more V momentum as they must do coming down from the faster-moving air aloft (Fig. 9.12). The instantaneous downward flux of V momentum is given by $-\rho w V$, where the minus sign ensures that downdraughts (negative w) contribute positively to the flux. Note that $-\rho w$ is the downward mass flux. The average downward flux of momentum is therefore given by the average of all the instantaneous values of the triple product. As mentioned in section 7.2, this flux is identical with the average horizontal shearing stress

$$\tau = - \overline{\rho w V} \tag{9.13}$$

Although the average value of the vertical mass flux is zero near a horizontal surface (because air does not form condensations or rarefactions so long as it flows more slowly than the speed of sound), the momentum flux can have quite substantial values when variations in w and V are negatively correlated, as they are in Fig. 9.15. We can look into this further by examining the right-hand side of eqn (10.13) more closely.

Contributions from density variations are so small relatively that ρ can be treated as a constant and removed from under the overbar. It is shown in appendix 9.4 that eqn (9.13) simplifies to

$$\tau = - \rho \overline{w V'} \tag{9.14}$$

where V' is the instantaneous deviation of V from its average value, and w is identical with w' because average updraught is zero. The term $\overline{wV'}$ is the *covariance* of vertical and horizontal winds, and is non-zero when their fluctuations are correlated (Fig. 9.15). Combining eqns (9.14) and (9.7) we see that

$$u_*^2 = - \overline{wV'} \tag{9.15}$$

which stresses the interpretation of u_* as a velocity related to the gustiness of winds.

These results show that if we describe air flow in terms of average rather than instantaneous values, we must allow for correlations between variations occurring within the averaging period otherwise we will miss potentially significant

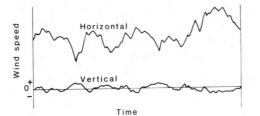

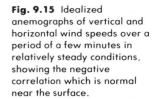

Fig. 9.15 Idealized anemographs of vertical and horizontal wind speeds over a period of a few minutes in relatively steady conditions, showing the negative correlation which is normal near the surface.

momentum fluxes, which is to say that we may ignore important forces. The equations of motion (eqn (7.12)) are always written in terms of average values, since it would be useless to apply them to instantaneous observations of quantities varying so rapidly and considerably as winds, and hence they need correction. A thorough treatment of the equations of motion in terms of averages and instantaneous deviations [56] shows that terms such as $\partial(\overline{wu'})/\partial z$ must be included along with viscosity terms such as $\nu \partial^2 u/\partial z^2$ to represent friction terms such as $\overline{F_x}$ in full in eqn (7.12). The extra terms are the gradients of terms such as $\rho \overline{wu'}$ which are known as *Reynolds stresses*, and their inclusion completes the treatment of the equations of motion begun in Chapter 7.

The Reynolds stress represented by eqn (9.14) is often the largest of the many possible components because $\partial V/\partial z$ is usually the largest shear, being the vertical gradient of horizontal wind. It can be measured by recording simultaneously the outputs from a conventional anemometer (measuring V) and a propellor on a vertical shaft (measuring w) and calculating their covariance — easily done now using a dedicated microchip. Each anemometer must of course be able to respond accurately to the smallest eddies contributing significantly to the stress, and this requires instruments with much faster responses than the normal robust and therefore rather sluggish instrument used for synoptic observation.

A very similar treatment shows that the vertical turbulent flux of sensible heat H is given by

$$H = \rho \, C_p \, \overline{wT'} \tag{9.16}$$

where $\overline{wT'}$ is the covariance of updraught and temperature, and the minus sign is absent this time because it is conventional to consider an upward heat flux to be positive. At night, turbulence persisting above a cooling surface maintains a negative covariance of updraught and temperature and hence a downward heat flux, whereas by day vigorous turbulence over strongly heated ground maintains a strong positive covariance and hence an upward flux.

The corresponding expression for the turbulent flux of water vapour is

$$F = \rho \, \overline{wq'} \tag{9.17}$$

where q is the specific humidity. Multiplied by the coefficient of latent heat L, this vapour flux becomes the turbulent flux of latent heat.

9.9 The surface boundary layer

Turbulence as outlined in the previous sections dominates the distributions and vertical transports of momentum, heat and matter such as water vapour and carbon dioxide throughout most of the atmospheric boundary layer (the laminar boundary layer being a spatially very small exception). In the lower parts of this turbulent layer the turbulence is so dominant that some striking regularities emerge. For example the horizontal components of the equations of motion (eqns (7.12a) and (b)) give rise to a simple balance along the flow between the gradients of the Reynolds stresses and the longitudinal component of the pressure gradient force

$$\frac{1}{\rho}\frac{\partial p}{\partial n} = \frac{\partial}{\partial z}(\overline{wV'}) = \frac{\partial}{\partial z}\left(\frac{\tau}{\rho}\right) \tag{9.18}$$

where n is a horizontal axis pointing upwind. The pressure gradient $\partial p/\partial n$ is usually dictated by the synoptic-scale pressure field, and consequently is of order 10^{-3} Pa m^{-1} (section 7.7). Ignoring vertical variations of air density ρ, the change $\triangle\tau$ in τ to be expected in a vertical depth $\triangle z$ is given by a simple form of eqn (10.18):

$$\triangle\tau \cong \triangle z \frac{\partial p}{\partial n} \sim 10^{-3} \triangle z$$

If τ has a value of 0.1 Pa (typical of fairly quiet conditions), then the change $\triangle\tau$ will be less than 10% of τ if $\triangle z$ is less than 10 m. In fact the uniformity of average momentum flux τ along a vertical from heights $\sim$ cm (clear of most small roughness elements) to $\sim$ 10 m is well confirmed by observation, and is one of the most striking regularities of the surface boundary layer, otherwise termed the *constant flux layer* because of the vertical uniformity of the vertical fluxes of heat and matter as well as of momentum. Complications arise because the upper limit of the surface boundary layer varies very considerably with time and position, especially over land. Using the 10% criterion again, the layer is only $\sim$ 1 m deep in nearly still nocturnal conditions when turbulence and τ have died away to low levels, but it may be $\sim$ 100 m deep when strong wind shear or convection encourage very vigorous turbulence. But though the change from one extreme to the other can be quite rapid, as from dawn to mid-morning for example, it is usually sufficiently slow for τ to be well-defined in ten-minute averages, and it is in these that the vertical uniformity of profiles is observed.

The relative simplicity of the surface boundary layer, and the fact that it is the part of the atmosphere in which we spend most of our outdoor lives, has attracted great attention since the study of the atmospheric boundary layer began in earnest early this century. In the absence of an approach from the first principles of the theory of turbulence, people looked for simple relationships involving readily measureable properties, in particular ten-minute average values. In the case of wind, the vertical profile of average wind consistently shows a characteristic shape with very strong wind shear $\partial V/\partial z$ near the surface and smaller values at increasing heights (Fig. 9.12). The simplest dimensionally consistent expression for the shear $\partial V/\partial z$ is of the form

$$\frac{\partial V}{\partial z} = \frac{A\,U}{z}$$

where A is a pure (i.e. dimensionless) number and U is a velocity characteristic of air flow in the layer. A great deal of experimental evidence in the atmosphere, and in wind tunnels and water flows, confirms that this simple relationship actually holds in the form

$$\frac{\partial V}{\partial z} = \frac{u_*}{kz} \tag{9.19}$$

where u_* is the friction velocity introduced in section 9.6 and k is an apparently universal constant with value close to 0.4 (experimental difficulties frustrate very accurate measurement) named after *von Kármán*. It is obviously reasonable that the velocity parameter on the right-hand side of eqn (9.19) should be u_* since this reflects the turbulent activity which dominates the layer. The further appearance of u_* is another example of its great significance.

Notice that eqn (9.19) is consistent with the shape of the mean wind profile, with shear increasing with decreasing height (Fig. 9.12). In fact infinite shear would be implied at zero height, but the surface boundary gives way to the laminar boundary layer (or to the layer penetrated by roughness elements such as grass stems) before this unrealistic extreme is approached. Nevertheless the presence of very strong shear at very low heights, together with the role of wind shear in maintaining turbulence (eqn (9.11) and discussion) is consistent with the observed presence of vigorous small-scale turbulence near the surface for so much of the time. Over land this is quenched by the strong temperature inversions which develop on still, clear nights, but over the sea these are inhibited by the great heat capacity of the surface layer of water, so that strong shear and turbulence are present nearly all of the time, each playing a role in raising and maintaining sea waves [58].

If we assume that τ and ρ are uniform along a vertical axis through the surface boundary layer at any particular location and time (realistic, though the uniformity of average wind stress τ is never very precise) then it follows that u_* too is uniform, and we can easily integrate eqn (9.19) to obtain an expression for the vertical distribution of average wind speed V:

$$V = \frac{u_*}{k}\ln z + B \tag{9.20}$$

where B is a constant. Such a simple linear relationship between average wind speed and the logarithm of height is so widely observed, at least to a reasonable approximation (Fig. 9.16), that it is known simply as the *log wind profile*. It implies (Appendix 9.5) that wind speed increases in equal steps for equal multiples of height increase, the speed increment for a doubling of height being $(u_* \ln 2)/k$, i.e. almost exactly $1.73\, u_*$. Because of the straight-line relationship in Fig. 9.16, extra-

Fig. 9.16 A nearly pure log wind profile (time 0835 CST) of mean horizontal wind over a prairie, including aerodynamic roughness length (5.4 mm) found by extrapolation to zero wind speed (dashed line through several cycles of the vertical scale). Curved lines show slight deviations observed in adjacent convectively stable (time 0435 — before dawn) and unstable (time 1035) situations. (Data from [56])

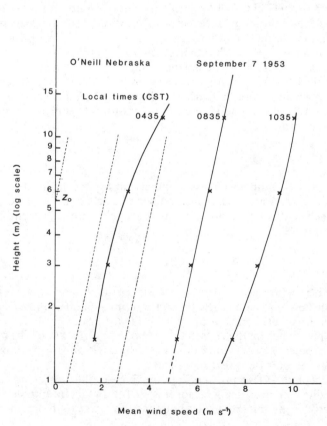

polation suggests that there is a non-zero height z_0 at which V is zero. This is called the *aerodynamic roughness length*, and it has a fairly well-defined value for any particular situation in which the surface does not deform much with increasing wind speed. It appears as the lower height limit in the particular form of eqn (9.20):

$$V = \frac{u_*}{k} \ln \frac{z}{z_0} \tag{9.21}$$

According to eqn (9.21) the average wind speed at any height in the surface boundary layer is completely determined by the characteristic length z_0 and the characteristic velocity u_*. A trivial rearrangement produces a general relationship between *normalized wind speed* V/u_* and normalized height z/z_0:

$$\frac{V}{u_*} = \frac{1}{k} \ln \frac{z}{z_0} \tag{9.22}$$

In recent studies of all parts of the atmospheric boundary layer, sophisticated use is made of such general, normalized relationships, most of them much more complex than eqn (9.22). It is possible to conduct quite general and subtle discussions in such terms, and hence derive relationships to be checked against observation. For example the logarithmic function on the right-hand side of eqn (9.22) is a particular case of the more general function $F(z/z_0)$ which applies to the surface boundary layer in a much wider range of atmospheric conditions.

There is no simple relation between the value of z_0 and the average height l of the surface roughness elements (e.g. lengths of grass stems in a meadow); this is fairly obviously because shape and distribution must also be important, but as a very rough guide $z_0/l \sim {}^1/_{10}$ in many cases. Because z_0 is typically very much smaller than l it follows that the actual shape of the wind profile must deviate from the idealized log profile well above a height of z_0, since we can hardly expect flow among the roughness elements to be similar to flow clear of their tops. It follows that the only way to measure z_0 for a surface is to observe the profile of average wind above it, using an array of anemometers on a mast, and to find z_0 by extrapolation as on Fig. 9.16. Since air flowing from above one surface to above another with a very different z_0 value adjusts only very slowly to the change in underlying surface (as a rule of thumb the depth of the fully adjusted layer is one hundredth of the *fetch* — the horizontal downwind distance from the boundary between the surfaces — Fig. 9.17), it is important that the mast be positioned with adequate fetch. In small-scale terrain like much of the British Isles, the shortness of homogeneous fetches means that the surface boundary layer is seldom fully adjusted, and that the log wind profile holds only approximately.

When roughness elements are tall and densely packed, as they are in mature cereal crops and forests, for example, heights must be measured relative to a datum

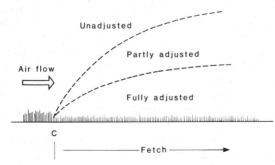

Fig. 9.17 Sketch of downwind growth of adjustment of a surface boundary layer to a change in surface roughness at C. The vertical is greatly enlarged, but is only notional since the partly adjusted layer should be ~ 10 times deeper than the fully adjusted layer.

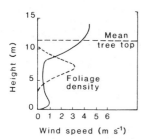

Fig. 9.18 Profile of hourly mean wind speed in a forest of Sitka spruce in the middle of a sunny day. Curvature represents profiles over open ground above the level of maximum foliage density. (After [54] and [59])

significantly raised above the ground to find a satisfactory fit with the log wind type of profile. If the required *zero plane displacement* is d, then

$$V = \frac{u_*}{k} \ln \frac{(z - d)}{z_0} \tag{9.23}$$

fits observed data quite well down to levels just above the crop or canopy. Wind flow within the crop or forest follows no general profile, being controlled by the vertical distribution within the particular vegetation in question (Fig. 9.18). Forests with dense crowns and nearly bare trunks below can even have a secondary wind speed maximum at bare trunk level.

Rearranging eqn (9.21) to find an expression for the tangential stress τ, we see that this stress is proportional to the square of the average wind speed V at any particular level.

$$\tau = \rho C_D V^2 \tag{9.24}$$

The constant of proportionality C_D is the *drag coefficient*, and depends only on the normalized height z/z_0 at which V is measured. The sensitivity of stress τ to V apparent in eqn (9.24) points to the importance of rare windy extremes in imposing high stresses on the surface, and thereby causing damage. Forests can bear scars for decades which were caused by only an hour or so of unusually high average winds with their associated even briefer gusts.

Strictly speaking the log wind profile is to be expected only when the air is neutrally stable for convection, otherwise a buoyancy term should appear in eqn (9.19) and its derivatives. In convectively stable situations there are slight but systematic deviations from the log wind profile, because the damped turbulence allows a greater wind shear aloft. In convective instability the opposite occurs, the wind shear aloft being reduced by the enhanced turbulent friction. However, these deviations from the pure log wind profile are quite small and are detected only by careful observation. A much more obvious result of the reduction of turbulence which occurs on most nights, and its increase during most daylit periods, is that at any particular location (i.e. any particular height above any particular surface) the average wind speed increases with the turbulence. This is apparent in eqn (9.21), where V is directly proportional to friction velocity u_* for fixed z and z_0, and u_* in turn increases as the square root of the turbulent momentum flux τ (which is the horizontal shearing stress). Clearly when turbulence is vigorous, as it is when enhanced by thermally driven convection, momentum from faster-moving layers aloft is transferred down to the surface boundary layer, speeding up the average wind there. At night this downward transfer may die away almost completely, if the surface cooling and stabilizing is strong enough, so that the surface boundary layer is brought nearly to a halt by friction with the underlying surface. The resulting diurnal variation in wind speed near the surface is one of the most familiar features of climate over land surfaces (Fig. 9.19). Over the sea, the diurnal rhythm in thermally driven convection is greatly muted by the large thermal capacity of the sea, so that the effect is weaker and much less closely tied to day and night. The extra momentum delivered to the surface boundary layer daily over land is of course brought down from higher levels, where the average wind speed is correspondingly reduced. Figure 9.19 shows this happening quite clearly at the 300 m level, well above the surface boundary layer. At such levels the result of enhanced turbulence in daylight is to extend upwards the effects of the frictional drag at the surface.

By multiplying each side of eqn (9.19) by u_* and rearranging, we find another expression for the horizontal shearing stress:

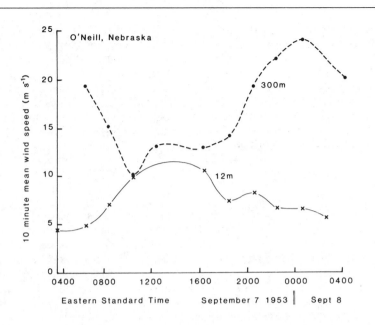

Fig. 9.19 Typical diurnal variation in mean wind speed at heights of 12 and 300 m above a prairie site. (Data from [56])

$$\tau = \rho \, (k \, u_* \, z) \, \frac{\partial V}{\partial z} \tag{9.25}$$

Comparing this with the equivalent expression for wind stress and shear in laminar flow (eqn (7.4)), it is apparent that $k \, u_* \, z$ plays the same role as the kinematic coefficient of eddy viscosity ν. However, unlike its molecular counterpart this *coefficient of eddy viscosity* K is not simply determined by the thermodynamic state of the air; it depends on both the turbulent intensity (through u_*) and the height z, and it is the increase of K with z which maintains the inverse relation between wind shear $\partial V / \partial z$ and height z which is so characteristic of the surface boundary layer.

Similar analysis leads to an equivalent expression for the turbulent vertical flux of sensible heat

$$H = - \, \rho \, C_p \, (k \, u_* \, z) \, \frac{\partial \theta}{\partial z} \tag{9.26}$$

where H is the sensible heat flux density in watts per square metre, and the minus sign is included because it is conventional to regard an upward heat flux as positive. Positive H occurs when there is a lapse of potential temperature with increasing height, i.e. a superadiabatic lapse rate (section 5.12). In this simple type of analysis the eddy coefficient for heat transport (the turbulent equivalent of the thermal diffusivity) is assumed to be the same as the eddy coefficient for momentum transport, $k \, u_* \, z$. More detailed analysis and observation suggest that this is not strictly the case, and that heat is transported more efficiently than momentum when the layer is convecting, while the opposite is true when the layer is convectively stable.

HEAT AND VAPOUR FLUXES

Just as τ, the average vertical flux of horizontal momentum, is nearly uniform throughout the depth of the surface boundary layer, so is the average vertical flux of sensible heat H, provided there is no stratifying mechanism at work, such as a shallow fog or haze layer, interfering with net radiant and convective heat fluxes near the surface. If H is nearly uniform along a vertical axis, then according to eqn (9.26) the vertical gradient of potential temperature must be inversely proportional to height z, which is consistent with the commonly observed concentration of strongly superadiabatic lapse rates close to strongly heated land surfaces (section 5.13), and a similar concentration of subadiabatic lapse rates close to cooled surfaces. We can estimate the lapse $\Delta\theta$ of potential temperature between heights z_1 and z_2 by integrating eqn (10.26) to obtain

$$\Delta\theta = \frac{H}{\rho \, C_p \, k \, u_*} \, \ln \left(\frac{z_2}{z_1} \right) \tag{9.27}$$

Substituting 100 W m^{-2} for H (corresponding to fairly strong surface heating), and 0.3 m s^{-1} for u_*, it follows that the potential temperature lapse between the base of the surface boundary layer and screen level (let us say heights 1 cm and 1.5 m respectively) is about 5 °C. Although eqn (9.27) is probably not very accurate in these conditions, because of the presence of significant buoyancy, the indication that temperature lapses of this magnitude can occur is well supported by observation. The temperature lapse switches over to a temperature inversion in the presence of significant surface cooling. This behaviour, together with the similar behaviour of the laminar boundary layer overlying the surface, indicates the marked increase of diurnal temperature variations between the modest values measured at screen level and the much larger values observed much closer to an exposed land surface (Fig. 9.20) — an increase which has a marked influence on the flora and fauna living there.

Similar analysis leads to the equivalent expression for the average vertical transport of water vapour by turbulence:

$$F = - \, (k \, u_* \, z) \frac{\partial \rho_v}{\partial z} \tag{9.28}$$

where F is the vapour flux density and ρ_v is the vapour density, and there are reservations about applicability and accuracy similar to those associated with eqns (9.26) and (9.25). Multiplied by the coefficient of latent heat L, this vapour flux becomes the flux of latent heat

$$H_1 = - \, L \, (k \, u_* \, z) \, \frac{\partial \rho_v}{\partial z} \tag{9.29}$$

Notice that to the extent that the eddy diffusion coefficients are identical in eqns (9.26) and (9.29) (though in principle buoyancy must enforce some differences

Fig. 9.20 Daytime and night-time profiles near the surface. The large diurnal range is typical of continental mid-latitudes in summer (Paisley, Ontario, 19 July 1962). Detail within 2 m of the surface is lost by the balloon observing technique. (After [55])

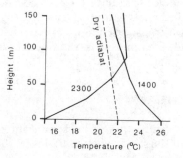

which are ignored in this simple treatment), they cancel in the expression for the Bowen ratio

$$\beta = H/H_1 = \frac{\rho C_p}{L} \left(\frac{\partial \theta}{\partial z} \Big/ \frac{\partial \rho_v}{\partial z} \right) \tag{9.30}$$

with the result that this very important climatological parameter can be estimated simply by comparing vertical profiles of mean temperature and vapour density. Although this must be severely complicated by the prevalence of large diurnal variations in buoyancy effects, especially over land, it is a measure of the difficulty of finding better methods that this approach has been used extensively to partition sensible and latent heat fluxes in major studies of the atmospheric energy budget, such as those discussed in Chapter 8.

9.10 Above the surface boundary layer

Rising above the surface boundary layer, turbulence gradually weakens in the diminishing shear, and eddies become larger and in some cases rather too structured to be considered turbulence proper (for example the vortex-ring type of thermal). The average horizontal stress τ arising from the Reynolds stresses reduces gradually from the value in the surface boundary layer, and eventually becomes insignificant in comparison with horizontal pressure gradient force. At such levels therefore the remaining Reynolds stresses have negligible effect on the balance between pressure gradient and Coriolis forces, and the quasi-geostrophic balance discussed in section 7.8 prevails. The lowest level at which this happens is known as the *gradient level*, and though seldom sharply defined by observation, the gradient level is considered to be the upper limit of the atmospheric boundary layer and the lower limit of the free atmosphere.

Rather subtle arguments suggest that the height of the gradient level above the local surface should be a modest fraction of the space scale u_*/f, where u_* is the friction velocity in the local surface boundary layer and f is the Coriolis parameter. Observations in the atmospheric boundary layer bear this out reasonably well, and show that the gradient level is ~ 500 m above the surface on many occasions, being much lower when the turbulence in the surface boundary layer is relatively weak (as on clear, still nights over land), and considerably higher when the surface boundary layer is unusually turbulent.

Consider the forces acting on a layer of air somewhere between the surface boundary layer and the gradient level. They will produce a net force opposing the motion of the layer because the turbulent drag forward maintained by the slightly faster air just above the layer in question is somewhat less than the turbulent drag backward maintained by the slower motion just below the layer (remember the shearing stress τ diminishes upward, as in Fig. 9.21). As discussed in section 7.11, this net drag pulls the air flow away from its geostrophic alignment with the isobars so that it has a component toward low pressure and is subgeostrophic in speed. This sort of behaviour was first analysed by Ekman in 1905 (actually in relation to the behaviour of surface layers of the ocean in the first instance), and his theory outlines the elegant exponential spiral of wind vectors depicted in Fig. 9.22. The cross-isobar angle α increases with decreasing height above the surface, giving rise to winds which *back* with decreasing height in the northern hemisphere (i.e. their

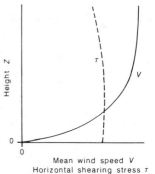

Fig. 9.21 Idealized vertical profiles of mean wind speed and shearing stress in and just above the surface boundary layer.

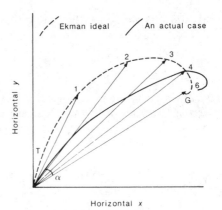

direction rotates anticlockwise when seen in plan) and *veer* with decreasing height (rotate clockwise) in the southern hemisphere. The Ekman spiral is confirmed in outline by observations, but the maximum value of α never reaches the 45° predicted on the strength of his over-simple treatment of turbulent diffusion, and seldom exceeds 30°.

The maximum value of the cross-isobar angle is reached at the top of the surface boundary layer and is maintained from there down to the surface because the relatively small differences between the large values of τ at different heights in the surface boundary layer (τ being large and nearly uniform throughout the depth of the surface boundary layer) are able to balance the pressure gradient and Coriolis forces without requiring significant turning of the wind vector with height. Of course the wind speed in the surface boundary layer falls away more and more rapidly with decreasing height (Fig. 9.21), in accordance with the log wind profile outlined in the previous section. In fact observation and theory indicate that at the 10 m level wind speeds are typically about 40% of the geostrophic value, the percentage increasing with increasing cross-isobar angle and decreasing surface roughness.

Although turbulence weakens above the surface boundary layer, the larger eddies aloft seem to be well able to maintain the upward fluxes of sensible and latent heat which power the great cloudy weather systems of the free atmosphere. When the whole atmospheric boundary layer is being stirred by buoyant convection, its upper parts are the site of rising, structured eddies often loosely entitled thermals. The precise nature of such thermals has not been determined yet by observation, since they are too large and mobile ($\sim$ 100 m and $\sim$ 1 m s^{-1} respectively) to be investigated by instruments on masts, and too small to be traversed by more than one instrumented aircraft at a time. Indirect evidence from birds and glider pilots who use their gentle updraughts to stay aloft with minimum effort can be interpreted either in terms of plumes fountaining from the strongly superadiabatic surface layer, or in terms of discrete ring vortices (like invisible giant smoke rings) floating up after disconnection from their surface origins.

However, the whole layer above the surface boundary layer is often layered to an extent which makes unrealistic both the maintenance of the Ekman spiral, and convective fluxes which are even remotely uniform in the vertical, especially over land where the diurnal changes can be so large. For example, during the destruction of the nocturnal inversion discussed in section 5.13, the deepening convecting layer is bounded above by the very stable remains of the inversion, which may reach down to within a few metres of the surface in the early morning. The resulting layering of turbulence is bound to lead to sharp vertical gradients of vertical fluxes of heat and momentum and consequent shaping of the profiles of temperature and wind speed. Layering is also produced by the differential flow associated with sea breezes and slope winds, both of which are important examples of processes which influence surface and boundary layer conditions very strongly over land on medium and small scales.

9.11 Airflow over uneven surfaces

A number of different processes give rise to patterns of atmospheric behaviour near the surface which have horizontal scales ranging from ~ 10 m to as much as 100 km. There are mobile patterns which are associated with shower clouds and systems of shower clouds which will be dealt with later (section 10.7), and there are patterns which are locked onto terrain features even though they may be quite short-lived. Nearly all locked patterns occur over or near land rather than water surfaces, although land-locked water often has its own distinctive effects. Consider a few examples of land-based effects.

SLOPE WINDS

When a sun-facing slope is warmed, a proportion of the buoyant forces generated can cooperate on the scale of the sloping surface itself to produce an overall up-slope (*anabatic*) wind. This may arise in conditions which are otherwise calm, or it may act to modify a larger-scale flow associated with prevailing weather systems. Anabatic winds are often quite gentle (~ 2 m s^{-1}) because the overall buoyant force is largely offset by the turbulent friction generated by the winds and by the convection which typically penetrates the anabatic layer. However, in extreme terrain much stronger winds can be generated.

By the same mechanism in reverse, nocturnal cooling on a sloping surface can generate down-slope (*katabatic*) winds which tend to be rather stronger and shallower than anabatic winds because of weaker turbulent friction with the over-lying air (Fig. 9.23) and with the underlying surface. Violent katabatic winds can sweep down snow-covered coastal slopes, triggered no doubt by some shift in the

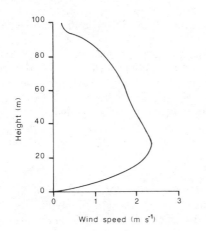

Fig. 9.23 Vertical profile of wind speed in gentle katabatic flow down a 40° slope on the Nordkette, near Innsbruck. (After Defant in [60])

315

Fig. 9.24 Arctic sea smoke near the Ross Ice Shelf, Antarctica. The shelf edge in the background is about 25 m high.

larger-scale air flow, and threaten small boats coasting round Greenland and Iceland for example; and the climatology of the coastal fringe of Antarctica is dominated by the katabatic gales which sweep off the central plateau in all but the summer season. The diaries of pioneer explorers of Antarctica are full of descriptions of the resulting blizzards, one of which cruelly trapped Scott and his south polar party only ten miles from the relative safety of a large supply dump.

The gentler katabatic flows of small-scale hilly country can produce ponding of cool air in hollows so that by dawn there may be quite sharp temperature contrasts between the lower and upper parts of the slopes (kept relatively warm by the continual shedding of chilled air during the course of the night). Relative humidities are highest in the coldest air, even if there is no stream or lake to contribute more by evaporation, and fog will therefore form first in the hollows and valleys. If there is a lake in the hollow, the ponding cold air may well become considerably colder than the water surface (because of the latter's much larger heat capacity — section 9.4), giving rise to *steam fog* as the warm surface air convects into the cooler air ponding just above it and is chilled to sporadic saturation by the resulting mixing. Katabatic flows onto the sea from glaciers and ice sheets in high latitudes have a similar effect, but the greater temperature contrast makes the resulting *arctic sea smoke* (Fig. 9.24) much deeper (~ 30 m) and more visually obvious than most steam fog.

In more extreme terrain and in the presence of very strong nocturnal or even seasonal cooling, the ponding of cold air can lead to *frost hollows* whose minimum temperatures may be tens of degrees below those of the surrounding area. On a larger scale, some of the lowest screen temperatures on Earth have been recorded in winter in the Siberian town of Verkhoyansk (− 51.9 °C monthly mean in December 1904), which lies in a valley just inside the Arctic circle surrounded by a horseshoe of mountains rising extensively above the 1000 m level.

SEA BREEZES

When coastal land is warmed by the morning sun, it can quickly become considerably warmer than the adjacent sea, especially if there is no strong overall wind

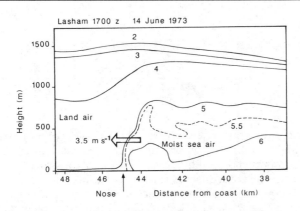

Fig. 9.25 The nose of a sharp fast-moving sea breeze front 45 km inland from the south coast of England, with isopleths of specific humidity in g kg^{-1}. The dry air aloft is subsiding gently in an anticyclone. (Redrawn from [61])

to minimize the contrast by rapid advection. In addition to the normal convection over land, there is in such conditions a tendency for the air there to rise relative to the adjacent cooler and denser air over the sea, in a weak but persistent large-scale buoyancy effect. In fact the uplift may occur by an enhancement of the convective updraughts compared with downdraughts over the coastal margins, encouraging enhanced development of cumulus and even cumulonimbus there. The rising air is replaced by the cool, moist air which is drawn inland over the coast as a *sea breeze*. The flow is seldom more than a few hundred metres deep, above which there is often a more gentle and diffuse seaward flow. The boundary between the sea and land air as the former flows inland can be very sharp, with marked contrasts in temperature, humidity and haziness being ascribed to the presence of a *sea breeze front* (Fig. 9.25). Sometimes there is a definite line of enhanced convection along the front which is welcomed by glider pilots, and lifts insects and the birds which feed on them.

In middle latitudes the most favourable conditions for well-marked sea breezes seem to occur in summer when there is a light synoptic-scale air flow along the coast with the sea on the left hand (in the northern hemisphere). The sea-breeze front may move tens of kilometres inland in such conditions before the waning of the sun in the late afternoon begins to remove the driving agency of unequal heating. Even so, a well-developed sea-breeze front may continue rolling inland under the influence of its established density contrast until late evening. When a sea breeze persists for more than a couple of hours there is usually a rotation of the breeze vector in the direction to be expected from the operation of the Coriolis force, which produces a veering in the northern hemisphere.

Even though well-marked sea-breeze fronts may occur on only a few days per year in any particular coastal region, the much more frequent sea-breeze tendency may have a very considerable effect on local climate, lowering temperature and reducing cloud and showery rainfall along the coastal margins as compared with more than a few kilometres inland. Indeed the resulting enhanced freshness and brightness of coastal climates has contributed greatly to the rise of the seaside resort in Britain and similar countries.

At night the sea-breeze effect tends to reverse, with the production of cooler, denser air over land leading to a seaward flow (a *land breeze*) which is known and used by coastal sailing craft.

HILL AND MOUNTAIN WAVES

When an airflow impinges on a hill or mountain, air is diverted horizontally and vertically by the obstruction. In the typical slight vertical stability (convective

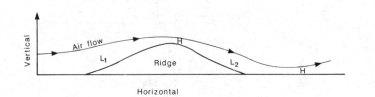

stability) of the lower troposphere, the forcible elevation of air passing over a hill, and of many higher layers of air, sets up vertical oscillations of the air about its undisturbed level, which continue downwind of the obstruction, often for substantial distances. The consequences for the free atmosphere are mentioned again in section 10.8, but at and near the surface the most important result is that wind speeds can be greatly enhanced or reduced at several zones on and downwind of a hill. A typical situation is depicted in Fig. 9.26, in which relatively high winds are induced by congestion of air flow near the crest of the hill, and again several kilometres downwind, with lower wind speeds where air flow is reduced by stagnation (position 1 on Fig. 9.26) or *separation* (position 2). In very windy weather the enhanced winds can become damaging: in a famous westerly gale over northern England in 1962, enormous damage was done to forests and to the city of Sheffield, in which over two thirds of all buildings were officially registered as having been damaged. All this damage occurred in regions in the lee of the Pennines where subsequent detailed analysis of the airflow in the gale indicated that wind speeds were enhanced by being at the base of a lee wave [62].

In some cases the influence of topography on airflow is just as might be expected: a valley which is nearly parallel to the prevailing wind has a channelling effect which raises wind speeds above values found in otherwise identical valleys lying across the prevailing wind. (Particular parts of a crosswind valley may of course be especially windy because of the presence of lee waves, but not the whole valley.)

CLOUD AND PRECIPITATION

Consider the flow of moist air over a hill or mountain. The enforced rise will be rapid (the updraught is the product of the surface slope and the horizontal wind speed) and therefore nearly adiabatic. Unsaturated air therefore tends to rise, preserving its lifting condensation level, and produce cloud at this level if the total rise is large enough.

In the particular case of a layer which is well stirred by mechanical or thermal convection from the surface up to cloud base and beyond, the level of cloud base over the surrounding low ground should be maintained over the hill, though there may well be more and thicker cloud cover over the hill in response to larger and stronger areas of uplift there (Fig. 9.27a). This is easily observed to be the case in hilly country, where the nearly uniform level of nearby and more distant cloud bases is particularly obvious when the observer has climbed close to cloud-base level, and is a nice confirmation that the combination of turbulent mixing and adiabatic eddy motion produces the same effect as wholesale adiabatic elevation.

In the presence of prolonged rainfall, however, it is normal to see low cloud blanketing the upwind slopes of hills at levels far below the base of the nimbostratus producing the general rain (Fig. 9.27b). It seems that air in the lower parts of what is the subcloud layer over the low ground is significantly moistened by evaporation from the wet surface and the falling precipitation, and has in conse-

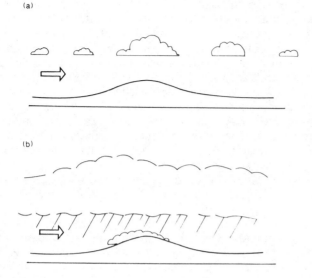

Fig. 9.27 Cloud enhancement over hills. (a) Cumulus enhanced by buoyant or forced updraught, but cloud base level is uniform throughout the well-mixed subcloud layer. (b) Thicker nimbostratus over the hill, with a collar of low cloud (hill fog) where dampened surface air is forced to rise.

quence a lower lifting condensation level which however does not become apparent until the air is forced to rise over the hill (though it may contribute to scraps of *scud* over the lower ground — section 10.2).

At higher levels, moist air may be raised to condensation giving humps of cloud outlining the distorted airflow, and in the middle and upper troposphere these may be particularly smooth and lens-like (*lenticular* — see section 10.8). If the moist layer is already cloudy, the cloud layer will be deepened by the enforced rise, as shown in the case of a shallow layer in Fig. 9.27b; and if there is a deep layer of nimbostratus, then it is deepened and its rate of precipitation enhanced over and slightly upwind of the hill in ways examined in section 10.8. There is often a corresponding reduction in cloud on the downwind side of the hill, and a reduction in precipitation which may produce a climatological *rain shadow* if the orography is extensive enough and the paths of prevailing weather systems well enough aligned. A pronounced zone of sinking flow on the steep lee of a hill can suppress low cloud which otherwise blankets the sky.

Note that the three examples of patterns mentioned above are patterns of flow. Clearly the patterns of radiant flux discussed in section 9.2 can have effects on small-scale climate which are equally important or even more so. Flow and radiation can interlock when the undulations of air in the low troposphere produce patterns of cloud which persist for hours or longer, shrouding some areas under cloud in the crests of waves over or downwind of hills, while leaving other areas in bright sun beaming through holes in the cloud cover maintained in the unsaturated troughs of the undulating airflow. The presence of cloud holes is most marked in the case of otherwise complete overcasts of low cloud such as stratocumulus or stratus. In idle boyhood moments the author noticed that his part of the city of Belfast was often bathed in sun amidst surrounding gloom as low cloud fell slowly down the steep lee slopes of hills to the west, dissolving and not reforming until the air rose suddenly and permanently above the condensation level again several kilometres downwind. Such local effects are commonplace in non-uniform terrain, and their influence on local deviations from overall weather patterns is often considerable, as necessarily observant locals such as farmers and coastal fishermen are usually aware.

319

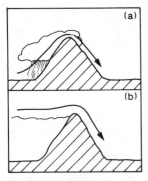

Fig. 9.28 Contrasting air flows associated with föhn conditions in the lee of mountains. In (a) there is saturated ascent upwind, whereas in (b) air slides over dammed air upwind of the mountain. In each case the descending air tends to warm dry-adiabatically.

The best known of all such local effects are the *föhn* winds of the European Alps, and the corresponding *chinook* winds on the eastern flanks of the Rocky Mountains of North America. They appear suddenly as unusually warm, dry and strong winds blowing down the lee slopes of the respective mountain massifs and continuing for as much as 100 km on to leeward. Once established they can continue for days, until a change in the large-scale pattern of weather and flow cuts them off. In winter the sudden warming can melt snow so rapidly that there is local flooding. In summer the desiccating winds can dry the surface and vegetation so much that there is serious risk of bush or forest fires. In the Italian Lake District in summer, the troublesome haze of the Po valley is swept away in a trice, replaced by unusually clear air, which persists throughout the föhn. Temperature rises range from a few °C to over 20 °C in some almost legendary chinooks. Relative humidities usually fall below 50% and have gone below 10% on occasion. Winds often exceed gale force, and can be magnified by local topography to damaging strengths.

Föhn and chinook events are conventionally and very commonly explained in terms of forced ascent and descent of air flowing over a mountain range which is reasonably high, and extensive both along and across the impinging airflow — an explanation first advanced by Hann in the infancy of meteorology (1866). The dryness and warmth of the descending flow is explained by invoking cloud and precipitation on the upwind side, and dry descent on the downwind side (Fig. 9.28a). In the simplest picture this would impose saturated adiabatic ascent and dry adiabatic descent on the air stream crossing the mountains, which could produce a resultant temperature rise of about 4 °C for every kilometre of intervening mountain height. Detailed case study confirms that this certainly happens on some occasions. Indeed the famous clarity of the notherly föhn in the Italian Lake District shows that much industrial and natural haze has been washed out of the air during cloudy ascent of the northern slopes of the Alps. However, there are other occasions when a föhn or chinook occurs when there is no precipitation on the windward side of the mountains. It seems that on such occasions the impinging airflow is largely dammed below the top of the mountain barrier, leaving the air above to slide over the top and descend on the lee side (Fig. 9.28b). The temperature rise on the lee side is then a measure of the stability (the increase in potential temperature with height) of the air to windward. Neither of these models accounts specifically for the considerable winds which are common in föhn and chinook, and for these we must look to the dynamics of lee flow as mentioned earlier in this section and again in section 10.8. Active research is continuing on these and other aspects of these interesting and often quite dramatic examples of the interaction of topography and meteorology [60].

9.12 Surface microclimate

VEGETATION

Conditions near the ground surface are strongly influenced by the presence of vegetation.

Short grass establishes the value of the aerodynamic roughness length z_0 and therefore scales the log wind profile in the overlying surface boundary layer. The presence of reasonably well-watered grass also ensures that the latent heat flux from the surface is relatively large (i.e. that the Bowen ratio is relatively small) and that the daytime temperature rise is relatively modest. The binding of the soil by grass roots prevents the production of dust storms, which are such a feature of arid lands where they temporarily but drastically transform conditions by decimating solar input. It seems possible that before the colonization of Earth's land areas by plants (a relatively recent event on the timescale of Earth history), dust storms were a major climatic feature, possibly even to the extent that they are still on the planet Mars, where hemisphere dust storms shroud the surface for weeks on end.

Long grass and dense shrubbery interfere with fluxes of solar and terrestrial radiation to such an extent that the radiatively effective surface is raised above the ground surface and spread over a significant height range, the spread being greater for more open vegetation. This means for example that it is the upper parts of the vegetation which are warmed during the day and cooled at night, and that these gains and losses of heat are only subsequently communicated downwards by turbulence, radiation and even by the falling of rainwater from wetted leaves. The upward communication to the overlying atmosphere is much the same as for an unvegetated surface after allowing for differences in the effective heat capacity of the active vegetated surface and the typical presence of a substantial flux of water vapour with its associated evaporative cooling. The effect on daytime and night-time temperature profiles in a stand of vegetation is outlined in Fig. 9.29 for the case of a cereal crop, but effects in other types of vegetation are similar in principle. The basic result of the elevation of the radiatively active layer above the ground surface is to make the microclimate of the layer of the region between the active layer and the ground much less extreme, elevating nocturnal temperature minima and depressing daytime maxima to produce a much smaller diurnal variation there. This moderate climate then encourages colonization by plants and animals which could not tolerate the more extreme conditions prevailing before vegetation.

The greatest effects of this type occur in forests, where the radiatively active layer may lie 10 m above the ground in temperate forests and several times this height in the great tropical rain forests which lie mainly in equatorial regions and are watered by precipitation from the intertropical convergence zone. The microclimate of the surface region beneath the dense, lofty canopies of tropical rain forests is familiar from the many films of wildlife there: the humid gloom, with its near absence of wind and diurnal temperature variation, contrasts sharply with conditions at and above the lofty canopy, and is inhabited by a whole range of living things which are irreversibly eliminated when the trees are felled on any substantial scale. Even the ground surface is then unable to cope with the change of conditions imposed by sudden exposure, being alternately baked by the sun and scoured by the deluges of rain from great equatorial cumulonimbus. Bearing in mind the considerable role played by this lush vegetation in the oxygen and carbon cycles of the Earth's atmosphere and surface (sections 3.2 and 3.4), the wholesale destruction of large amounts of tropical rain forest in the present century seems likely to be remembered as one of the classic examples of unenlightened self-interest.

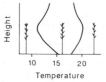

Fig. 9.29 Idealized daytime and night-time temperature profiles in a barley crop. (After [51])

FOG

Fog forms when air close to a surface becomes slightly supersaturated and produces a layer of cloud in contact with the surface. The steam fog already mentioned is the only type of fog which is formed over a relatively warm surface; all other true fogs are formed over relatively cool surfaces. (*Hill fog* is not a true

fog, being simply cloud which is low enough to envelope the upper parts of hills or
mountains).

Radiation fog forms in shallow layers, usually only a few metres deep, when
radiative cooling of a ground surface cools the overlying air below its dew point. It
often appears shortly after dusk on a relatively windless evening, under clear or
nearly clear skies, over land which is fairly moist. In these conditions the air in the
first few metres above the surface is almost still, and the fog when it forms outlines
what is virtually a huge laminar boundary layer (Fig. 9.30). Local inhomogeneities
may stratify the humidity and give rise to horizontal leaves or fingers of denser fog
which enhance the laminar appearance. Over uneven surfaces radiation fog tends
to form in hollows, as they fill with dense, moist air from neighbouring slopes.
Once formed, the fog layers tend to persist by direct cooling of their top surfaces by
net long-wave radiation, and thicken gradually, but unless they grow to a substantial depth (10 m or more) they are easily removed by an increase of wind, or by
solar heating next day. Fog formation shields the underlying surface somewhat
from further radiative cooling, though very gentle convection begins to spread the
top surface cooling throughout the thicker fog layers.

Advection fog occurs when relatively warm air is chilled to saturation by over-
running a sufficiently cool surface. If the air is close to saturation to start with, then
only a little further cooling is necessary to form a fog layer which may be tens of
metres thick. Pulses of relatively warm, moist air often move over land in autumn,
winter and spring in mid-latitudes, as part of air movements in weak, slow-moving
synoptic-scale weather systems. In autumn and winter particularly, the land is
usually much cooler than the sea, so that the invading air is cooled quite sharply.
Even if this is not sufficient, radiative cooling under clear night skies often
completes the production of saturation and fog. Advection fog over land is often
produced jointly by advection and radiation in this way. Fog layers produced
wholly or in part by advection are usually substantially thicker than those produced
purely by radiation, and this has important consequences.

The top surface of a substantial fog layer replaces the shrouded underlying
surface as the new radiatively effective surface. This does not affect the output of

Fig. 9.30 Radiation fog
about 2 m deep forming in a
meadow about dusk.

terrestrial radiation much, since thick fog behaves like a blackbody in much the same way as any terrestrial surface. However, it transforms the response to any available solar radiation, since fog behaves like any cloud in back-scattering strongly and absorbing very weakly. On a clear night therefore the top of the fog layer loses heat rapidly by net emission of terrestrial radiation, but this cooling is spread throughout the fog layer by the gentle convection which it encourages, so that the sharply concentrated surface cooling typical of an exposed land surface is spread out vertically, and somewhat reduced by the addition of the heat capacity of the foggy air (a 20 m layer of air having much the same heat capacity as a centimetre layer of soil). The sharp temperature inversion which would be observed over an exposed land surface is therefore lifted to the fog top and weakened significantly. In late autumn and winter, the weak sun when it rises may be incapable of removing the fog layer by warming, the direct warming being minimal and the indirect warming of the underlying surface being very small if the fog layer is deep enough. In fact all the daylight may do is reveal the gently convecting fog layer to an overflying observer (Fig. 9.31). Once a substantial fog layer is established in these conditions it tends to maintain itself and the cool layer of air in which it persists, until the synoptic situation allows increasing wind to disperse the fog by shear-driven turbulent mixing with the overlying air.

In industrial or urban regions with significant numbers of short chimneys, the smoke and gases emptied into the fog layer (which is usually bounded above by a weak inversion maintained by radiative cooling of the fog top) can rapidly produce unpleasant and even dangerous concentrations of pollutants. London has been prone to such smoky fogs (*smogs*) for centuries, in fact since the burning of coal in domestic open fires became widespread in the sixteenth century. The old local name for them is *pea soupers*, which nicely describes the greeny-brown obscurity in which they enveloped buildings and people. The dirtying and corroding effects on buildings have long been obvious (St Paul's Cathedral was already blackened before its completion in the early eighteenth century), but the effects on people

Fig. 9.31 Aerial view of a convecting fog blanket produced by advection and radiative cooling. The spire of Salisbury Cathedral is 123 m high.

323

were largely ignored until it became clear from hospital records that a spectacular but not untypical smog which lasted for several days in the winter of 1953 had killed several thousand people made vulnerable by age or chest infection. The ensuing outcry led to the introduction of controls on the production of smoke by inefficient combustion of fuels which have transformed the appearance of British cities in subsequent decades. The numbers of foggy days in cities have also dropped, showing that fog formation was being positively encouraged by the availability of condensation nuclei in smoky air.

Advection fog also forms over the sea. The fogs of the Grand Banks fishing grounds off Newfoundland (graphically described by Rudyard Kipling in the novel *Captains Courageous*) are formed by the advection of warm, moist air from the Gulf Stream over the cold, fish-laden waters of the Labrador Current. Fogs form in summer in the coastal waters round the British Isles when warm, moist air from nearby land is cooled as it flows over the cool sea. This sea fog or *har* plagues the eastern coasts of Britain, particularly in fine weather, when light easterlies associated with an anticyclone maintain large sheets of fog over the North Sea, some of which drift onto the British coasts. The fog is dispersed by flowing over a few kilometres of sunlit land, but the coastal strip may remain shrouded and chilled for days while the rest of the country is basking in the sun.

If the overall wind speed in and above a deep fog layer begins to increase, the fog-top inversion may survive and rise while the fog layer is deepened by stirring. In these circumstances, the lowest part of the layer may become slightly unsaturated, leaving a layer of stratus cloud, which in the context is simply lifted fog, overlying a shallow subcloud layer. Something like this can occur in the poleward extremity of the warm sector of an extratropical cyclone (section 11.3).

Appendix 9.1 Direct sunlight on projections

Figure 9.3 shows a beam of direct sunlight with solar elevation angle E impinging on a symmetrical projection with base angle α. From the construction it is apparent that the beam makes an angle $\alpha + E$ with the plane of the slope. Since the beam makes an angle E with a horizontal (by definition of solar elevation angle), the ratio RI of the irradiance of the sloping and horizontal surfaces is given by the sine rule for oblique irradiance (appendix 8.2)

$$RI = \frac{\sin(\alpha + E)}{\sin E}$$

Note that for any chosen E, this expression confirms that the maximum slope irradiance occurs when $\alpha + E = 90°$, ie. when the beam is normal to the slope. Values of RI are plotted in Fig. 9.4 for three representative slopes and a range of solar elevations.

To find an expression for the absolute irradiance AI of the slope, we can use eqn (8.7) (appendix 8.2) for the horizontal irradiance HI at the base of an atmosphere with zenith transmissivity τ_{90}, together with the identity

$$\begin{aligned} AI &= HI \times RI \\ &= S \sin(\alpha + E)\, \tau_{90}^{\ \operatorname{cosec} E} \end{aligned}$$

where S is the solar constant. For simplicity, values plotted on Fig. 9.4 are for AI/S. All are calculated for the case where zenith transmissivity is 0.8, which corresponds to a very clear sky (appendix 8.2).

Appendix 9.2 Daily insolation

For many purposes in climatology and botany it is useful to have the daily total solar input. If the zenith transmissivity is steady throughout the day, or if, more realistically, several more variably sunny days are averaged, the profile of solar horizontal irradiance from dawn to dusk lies close to half of a sinusoidal curve, and the daily total insolation can be found by standard integration. Small fillets at dawn and dusk are ignored in the process, but the error is small compared with others, chiefly those arising from variations in zenith transmissivity.

Figure 9.32 shows half a sinusoid fitted between dawn and dusk to represent the daily profile of horizontal irradiance I. The formal expression for such an ideal I is

$$I = I_{\mathrm{m}} \sin \left(\frac{\pi t}{N} \right)$$

where I_{m} is the maximum irradiance (at noon), t is time measured from dawn, and N is the time interval between dawn and dusk. This can be justified by checking the I values it gives at dawn, midday and dusk. The daily insolation DI is found by integrating the expression for I between dawn and dusk.

$$
\begin{aligned}
DI \quad &= \int_0^N I \, dt \\
&= I_{\mathrm{m}} \int_0^N \sin \left(\frac{\pi t}{N} \right) dt \\
&= \frac{2 N I_{\mathrm{m}}}{\pi}
\end{aligned}
$$

In the case of a sunny summer day in Britain, N may be 6×10^4 s (over 16 hours) and I_{m} 650 W m^{-2}. It follows that DI is 24.8 MJ m^{-2}. Note that the daylight average irradiance (DI/N) is $2/\pi$ times the noon maximum.

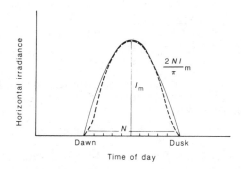

Fig. 9.32 Idealized symmetrical diurnal time sequence (dashed) of surface insolation. The solid line is a fitted half-sinusoid.

Appendix 9.3 Sunlight in water

Adapting the method of appendix 8.2 to the penetration of solar irradiance I into water of increasing depth z, we have an expression for the change dI produced as sunlight passes a little further into the depths:

$$dI = - k \, dm \, I$$

where k is the absorption coefficient (in a less simple model it varies with wavelength), dm is the mass of water per unit horizontal area in the little range of depth containing dm, and the minus sign ensures that I reduces with increasing depth. Replacing dm by $\rho \, dz$, we find a relationship between dI and dz which can be integrated in the simple case where k and ρ are uniform to give an exponential decay of irradiance with increasing depth.

$$I = I \exp (- k \, \rho \, z)$$

where I is the irradiance at the water surface.

According to appendix 4.4, the exponential and decadal scale depths are respectively $(k \, \rho)^{-1}$ and $2.3026/(k \, \rho)$.

Appendix 9.4 Turbulent momentum transport

According to eqn (9.13), the average vertical downward flux of horizontal momentum through horizontal unit area is given by

$$\tau = - \overline{\rho w V} \tag{9.31}$$

where the overbar represents averaging over a chosen time period (often 10 minutes) and the group of terms under it represents the instantaneous momentum flux density. Since variations in density are proportionately very much smaller than those in $\overline{w}$ and V, we can treat it as a constant and remove it from under the overbar. The instantaneous horizontal downwind speed can be written as the sum of the average speed and the instantaneous deviation from it:

$$V = V' + \overline{V}$$

The vertical speed can be analysed in the same way, but w is virtually zero in most realistic conditions in the surface boundary layer, so that

$$w = w'$$

Substituting these in eqn (9.31) we find

$$\tau = - \rho \, \overline{w' (V' + \overline{V})}$$
$$= - \rho \, [\overline{w' \, V'} + \overline{w' \, \overline{V}}]$$

Since V is constant in the averaging period by definition, the second pair of terms in the bracket includes $\overline{w'}$, the average of the deviations of w from its average, which is zero by definition. Hence

$$\tau = - \rho \overline{w V'} \tag{9.32}$$

dropping the unnecessary prime from the updraught speed.

Appendix 9.5 The log wind profile

According to eqn (9.20) in the neutrally buoyant surface boundary layer we have

$$V = \frac{u_*}{k} \ln z + B$$

If we choose two heights with ratio X

$$z_2 = X z_1$$

then it follows from the log wind profile that the difference in wind speed is

$$V_2 - V_1 = \frac{u_*}{k} \ln X \qquad (9.33)$$

Since k is a universal constant, and u_* is constant for a particular profile, we see that the difference in mean wind speed on any particular profile depends only on the choice of X. If we choose a series of heights in geometrical progression then it follows that the wind speeds will be in arithmetical progression. As an example, consider the difference between wind speeds at 2 m (head height) and 10 m (standard anemometer height). According to eqn (9.33)

$$V_{10} - V_2 = 1.61 \frac{u_*}{k}$$

showing that in a gale (17 m s^{-1} at 10 m height) with friction velocity $u_* = 2$ m s^{-1} (a reasonable value), the wind speed at head height (2 m) is only 9 m s^{-1}. Notice that in the same conditions the wind speed at 50 m (near the masthead of a large sailing ship) is 25 m s^{-1} (49 knots).

Problems

LEVEL 1

1. Why is the laminar boundary very significant to an elephant despite the gross discrepancy in their volumes?
2. If the solar elevation is 40°, what can you say about sunlight falling on a slope of 45° (a) facing towards and (b) facing away from the sun?
3. If a typical incident solar ray makes three successive impacts on the interior of a stand of vegetation before escaping back to the atmosphere, and each element of vegetation surface has an albedo of 0.5, what is the gross albedo of the stand?
4. Why should there be less hoar-frost round the bole of the leafless tree than in the middle of a field on the same occasion?
5. Which of the following usually contributes most to the very large heat capacities of lakes and seas: the large specific heat capacity of water; its turbulent mixing; its transparency to sunlight?
6. While illuminated by the sun, ground passes less sensible heat into the atmo-

sphere when it is saturated than when it is dry. Suggest two reasons for this.

7. Why would it not be helpful for synoptic observers to report winds averaged over 10 s instead of 10 mins?

8. A complex computer model of the nocturnal boundary level at an open site in winter is able to forecast values for the Richardson number near the surface on the basis of conditions at the previous dusk. What values would you look for to give a warning of ground frost?

9. In a pure log wind profile, how do you expect (a) the wind speed and (b) the wind shear to vary at equal multiples of height?

10. In a simple model of the temperature profile over heated ground, how should the temperature gradient at a fixed level close to the ground vary with (a) convective heat flux and (b) variance of wind speed?

11. Because of the Ekman spiral, the wind direction at 10 m needs to be altered in a consistent sense to represent the direction of the gradient wind. What is this sense in the northern hemisphere?

12. In sloping terrain, why are katabatic winds more common by night than by day?

13. People living near hills often remark that in overcast conditions the sky is especially threatening in the direction of the hills. Why should this be so?

14. Why should fog often gather in hollows and valleys, leaving even very low hills protruding into the clear air?

LEVEL 2

15. A beam of direct sunlight with flux density 1000 W m^{-2} falls on several surfaces. Given that the angle of solar elevation is 35°, find the insolation on (a) a surface tilted toward the sun at an angle of 30° to the horizontal, (b) a surface tilted away from the sun at the same angle, and (c) a horizontal surface.

16. Calculate the direct-beam insolation values for the surfaces and solar elevation of problem 17 on an occasion when the sky was cloudless but rather hazy, having a zenith solar transmissivity of 0.65.

17. Given that problems 17 and 18 represent conditions at noon on a day in which the clarity of the atmosphere remained uniform throughout the full 9 hours of daylight, find the total insolation on a horizontal surface for the day.

18. Find the angles of solar elevation at noon in midwinter and midsummer at the Arctic (or Antarctic) circle, and at latitudes 55° and 25°.

19. Using a finite-difference form of the Richardson number, estimate the difference in wind speed across the layer between 2 and 3 m above the surface, which would be just sufficient to initiate turbulence despite a temperature inversion of 2 °C across the same layer.

20. In a very simple model of wind variations in the surface boundary layer, horizontal wind speeds V are 3 m s^{-1} when there is a downdraught (of 1 m s^{-1}), and they fall to 1 m s^{-1} when there is an updraught w (again of 1 m s^{-1}). The up- and downdraughts alternate regularly and account for all the time in the observation period. Sketch V and w on a common graph against time, and calculate their covariance and hence the friction velocity and average downward eddy flux of momentum (assuming air density to be 1.2 kg m^{-3}).

21. The average wind speeds at 8, 4 and 2 m height on a certain occasion are 7, 6 and 5 m s^{-1}. Find the implied roughness length of the underlying surface, the wind speed at 10 m, and the friction velocity.

22. In a certain föhn event, the air descending the lee slopes is observed to be up to 5 °C warmer than air at the same level on the upwind slopes. Find the loss of vapour content (specific humidity) implied by an adiabatic model of the ascent and descent (*Hint*: consider eqn (5.21)).

23. In problem 24, given that the air at the base of the upwind slope was as specified in problem 5.15, find the implied atmosphere pressure at the crest of the ridge, and convert this to a ridge height above sea level (1010 mbar) in kilometres using the rule of thumb that pressure lapses by 1 mbar per 10 m.

LEVEL 3

24. Discuss the factors which contribute to the large seasonal range of climate observed in a continental site at latitude 40° in the rain shadow of a range of hills.

25. A clear calm night has established a marked temperature inversion near the surface. Now the wind begins to increase in response to an approaching synoptic-scale weather system. Describe what you would observe with simultaneous measurements by fast-response thermometer and vertical-propellor anemometer before and after turbulence breaks out. What will happen to the average air temperature at the same time?

26. Estimate the solar, terrestrial and net radiative balances for midsummer's day on the Arctic circle, when the surface solar albedo is 0.2, the zenith transmissivity for solar radiation is 0.7, the surface temperature remains constant at 10 °C, and the surface long-wave emissivity is 0.95.

27. Compare and contrast the climatic environment of a small crawling insect, an Irish wolfhound and a man, each living in a mid-latitude continental prairie. Compare the experiences of the man and dog, given that they have the same body weight.

28. Find the correction factor to be applied to wind speeds measured at 2 m to bring them to the equivalent value at 10 m, assuming a log wind profile and an aerodynamic roughness of 1 cm over the surrounding area. Note that this is a useful procedure to estimate synoptic standard wind speeds when there is no anemometer at 10 m. Consider the likely sources of error in such estimates.

29. The relationship between potential temperature lapse and sensible heat flux density H in eqn (9.27) rather unhelpfully contains the friction velocity u_*. However, u_* can be replaced using the log wind profile. Do this and derive the useful relationship

$$H = \rho C_p\, C_D\, \Delta\theta\, \Delta V$$

where $C_D = [k/\ln(z_2/z_1)]^2$, which is the drag coefficient C_D in eqn (9.24).

Derive a similar expression for the latent heat flux, and use the two to show that the expression for the Bowen ratio is

$$\beta = \frac{\rho\, C_p}{L}\, \frac{\Delta\theta}{\Delta\rho_v}$$

which is the most widely used expression for the Bowen ratio.

30. Give an account of ways in which local meteorological conditions in hilly terrain can be influenced by the terrain.

CHAPTER 10

Smaller-scale weather systems

10.1 Introduction

We have seen how the structure of the atmosphere both controls and is maintained by atmospheric behaviour, and that such interaction is especially vigorous in the troposphere. Much of this behaviour appears to us to be a series of events related to types of surface weather: cumulus clouds blossom briefly and fade, showering the surface with precipitation if they are big enough; depressions develop and trail their attendant fronts across thousands of kilometers of middle latitudes, while neighbouring areas are covered by gently moving and developing anticyclones; tropical cyclones form over warm oceans and drive westward, scourging the surface with tremendous winds and rains before turning polewards and subsiding to more normal intensities. Since such weather-related events have quite characteristic patterns of behaviour and structure, we will regard them as systems, to be examined descriptively, much as living organisms are described and catalogued, and also mechanistically, to see if we can identify their functional parts and how they operate in terms of the basic physical and meteorological processes which we believe underlie all such behaviour.

There is a wide range of tropospheric weather systems, differing in scale and type. In this chapter we will concentrate on those which are not large enough to appear unambiguously on a weather map, although they may individually and quite strongly influence conditions at one or two adjacent stations on any particular map. Such systems are called sub-synoptic in scale since they do not appear individually on synoptic-scale maps, although populations of individual systems may well dominate large areas of any particular map. The shower cloud (cumulonimbus) is a typical example. The sub-synoptic range is so wide that it is often subdivided into micro-scale, small-scale and meso-scale. Micro-scale refers typically to systems no larger than large individual eddies in the atmospheric boundary layer, say ~ 100 m. These affect surface conditions mainly through the turbulent variability described in the previous chapter and will not be discussed further. Small-scale refers usually to systems no larger than a single large cumulonimbus, say 10 km in both horizontal and vertical dimensions (though the anvil of a mature cumulonimbus may stretch up to 100 km downwind of the main body in the presence of strong wind shear), and has been used in this way in Chapter 7. *Meso-scale* usually refers to systems or patterns of systems intermediate in scale between the small and synoptic scales, which means between ~ 10 km and ~ 100 km in the horizontal according to my summary. However, none of these sub-ranges of scale is well defined in meteorological usage, and the meso scale in particular is used especially loosely.

The following sections describe and simply analyse some of the most important types of small and meso-scale tropospheric weather system.

10.2 Cumulus

One of the most common types of cumulus is the small *cumulus* which often begins to appear in previously clear skies over land in mid-morning, after the stable layer left by the previous night's cooling has been eliminated by solar heating (section 5.13). This is often called fair-weather cumulus because it is typical of fine summer weather, though it is a misleading title on those quite common occasions when the clouds subsequently develop enough to give a showery afternoon. Some features of the appearance and behaviour of cumulus humilis have been mentioned already in sections 6.4 and 6.9 in discussions about the sub-cloud layer and about droplet-scale mechanisms. In the present section we will summarize and discuss these and other aspects of cumulus, mainly as they are apparent to the unaided ground-based observer.

The individual cumulus has a very characteristic shape, and is usually just a few tens of metres across when it first becomes clearly established (Figs 6.3, 6.11 and 10.1). The nearly flat base corresponds reasonably to the lifting condensation level of the well-mixed sub-cloud layer when observations are available (section 5.11), and the quite obvious consistency in base level across an extensive field of such cumulus (Fig. 10.1) supports this identification. As mentioned previously (section

Fig. 10.1 Cumulus with characteristic flat bases and humped upper parts.

Fig. 10.2 A small cumulus seen at close range, showing its ragged base tilted slightly upwards in the direction of drift.

6.5) the cloud base is often a little higher than the lifting condensation level deduced from screen level measurements, but the discrepancy is so consistent on any particular occasion that it seems likely to arise from a too-simple model of the lowest part of the well-mixed layer, rather than from inaccurate measurement. Closer inspection (Fig. 10.2) usually reveals a raggedness of cloud base on a scale of metres, and a slight but consistent upward tilt in the ambient downwind direction. The first clearly reflects the turbulent inhomogeneity of the air in and below the cloud, but the second seems to indicate a mechanism on the scale of the cloud itself which is still unknown.

The sides and top of the individual cumulus are characteristically domed and knobbly. The domed shape suggested the generic name 'cumulus' (Latin for hill) to the pioneers who were naming cloud types in the early nineteenth century, in the wake of the latinate Linnaean classification of living organisms, and is clearly consistent with the rising of an extensive, rounded parcel of buoyant air above its lifting condensation level. The knobbly appearance, which often closely resembles the heart of a ripe cauliflower, again reflects the presence of turbulence on scales much smaller than the cloud itself.

If an individual cumulus is observed for even a little longer than the usual casual glance, it becomes apparent that it is very dynamic: the dome swells visibly in the space of a minute, and the knobs protrude and retract even more rapidly — both clearly consistent with the operation of mechanisms already mentioned.

The rate of rise of the top of the cloud is somewhat less than the maximum updraught speed inside the cloud, because the rising mass of air continually tends to turn itself inside out like a rising smoke ring (technically a *ring vortex*). Updraught speeds seldom exceed 2 m s^{-1}, because the weak buoyancy is almost exactly offset by drag. Drag arises obviously as the rising air forces its way through its surroundings, and less obviously but equally importantly because of the *entrainment* (section 7.12) of previously static ambient air.

The seething detail of the knobs in particular is suggestive of intense mixing between the swelling cloud and its cloudless environment, and careful observation confirms that the net effect of such mixing is to extend the cloudy volume quite rapidly by entrainment and saturation of previously cloud-free air. This entrainment is in addition to the tendency of the rising air simply to displace the surrounding air, and is a crucial feature of the whole cumulus family (including cumulonimbus). The sharpness of the boundary between cloudy and cloud-free air shows that cloud droplets form and disappear so rapidly that any individual micro-scale eddy is either densely clouded or quite clear depending on whether it is saturated or not. The implied high speed of response (of activation and deactivation — see section 6.9) is made possible by the very small size of the droplets: $\sim$ 10 μm.

A particular cumulus usually continues growing horizontally and vertically for five minutes or so, behaving as described and drifting in the ambient wind. It may then be over 100 m broad and somewhat less deep, and may have drifted several kilometres downwind of its point of origin. After another few minutes it will often seem to lose vitality (notice how the correspondences with living organisms keep appearing), stop growing, lose sharpness of outline, and begin a process of dissolution which is completed in only a few more minutes, leaving the air as clear as it was initially. During dissolution the cloud usually becomes less obviously white, and its edges less distinct, which is consistent with selective evaporation of the large numbers of smaller cloud droplets as the marginally saturated cloud continues to entrain unsaturated air, despite its waning vigour. The largest droplets last longest, so that toward the end of the cloud's recognizable existence it becomes a dull (the few larger droplets scatter sunlight relatively poorly), diffuse mass, heading for extinction by a process which could be called terminal detrainment.

Two factors may affect the shape of the cloud quite considerably at maturity. If the effective lifting condensation level lies only a few tens of metres below the base of a convectively stable layer (a temperature inversion is an extreme example), then the cumulus will be vertically stunted by the loss of buoyancy in air rising into the base of the stable layer, and will show a pancake shape throughout its life. This is typical of the shape of cumulus forming beneath the base of the subsidence inversion of an anticyclone (see section 11.5). Indeed the continual formation of such flattened cumulus in anticyclones often maintains large areas of *stratocumulus*, in which the individual clouds are so closely packed that they form a sheet patterned on a scale of $\sim$ 100 m by narrow gaps between adjacent clouds, and the underlying

Fig. 10.3 Stratocumulus, forming nearly total overcast.

surface is very significantly shaded from the sun in consequence (Fig. 10.3). In heavily industrial regions the trapping of smoke and other pollution under the same stable layer can give rise to quite gloomy and noxious conditions at the surface.

The other factor affecting cumulus shape is the presence of wind shear (i.e. vertical shear of horizontal wind) in the layer containing the cumulus. Such shear is never as marked as in the surface boundary layer, but it can nevertheless be strong enough to make the cumulus lean quite noticeably. Though the shear at cloud level need not parallel the wind there, which in turn need not parallel the surface wind and shear, in fact these shears are often nearly parallel, with the result that cumulus are seen to lean nearly enough down the direction of the surface wind and shear. In very windy conditions the shear may be so strong as to inhibit the orderly development of cumulus, producing a ragged, heavily sheared cloud known as *scud* or *fractocumulus* when it is accompanied by rainy cloud at higher levels, as is often the case in strong winds (Fig. 10.4 and section 9.11).

In the trade-wind zones there is a large population of cumulus, much of it taller than small cumulus because of the rather different stratifications of temperature and humidity which prevail there, which tend to lean backwards against the surface winds (Fig. 8.13) in what is apparently the wrong way, at least by comparison with observation at higher latitudes. The reason for this is that the trade-wind flow is quite shallow, extending no more than a few kilometres above the oceans, where their reliability was once so widely exploited by sailing ships. For example, in the northern hemisphere the characteristic north-easterly flow fades rapidly with height and is replaced by a westerly flow in the middle of the low troposphere, which tends to persist and increase with height (Fig. 10.5). The trade-wind cumulus are therefore rising through a layer in which wind shear and near-surface winds are nearly opposed in direction, and their upper parts are therefore sheared backwards in relation to their bases, unlike the case of such clouds at higher latitudes, where shear and flow usually have much the same direction.

Fig. 10.4 Fractocumulus (scud) silhouetted against an advancing clearance of overhanging frontal nimbostratus.

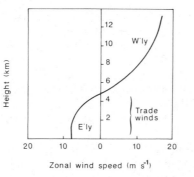

Fig. 10.5 Vertical profile of zonal wind component in and above winter trade winds. In summer the upper westerlies may be weak or replaced by easterlies. (After [27])

10.3 Cumulus development

Unless there is a convectively stable layer aloft, some small cumulus seem to avoid the dissolution described in the previous section, and carry on growing until they become elevated hills and mountains of cloud (Fig. 10.6), likely to produce precipitation and therefore by definition become cumulonimbus. Let us consider the process of development of such *cumulus congestus* (swelling cumulus) from small cumulus, and the factors which encourage it on some occasions.

The updraughts in small cumulus, and in the cloud-free thermals which feed them from beneath, are rather weak, seldom exceeding 2 m s^{-1}. As mentioned in sections 7.12 and 10.2, this is consistent with a condition of near balance between buoyancy and the combined effects of form and entrainment drag. The buoyancy is weak because the temperature excess in a thermal is usually less than 1 °C, and is partly offset by the weight of cloud water borne by the updraught above cloud

Fig. 10.6 Cumulus congestus, showing typical mountainous shape.

base, and the temperature excess is small because of the weakness of solar warming of the terrestrial surface, and because of the readiness of atmospheric turbulence to dissipate temporary inhomogeneities. In these conditions the loss of buoyancy associated with the evaporative cooling which occurs all over the upper surfaces of a cumulus, and continues throughout its life, is potentially very important (appendix 10.1). Indeed, its tendency to make the flanks of a cloud sink rather than rise is often quite clearly visible on larger clouds. Cloudy updraughts which can avoid or delay the onset of such destruction of their buoyancy have a considerable advantage over their fellows.

Suppose that conditions are especially favourable for the formation and persistence of cumulus: because the temperature stratification aloft is neutral or even slightly unstable (see section 5.12), the ambient air above cloud-base level is sufficiently moist to minimize evaporative cooling, and there is an adequate supply of buoyant moist air rising from the surface layers. The instability can be induced by large-scale factors such as the warming of surface layers by the sun or by advection over a warmer surface, or by the cooling of upper layers by the differential advection which is such a feature of flow in extratropical cyclones (section 11.4). Indeed in the latter there are effective mechanisms involving large-scale vertical motion too. Gentle but persistent wholesale uplift can reduce stability by raising a layer whose lower parts are sufficiently more humid to saturate before the upper parts, as exemplified in section 10.6. Another effective mechanism is the differential uplift associated with vertical stretching and synoptic-scale horizontal convergence (section 7.13) whose effect can be seen by raising and stretching a layer with a sub-adiabatic lapse rate on a tephigram.

In such conditions many small cumulus form, and some are reinvigorated by new thermals rising into their bases before their initial impetus is lost. These new thermals are able to rise through the pre-existing cumulus without evaporative loss of buoyancy until they emerge from the protective body of the cloud, and obviously tend to reach this stage with greater buoyancy than the pioneering thermals which first produced the cloud in the teeth of evaporative destruction of buoyancy. Cumulus receiving such additional buoyancy must therefore be considerably advantaged over their contemporaries, who have to make do with a single initial input, and tend to grow larger and live longer as a result. Indeed they are likely to receive repeated additions of buoyancy from beneath cloud base by the same mechanism, each new influx being protected from destructive evaporation by rising through successively larger depths of accumulated cloud before temporarily taking the brunt of pushing the cumulus congestus further up into the unsaturated environment. The end result is that a fraction of the total number of cumulus-type clouds grows much larger than their fellows in these conditions, often extending up into the middle troposphere ($\sim$ 5 km above sea level).

The size of the fraction becoming cumulus congestus in any particular situation varies considerably with the temperature and humidity stratification of the lower and middle troposphere, as does the average and maximum heights reached by the clouds, and these in turn are affected by the mechanisms already cited. Note that the convergent mechanism can be enhanced by frictional convergence in the planetary boundary layer of a low-pressure system, as is believed to be important in tropical cyclones (section 12.3). When sufficiently encouraged, cumulus congestus may outweigh small cumulus in terms of cloud volume, but are usually outnumbered by the small cumulus which continue to coexist.

Careful observation of individual cloudy thermals by time-lapse photography shows that as they rise they tend to expand as if bounded by an imaginary cone, whose vertex has a semi-angle of about 15° (Fig. 10.7). From this and from laboratory work with similar convection in water, as well as from a variety of

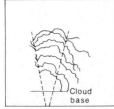

Fig. 10.7 Outlines from a sequence of photographs of a cumulus congestus taken at one-minute intervals. Over the whole sequence the chosen cloudy thermal grows from being a small lump on the edge of the congestus to become its main turret. (Redrawn from [63])

theories, it is known that conical expansion is a characteristic feature of such pene-trative convection, and one which results from net entrainment. The familiar shape of the congestus, on the contrary, shows a marked narrowing with height (Fig. 10.6). The reason for this inversion in the aggregate of the behaviour of the individual cloudy thermals lies again in the role of the established body of cloud as protector of new cloudy thermals: those on the flanks of the cloud suffer much more evaporative dissolution than those in its core, so that the congestus as a whole converges upward despite the tendency of its individual constituents to diverge. The sacrifice of the flanks, however, allows the central core to rise relatively effi-ciently into the middle and upper troposphere, a virtual prerequisite for the production of cumulonimbus.

10.4   Cumulonimbus

Though occasionally a few spots of rain may reach the surface from a small cumulus congestus, and low stratocumulus may produce slight drizzle, it is a familiar fact that most showers fall from clouds which have already grown to a very substantial height, and hence warrant the title cumulonimbus. Indeed it is the dark bases of such tall clouds, shadowed by efficient backscatter of sunlight by the thick cloud above, which give the clouds their nimbus title and are correctly regarded as threatening rain.

The reasons for the association of cloud thickness with capacity for precipitation follow directly from the discussion of section 6.11 and problem 6.8. It takes a considerable length of time, probably at least 20 min, for significant numbers of cloud droplets (or crystals) to grow selectively to the millimetric size at which their fall speeds are big enough to outweigh updraught speeds, and they are able to survive evaporation in the sub-cloud layer. Only fairly substantial clouds last as long as this, and even if a shallow cloud managed to avoid dissolution, fall speed and updraught would have to be very nicely balanced for some minutes to allow precipitation to develop in such small clouds.

Most showers fall from cumulonimbus which have very recently developed from cumulus congestus. Where cloud base is reasonably high and the shower is viewed against a pale background sky, the virga or long trails of precipitation reaching down from cloud base to surface can be visually very striking (Fig. 6.17), par-ticularly if they are curved by wind shear, as is often the case in the blustery winds which accompany showers in cool equatorward airflows in middle latitudes. The visible thinning of the precipitation arises as the smaller precipitation droplets are evaporated to extinction before the larger, the great sensitivity of extinction distance to droplets size (Appendix 6.4) ensuring that selective evaporation is always significant. If the air is so cold that even the low troposphere is below 0 °C, the precipitation will fall as snow, which has a characteristically pale and diffuse appearance beloved of painters such as Turner (Fig. 10.8).

As a cumulonimbus drifts along with the wind over land, it leaves a trail of precipitation (rain or snow, often accompanied by hail) which is as wide as the active core of the cloud (usually no more than a few kilometres) and as long as the distance the cloud is able to travel during its active lifetime. Since this life often lasts about an hour, the length of the affected strip may vary from only a few kilo-metres in very light winds to more than 50 km in a gale. Comparing clouds with

337

Fig. 10.8 Winter cumulonimbus precipitating snow in north-west England.

similar precipitation production and lifetime but different drift speeds, we see that the conventionally measured rainfall on any part of the strip is inversely proportional to the total length of the strip and is greatest when the shower is stationary. Showers occurring in very weak airflow are therefore particularly sporadic and heavy in the affected areas, but even in a vigorous airflow the proportion of surface area subject to showers may be quite small. The treatment of such uneven distributions in observation and forecasting has been mentioned in section 2.8. Typical rainfalls (or the equivalent rainfalls of snow and hail showers) from showers of normal intensity are observed by standard (stationary) rain gauges to range from about 1 to 10 mm. (Unusually intense storms are discussed in section 10.6) Though such totals are quite modest, and compare with rainfalls from weak to moderate fronts, they usually fall in a small fraction of an hour and therefore correspond to short bursts of moderate to heavy rain, with rates of rainfall $\sim$ 30 mm h^{-1}, contrasting with the sustained periods of mainly light rain which are typical of large areas of frontal rainfall. In weak airflows the prolongation of such falls in limited areas can easily overload the natural drainage there, causing localized flooding.

It often seems that a cumulus congestus, having taken 20 min or so to reach the point of becoming a cumulonimbus by building slowly up to a vertical extent of several kilometres, begins to grow much more conspicuously as precipitation begins. As mentioned in section 6.11, in middle latitudes this transformation seems to be associated with widespread glaciation of the upper parts of the cumulonimbus. The rapid release of latent heat of fusion as large amounts of supercooled water cloud freeze must tend to enhance buoyancy and further growth (appendix 10.1), as must the loss of some of the dead weight of cloud by precipitation. It is certainly often apparent that as precipitation begins to fall from the cloud's base, its top begins to grow more rapidly into the upper troposphere, often by sprouting a cloud tower which at least initially is considerably narrower than the pyramidal heap of congestus beneath. Rates of rise of 5 to 10 m s^{-1} are observed to be quite normal, and are consistent with typical updraught speeds in the cloudy interior measured and estimated by a variety of direct and indirect methods. Soon veiled by glaciation, this tower is often quickly stretched very asymmetrically by strong wind shear in the high troposphere, producing the anvil-shaped top which is so characteristic of the mature cumulonimbus (Fig. 10.9).

Fig. 10.9 Distant summer congestus and cumulonimbus with nearly symmetrical anvils forming over sunlit western slopes of the Pennine chain.

The top of the anvil is usually fairly flat, indicating that the updraughts are being halted at a fairly definite altitude which is often observed to coincide with the tropopause. The situation is sketched in Fig. 10.10, which shows how the updraughts are inhibited by the convectively very stable low stratosphere. A relatively simple treatment (Appendix 10.2) confirms observation by aircraft that updraughts can overshoot their ultimate equilibrium level by several hundreds of metres on account of their vertical momentum before sinking back and spreading out. Sometimes there is a stable layer some way below the tropopause which serves as the upper limit of cumulonimbus development. Whichever is the case, the collision between the towering updraughts and the stable layer feeds large volumes of ice cloud laterally into the anvil — a flux which is maintained throughout the active life of the cloud. In the course of an hour's vigorous activity, huge amounts of ice cloud may be disgorged in the high troposphere. Since anvil cloud is relatively slow to evaporate, because vapour-density differences between the cloud and the cloudless ambient air are very small (even when the ambient air is very dry, as can be seen by extrapolating Fig. 6.2 to the very low temperatures typical of the high troposphere), the resulting anvil may stretch many tens of kilometres down the direction of the wind shear. The lower edge of the anvil rises progressively upwards with increasing distance from the main cumulonimbus tower as smaller and smaller ice crystals sink more and more slowly into the underlying clear air (Fig. 10.10). Sometimes the undersurface of the anvil near the tower may hang down in smooth breast-like shapes known as *mammatus* (Fig. 10.11), which indicate the presence of pockets of negative buoyancy probably driven by the evaporation of anvil material into the dryer air of the middle troposphere.

Like the much smaller cumulus, cumulonimbus have a fairly clearly defined life cycle. Initial growth as cumulus congestus is followed by the transition to cumulo-

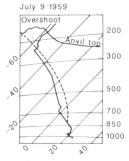

Fig. 10.10 Tephigram of conditions in the troposphere on the occasion of an intense cumulonimbus (a severe local storm) in southern England. Radar and aircraft observation showed tower penetration of the stratosphere and anvil tops at the levels sketched (horizontal structure is unrelated to the tephigram). The dashed line is a saturated adiabat. (After [42])

Fig. 10.11 Mammatus on the underside of the anvil of a summer cumulonimbus in southern England.

nimbus, frequently accompanied by rapid extension into the high troposphere. Precipitation falls in intensive shafts through the interior of the cloud body and into the clear air beneath cloud base. As these develop, air in their vicinity begins to fall, partly through frictional drag, and partly by chilling of unsaturated air by evaporating precipitation (section 5.11). When they reach the underlying surface these downdraughts are felt as the chilly gusts which usually accompany all but the slightest showers. Eventually the updraughts are overwhelmed by the downdraughts and the cloud begins to die, essentially as a result of its earlier success in producing cloud and precipitation. The main trunk of cloud dissolves into a temporary veil of precipitation, which looks much less solid than cloud because its consists of a relatively very small population of very large droplets and particles. After another 10 min nothing may be left of the recent glistening mountain of cloud apart from the tuft of anvil drifting and slowly dissolving in the high troposphere and the long track of surface which has received precipitation — symbols of the vertical exchange which underlies all convection.

Many cumulonimbus are associated with hail and thunder. As mentioned in section 6.10, hail is produced by the relatively rapid accretion of ice and supercooled water (mainly the latter) onto a falling frozen embryo. The growing hail stone soon has a relatively high fall speed through the ambient cloudy air, and in fact is able to grow to a substantial size only if the updraught of cloudy air is strong enough to hold it inside the body of the cloud for some minutes. However, cumulonimbus are typified by their strong updraughts, and very large cumulonimbus by their very strong updraughts — often exceeding 10 m s^{-1}. Hail is therefore typical of larger cumulonimbus, though in the British Isles comparatively weak winter cumulonimbus often produce little conical pellets of spongy hail (*graupel*). The association of maximum particle size with updraught strength explains why the largest hail fall from summer cumulonimbus in middle latitudes, the potential for the buoyancy which powers all such convection being greatest then. Hailstones more than a few millimetres across remain unmelted even after falling through a kilometre or more of warm air and often appear as a narrow shaft of hail in or beside the core of heaviest rain (some of which is probably melted smaller hail). In lower latitudes the low troposphere is so warm that even quite large hail is melted before it reaches the surface.

The same strong updraughts which encourage the growth of hail usually separate electrical charge efficiently enough (section 6.12) to cause lightning within the cumulonimbus and between its lower parts and the underlying surface. Most substantial cumulonimbus produce thunder in this way, and even those small ones which do not thunder audibly produce crackles of radio noise which indicate that weak sparks are jumping between regions of separated charge. The frequency of lightning strokes, like the maximum size of hail, is a measure of updraught strength, which in turn is a measure of the vigour of the parent cumulonimbus.

Civil aircraft are normally routed to avoid cumulonimbus because of the discomfort and even danger of flying quickly through alternating up- and downdraughts. Sudden downward accelerations in particular (widely but quite misleadingly attributed to the presence of air pockets) can throw coffee cups, and occasionally even unstrapped passengers, upward relative to the aircraft's interior — a potentially dangerous example of how an unexpected acceleration seems like a force in the opposite direction. In a very few cases such loads have caused structural failure of the aircraft and subsequent catastrophe. Although a few accidents have been attributed to lightning strikes on aircraft, this risk is relatively small, since the aircraft tends to keep close to the electrical potential of the ambient air. Aircraft icing is potentially a more important hazard (section 6.13). Larger aircraft carry a small radar which can detect the shafts of heavy precipitation which are associated with the active cores.

Consider a typical sequence of events experienced at a point on the ground lying in the path of a well-developed summer cumulonimbus. The air is warm and humid, both factors contributing to the convective potential of the situation. The sky becomes gradually overcast as the anvil of the advancing cloud reaches overhead and beyond, though the details are often visually very unclear if the low troposphere is hazy, as it often is in such conditions. The wind, which may have been blowing gently in much the same direction as the storm movement, dies as a persistent draw of air toward the base of the updraught sets in and locally offsets the overall airflow near the surface. (Sometimes there is no obvious relation between surface flow and storm motion, which is essentially with the airflow several kilometres above the surface, and occasionally the two are opposed.) As yet there is no obvious evidence of the storm: there is no precipitation and usually no evidence of thunder or lightning unless it is already after dusk, but there is widespread awareness of the 'lull before the storm', which is confirmed as the main cloud mass begins to loom overhead. The depth of cloud is such that it may become dark enough to trigger automatic street lights.

Quite suddenly there is a sound of wind, and nearby trees begin to move a few seconds before the squall of downdraught reaches the observer. If there is time to notice it before the onset of precipitation, the squall is obviously cooler than the sultry air it has replaced. The rain begins very suddenly, almost as if switched on, and may be heavy and contain bursts of small hail. As the same time the sky overhead begins to lighten considerably, giving the appearance of a pale grey vault. This is the tall column of downdraught where the cloud has been largely replaced by the much less optically dense veil of precipitation. This too is the region where the electrical activity is most obvious, with lightning strikes to ground, and less distinct flickering and rumbling from lightning in the middle troposphere. The sky lightens even further as the rain begins to ease off, and the squall winds subside. As the storm passes away downwind it may become clearly visible for the first time as a glistening mountain of cloud. The deposited carpet of cool, fresh downdraught remains, to be warmed again by the sun if there is enough of the day left.

10.5 Clouds and their environment

So far we have distinguished between cloudy and cloud-free air as if they were as separate as they appear to be visually. But in sections 5.12 and 5.13 we considered how convection, like any mixing, tends to produce an environment which is unaltered by further convection. To the extent that this tendency is realized, therefore, the ambient air around thermals below cloud base, and around the cloudy thermals above, is not independent of the buoyant masses which are currently active — it has probably already been processed by many such thermals through successive entrainment and detrainment. If the individual thermals are scarce and weakly interactive with their surroundings, the ambient air may wait so long between successive interactions that it is seriously out of equilibrium with the thermals, and the distinction between thermal and environment may be quite marked in consequence. But in fact most cumulus types of convection are so vigorous that thermals and environment are only slightly out of equilibrium. This is why the temperature excesses inside clouds are usually so small, less than 1 K in cumulus humilis and usually less than 5 K in even quite substantial cumulonimbus. It is also why the various types of cumulus convection are associated with characteristic vertical profiles of temperature and humidity which we will now briefly summarize.

Figure 10.12 depicts typical vertical profiles of potential temperature in the presence of various types of cumulus convection. The vertical scale has been made roughly proportional to the logarithm of height to enlarge the details of structure in the lower troposphere which would otherwise be inconveniently small. Remember that conversion from potential to actual temperature would simply tilt the whole towards the top left of the diagram by about 10 K km^{-1} (section 5.6). The profiles represent conditions outside the cloud masses which, remember, occupy only a small fraction of the total horizontal area at any level above cloud base. They therefore represent what is almost always recorded by radiosondes in these conditions, since only a small minority of sondes actually ascend through any substantial depth of such cloud. Conditions inside a cloud of the cumulus family are always very

Fig. 10.12 (a) Vertical profile of ambient potential temperature in the presence of cumulus and flattened cumulus, with a saturated adiabat shown dashed. Dotted portions correspond to flattened cumulus. (b) As (a) but with vertical scale considerably reduced to accommodate the full height of a cumulonimbus. (Based on [40])

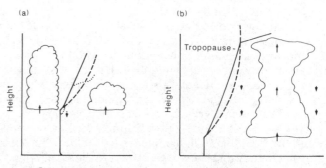

close to saturation, and are therefore moister than the air outside the cloud, but are usually only a very little warmer, as discussed above.

Profile (a) in Fig. 10.12 represents the conditions accompanying small cumulus, either in the early stages of a day over land which will later give rise to cumulus congestus, or (dotted) confined beneath a subsidence inversion, or similar convectively stable layer, a little above cloud base. The sharply superadiabatic layer (sharp lapse of θ) typical of the surface boundary layer over land (section 9.9) is apparent, as is the nearly adiabatic layer reaching up to cloud base. Careful measurements suggest that this layer is not exactly dry adiabatic (uniform θ), in that θ lapses by ~ 1 K in the first few hundred metres above the surface boundary layer, and recovers by the same amount up to cloud base. If this is a general result, it must arise from details of the way in which thermals vary with height, which are still neither clearly observed nor understood. Near cloud base the profiles differ somewhat, depending on whether we are beneath a cloud or a gap between clouds. Rising into cloud base the temperature profile converts from a dry adiabat to something close to the appropriate saturated adiabat, as expected (section 5.10). However, rising toward the level of cloud base whilst beneath a clear gap between clouds, θ tends to begin to rise before cloud base is reached, for reasons which are no doubt complex in detail, but which in principle depend on a very important aspect of cloudy convection which is worth brief attention.

The cloudy updraughts are only one part of an unsteady convective cycle which must include downdraughts as well. Some of the downdraughts occur within the clouds, as mentioned in descriptions of cumulonimbus, but most of the sinkage occurs in the clear air between clouds. These up-and-down motions must maintain a close balance of mass flux (only very slightly biased by the presence of any large-scale convergence or divergence — section 7.13). But the rising air tends to follow a saturated adiabat, whereas most of the sinking air must tend to follow a dry adiabat. Air sinking dry adiabatically after cloudy ascent will as a result tend to become warmer than the rising cloudy air. In the main this tendency is held in check by the frequency and vigour of the interactions between clouds and environment,

Fig. 10.13 Cumulus congestus, probably kept unusually narrow by strong convective instability. The sharp contrast between conditions in the cloudy rising air and the clear ambient air is visually obvious, as is the large area of the intervening mixing zone.

343

mentioned already. However, it seems that the interactions at around cloud-base level are sufficiently weak to allow slightly warmer air to extend a little way down into the sub-cloud layer between clouds.

Above cloud base the potential temperature profile deviates significantly from the expected saturated adiabat, tending consistently to lower temperatures (to the left in Fig. 10.12), but still maintaining a rise of potential temperature with height. This too is a consequence of the different adiabatic processes being followed by the rising and sinking air. If both branches followed the dry adiabatic, then according to the general law of mixing (section 5.12) the equilibrium profile approached by continued mixing of this type would be a dry adiabat, as is very nearly the case in most of the sub-cloud layer. If both branches followed the wet adiabatic process, then that would be the equilibrium profile. Since the branches differ, the equilibrium must fairly obviously lie somewhere between the dry and wet adiabatic profiles, at a position determined by details of the mixing effected by any particular population of clouds, including for example the vertical profile of rates of entrainment and detrainment. This is borne out by observation, though incompleteness of current theory and observation prevents detailed matching. Observed environmental lapse rates in layers populated by small cumulus often lie about one third of the way towards the dry adiabatic from the relevant saturated adiabatic, the fraction decreasing with increasing cloud cover. As already mentioned, temperatures inside the clouds differ very little from those outside. In fact some small cumulus seem to be capable of operating on zero temperature excess, relying on their excess water vapour to maintain buoyancy.

Above the layer of small cumulus the potential temperature profile typical of cloudy convection may extend through a deep layer populated by cumulus congestus, or potential temperature may rise sharply, indicating the presence of a convectively stable layer (dotted line on Fig. 10.12). In the latter case the cloudy updraughts lose buoyancy very quickly as they rise into the warmer ambient air. Terminal detrainment quickly follows the cessation of upward motion, but a little exchange of air between the convecting and stable layers occurs in the process.

When the stable layer is maintained by subsidence of the lower middle troposphere in an anticyclone, it may persist for days in a type of dynamic equilibrium with the underlying convection. If convection has the advantage, as may become the case if the air moves over warm land or warm water, the convecting layer eats upwards into the stable layer, but at a rate which diminishes as the erosion proceeds, since this tends to sharpen the 'elbow' in the temperature profile and hence increase the stability of the immediately overlying air. If the subsidence has the advantage, either through weakness of convection or enhanced anticyclonic vigour, then the convecting layer is squeezed downwards into a progressively shallower layer, producing stratocumulus as mentioned in section 10.1, and even leading to the extinction of all cloud if the base of the stable layer sinks below the lifting condensation level. The dynamic interaction at the base of the stable layer exchanges sensible heat and water vapour between the subsiding and convecting layers, the subsiding layer being cooled and moistened and the convecting layer being dried and heated as a result. Estimates have shown that an appreciable fraction of the heat input into the bottom layers of an anticyclone over land is brought down from the subsiding air in this way, most of the remainder coming through direct absorption of sunlight by the ground.

Profile (b) in Fig. 10.12 represents conditions accompanying cumulus congestus and cumulonimbus convection. The details up to levels reached by small cumulus are essentially the same as in profile (a), which is consistent with the observation that large cumulus are normally accompanied by small cumulus, though not the stunted type associated with anticyclones. Above these levels the potential

temperature profile remains intermediate between the dry and saturated profiles, for reasons already discussed, but tends in the upper troposphere to return
toward the saturated adiabat through conditions at cloud base. In fact the equilibrium buoyancy level indicated by drawing a saturated adiabat upwards from
the lifting condensation level associated with screen level conditions often gives
a rough but observably reasonable value for the level of cumulonimbus tops
(Appendix 10.3).

The tendency for ambient temperature profiles to recover toward the saturated
adiabat from the lowest limit of the convection no doubt reflects the concentration
of detrainment and dissolution by cumulonimbus, especially in the upper troposphere. It is as if the middle troposphere were partly bypassed by the deep convection currents linking low and high troposphere, so that the environment in these
intermediate levels is influenced only at second hand by the tumult of small
cumulus rapidly processing the low troposphere and the scattering of cloudy
towers reaching toward the top of the troposphere to disgorge their fans of icy
anvil. The intertropical convergence zone is densely populated by many of the
largest cumulonimbus on Earth, and it is significant that these are often technically
termed *hot towers*, on account of their role in piping latent and sensible heat from
the warm surface (most of it ocean) to the vicinity of the equatorial tropopause
about 15 km above. In fact such hot towers collectively represent the ascending
branch of the Hadley circulation outlined in section 4.7.

The existence of a layer of minimum interacting between updraughts and their
middle-troposphere environment is associated with a basic property of entrainment and detrainment: that it is least effective in proportion to the mass of rising
air when updraughts are largest. Basically this is yet another example of the importance of the decline of surface-to-volume ratio with increasing scale which applies
to phenomena as diverse as the heat economies of mice as compared with
elephants, and the slowing rate of growth of cloud droplets growing by condensation (section 6.10): the mass of a thermal is proportional to its volume whereas its
rates of entrainment and detrainment are proportional to surface area. The
insulating effects of large scale are confirmed by more technical arguments which
follow from an extension of the treatment of entrainment begun in section 7.12.
These indicate that small cumulus should mix themselves out of existence in the
short time taken for their tops to rise a few hundreds of metres, as indeed is usually
observed, whereas updraughts more than a kilometre or so in breadth should be
able to rise through the full depth of the troposphere without losing more than a
fraction of their mass by mixing with the environment. The advantage of scale is
apparent too in the mechanism outlined for the development of cumulus congestus
(section 10.2).

An inevitable consequence of such minimal interaction is that it encourages
maximum differences between clouds and environment: temperature excesses in
cloudy interiors are largest in the mid-troposphere and often amount to several K in
cumulonimbus. Buoyancy is thereby enhanced and this together with the reduction
in entrainment drag per unit mass (section 7.12) encourages strong updraughts.
Compared with the fretful inefficiency of small cumulus, cumulonimbus, are relatively very efficient in all that they do, and are capable of producing quite vigorous
weather locally, as described in the previous section. If conditions are such that
cumulonimbus produce weather which is so severe as to be locally damaging at the
surface, and this applies to a small but significant fraction of the total, then the
cumulonimbus are called *severe local storms*.

10.6 Severe local storms

Certain land areas are seasonally subject to unusually intense cumulonimbus when
intensity is judged by rates of precipitation, maximum size of hail, squally wind
speeds, and electrical activity. Such storms also quite often produce one or more
tornadoes (Fig. 10.14): intense localized vortices extending from cloud base to
surface, where they produce damaging winds (but not always the devastation
reported by the media — see later in this section).

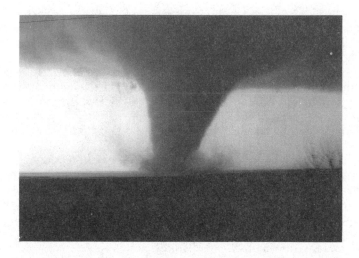

Fig. 10.14 A powerful
tornado in the United States
(10 April 1979) showing the
parent cloud base tapering
downwards into the vortex
core. The great breadth of the
cloudy core shows that the
pressure in its centre is very
low. Near the ground, violent
winds maintain a plume of
centrifuging debris.

Severe local storms have been studied quite extensively in recent years because of
the very considerable damage they can do. Much of this study has been carried out
in North America, partly because of the considerable incidence and severity of
storms there, but also because that particularly affluent and sophisticated population demands detailed forecasting of such events, and supports a greater *per capita*
number of forecasting and research meteorologists than anywhere else in the
world. However, it is important to realize that severe local storms occur in many
regions scattered throughout the world, including the British Isles, Europe and
European Russia, north India, north Indochina and China, with many of the
features mentioned below in the North American context.

Despite gross differences of geography and climate between different affected
regions, it seems that there are two features which are usually present when such
storms occur. The first is a mechanism for accumulating and then fairly suddenly
releasing substantial convective instability. Without such a temporary bottling up
of energy, the atmosphere releases its instability as soon as it appears, in weak,
unspectacular cumulonimbus, which do little more than contribute a little to local
precipitation. The second feature is the presence of significant wind shear distributed through a considerable depth of the troposphere. Though much less
obviously relevant than the accumulation and release of convective potential, wind
shear appears to play a crucial role in making storms especially vigorous and long-
lasting, and therefore severe, and it does so in two quite distinct ways which
correspond to two different types of storm — the *multicell* and the *supercell*
storms. Let us consider these two features, and the ways in which they are believed
to contribute to the appearance of severe local storms.

The accumulation and sudden release of convective instability occurs in different ways in different geographical regions, and usually depends on quite specific local distribution of sea, land and topography, though the release of conditional instability (section 5.12) is often a common feature. Consider, for example, the severe local storms which occur in the central United States in the spring and summer. These are amongst the world's most spectacular storms, and often occur ahead of weak cold fronts which have crossed the Rocky Mountains and are moving slowly across the Great Plains. A typical situation is depicted in Fig. 10.15, which shows a weak cold front orientated from roughly south-west to north-east. The associated jet stream is similarly orientated (according to the thermal-wind relationship — section 7.10) and provides the necessary wind shear throughout the troposphere in a fairly broad zone on either side of the jet core. As will be outlined in the next chapter, the flow of air in the vicinity of fronts is much more complex than is apparent on either horizontal or isobaric surfaces. There is always a three-dimensional interlacing of airflows from source areas which can be hundreds or even thousands of kilometres apart, and which look quite unconnected with the system on purely horizontal or isobaric charts.

In the case depicted, the air in the low troposphere well ahead of the advancing cold front is flowing northwards from the Gulf of Mexico, and is consequently very warm and humid. Such air is potentially very good convective fuel, on account of its high wet-bulb potential temperature, but the potential would be frittered away in rather ordinary cumulonimbus convection if its release were not temporarily inhibited by the presence of an overlying 'lid' of warmer air which confines any convection in the warm, Gulf air below a convectively stable interface between the two layers (Fig. 10.16). The warmer air is dry and seems to originate over the arid highlands of New Mexico and Mexico to the south-west, where the strong sun shining on the dry, elevated surfaces produces air with relatively high potential temperature, but low wet-bulb potential temperature. This air is then drawn into the confluent sandwich of air streams which is so typical of the vicinity of fronts, where it overlies the Gulf air.

The release of the convection potential of the troposphere as a whole occurs when the underlying moist Gulf air becomes warm enough to convect through the last vestiges of the weakening stable layer of the lid. This warming seems to come

Fig. 10.15 Airflows associated with outbreaks of severe local storms in the southern central United States. Warm moist air from the Gulf of Mexico is labelled G, and hot dry air from the Mexican Highlands is labelled M. (After [42])

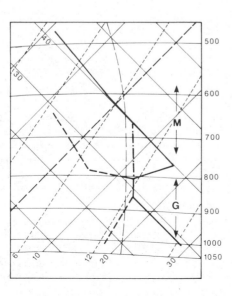

Fig. 10.16 Tephigram of Gulf and Mexican air streams at position X in Fig. 10.15. The chained line shows how the saturated Gulf air would rise in general uplift. (After [27] and [64])

about partly as a result of solar heating of the Gulf air as it over-runs the warmed Plains in the middle or latter part of the day, and partly as a result of the general uplift of the whole sandwich of airflows in the lower half of the troposphere which is associated with persistent large-scale convergence in the vicinity of the frontal system (sections 7.13 and 12.1). The general uplift begins to warm the moist Gulf air faster than the dry Mexican air as soon as cloud begins to form in the Gulf air, since the latter will then tend to rise wet-adiabatically while the unsaturated Mexican air continues along a dry adiabat (Fig. 10.16). In fact as the lid fades away, the steep temperature lapse rate of the Mexican air directly surmounts the cloudy Gulf air beneath, comprising a composite layer which is convectively extremely unstable. Clouds begin to tower up into the Mexican air, accumulating buoyancy at least to the level of the upper middle troposphere, as they rise and become progressively warmer than their immediate environment, until the most favoured reach the tropopause. These then offer efficient paths for the rapid ascent of more warm, moist Gulf air into the high troposphere, with consequent copious production of cloud, rain, hail and thunder.

It is now that the presence of wind shear through a deep layer begins to have its effect, achieving rather similar results by interestingly different means in the cases of the multicell as compared with the supercell storms.

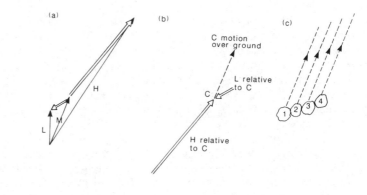

Fig. 10.17 (a) Horizontal wind vectors in the low, middle and high troposphere (L, M and H as observed from the surface) on an occasion such as shown in Fig. 10.15. The vector difference between M and H agrees in direction with frontal alignment through the thermal wind relation. (b) The same winds relative to a storm cell C moving with the middle troposphere flow. (c) Consistent development of new cells on the right flank pulls apparent storm motion to the right of individual cell motion.

In the *multicell* storm, the wind shear between lower and middle levels (somewhat above the disintegrating lid) is outlined in Fig. 10.17 for the situation described above. The individual cumulonimbus tends to move with the flow at middle levels, so that we can find the airflow at low levels (and any other level of interest) relative to the travelling cumulonimbus by vectorially adding an overall movement which exactly opposes the middle-level flow. When this is done in the present case we find that, from the viewpoint of an observer riding on an individual cumulonimbus, the low-level flow of warm, moist air flows in from a direction which is distinctly to the right of the direction of motion of the storm. Here the humid air rises in the powerful updraught, producing the great column of cloud and consequent precipitation and electrical activity which constitutes the currently active convective cell. However, as the inevitable downdraught begins to choke the updraught, the inflowing surface air begins to develop a new updraught on the right flank of the previous one, climbing over the spreading flood of downdraught. This new cell flourishes at the expense of its parent until it too is replaced by another on its right flank. The sequence is depicted on Fig. 10.17, and has no obvious end for as long as there is potentially warm and cool air to be vertically exchanged by convection. Although close inspection shows the sequential development of new cells, the overall impression is that there is a single, powerful storm which is moving distinctly to the right of the flow at middle levels. The significance of the title multicell is obvious. The sequence may run on until evening cooling of the surface air reduces its convective potential, or it may break down earlier because some local topographical feature inhibits the development of a new cell. Many quite ordinary cumulonimbus appear to have the multicell structure too, but when the convective potential is high enough to produce severe storms, the multicell structure ensures that the severity is persistent and distributed over substantial tracts of surface.

In the *supercell* storm the prevailing deep layer of wind shear allows a single powerful convective cell to become organized into an essentially steady, though highly dynamic, state. The detailed dynamics of the development and maturity of these storms are not well understood, and are far beyond the level of an introductory text, but it is possible to see one basic principle at work as follows. The downdraughts associated with the falling and evaporating precipitation originate in the upper middle troposphere, where the air is moving much more quickly than the air near the surface, on account of the presence of pronounced shear. The downdraught air tends to maintain its initial momentum as it sinks down to lower levels, flooding north-eastwards in the particular case of the situation outlined above and in Figs. 10.15 and 10.16, and thereby helping to 'shovel' the warm, moist air into the base of the updraught, which for the same reason adopts a corresponding slope backward relative to the storm's direction of motion (Fig. 10.19).

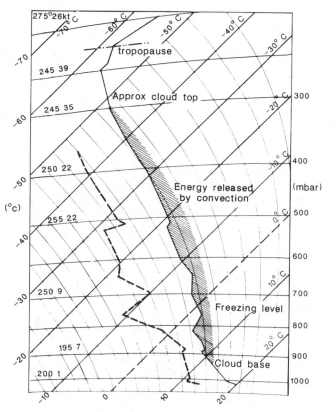

Aughton 1200Z 19 May 1989

Fig. 10.18 The troposphere on the occasion of unusually heavy rainstorms near Halifax (north-central Britain). The area of the shaded zone on the tephigram is a measure of the potential energy available for conversion in air rising wet-adiabatically from cloud base (remaining strong buoyant up to the 300 mbar level). A conglomeration of cells of heavy precipitation was observed by radar moving slowly over several hours. (After [65])

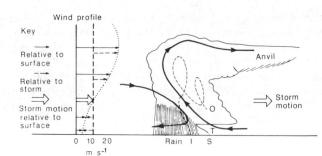

Fig. 10.19 Vertical section through a supercell storm showing updraught and downdraught, inflows and outflows, as observed from a frame moving with the storm (i.e. with the flow in the lower middle troposphere). The dashed line indicates the path of a large layered hailstone from effective origin O to impact I on the surface. The squall or gust front S is the leading edge of the cool downdraught, and tornadoes are most likely in the region T. (After [27] and [42])

The result is that in at least the lower half of the troposphere the rising and falling airflows are adjacent and parallel, rather than entangled as they tend to be in the more vertical, ordinary and multicell cumulonimbus, and are tilted so that the updraught overhangs the downdraught. The burden of precipitation no longer falls back through the parent updraught; instead it falls into the downdraught, enhancing it by drag and evaporative cooling. The self-choking tendency of more vertical cumulonimbus is therefore avoided, and the single supercell could, on this simple view, continue to operate for as long as the convective potential of the overall situation was maintained.

We might expect the supercell storm to travel with the air in the middle troposphere, as is the case with ordinary cumulonimbus and the individual cells of the multicell storms, but this is apparently not generally the case. The movement is often distinctly to the right of the mid-tropospheric flow, but leftward motion has also been observed, and neither of these tendencies is properly explained. Whatever is the storm motion, it is the flow relative to the storm which is nearly steady and which is depicted in a simple two-dimensional vertical section in Fig. 10.19. The three-dimensional pattern of relative flow in the case of a right-moving storm is depicted in plan view in Fig. 10.20, where it is apparent that the single supercell is maintaining a configuration of flow and storm motion which is similar to that of the composite multicell.

In the space of the several hours which is the normal lifetime of supercell storms, their relatively rapid movement over the ground means that they can cover many tens of kilometres, leaving a carpet of precipitation, chilled air and more or less serious damage.

The separation of updraughts and downdraughts is obviously consistent with the tremendous power of supercell storms. Updraughts become horizontally very large (up to 10 km), and the minimization of drag and maximization of buoyancy which follow from the advantages of scale ensure that they also become very strong. Vertical speeds of as much as 50 m s^{-1} (100 mi/h) are believed to occur occa-

Fig. 10.20 Plan view of a right-moving supercell storm, showing low-level inflow and climbing flow (dashed), and high-level flow and outflow (solid). Concentric envelopes contain areas of cloud, rain (R) and hail (H). The vault V is largely free of radar echo because the updraught there has few precipitation particles though it is full of dense cloud. A hook-shaped radar echo is often observed beside V just before tornadoes appear in the same area. (Redrawn from [8])

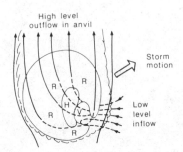

sionally, on the basis of some exceedingly bumpy traverses by military aircraft and dramatic inadvertent ascents by gliders, and from indirect evidence such as the height of updraught overshoot into the stratosphere (Appendix 10.2). Such powerful updraughts maintain high rates of production of cloud and precipitation, and permit the growth of giant hail (section 6.10), whose size provides further indirect evidence of updraught strength, through the inferred fall speeds (Fig. 6.10) which obviously must be exceeded by updraught speeds throughout much of the life of the growing stone. In fact the onion structure observed in some giant hail suggests a vertical oscillation of the growing stone as depicted in Fig. 10.19 — the stone being first carried aloft in the great updraught, then falling out of the forward overhang of the anvil back into the updraught again, and repeating this cycle until it is big enough to have a fall speed substantially greater than the core of the updraught, through which it finally falls to the surface. The large variations in temperature and cloud conditions between the extremes of the looping path are consistent with the alternating zones of clear and opaque ice, as mentioned in section 6.11. Giant hailstones five or more centimetres across can clearly be very damaging, but they are produced in such small numbers by the ultimate example of selective growth that reports of injury are rare.

The very rapid convergence of air at the base of such powerful updraughts can generate tornadoes by concentrating residual synoptic or mesoscale vorticity (by the spinning-skater effect — section 7.14). According to section 7.15 the fall of pressure in a horizontal traverse to the core of a tornado can be as much as 100 mbar. Since this is comparable with the vertical lapse of pressure in the first kilometre of the atmosphere, the base of the cumulonimbus cloud is often drawn right down to the surface, as air spirals violently into the axis of the vortex before ascending, to form the sinuous trunk of cloud which is typical of the mature tornado (Fig. 10.14). Air whirls around the tornado vortex with tangential speeds probably exceeding 100 m s^{-1}, though direct measurement is virtually impossible in such extreme and localized conditions – the violent core of the vortex being seldom more than a couple of hundreds of metres across. Most buildings suffer severe damage in such conditions, with roofs taken off and unframed walls demolished. Those stout enough to remain intact at the first onslaught may succumb to a tendency to explode if they do not leak internal pressure fast enough to match the sudden fall of external pressure as the vortex core passes over. The dislocated components are then whirled away by the fierce winds just outside the vortex core, contributing to the cloud of fast-moving debris which continually sprays tangentially out of the foot of the tornado and is potentially lethal to any exposed animal or man (Appendix 7.7). Fortunately, the conditions for the maintenance of a tornado are only marginally attained in even the most vigorous storm, and the majority of individual vortices seldom remain at full damaging strength for more than 20 min, but that is cold comfort to anyone whose farm or house has been traversed in that brief, violent life.

The downdraught too can attain great strength through separation from the updraught, and through the effects of the very heavy precipitation, and it is not unusual to record gusts of hurricane strength (> 33 m s^{-1}) where fallen air spreads out at the surface. Though much less intense than tornadoes, these winds can cause damage to crops (which can become irreversibly *lodged* on the ground by being battered down when wet) and buildings over a much wider path than that of a tornado.

The various types of very severe weather mentioned above are not confined exclusively to supercell storms, since many very severe storms are of the multicell type. However, the efficiency of the supercell structure means that virtually all supercell storms are especially severe.

The combination of hurricane-strength winds, torrential rains, giant hail, intense thunder and lightning, and the possibility of tornadoes, makes the severe local storm one of the most damaging types of weather system, and this together with the intrinsic interest of such powerful systems attracts the attention of research scientists and forecasters. The North American examples are amongst the most spectacular, and include a very considerable incidence of tornadoes, as news photography confirms each year when towns or villages are unlucky enough to be traversed by them. Because of this, there are well-developed ways of forecasting the particular conditions known from long experience there to be likely to produce very severe storms, and of monitoring weather radars for the tell-tale signs of the imminence of tornado formation (in particular the hook echo in Fig. 10.20). When these are seen, local broadcasting is interrupted to give very specific local warnings to areas at risk, which are repeated and amplified if tornadoes are actually sighted. Little can be done to protect against damage, but taking cover in storm cellars or sheltered rooms in solidly constructed houses very considerably reduces the risk of injury by flying debris.

Although less spectacular than the tornado, one of the most expensive types of associated damage is that done to crops by the deluges of moderate-sized hail which often form in the core of each precipitation shaft (Fig. 10.21). Such stones are only a tiny fraction of the weight of giant hail, but they are nevertheless heavy enough to strip the head from a mature cereal plant, and numerous enough to reduce the yields from affected areas almost as if they had been visited by locusts. In fact the economic danger to farmers in parts of the United States which are prone to such damaging hail storms is such that it is prudent to insure against hail damage, and one enterprising meteorologist has investigated the distribution of hail from a series of storms by plotting the locations of fields subject to insurance claims for hail damage.

In some countries affected by similar storms (Russia and northern Italy for example) there have been persistent but largely inconclusive attempts to reduce hail damage by firing explosive rockets or shells into threatening clouds in the hope of

Fig. 10.21 A maize crop destroyed by a shower of centimetric-scale hail. Smaller stones lack the damaging fall speeds (Fig. 6.10), and larger ones cannot fall densely.

inhibiting the growth of hailstones to the damaging size. Although such efforts probably help psychologically more than they do physically, there is some evidence that they can occasionally be effective. The explosive dispersal of freezing nuclei has also been tried and at least has the merit of being useful in principle, being an attempt to use cloud seeding to increase the numbers and therefore decrease the sizes of the hail by deliberate exaggeration of techniques used elsewhere for precipitation enhancement (section 6.13).

In different regions, different aspects of severe surface weather are accentuated at the expense of others. In most, the electrical activity causes damage to power lines etc. In many areas it is the torrential rainfall as well as the hail. In North America it is the hail and tornadoes. There is strong evidence that devastating tornadoes are extreme examples of a spectrum of vortices. Patient work by volunteers in Britain [95], sifting eye-witness accounts and inspecting damage, shows that tornadoes capable of causing substantial damage, rather than devastation, are quite widespread in southern Britain, and probably in Western Europe (and probably, by inference, in most areas visited by severe local storms). In some fairly dry regions like central northern India, which are nevertheless subject to cumulonimbus vigorous enough in at least some respects to be classed as severe local storms, the downdraught is the most significant feature for surface conditions. This arises because of strong evaporative chilling of rain falling through the very deep sub-cloud layer. This can even entirely remove the rain, like torrential virga tapering to nothing (section 6.10), leaving the surface dry, but subjecting it to a blast of cold air which raises vast dust clouds from the arid land.

10.7 Convective systems

We have already had two examples of the way in which individual cumuliform elements can combine and cooperate to produce more extensive and persistent systems. Individual cloudy thermals tend to combine to form the aggregate which we call cumulus congestus when there is sufficient convective instability through a substantial fraction of the troposphere (section 10.3). And in the presence of wind shear through a similar fraction, individual cumulonimbus can cooperate to produce the multicell sequence (section 10.6). There are other examples of convective cooperation which produce distinctive systems of cumuliform elements on scales ranging from ~ 100 m (small scale) to more than 1000 km (synoptic scale), and these we will now briefly and selectively review.

Photographic reconnaissance by aircraft over warm seas shows that small cumulus often appear in remarkably straight, parallel lines called *streets* (Fig. 10.22), which can individually run downwind for as much as 100 km. Lateral separation of adjacent lines is about 1 km. Over land, such lines are much more irregular and almost always run downwind from a sun-facing slope or some other especially effective source of thermals, but the ocean surface is normally too homogeneous to suppose that there could be a corresponding origin for maritime streets. Although the mechanism is still not properly understood, it is clear that maritime cumulus streets are a symptom of a larger-scale dynamic structure in the convecting layer which has the effect of encouraging convection along long parallel lines, and discouraging it along adjacent intervening lines. Given the marginal buoyancy of small cumulus, the encouragement and discouragement need be only very slight to

Fig. 10.22 Cumulus streets aligned with trade winds.

produce a very marked degree of alignment in what would otherwise presumably be a random cloud distribution. There is some evidence that opposing wind shears above and below a low-level wind maximum plays a significant role, with repulsion between the opposing horizontal vorticities of rising and sinking air currents encouraging their lateral separation [66]. Observation of soaring sea birds suggests that similar streets of thermals are quite widespread in the maritime boundary layer even when there are no cumulus aloft, and this has been supported by recent observation by high-power radar which is able to detect cloudless thermals by backscatter from their small-scale turbulence.

Satellite pictures of showery air streams over oceans in middle latitudes almost always show a characteristic distribution of ring-like patches or cells (Fig. 10.23) which was quite unsuspected before there were any weather satellites, since the cells are too small to be resolved by the very incomplete synoptic network there, and too large to be coherently viewed by normal aircraft. Each cell seems to consist of an extensive cloudless area surrounded by a roughly hexagonal ring of well-developed cumulus, including congestus and cumulonimbus, and the whole showery zone is made up of a close-packed mesh of such cells. Cells range in diameter from about 20 to 200 km, depending on larger-scale meteorological conditions, especially the depth and instability of the convecting layer. Similar cellular structure is observed on the western margins of the great maritime subtropical anticyclones, where gradually destabilizing air flows polewards and eastwards. All such cells are termed *open* on account of their open centres, and to distinguish them from the related but contrasting closed variety.

Oceanic areas covered by extensive stratocumulus are observed to be patterned by *closed* cells of a somewhat smaller horizontal scale and a much smaller vertical scale than open cells. These consist of extensive pancakes of thicker stratocumulus bounded by roughly hexagonal margins in which the stratocumulus is fragmentary or absent. Again the whole zone is covered by a fairly close-packed mosaic of these mesoscale elements, each of which in turn consists of an assembly of small-scale

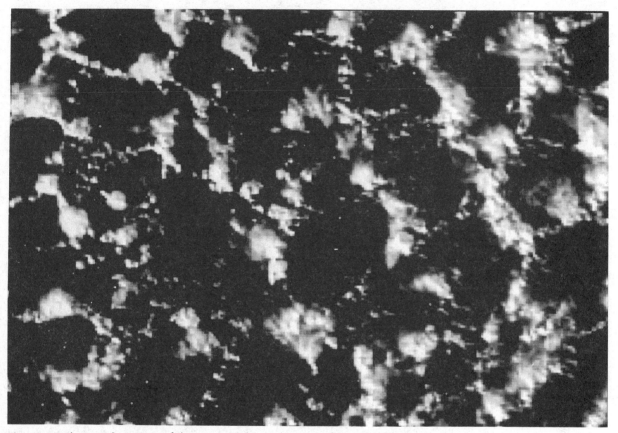

Fig. 10.23 Close-up of open cells of showery convection in the centre of Fig. 2.12 (55° N, 20° W). Cells are 20 to 30 km across in this zone.

stratocumulus elements. Closed cells are most often observed on the eastern flanks of maritime anticyclones, where the subsidence inversion is relatively low and pronounced, and where the ocean surface is often kept cool by upwelling of cold water from the ocean depths.

Open and closed cellular convection is also observed on a scale of centimetres in laboratory experiments pioneered by Bénard at the end of the last century. Small tanks of water heated gently from below are observed to form closed cells (Fig. 10.24), while air in the same situation forms open cells. In pioneering theoretical studies of convection, Rayleigh was able to show that the crucial difference between the two types is the vertical profile of viscosity: in the air tank the viscosity is largest near the warm base, while in the water tank the viscosity is smallest there, because of the opposite dependence of viscosity on temperature in gases and liquids. On the relatively enormous scales of atmospheric cellular convection, the effects of molecular viscosity are overwhelmed by eddy viscosity (section 7.2), but the fact that the Rayleigh number regimes (eqn (10.12)) are similar, if molecular viscosity is replaced by realistic estimates of eddy viscosity in the atmospheric case, is highly suggestive of a dynamic similarity between the two very different scales. Rayleigh numbers then lie in the range known to be associated with *cellular* convection (laminar and regularly patterned), rather than the range of larger values > ~ 50 000 known to be associated with *penetrative* convection (turbulent and

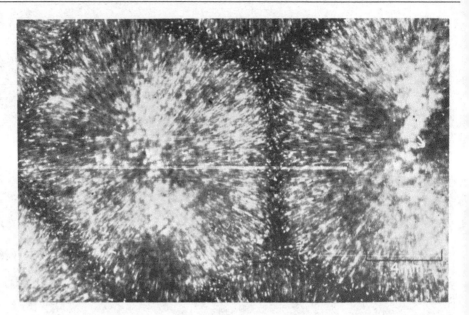

Fig. 10.24 Closed convection cells in a small tank of water heated gently from beneath. Motion is revealed by reflection streaks from aluminium powder in a time-exposure photograph. (From [57])

chaotic). Now atmospheric cellular convection is far from laminar on the small scale, since the convective elements are the intrinsically turbulent cumulus, but on the mesoscales of the cells themselves the turbulence is so small in scale for the individual eddies to be indistinguishable. Moreover the vertical profiles of eddy viscosity resemble their molecular counterparts: in open cells there is an upward lapse of eddy viscosity from a maximum close to the relatively warm sea surface (the normal state of affairs in and above the turbulent boundary layer); while in the case of closed cells, radiative cooling of the upper parts of the stratocumulus layer produces negative buoyancy which encourages an elevated layer of enhanced turbulence. It seems likely that the small laboratory models with their associated simplified (but far from simple) dynamics isolate the meso-scale dynamics which gently but persistently constrain the cumulus elements to form these large patterns of cells in the atmosphere [67].

Relatively recent detailed studies of precipitation distribution in fronts show that there are quite pronounced patterns on the meso-scale (Fig. 10.25). Bands of much more intense precipitation are embedded in the synoptic-scale areas of frontal precipitation, often aligned fairly closely with the nearest front. Although these bands are too small in scale to be resolved by the synoptic surface network, their presence is consistent with the large variations in rainfall rate which are such a familiar feature of most fronts (Fig. 2.5). Weather radars show that these bands coincide with bands of cumulonimbus (whose shafts of precipitation show up as vertical stripes of radar echo on a range–height display) embedded in the frontal nimbostratus. The instability producing these arises from the differential advection which is typical of flow in the vicinity of fronts (Section 11.4), in which air differing streams are brought together in a complex vertical sandwich, as mentioned in the previous section. Direct instability (vertical lapse of potential temperature) does not occur, but if a dry stream should flow over a saturated stream with much the same potential temperature but higher wet-bulb potential temperature, then the composite layer is unstable for cloudy convection, and clouds will mushroom up from the saturated layer. This is very similar to the

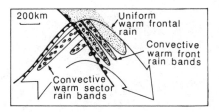

Fig. 10.25 Meso-scale patterns of enhanced precipitation in and near the fronts of a mid-latitude depression. The broad arrow is the warm conveyor belt (section 11.4) of ascending air in which bands of moderate rain are embedded (light stipple), which in turn contain patches of heavy convective rain (dark stipple). (From [68])

mechanism cited in the previous section, but here the instability may be quite gentle, adding a cumuliform pattern to an existing substantial stratiform system (Fig. 10.26). Though the cumulonimbus are often quite feeble, they are nevertheless capable of enhancing the precipitation rate severalfold at the surface, either directly or by interactions between their precipitating ice crystals and underlying layers of supercooled water cloud associated with the ambient nimbostratus. In cold fronts the embedded cumulonimbus tend to be much more vigorous, producing bursts of torrential rain and hail, vigorous gusts and lightning. The alignment of meso-scale rainbands parallel to the ambient or nearby fronts (for some bands form outside the frontal cloud masses) presumably arises because the airflow at middle levels, in which the convection is embedded, is similarly aligned by the thermal wind effect, but the precise mechanism is not well understood.

One type of linear alignment of cumulonimbus has been known about for a considerable time, because it is big enough to show clearly on synoptic weather maps, and because the cumulonimbus are often extremely vigorous, often being severe local storms. This is the *squall line* of the central and eastern United States. Typical conditions for the formation of a squall line are identical with those already described in section 10.6 and Fig. 10.15 in relation to the formation of severe local storms in the same region. On this scenario the cumulonimbus begin to erupt along or just ahead of the cold front which is moving from the west or northwest. By some sort of dynamic cooperation the cumulonimbus become very clearly and closely aligned, to the extent that the very blustery leading edges of their squalls merge into a nearly continuous line, giving the phenomenon its title. The line may be several hundreds of kilometres long, acting like a very sharp forward extension of the cold front. However, the ability of the advancing wedge of cool, dense air to help 'shovel' the warm air ahead up into the updraught seems to encourage the squall line to advance faster than the parent cold front, so that it becomes a detached line of powerful cumulonimbus. Over the day or two in which the line squall is identifiable it may advance several hundreds of kilometres into the warm sector as a narrow band of intense rain and hail (Fig. 10.27). Forecasters are particularly interested in the development and progress of line squalls because of their tendency to be or become sites of very severe weather, including tornadoes. In fact it seems that the updraughts and downdraughts can become separated in much the same way as in the supercell storm (section 10.6). Lines of cumulonimbus appear in

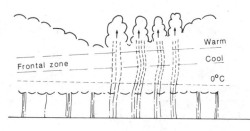

Fig. 10.26 A vertical section across a warm front, showing embedded weak vertical convection whose precipitating ice and snow crystals seed the underlying cloud sheet. (After [69])

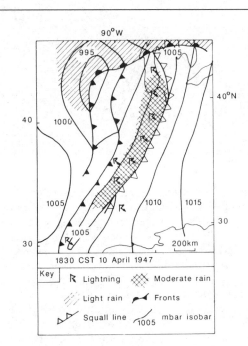

90°W

40°N

30

200km

1830 CST 10 April 1947

Key | R Lightning | Moderate rain
Light rain | Fronts
Squall line | 1005 mbar isobar

Fig. 10.27 A classic case of a North American squall line, extending over 1500 km from Texas to Michigan and lying between 200 and 500 km ahead of a doublet of cold fronts. (After [27] and [70])

other parts of the world (west Africa and northern India for example), and although their mode of formation varies from place to place, and few are as energetic as the American line squalls, they all seem to derive some dynamic advantage from their linear cooperation.

There is considerable observational evidence to show that *tropical cyclones* can be regarded as spiral assemblages of very vigorous cumulonimbus clustered round the relatively quiet eye. The characteristic spiralling arms and the massive wall which surrounds the eye are composed of very large and vigorous cumulonimbus. The violent winds spiralling inwards toward the eye wall feed the powerful updraughts of the cumulonimbus, which in turn sustain the very high rainfall rates and the copious production of ice cloud which fans out aloft to produce the dense white shield observed by satellite (Fig. 12.10). Other types of tropical weather system too seem to be cooperating assemblages of cumulonimbus, although the modes of cooperation are largely unknown. Tropical weather systems in general and hurricanes in particular are outlined in sections 12.3 and 12.4. Even the vast intertropical convergence zone is in essence a planetary-scale girdle of very tall cumulonimbus linearly constrained (and constraining, since as usual there is no clear chicken or egg) by the Hadley circulation (Fig. 8.14).

10.8 Atmospheric waves

There are many examples of atmospheric waves on the small and meso-scales which are at least superficially similar to waves on the ocean surface.

Acoustic echo sounders have been used in recent years to probe the atmospheric

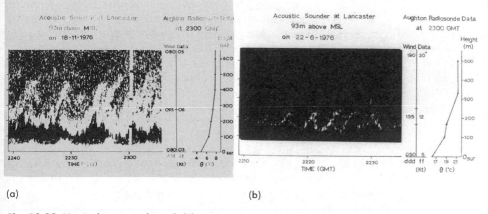

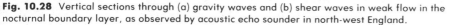

(a) (b)

Fig. 10.28 Vertical sections through (a) gravity waves and (b) shear waves in weak flow in the nocturnal boundary layer, as observed by acoustic echo sounder in north-west England.

boundary layer by detecting backscatter from variations in air density on the scale of half the acoustic wavelength used ($\sim$ 10 cm). Individual variations are associated with micro-scale turbulence, but they are patterned by events on a somewhat larger scale. At night they are seen to be almost incessantly patterned by events which are clearly wavelike (Fig. 10.28a). Activity like this was largely unsuspected before acoustic sounders enabled effectively continuous monitoring of the nocturnal boundary layer, but it has now been observed to be typical of overland convectively stable boundary layers from the tropics to the poles.

Many of the patterns are obviously *gravity waves*, in which vertical displacement from equilibrium is restored by gravity, with subsequent under- and overshooting of the equilibrium level giving rise to a series or *train* of waves. As such they resemble surface water waves, but actually they are best compared with *internal* water waves which are observed under water on horizontal boundaries between water of slightly different densities. In the atmospheric case the boundaries are always more or less spread out in the vertical by mixing, but the dynamics are essentially the same. A vertically diffused layer in which density lapses with increasing height (the opposite would of course be convectively unstable) is simply a convectively stable layer, of the type discussed in section 5.12. It can be shown (appendix 10.4) that if a parcel of air is displaced adiabatically and vertically in this layer and then released, it will oscillate about the level of neutral buoyancy with what is called the Brunt–Väisälä frequency $N/2\pi$, where

$$N = \left(\frac{g}{\theta} \frac{\partial \theta}{\partial z} \right)^{1/2} \tag{10.1}$$

and $\partial\theta/\partial z$ is the vertical gradient of ambient potential temperature, which expresses the degree of convective stability of the layer. This expression confirms what is intuitively reasonable: that the frequency of oscillation increases with stability, just as the rate of vibration of a spring-mounted mass increases with the stiffness of the spring.

We can easily imagine how the waves in Fig. 10.28a come about. Nocturnal cooling of the land surface established the stable layer, and airflow over humps and hollows in the surface give the air its initial vertical push. The waves then propagate through the air, just as waves propagate over the water surface, and produce the patterns shown by raising and lowering a slightly turbulent layer above the acoustic

359

sounder. The speed of wave propagation over the sounder is the algebraic sum of the *celerity* of the waves through the air, and the speed of airflow. If the two speeds were exactly equal and opposite, as might be in the case of waves rippling upwind from a fixed obstacle, then we would have standing waves, which could be seen only if we had several acoustic sounders operating instantaneously along the flow.

A fundamentally different type of wave is apparent in Fig. 10.28b, again from acoustic sounding of a nocturnal boundary layer. Although it may slightly resemble a breaking sea wave, this is a misleading coincidence of appearance, since these interlocking wave shapes are believed to result from the *Kelvin–Helmholtz* instability of a strongly sheared layer. Such shearing instability is responsible for the flapping of a flag, which is encouraged by the difference in wind speeds either side of the cloth, and restrained by the stiffness and strength of the cloth. In the atmospheric case the sheared layers are horizontal and the restraining force is the convective stability of the sheared layer. A dynamic analysis of the balance between rippling by shear and smoothing by convective stability is beyond the scope of this book, but yields as a critical parameter the Richardson number *Ri* quoted in the discussion of turbulence in section 9.8:

$$Ri = \frac{g}{\theta} \frac{\partial \theta}{\partial z} \bigg/ \left(\frac{\partial V}{\partial z} \right)^2 = \left(N \bigg/ \frac{\partial V}{\partial z} \right)^2$$

When *Ri* falls below 0.25 (according to the detailed theory) then the convective stability is unable to restrain the tendency to ripple, and the shear waves grow in amplitude to the point where orderly flow breaks down in turbulence. The whole process has the effect of converting some of the kinetic energy of the original orderly flow into turbulent kinetic energy (and ultimately heat), and considerably deepening the originally shallow sheared layer. The prevalence of the interlocked wave patterns in the nocturnal boundary layer strongly suggests that strongly sheared layers spend a considerable proportion of their time in the process of growing these shear waves, which are quite distinct from the gravity waves considered earlier, and are much shorter in wavelength. There is some interconnection between the two types when conditions encourage large-amplitude gravity waves, which concentrate shear at their crests and troughs which then generates shear waves and turbulence.

Gravity and shear waves play a very considerable role in the activity of stable boundary layers, the shear waves in particular being an important link in the production of turbulence which transports momentum, heat and material (water vapour, etc.) to or from the cool surface. However, each type is also important at much larger scales and heights in the atmosphere.

As mentioned already (section 9.11), airflow over hills and mountains often maintains standing waves in the lee of the obstruction. These are basically the same as the much smaller standing waves mentioned above. The theory of such waves is quite well developed but is subtle in principle and complicated because of differences arising from the vast range of conditions of atmosphere and terrain. A number of points can be made in relation to the outline in Fig. 10.29.

1. The dominant wavelength λ induced by an obstacle is given approximately by assuming that air flows through the steady waves at the undisturbed wind speed *V*, swinging up and down at the Brunt–Vaisala frequency $N/2\pi$. Since

    $$V = \text{frequency} \times \text{wavelength}$$

    we have

    $$\lambda = 2\pi \frac{V}{N}$$

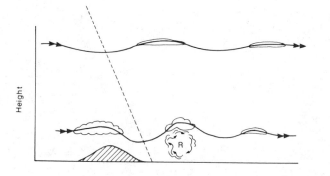

If we substitute reasonable values for the stability of the air flow (appendix 10.4), corresponding to a lapse rate of about 75% of the dry adiabatic value, then we have the useful rule of thumb that the dominant wavelength in kilometres is related to the average wind speed in metres per second by

$$\lambda \cong 0.6V$$

the fraction (and wavelength) decreasing with increasing stability. Typical wavelengths of ~ 3–20 km follow from typical wind speeds.

2. As implied by the absence of terrain profile in the above, the wavelength of the lee waves is independent of terrain shape (i.e of any 'wavelength' there), but this is not the case for the amplitude, the vertical oscillations in the waves. The largest-amplitude waves are produced by terrain whose wavelength along the flow is considerably smaller than the dominant lee wavelength, but whose wavelength across the flow is very long. A hill or range which is very long in the direction of flow is ineffective because the air merely climbs slowly up one side and sinks as slowly down the other, without further vertical motion. And a hill which is short in the cross-flow direction is easily bypassed by lateral airflow. Amplitude obviously tends to increase with terrain height, but quite small hills can be surprisingly effective if other conditions are optimum.

3. For a given terrain, the largest-amplitude waves arise when there is a significant increase with height in wind speed toward the barrier, and a significant decrease in stability. Largest amplitudes often appear in the vicinity of a stable layer and may considerably exceed the height of the mountain or hill at heights far above hill or mountain top.

4. Motion is everywhere smooth and wavelike except in the immediate lee of a sharply scarped hill, and under the first wave crest downwind where there may be a violently turbulent *rotor* in which trapped air is continually rotating.

5. For subtle dynamic reasons the axes of the wave crests and troughs shift upwind with increasing height throughout the region in which the wave train is exporting energy upward. This is believed to be significant on the large scale in communicating surface drag to the middle and upper troposphere, just as the wake of tied waves increases the drag on a boat and spreads its equal and opposite effect through a large water mass.

Hill and lee waves become conspicuous when a layer of air is lifted above its lifting condensation level in the wave crests, producing lenticular clouds whose smoothness and stability amid the flowing air nicely depict the wavy nature of the flow (Fig. 10.30). The dependence on atmospheric stability often enhances the number and size of lenticular clouds at dusk, producing clouds kilometres above

361

Fig. 10.30 Lenticular clouds over the Yosemite National Park, USA. The larger clouds (middle right) are at a lower level and they and the edges of the clouds above them show that the humidity structure of the middle troposphere is finely structured in the vertical.

hills which are only a few hundred metres high. Updraughts and downdraughts in large-amplitude wave clouds are sufficient to make flying conditions uncomfortable and even dangerous. The danger can be compounded by severe turbulence in the vicinity of a steep hill or rotor, and many aircraft accidents have been attributed to such encounters, including the fatal crash of an airliner which broke up while allowing its passengers a view of Mt Fujiyama in the lee of that steep, conical mountain (height 3.8 km).

Lenticular clouds have been noted for at least as long as they have been specifically named, but only with the coming of the satellite has it been clear that vast trains of lee waves are the norm in the vigorous cloudy airstreams associated with extratropical cyclones (Fig. 10.31). These would of course be invisible from the ground since they are embedded in thick nimbostratus, but they most probably give rise to at least some of the strong patterns of rain and snowfall long observed in hilly terrain. Detailed mechanisms are complex and difficult to confirm, but it is clear that cloud production is enhanced on the upwind side of a wave crest. If the wavelengths are long enough, like the ones apparent on satellite pictures such as Fig. 10.31, then precipitation too may be enhanced before the cloud is thinned in the lee of the wave crest. A related possibility is that precipitation falling through the denser cloud on the upwind side is able to grow much more rapidly by collision and coalescence, or that ice crystals falling through air suddenly forced toward water saturation by uplift grow rapidly by the Bergeron–Findeisen mechanism (section 6.11). Mechanisms such as these are believed to be involved in the enhancement of rainfall on the upwind flanks of hills and mountains, and its depletion on the near downwind side, but pictures such as Fig. 10.31 suggest that the influence could be much more pervasive, though hidden amid the inherent structure and variability of weather systems. It seems clear from some detailed studies of rainfall patterns that they may at least trigger the formation of some of the meso-scale rain bands mentioned in the previous section.

Shear waves too are important at much greater heights than the boundary layer in which they were noted above. In the strongly sheared zones below the core of a jet stream in particular, the Richardson number may be held at about 1 throughout substantial volumes of the high troposphere by the large-scale structure. Local enhancements of shear or reductions in stability may then produce sheets of turbulence far above and completely detached from the turbulent boundary layer. When first encountered by high-flying aircraft, these were judged most surprising in cloudless air, where there was no obvious convective source of turbulence, and became known as *clear-air turbulence* (CAT), but they are also found in the cloudy parts of similarly sheared zones. Turbulence of small scale and amplitude is wide-

AVHRR CHAN 2 21/8/80 42 - 324

Fig. 10.31 Satellite view of north-westerly flow over the northern British Isles (i.e. top left to bottom right), showing extensive wave trains associated with hills and mountains there. There appear to be some large closed cells in the south-west of the picture. Infra-red image from sun-synchronous NOAA-6, 0836 Z, 21 August 1980.

363

Fig. 10.32 Billow clouds in strong wind shear in the high troposphere downwind of Mt Teide, Canary Islands. Continued observation shows that each wave has only a limited life before it breaks up in turbulence. (Reproduced by permission from *Clouds and Storms: the behaviour and effect of water in the atmosphere*, by F.H. Ludlam, Pennysylvania State University Press, University Park, USA, 1980)

spread in the high troposphere, as may be judged on commercial flights by looking out for slight but persistent vibration, which feels as if the aircraft were taxiing over soft cobblestones. CAT can be much more violent on occasion, presumably as the aircraft encounters large shear waves or the large eddies which are their immediate aftermath, and the problem is sufficiently serious to have encouraged considerable study of ways avoiding severe CAT by forecasting risky zones, or detecting current CAT by special radar. The problem remains, however.

Shear waves occasionally produce clouds which pick out the crests of the turbulent wavelets arranged across the shear (Fig. 10.32). These are sometimes called *billow clouds*, which is unfortunate since breaking sea waves have no equivalent of the shear which is their crucial ingredient. Shear waves and CAT are important on a global scale because they ensure a distribution of frictional destruction of the energy of large-scale airflows through substantial volumes of the troposphere, rather than have it concentrated entirely in the turbulent boundary layer, or communicated from there by gravity waves.

Appendix 10.1 Evaporative loss of buoyancy

Consider a parcel of unit mass containing air, vapour and cloud, and suppose that some of the cloud water in the parcel evaporates, reducing the parcel's cloud content and increasing its vapour content. If the process is adiabatic, the heat needed to evaporate the cloud water must come from the parcel, chilling the mixture of air, vapour and cloud.

The heat needed to evaporate a mass m of water is simply the product of the mass and the specific latent heat L of evaporation. The fall $\triangle T$ in temperature is given by

$$\triangle T = (L\,m)/(\text{parcel heat capacity})$$

In full detail the parcel's heat capacity is the sum of the heat capacities of the air, the vapour and the cloud water, and is complicated by the exchange of water substance from water to vapour during the cooling. However, vapour accounts for such a small proportion of the total parcel mass (exceeding 1% only in the warmest and most humid parts of the troposphere), and cloud accounts for even less, that we can closely approximate the parcel heat capacity by assuming that the whole parcel is dry air, using the specific heat capacity C_p at constant pressure. This underestimates the actual heat capacity a little, since the specific heat capacity of cloud water is over four times, and that for vapour is nearly twice the value for dry air, but the small specific masses mean that the net underestimation is only a few percent at most. Note the use of C_p rather than C_v, since the former is relevant to all meteorologically relevant processes (sections 5.3 and 5.11).

Because we are assuming unit parcel mass

$$\triangle T = \frac{L\,m}{C_p} \tag{10.2}$$

Inserting values for L and C_p we see that the parcel temperature falls by about 2.5 K for each gram of cloud water evaporated in a kilogram parcel. Realistically we might expect to have a cloud water content of several g per kg, but on the edge of a cumulus this would evaporate only if the cloudy parcel were mixed with an unsaturated, cloudless parcel of much the same mass. The original calculation therefore still gives a reliable order of magnitude for the cooling effect. Since temperature excesses in cumulus are only ~ 1 K, evaporative chilling is clearly potentially very significant.

The negative effect of such chilling on parcel buoyancy is offset by two effects: the reduction of the burden of cloud water reduces the parcel density, and the increase in the relatively low-density vapour component (molecular weight only 18/29 that of air) decreases it further. A detailed examination of these effects on parcel density shows that they are equivalent to a warming of about 0.43 K for each g per kg cloud water evaporated at typical low troposphere temperatures, and slightly less at lower temperatures. The net effect on parcel density of all of these effects (chilling by evaporation, loss of cloud burden, increase of low-density vapour content), is therefore equivalent to a cooling of about 2.1 K for each g per kg evaporated, which is still very considerable, and readily accounts for the losses of buoyancy visible on the sides of cumuliform clouds.

The process described above is entirely reversible, showing that the condensation of 1 g per kg of water vapour to form cloud warms the parcel by 2.5 K and decreases its density as if by a simple warming of 2.1 K, with obvious relevance to the production of buoyancy by cloud formation.

The chilling or warming which respectively follow melting of ice cloud or freezing of water cloud can be found from eqn (10.2) by treating L as the latent heat of melting of ice (or freezing of water). The temperature changes are only about one seventh of those associated with evaporation or condensation since this is the ratio of the latent heats. There is no offsetting correction to apply because the vapour content is unchanged and the slight difference in density between water and ice has quite negligible effect on the parcel density as a whole.

Direct sublimation from ice to vapour involves an L which is the sum of the Ls for melting and evaporation at the temperature involved. Chilling by sublimation would therefore be especially destructive of buoyancy if the very small differences in vapour density between cloud and environment at the low temperatures involved (below − 25 °C) did not make the process very slow. Corresponding warming by direct condensation from vapour to ice, such as is envisaged in the Bergeron–Findeisen process, is correspondingly effective in producing buoyancy, and is quite fast for reasons discussed in section 6.11.

Appendix 10.2 Convective overshoot

Figure 10.33 represents an idealized version of the vertical temperature profile in the vicinity of the tropopause, in which an isothermal stratosphere surmounts a dry adiabatic high troposphere.

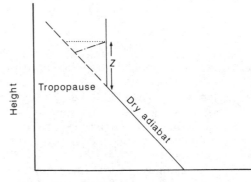

Fig. 10.33 Idealized temperature — height profile around the tropopause. Solid lines depict ambient temperatures and the dashed line traces an overshooting air parcel. Continued overshooting often maintains a temperature inversion below the isothermal stratosphere, enhancing the sharpness of the tropopause, and gradually raising its altitude.

Consider an air parcel in the updraught of a cumulonimbus which is neutrally buoyant as it rises through the dry adiabatic layer (remember that saturated and dry adiabats are nearly parallel in the very cold high troposphere). As it rises above the tropopause the parcel continues to cool dry adiabatically and so becomes colder than the surrounding stratosphere. At height z above the tropopause the temperature deficit of the parcel will be $z\,\Gamma_d$. According to Appendix 7.11 the resulting net downward buoyant force per unit parcel mass will therefore be $g(z\,\Gamma_d)/T$, which as expected increases with increasing penetration of the stratosphere. This force opposes the continued rise of the overshooting parcel and eventually brings it to a halt before sending it back down again.

Ignoring frictional and entrainment drag, the vertical component of the equation of motion is

$$\frac{\mathrm{d}w}{\mathrm{d}t} = -g\,\frac{z\,\Gamma_d}{T} \tag{10.3}$$

To solve this equation for the variation of w with z, note that

$$\frac{\mathrm{d}w}{\mathrm{d}t} = \frac{\mathrm{d}w}{\mathrm{d}z}\frac{\mathrm{d}z}{\mathrm{d}t} = w\,\frac{\mathrm{d}w}{\mathrm{d}z}$$

Rearranging eqn (10.3) to separate w and z terms and integrating, we find

$$\int w \, dw = \frac{g \, \Gamma_d}{T} \int z \, dz + A$$

Ignoring relatively slight variations in T, we can evaluate the integrals from the tropopause ($z = 0$, updraught w) to the level at which the speed of updraught has been reduced to zero (height z, $w = 0$), to find the height of penetration of the stratosphere in terms of the updraught speed w at the tropopause.

$$z = \left(\frac{T}{g \, \Gamma_d} \right)^{\frac{1}{2}} w = \left(\frac{T C_p}{g^2} \right)^{\frac{1}{2}} w$$

Substituting for C_p and g, and assuming a realistically low value of 220 K for T, we find (in SI units)

$$z \cong 48w$$

which would suggest that an updraught of 10 m s^{-1} should penetrate nearly 500 m into the stratosphere. This is bound to be an overestimate, since drag must tend to reduce the penetration, though the reduction cannot be reliably estimated. Observations of the tops of severe local storms indicate penetrations of more than 1 km on occasion, which implies updraught speeds greater than 20 m s^{-1}. After coming to a halt, the air sinks back to the top of the troposphere, and after some oscillation (damped by drag) settles there.

Appendix 10.3 Cumulonimbus tops

Figure 10.34 represents a sounding on a showery autumn day in north-west England. The showers were forming over a wide area of the northeastern Atlantic, as a cool north-westerly airstream over-ran the relatively warm sea (which is warmer there than even the western margins of the land throughout the winter). The sounding therefore portrays an air mass which is well adjusted to the showers it contains.

The cloud base at Aughton is very low, and the situation at low levels is complicated by a weak convectively stable layer which represents the influence of local land-based cooling (sunset was over an hour earlier). Over the slightly warmer sea, the lifting condensation level was probably as depicted. On the simplest adiabatic model of the showery convection, the rising air follows the saturated adiabat from the lifting condensation level, and is therefore warmer than the ambient air mass up to the level marked CT, where the saturated adiabat and environmental temperature profiles re-cross. On this model the air accelerates to produce a maximum updraught speed at CT, then overshoots and oscillates about that level before detraining there, as discussed in appendix 10.2.

In fact the inevitable entrainment must dilute rising air so that with increasing height it drifts to the cold side of the saturated adiabat, and widespread glaciation and consequent release of latent heat (Appendix 10.1 — a little earlier the evening sky was full of fresh and decaying anvils) must tend to reverse this trend in the

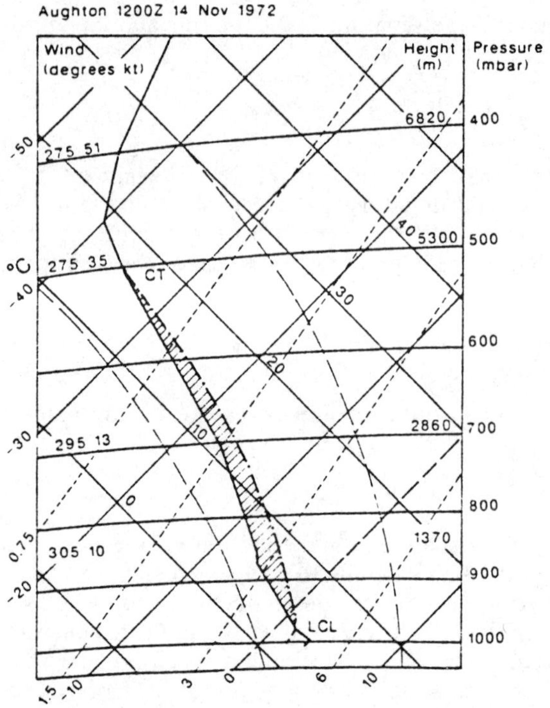

Aughton 1200Z 14 Nov 1972

Fig. 10.34 Tephigram for a showery day. The solid line traces the radiosonde temperature profile, and the dash-dotted line shows the construction used to estimate the altitude of cumulonimbus tops.

upper part of the convecting layer, with the result that the level predicted by the too-simple model agrees reasonably with observed heights of the tops of the highest cumulonimbus. The presence of both entrainment and frictional drag keeps the updraught speeds well below the frictionless values (given by the hatched area on the sounding and having value 20 m s⁻¹ on this occasion), and ensures that they are strongest in the upper middle of the convecting layer, because of the entrainment minimum there, rather than at the top.

Appendix 10.4 Oscillations in a stable atmosphere

Figure 10.35 depicts a convectively stable layer of air in which a parcel is displaced vertically and dry adiabatically from its original position of neutral buoyancy at O. According to Fig. 10.35 a raised parcel is cooler than its surroundings and consequently negatively buoyant, whereas a depressed parcel is warmer and positively buoyant. If a parcel is released from any initial displacement it will therefore accelerate towards O, overshoot, slow to a halt, accelerate back toward O again, and continue oscillating about O until the motion is damped out by drag, leaving the parcel at O. Let us examine the dynamics of such oscillation, slightly changing and generalizing the method used in appendix 10.2.

If the sub-adiabatic lapse rate of the stable layer is Γ, and the dry adiabatic lapse rate of the profile traced by the parcel is Γ_d, then the temperature excess of the parcel at any height z above O is given by $z(\Gamma - \Gamma_d)$. Since $\Gamma < \Gamma_d$, this expression shows that upward displacement (positive z) produces a temperature deficit, while downward displacement (negative z) produces a temperature excess, consistent with Fig. 10.35. Note that this is a more general form of the expression used in appendix 10.2, where Γ was zero since the stable layer was isothermal. Algebraic manipulation shows that the expression for parcel temperature excess can be simplified to $-z\,d\theta/dz$, where $d\theta/dz$ is the potential temperature gradient of the stable layer, and θ is the constant potential temperature of the parcel. There is only one temperature gradient in this form of the expression for the parcel temperature excess because the potential temperature gradient corresponding to dry-adiabatic parcel motion is zero.

We can use either of these expression for the parcel temperature excess to derive the buoyant force per unit parcel mass, as in appendix 10.2, and hence write down the vertical component of the frictionless equation of motion:

$$\frac{dw}{dt} = - \frac{g}{\theta}\,\frac{d\theta}{dz}z$$

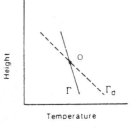

Fig. 10.35 Temperature–height diagram of vertical oscillation of an air parcel in a convectively stable environment.

Since $\dfrac{dw}{dt} = \dfrac{d^2z}{dt^2}$, the equation can be written in the form

$$\frac{d^2z}{dt^2} = -N^2 z \qquad (10.4)$$

where $N^2 = \dfrac{g}{\theta}\,\dfrac{d\theta}{dz}$

If the amplitude of the oscillation is small enough to allow the oscillating parcel to remain within a layer with uniform $d\theta/dz$, then N is a constant, and eqn (10.4) becomes a perfect example of the equation for *simple harmonic motion* — the idealized oscillatory motion which governs small-amplitude swinging of pendulums, and serves as the model for many natural and man-made oscillating systems. Standard procedure, or simple substitution, shows that solutions are of the form

$$z = A \sin Nt \qquad (10.5)$$

where A is the amplitude of the oscillation (the maximum distance the parcel moves above or below O), which, however, remains undetermined unless we define the parcel's starting speed, direction and height z, as was done in appendix 10.2. By contrast N is fully determined by the conditions already described (crucially by the convective stability of the ambient layer). According to the behaviour of sine functions, eqn (10.5) describes an oscillation which completes one cycle in a time period $2\pi/N$, which means in other words that the frequency of oscillation about O is $N/2\pi$. This is the *Brunt–Väisälä* frequency of adiabatic oscillation in a convectively stable layer, and it is apparent that the frequency increases with stability. Substitution of values shows that the isothermal layer considered in appendix 10.2 has a Brunt–Väisälä period of almost exactly 5 min. (Indeed it should now be clear that the analysis in appendix 10.2 was of one half of a Brunt–Väisälä cycle, with special interest in the amplitude rather than the frequency of the oscillation.) Less stable layers have longer periods (lower frequencies), tending to infinity as the layer approaches neutral convective stability (a dry adiabatic lapse rate).

Problems

LEVEL 1

1. Soaring birds and gliders are usually observed to be continually turning. Why do they need to do this?
2. Given that the top of a small cumulus rises at half the speed of the fastest updraught within it, estimate the typical speeds of such updraughts from the simple observation that many small cumulus grow to a depth of 100 m in 5 min.
3. Fractocumulus often appears at lower levels than any cloud observed before the arrival of the overlying nimbostratus. Suggest a reason for such behaviour.
4. In a certain north-westerly flow in the northern hemisphere, showers are observed to arrive in small groups separated by clear periods which range from 5 min to an hour. Interpret these observations in terms of the motion of a regular pattern of open cells, and deduce their breadth given that the pattern is moving at 15 m s^{-1}.
5. As a cumulonimbus slides overhead it becomes very dark. Where has the light gone?
6. An aircraft flies through a train of billows with a fairly uniform wavelength of 1 km at an air speed of 250 m s^{-1}. Given that the billows are moving with the ambient air, find the time period between successive jolts of the aircraft.
7. What is the minimum downward acceleration of an aircraft which can begin to throw passengers up out of their seats?

LEVEL 2

8. A certain cumulonimbus is moving at 20 km h^{-1}, and is raining at an even rate over a zone which is 2 km long in the direction of motion. Given that 5 mm of rain is measured by all gauges passed over by the full length of the shower, find the rainfall rate in mm per hour during the cumulonimbus' life.
9. Find the increase in temperature of air in which 1 g kg^{-1} of water substance behaves in each of the following ways: (a) water cloud condenses from vapour; (b) supercooled cloud water is frozen; (c) ice cloud condenses from vapour.
10. Using the tephigram in Figs 5.5 and 5.6 and referring to Fig. 10.15, trace the thermodynamic path of air moving adiabatically from a surface source in the Mexican Highlands (temperature 30 °C, pressure 900 mbar) to the 800 mbar level over the position X in the United States. Compare this with air which rises in dry and saturated convection from a surface source at X with surface pressure 1000 mbar and temperature and dew-point 26 and 23 °C respectively. What do you conclude about convection extending to greater heights?
11. Find the time period of vertical oscillation (the Brunt–Väisälä period) of an air parcel embedded in air with a temperature inversion of 1 °C per 50 m.
12. The core updraughts of some cumulonimbus are observed to overshoot the tropopause by 500 m, penetrating a low stratosphere which has a temperature inversion of 5 °C km^{-1}. Find the implied updraught speed at the tropopause in the absence of drag.
13. A large area of wave clouds is seen in the high troposphere by a satellite on a certain occasion, with a coherent wavelength of about 50 km apparent for

many wavelengths downwind from a hilly source region. Using the rule of thumb for lee wavelengths, estimate the wind speed in the high troposphere. How would this compare with a more stable atmosphere?

LEVEL 3

14. Consider the likelihood of a report of 8 octals of cumulus humilis being accurate.
15. Discuss in detail the role of condensation and evaporation in creating and destroying buoyancy in small cumulus.
16. Before the outbreak of a summer thunderstorm, the oppressive air has temperature 30 °C and dew point 24 °C. In the cool squall accompanying the main precipitation, these fall to 22 and 19 °C respectively. By following each up to the 700 mbar level on a tephigram (Fig. 5.5), at first dry adiabatically and then wet-adiabatically, find the implied temperature difference there. This is clearly a measure of the buoyancy driving the cumulonimbus. What surface observations give us this temperature difference most directly?
17. A certain cloudy air parcel is 1 °C warmer than ambient cloud-free air. If the cloudy and cloudless parcels mix in equal masses, and 1 g kg^{-1} of cloud water is evaporated in the process, find the resultant temperature of the mixed air compared with the ambient air. Find the effective temperature of the mixed air as far as buoyancy is concerned.
18. Discuss the microphysical processes at work in the upper parts of a cumulonimbus as it glaciates and produces an anvil, mentioning their relevance to buoyancy and precipitation.
19. Use the method of problem 6.20 to consider the growth of a giant hailstone of radius 5 cm in a cloud with 10 g kg^{-1} of water substance in collectable form, explaining how such stones can grow in a cloud which is seldom more than 12 km tall.
20. Describe and discuss the tendency of cumulus of all types to appear in non-random patterns.
21. A passenger aircraft flies from vertically static air into a downdraught of 20 m s^{-1}. Given that the aircraft adopts the new ambient motion in a horizontal distance of 500 m, and the plane is flying at 300 m s^{-1}, find its implied downward acceleration and describe the sensations and behaviour of the passengers. Given that the pilot finally zeros the aircraft's sink rate (by climbing through the sinking air) just as the aircraft flies out of the downdraught into a 20 m s^{-1} updraught, repeat your calculations and descriptions.
22. Surface air contaminated by the Chernobyl reactor fire reached Northern England on 3 May 1986 just as thunder showers were breaking out near the centre of a slow-moving weak depression. Assuming convective updraughts of 3 m s^{-1} 1000 m above the surface, use the method of problems 7.23 and 24 to find the associated convergence in this sub-cloud layer and estimate the reduction in horizontal area in 30 min. Consider the possible fate of contaminated aerosol particles in the converging layer.

CHAPTER 11

Large-scale weather systems in middle latitudes

11.1 Historical

Although people have known for centuries that the great storms of middle latitudes must be very large, much larger than can be appreciated from any single viewpoint on the Earth's surface, that knowledge has remained fragmentary and elusive until comparatively recently. While the European climate deteriorated from its thirteenth-century optimum, there were several storms which caused very widespread damage and loss of life by coastal flooding. Given the primitive communications, awareness that each of these disasters was caused by a single very large storm developed only in retrospect. The conspicuous fall of atmospheric pressure before the arrival of a gale was an added straw in the wind available from the mid-seventeenth century, but it took the journalistic effort and skill of Daniel Defoe to compile the first comprehensive account of the effects of a very severe gale, the great storm which struck southern Britain on 26 November 1703, and that account was completed a year after the event [71]. On 21 October 1743 Benjamin Franklin was prevented from seeing an expected lunar eclipse in Philadelphia by the sudden arrival of a north-easterly gale and associated cloud. He was surprised to learn later that the eclipse had been visible at Boston, 300 km to the north-east, and that the gale did not arrive there until the next day. His curiosity aroused, he wrote to people living along the connecting path and was able to show that the storm apparently moved from Philadelphia to Boston, against its constituent winds [72]. This is reckoned to be the first recognition that storms did not simply blow along with their own winds. During the next hundred years patient reconstruction of scattered observations of a few storms gradually revealed the great wheel-like (*cyclonic*) distribution of winds around a low-pressure centre which is now such a commonplace feature of weather maps, and confirmed that these wheels often moved as quickly as Franklin's storm. (Note that the term cyclonic is now used technically to mean wheel-like rotation anticlockwise in the northern hemisphere and clockwise in the southern, which is the observed sense of rotation around low-pressure centres. The opposite rotation, observed round high-pressure centres, is called *anticyclonic*.) Piddington and others concentrated such behaviour into an empirical law of storms which applied in low latitudes as well as high, and which worked well enough to allow sailors to plot a safe course in the vicinity of a potentially dangerous storm.

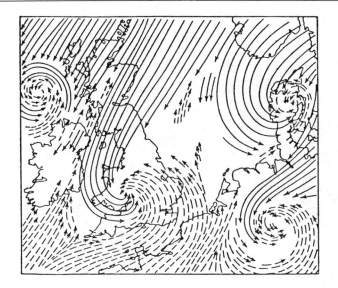

Fig. 11.1 Fitz Roy's picture of contrasting warm (dashed) and cold (solid) flows in cyclonic disturbances around the British Isles. (From [73])

Until the development of the electric telegraph in the middle of the nineteenth century, there was no point in trying to organize networks of people to observe and possibly forecast the movement of storms over extensive areas, because many storms moved and changed at least as quickly as the fastest available means of communication — the galloping horse. However, the arrival of the telegraph encouraged the formation of early synoptic networks (section 2.8), and quickly established and extended the picture of cyclonic storms, especially those found in middle latitudes. In Britain the young Meteorological Office flourished briefly under Admiral Fitz Roy, who years earlier had captained the Beagle as it carried Charles Darwin around the world. Fitz Roy was a good organizer and a perceptive observer (despite his fanatical opposition to Darwin's later theories), and Fig. 11.1 shows that he realized that the contrast between warm and cold *air masses* was an essential feature of the mid-latitude (i.e. extratropical) cyclone. There was a considerable delay in recognizing that these contrasts were concentrated in narrow bands of especially active weather, so that the modern picture of the extratropical cyclone and its associated fronts did not really emerge until the rise of a small group of gifted observers, analysts and theorists in the Norwegian meteorological service between 1910 and 1930. Cut off from many European observations during the First World War, they looked closely and perceptively at what observations they had and derived what has become known as the *Norwegian cyclone model* (Fig. 11.2).

Although the Norwegians made good use of visual observations of middle and high clouds to define behaviour throughout the troposphere, their model depended almost entirely on the surface synoptic network (section 2.9). But as aircraft became more numerous and adventurous, the need for upper-air observations grew. The Second World War triggered the establishment of radiosonde stations (section 2.10) on a wide scale, and these quickly confirmed the presence of a jet stream in the upper troposphere in the vicinity of each well-marked front, and indeed of the huge circumpolar vortices of fast westerly flow in the middle latitudes of each hemisphere — features which had been known about previously only rather haphazardly. The whole composite picture of structure in the low, middle and high troposphere was magnificently confirmed as soon as meteorological satellites began to operate in the early 1960s, but their capacity for high resolution over wide fields of view also revealed the presence of much meso-scale structure which had

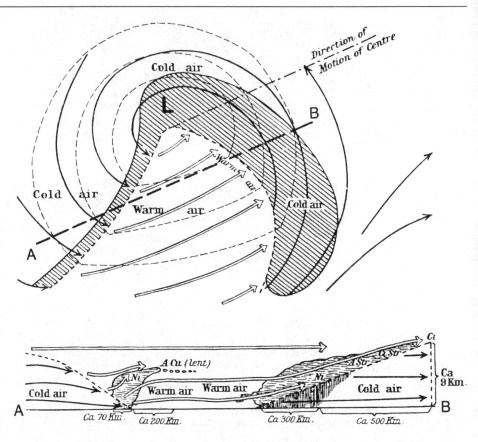

Fig. 11.2 The classical Norwegian cyclone model after Bjerknes and Solberg [73]. The top part is a plan view of surface conditions, with airflow, isobars, fronts, and areas of cloud and rain. The bottom shows a vertical section from A to B, showing flow in the plane of the section, cloud and precipitation. Subsequent amendments have added the familiar triangles and humps on the cold and warm fronts respectively, raised the tops of the deepest cloud masses, added jet streams aloft (Fig. 11.10) and altered cloud type abbreviations (see appendix 2.3).

slipped through the rather coarse synoptic-scale observation network. Despite much dedicated effort by meteorologists, the meso-scale structure revealed by satellites and weather radars is still rather poorly integrated with synoptic-scale structure encapsulated in the Norwegian cyclone model, but the synoptic-scale picture of events in both low and middle latitudes is now established beyond dispute.

11.2 Extratropical cyclones

ORIGINS

The weather map in Fig. 11.3 shows a typical situation in the northern Atlantic: a front is strung out from west to east across several thousand miles of ocean, and along it are several large weather systems in various stages of development, each marked by a wave-like distortion of the front. This is a *family* of extratropical cyclones, each of which has a characteristic common structure and life cycle. In fact since each moves eastwards along the extended front during its life cycle, the sequence from west to east on Fig. 11.3 corresponds to stages in that common cycle.

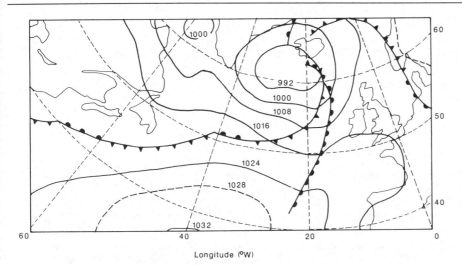

Longitude (°W)

Fig. 11.3 A family of depressions in the north Atlantic at 1200 Z on 6 July 1972. Like all actual cases it departs from the ideal, there being no really mature member between the old occluding system and the young waves to the west. Fronts in the far north-east are debris of the earliest members of the family. Notice the ridge of high pressure over the British Isles ahead of the warm front.

The front which links these depressions (using their familiar name) is called the *polar front* in the Norwegian terminology which has become universal, and is a region of relatively sharp temperature contrast separating the warm air of low latitudes from the cold air of high latitudes. Various factors can sharpen the background meridional temperature gradient imposed by the unequal distribution of solar input (section 8.7), distribution of land and sea being one of the most effective in the northern hemisphere. For example, in the north Atlantic in winter there is a particularly sharp contrast between the cold *polar continental* air in the interior of North America, and the warm *tropical maritime* air over the west Atlantic warmed by the powerful poleward flow of very warm water in the Gulf Stream. Although such air masses are clearly defined only in the lower troposphere (since the upper-troposphere westerlies flow freely by aloft) they are a useful descriptive concept, because they define surface atmospheric conditions, and because the strong thermal contrasts between them are observed to be favoured sites for *cyclogenesis* — the birth of depressions. The north-western Atlantic breeds depressions which drive across the Atlantic to arrive in a mature state in the western approaches to Europe, and the north-western Pacific and the northern Mediterranean (in winter) act similarly to spawn depressions is the mid-latitude westerlies (Fig. 11.4).

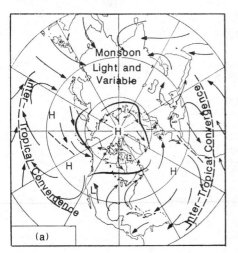

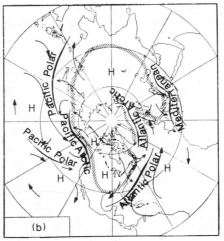

Fig. 11.4 Zones of cyclogenesis in the northern hemisphere in (a) summer and (b) winter. Note the northward migration in summer and the associated collapse of the Mediterranean zone, on which the local tourist industry depends. (After [72])

375

The Norwegian school believed it was the polar front itself which generated depressions through a wave-like instability, like a huge version of the waves on a rippling flag. Though current theory does not support this view in detail (section 11.6), it is such a persuasive interpretation that it is widely used as a descriptive device. Whatever the details, it is clear that cyclogenesis depends on the juxtaposition of contrasting air masses.

LIFE CYCLE AND STRUCTURE

The stages in the life cycle of a depression are shown in Fig. 11.5 — another classic diagram from the Norwegian school. In the first stage (a) the pre-existing front begins to develop a large elongated mass of deep stratiform cloud in response to persistent uplift throughout the troposphere. As low-level air converges, to replace the rising air, the planetary vorticity f is concentrated (section 7.14) and the air begins to rotate cyclonically, twisting the warm air polewards to the east of the centre and the cold air equatorwards to the west and sharpening the horizontal temperature gradients there, especially in the low troposphere.

These developments continue in the second stage (b), by which time large areas of continuous precipitation have appeared close to the deforming front. Satellite pictures (Fig. 11.6) clearly show the large area of middle- and high-level cloud, often with a spume of cirrostratus fanning north-eastward over the crest of the frontal wave. The rightward curvature of this high cloud shows that the flow there is twisting anticyclonically in strong divergence above the main uplift (Fig. 7.29). By now sea-level pressure is falling sharply near the wave crest as divergence in the upper troposphere removes air from the depression centre faster than convergence in the low troposphere replaces it.

Conventionally the fronts in Fig. 11.5 are marked in their surface positions. In fact the regions of strongest horizontal temperature gradient (the *frontal zones*) slope upwards at such very small angles (Figs 7.16 and 11.2) that in the upper troposphere they lie hundreds of kilometres poleward of their surface positions. As the depression develops, the frontal zones become more and more sharply defined, especially in the low troposphere to the west of the crest of the frontal wave. Temperature gradients ~ 5 °C per 100 km are observed where the cold polar air undercuts the warm tropical air at the *cold front*, forming a very sharp thermal boundary between the *warm sector* and the cold air to the north-west. The *warm front* acts as a weaker eastern boundary.

Development of the depression continues with further growth in amplitude of

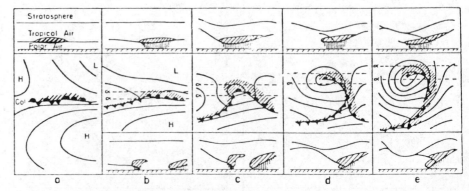

Fig. 11.5 The life cycle of an extratropical cyclone in the northern hemisphere as depicted by the Norwegian school. (Modified from [75])

Fig. 11.6 Infra-red satellite picture of young and old depressions in the eastern Atlantic. A young system still in the open-wave stage lies over Spain and France, moving north-east along the front trailing from the old occluded system whose comma-shaped cloud pattern is centred over Scotland. Notice the large meso-scale swirl of cloud to the west of Ireland. (1917 Z 14 December 1980)

the frontal wave and falling of the pressure minimum at its crest. Figures 11.2 and 11.5(c) are conventionally regarded as showing the beginning of the mature stage of the life cycle. The minimum surface pressure may now be tens of millibars below the undisturbed value, and surface winds may reach gale force over much of the area of cyclonic flow. Large masses of nimbostratus blanket the warm and cold frontal zones, giving substantial falls of rain or snow to the surface, depending on season and latitude. An extensive flange of altostratus and cirrostratus reaches

377

polewards and eastwards ahead of the warm front, enlarging the 'spume' first noted in Fig. 11.6. Associated with the well-marked warm and cold frontal zones, there are powerful cores of the westerly *polar front jet stream* in the high troposphere. The whole situation is described more fully in section 11.4.

In the early stages the meridional extent of the frontal wave (its north–south 'height') was smaller than its zonal wavelength. Now however the two are comparable and the warm sector is narrowing quickly, like an ocean swell steepening in shoaling water (but remember the cyclone 'wave' is *not* a vertical gravity wave like a swell). This narrowing leads to the *occlusion* of the warm sector, as the cold front overtakes the warm front and forms a composite or *occluded* front (Figs 11.5(d) and 11.7). During occlusion the still-deepening low-pressure centre moves polewards and westwards from its previously central position in the depression, drawing out the occluded front which connects it to the rump of the warm sector. The jet-stream core is still aligned with the trailing cold front to the west, but not with the occluded front, which it cuts somewhat poleward of the warm surface air (Fig. 11.18). This lack of alignment shows (through the thermal-wind relation) that the thermal contrast across the occluded front is much less than it was in the cold front. There is still some contrast however, since the fresh polar airstream west of the occluded front is often colder than the 'older' cool air ahead of it.

After occlusion the depresssion progressively loses the marked temperature contrasts which were such important features of its growth to maturity. In fact the centre of the circulation is now predominantly cold (giving the term *cold core*), and spinning well to poleward of the nearest warm air mass. The occluded front may continue to maintain widespread cloud and precipitation for several days, during which its poleward end may spiral round the low-pressure centre, producing a striking 'comma' shape on satellite pictures (Fig. 11.7) — about the only time that the low-pressure centre, such an obvious feature in analysed weather maps, appears clearly in satellite pictures. The cyclonic rotation in the low troposphere may persist similarly, maintaining strong winds for several more days, but it is like a vast eddy shed after vigorous formation, inexorably slowing through friction and filling as convergence below exceeds divergence aloft (Fig. 7.29).

FAMILIES OF DEPRESSIONS

The eastward and poleward motion of the depression from birth to occlusion is not well represented in Fig. 11.5. The open waves (a) and (b) in particular may move and develop rapidly in the prevailing westerly flow, testing the skill of forecasters, especially in areas of sparse data like the central Atlantic. Eastward motion continues through maturity but dies away after occlusion, leaving the frontless swirl of Fig. 11.5(e) turning about a nearly fixed centre. However, the residue of the cold front, which was such a prominent feature of earlier stages, lies like an uneven string unwound from the equatorward rim of the depression as it rolled eastwards. This residual front still contains considerable contrast between polar air flowing equatorwards down the western side of the depression and the tropical air at lower latitudes (often the poleward flank of a subtropical anticyclone). This thermal contrast may encourage further cyclogenesis, beginning to wave the front again as in Fig. 11.5(a) and passing through the same life cycle along the track of the trailing front to merge with its decaying parent after occlusion. The second wave in turn leaves a trailing front which can encourage the growth of a third, and so on through a family of six or more depressions, all decaying into

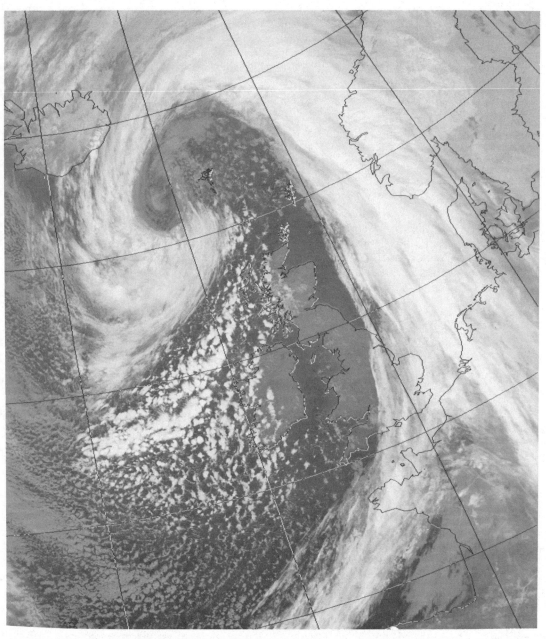

Fig. 11.7 Infra-red satellite picture of an occluded depression. The warm sector remnant lies over France, with an extensive occluded frontal cloud mass lying meridionally over the North Sea, southern Norway and the Norwegian Sea. The great westward hook of cloud surrounds the low-pressure centre at sea level, and lies to the north of the mottled pattern of shower clouds in the cool westerlies. (0910 Z 4 January 1980)

and re-invigorating the same final swirl begun by the founding parent. The sequence comes to an end when no further cyclogenesis occurs on the trailing front, often as a ridge of high pressure begins to build there, extending polewards from the local subtropical anticyclone. The old vortex, cut off from further invigoration, finally dwindles and dies, leaving little trace of the enormous weather systems which have troubled a considerable fraction of the middle latitudes for two weeks or more.

11.3 The mature depression

Let us examine in more detail the structure of a mature extratropical cyclone. As shown in Fig. 11.5 (c), the mature system has well-developed cold and warm fronts separating the warm sector from cold air to the east and even colder air flowing equatorwards to the west. The structure is summarized in the classical plan and vertical cross-section of the Norwegian Cyclone Model published in the 1920s (Fig. 11.2a and b).

PRESSURE AND WIND

The low-pressure centre lies close to the apex of the surface fronts, and the isobars form a large pattern with many closed isobars when they are drawn at intervals of one or two millibars, as is normal practice. The isobars curve round fairly smoothly in the cold air but run almost straight across the warm sector. The pattern of fronts and their associated weather usually moves parallel to the warm sector isobars — a useful rule of thumb in forecasting. The isobaric curvature absent from the warm sector appears to be concentrated at the two surface fronts, especially at the cold front where a deep kink in the isobars is often observed. These sharp kinks are hydrostatically consistent with a boundary between cold dense air and warm less dense air (Fig. 11.8) which slopes up from the position of the surface front. Moving horizontally from the surface front into the cold air, the increasingly deep layer of denser air produces a relatively sharp increase of surface pressure. In the presence of the overall fall of pressure towards the low-pressure centre, this sharp pressure gradient appears as a sharp clockwise turn of the isobars as we go from the warm sector into the cold surface air across the surface cold front. Of course there are no actual discontinuities of air density in the real atmosphere, but the boundaries can be quite sharp on the synoptic scale, particularly at the cold front.

Fig. 11.8 Horizontal pressure patterns near a front. The solid lines in (a) show the isobar pattern associated with uniform westerly geostrophic flow in the northern hemisphere. These change to the dash-dot pattern when we add a cold front as shown (b), to accommodate the associated wedge of denser air.

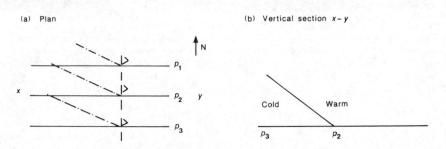

(a) Plan

(b) Vertical section x–y

Associated geostrophically with these kinks in the isobars, the winds too must turn sharply clockwise as we go across the surface cold front from its warm to its cold side. Clockwise rotation of wind direction is known as *veering*, whereas anti-clockwise rotation is termed *backing*. In the particular layout in Fig. 11.2(a) the geostrophic wind veers from azimuth 260° to about 310° in going across the surface cold front, which is roughly equivalent to a veer from a south-south-westerly to a north-westerly in compass directions. Of course surface winds are not accurately geostrophic because of friction (section 7.11), but although reduced and backed by the effects of friction, they too must veer sharply across the cold front. In fact the surface position of a vigorous cold front is often the site of vicious squalls from cumulonimbus embedded on the edge of the nimbostratus, in which the sharp veer can be concentrated into a few hundred metres on the edge of the squalls. This behaviour was well known to sailors in square-rigged sailing vessels, who often had to go aloft to reset the angle of the spars, and to reef in the larger sails, as the squally wind veer came upon them. Struggling with heavy wet canvas cracking like gunshot under the strain of winds gusting well above gale force while lashed by bursts of associated rain and hail (and lightning — see section 6.13) would be highly uncomfortable in any situation, but 30 m above a deck heaving in the confused sea produced by the rapid veer of wind, it was an unforgettable and highly dangerous experience which exacted a heavy annual toll of death and serious injury for the centuries in which such ships were widely used. By contrast the wind veer associated with warm fronts (as we cross from the cold to the warm side of the surface front) is usually smaller and more gentle. Such veers of wind and changes in pressure tendency are commonplace at fixed observation positions as fronts drive across from the west; and the sharp rise of pressure after the surface passage of the cold front is a useful telltale which often enables it to be positioned accurately on a synoptic chart.

CLOUD AND PRECIPITATION

Figure 11.2(b) represents a vertical section through the warm sector and its bounding fronts, along the line A–B in the plan view (a). Working westwards from its eastern boundary we have the sequence of observations and weather as it would be experienced at a fixed geographical location as the structure moves across from the west. Firstly there is fair weather in the little ridge of high pressure which usually precedes the warm front, with stunted cumulus if there is any convection at all. Then the outlying edge of the frontal cloud sheet is heralded by the appearance of hooked cirrus (mare's tails), produced by patchy formation of ice cloud in the high troposphere (Fig. 2.7). The motion and appearance of these clouds is quite informative. Their progression across the sky indicates the speed of approach of the depression while their individual movement along the overall pattern reveals the relative flow of the air in the high troposphere. According to the thermal wind relation (section 7.10) this air should move so that cooler air is on the left looking downwind in the northern hemisphere. The flow should therefore be out of the paper in Fig. 11.2(b), which is from right to left when viewed by an observer still ahead of the approaching front, watching the hooked cirrus as they invade the sky. This is readily confirmed by observation, since the mare's tails are usually unobscured by lower cloud. The curved hooks are formed as streamers of the larger ice crystals in the cloud fall into the slightly lower and slower air.

The hooked cirrus usually gives way to a film of cirrostratus which fills the high troposphere and slightly diffuses the sunlight (Fig. 11.9). The shallow layer of

Fig. 11.9 The 22° halo around the sun indicating the presence of a thin veil of cirrostratus which is often a forerunner of an approaching warm front.

highly regular, hexagonal ice crystals can produce by refraction a faint but striking halo around the sun, whose angular radius is 22° [76]. The cirrostratus thickens steadily as its bottom edge lowers, and by the time this has sunk down to middle levels (below the 7 km level technically) the sun's position is barely visible. The cloud layer now consists of cirrostratus and altrostratus, though quite often only the altrostratus is visible from the surface. Aircraft observations suggest that such thick layers are usually built up of several thin layers rather than a single thick one. When seen over hilly land, there are often darker patches where layers of moist air in middle levels form lenticular clouds in standing waves over the hills (section 10.8). Shortly after this the cloud thickens and darkens to nimbostratus and the precipitation begins. Nearer the surface position of the front most of the lower troposphere too is filled with cloud, so that there is a multi-layered sandwich of cloud filling most of the troposphere and producing gloom and moderate or heavy precipitation at the surface.

The precipitation consists of large ice crystals sifting down through the very gentle uplift of air in the front. If the air in the low troposphere is cold enough, no more than a degree or so above freezing, then the precipitation will fall as snow. This tends to happen in winter if the depression is at a fairly high latitude or in a continental interior in middle latitude, or of course if the surface is high enough above MSL. In a maritime climate such as Britain's, even in winter the precipitating snow normally melts at least several hundred metres above sea level and falls as rain on low ground. The large wet flakes in the melting zone produce a strong horizontal echo on weather radars (Fig. 6.27) known as the *melting band*. There are usually large quantities of supercooled water cloud in the lower troposphere, some of which is seeded by ice crystals precipitating from above to form meso-scale patches of heavier rain at the surface (section 10.7). Apart from these, precipitation in warm fronts tends to be more steady than in any other type of weather system.

As precipitation falls into the sub-cloud layer it tends to moisten the layer by continual evaporation. Although there is no sun in these conditions to drive convection, the strong winds and their associated shear maintain mechanical convection (section 9.8) in low layers. The tumbling masses of moistened air reach their lifting condensation levels here and there to produce the torn cloud fragments called *fractocumulus* (Fig. 10.4) which are characteristic of such conditions. Sometimes rain falling into cold air overlying the surface at the narrow end of the wedge of cold air near the surface front will saturate the cold air to produce *frontal fog*.

The vertical cross-section (Fig. 11.2b) is probably unrealistic in indicating that the upper surface of the warm frontal cloud mass descends toward the warm air, though such is the respect for the Norwegian model that the feature is reproduced faithfully in many texts. Aircraft and satellite observations show that the upper surface is fairly horizontal across the front, although there is some meso-scale sculpting which can cast dramatic shadows in oblique illumination. The frontal cloud mass is therefore wedge-shaped in section with maximum thickness close the surface front.

THE JET STREAM

The Norwegians originally had little upper-air data, and had to rely on visual observations from the surface. Though they knew that there were high winds in the upper troposphere near fronts, they had almost no direct measurements of them, and consequently did not emphasize this feature. However, the radiosonde network has changed all that, and the polar-front jet stream is now an integral part of the model, and one which is of great practical importance to the large amount of air traffic which is routed at these levels. The core of the jet stream is presented in Fig. 11.10 together with an outline of the patterns of Fig. 11.2. Notice that the jet cores lie in the warm air (although at these levels there is nothing like the thermal contrast typical of the low troposphere) roughly vertically above the positions of the frontal zones in the middle troposphere, as is to be expected from the thermal wind relation. Core speeds of as much as 75 m s^{-1} are quite typical in vigorous cold fronts. The hooked cirrus ahead of the warm front is moving on the eastern flank of the jet core there: hence the visible equatorward motion noted earlier.

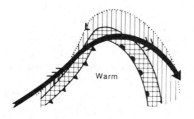

Fig. 11.10 An idealized picture of a mature extratropical cyclone (similar to that in Fig. 11.2) showing plan positions of the centre of low pressure (L), the warm and cold fronts, areas of stratiform cloud (hatched) and precipitation (cross-hatched), and the axis of the associated polar-front jet stream centred on the 300 mbar level, with barbs in regions of highest speed.

It is tempting to imagine from the appearance of Fig. 11.2(b) that the great wedge of warm frontal cloud is produced by air creeping up the frontal slope in the plane of the frontal cross-section. Though there certainly is some motion in that plane, it is much weaker than the components of motion perpendicular to the plane (i.e. parallel to the front) — the jet core being the extreme example. It is flow up the slight but very extensive slopes *along* the fronts which begins and maintains the production of cloud and precipitation. In the case of the cold front the air is climbing as it travels polewards (into the page in Fig. 11.2), whereas in the warm front the air in the upper troposphere is climbing as it flows equatorwards (out of the page). In fact the three-dimensional motion of the air is quite difficult to deduce accurately from observation, and is very complex insofar as it is known, as will appear in the next section.

WARM SECTOR AND COLD FRONT

Continuing with the westward section of Fig. 11.2, we have now reached the warm sector. This is often largely free of substantial cloud sheets, at least clear of the poleward extremity where stratus and even fog may form as the humid air is carried polewards over colder and colder surfaces (a classical example of advection fog). Being a tropical air mass, the warmth of the warm sector can be quite striking at and near the surface. In winter this warmth is an important source of heat for middle latitudes, and is an obvious example of the warm air advection considered more generally in section 8.12. In summer the warmth may become positively balmy when the sun adds to the intrinsic warmth of the air by heating the underlying land surface. However, on some occasions there may be much more than the quite typical scattering of small cumulus, even to the extent of there being patches of nimbostratus and cumulonimbus. Convective instability often increases on the western side of the warm sector (the North American squall line being an extreme example — see section 10.7). When winds are strong, as they often are especially in winter, the pleasant balmy feeling of the warm sector is spoilt for us, though the air temperature may be just as high. The temperature contrasts between the air before, in and behind the warm sector can obviously be masked or accentuated by local cooling or heating, especially overland. The dew-point temperature is a much more secure indicator of air-mass differences, since it is unaltered by simple isobaric heating or cooling (section 5.5). The dew point is highest in the warm sector, indicating its origin in the humid warmth of lower latitudes.

The cold front may arrive with great vigour, with heavy bursts of precipitation, squally winds and thunder, and the wind veer and pressure kick mentioned already. On other occasions it may approach much less dramatically, though the sky generally fills with cloud much more rapidly than it does with a warm front. Precipitation rates are characteristically higher and more variable than in warm fronts, consistent with there being quite vigorous cumulonimbus embedded in the nimbostratus, often in groups as mentioned in section 10.7, and with associated hail and thunder.

The cold frontal cloud mass is usually considerably narrower than the warm front because of the steeper frontal zone associated with the stronger temperature contrast in the lower troposphere (section 7.10). Its unpleasant weather therefore does not last too long, though note the different behaviour in young depressions mentioned below. The clearance at the back edge of the frontal cloud mass can be correspondingly much more rapid than its counterpart, the advance of the leading edge of the warm front, and occasionally offers a panoramic view of a nimbostratus edge stretching from horizon to horizon as it retreats from the observer (Fig. 11.11) — one of the few occasions when a surface-based observer gets a sense of the vast horizontal scale of large weather systems. Facing the retreating edge, small-scale structure such as aircraft condensation trails, etc. embedded in the upper part of the edge moves quickly from right to left with the jet stream there. The observer is now in the cold polar air behind the front, which is kept in a convectively unstable state by equatorward flow over progressively warmer surfaces.

Fig. 11.11 The clearance behind a cold front moving south (left) over the Irish Sea in late autumn. The overhanging flange of high and middle cloud reveals the backward slope of the frontal zone (Fig. 11.2, bottom). Small cumulus are growing in the cold air as it flows south over the relatively warm sea.

These north-westerlies (as they usually are) are therefore showery, with cumulonimbus hurrying along singly or in groups in the brisk flow of cool clear air. Sometimes there are larger areas of merged cumulonimbus or nimbostratus-like segments of cut-off occluded front which give short bouts of unpleasant weather, and of course it may become very cold in winter as the winds veer gradually to become northerlies.

DIVERSITY OF WEATHER

If this depression is not the last member of a family, the showers will gradually die away as they are suppressed by the ridge of high pressure preceding the warm front of the next wave depression advancing from the west. The observer is now about to repeat the sequence of observation and experience recounted above as another system moves across, but of course no two systems are the same, so that the next one may differ very considerably in intensity, path, speed, and maturity at time of passage. The orientation of the fronts relative to their direction of motion in particular can have a dramatic effect on the conditions experienced at any particular geographical location. In the mature state the fronts make quite large angles with their direction of propagation, with the result that they pass breadthwise across any location, and so do not last very long if they are moving quite quickly, as they usually are. However in their young state, when the frontal wave is still very open (see Fig. 11.5(b)), they tend to move nearly lengthwise, with the result that most of the adjacent surface is unaffected by cloud and precipitation while an unlucky stretch passes under almost the entire length of both the warm and cold fronts, experiencing therefore one or more days of nearly incessant rain or snow which may fade and re-intensify as the slightly buckled fronts meander laterally overhead.

A location in the vicinity of the great terminal vortex of the depression family, which may persist with little motion other than rotation for a week or two, may suffer conditions which have very little to do with the sequence associated with the mature depression. If the location is to the east of the centre of the old low, the winds are persistently southerly, and periods of rain and cloud may recur as the debris of old fronts pass over, with little thermal contrast or organized cloudy structure. If the location is to the west a persistent showery northerly flow is experienced, with longer periods of rain or snow as frontal debris swirls round from the east. Regular inspection of weather maps shows that a huge variety of conditions and sequences of conditions can arise depending on the orientation of the depression track. This has been assumed in Fig. 11.5 to be nearly westerly, but in fact it may have any orientation, though easterly motion is rare and usually sluggish. The classic pattern and motions depicted in Fig. 11.2 should therefore not be allowed to limit the actual range of behaviour, and indeed the Norwegians themselves were most careful and flexible in applying their model to the analysis of actual situations, as are all professional forecasters.

Note that the descriptions in this and the previous section have been written for the northern hemisphere. It is a useful exercise to re-examine them to see how they should be changed to apply to the southern hemisphere, bearing in mind that the direction of map rotation about the local vertical, and all related phenomena, reverses there. Figure 11.12 outlines a depression and its fronts in southern hemisphere mid-latitudes.

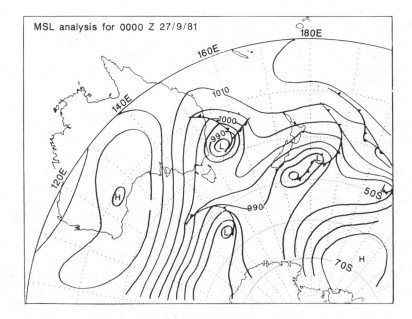

Fig. 11.12 Sea-level pressure at 0000 Z, 27 September 1981, showing a quadrant of the southern hemisphere with high-pressure zones in the subtropics and high latitudes and depressions in mid-latitudes. Airflow is anticlockwise round the anticyclones and clockwise round the depressions.

SEA WAVES AND STORM SURGES

The great wheel of strong winds of a mature depression can have obvious and familiar effects on underlying surfaces. Trees, buildings and crops are liable to damage in the more vigorous storms, the damage increasing rapidly as the average winds rise above gale force (about 19 m s⁻¹) and the strongest gusts rise much higher. The sea surface too becomes increasingly disturbed, with waves increasing in height, wavelength and celerity as the winds strengthen, and more and more crests break into seething turbulence. This response of the sea, and the difficulties it posed for mariners, provided early motivation for the development of meteorology, but though modern ships are much better able to withstand an angry sea than the relatively frail and complex sailing ships of earlier times, they are by no means immune. At the end of January 1953 a vigorous depression was passing to the north of the British Isles. The strong winds raised mountainous seas in the narrow North Channel between Scotland and Northern Ireland just as the car ferry *Princess Victoria* was making her way westwards. An unusually large wave struck the vessel's stern and buckled the car loading door there. The engine room flooded and the ship drifted helplessly to the south-west, calling for assistance. Sea conditions were so bad that despite the best efforts of lifeboats and several searching ships, the ferry was still alone several hours later when she foundered in broad daylight within a few miles of the nearest coast with heavy loss of life. The coxswain of one searching lifeboat claimed that his boat was thrown bodily out of the water by the steep high waves typical of storms in confined seas.

Coastlines are very significantly shaped by short periods of unusually violent waves, and the scouring currents they indirectly drive, and this role of the great storms has often exposed unwisely sited coastal settlements to discomfort, damage and even disaster. Some of these difficulties also arise from an interaction between wind and water which is much less obvious, but even more potentially damaging, than the great wind-driven waves of a storm. This is the so-called *storm surge*.

As a vigorous depression moves across a body of water, the sea over a considerable area may respond in ways which produce substantial alterations to the water level expected on the basis of the closely predictable astronomical tide. There are two types of response: a static one in which the sea surface behaves like an inverted water barometer, and a dynamic one in which shallow water in confined zones is set in vast oscillation by the moving and changing tangential wind stress. The static effect is relatively simple in principle. If an area of sea surface is subject to lower atmospheric pressure than its surroundings, then the surface there will rise up and the surrounding surface sink down until there is uniform total pressure on the highest horizontal surface which is below the water surface everywhere. If the area of low pressure is much smaller than the surrounding area of normal atmospheric pressure, then we can simply concentrate on the elevation of water in the region of low atmospheric pressure, and ignore the surrounding compensation. We can use the hydrostatic approximation in the water to find the elevation of the sea surface which accommodates the fall in atmospheric pressure. It is easy to show that because the density of water is 1000 kg m^{-3} a fall of 1 mbar in atmospheric pressure should support an elevation of almost exactly 1 cm in water level. Since minimum pressures in vigorous systems can be depressed more than 50 mbar, we see that water levels can be raised by more than half a metre — a significant value to those in charge of coastal defences, and likely to cause increasing concern as average sea levels rise because of global warming (section 4.10).

The dynamic effects can be even larger. If, for example, a northerly gale blows into the constricted North Sea between North Europe and the British Isles, such as may happen as a vigorous depression moves eastwards over Scandinavia, then the large area of southward tangential stress sets up an insistent water movement which may subsequently affect the whole region of coastal waters down to the coast of France. The water motion is actually a ponderous wave of extremely long wavelength, like the slopping movement of water in a shaken wash-basin, and like that slopping it can become very large if there is resonance between the imposed shake and the natural period of the water basin. Such time periods are many hours long in substantial water bodies like the North Sea, and in that time the turning of the Earth imposes a movement to the right in the northern hemisphere, just as it does on large-scale atmospheric flow. The result is that in the case of the North Sea, the

Fig. 11.13 Heavy seas at
Morecambe (north-west
England) during the storm
surge of 26 February 1990,
which caused the damage
shown in Fig. 4.16(b) and
extensively flooded Towyn
(North Wales). The huge burst
of water is caused by collision
between incident and
reflected waves, from which
high winds whipped water
inland over the sea wall and
road.

surge of highest water moves south along the east coast of Britain before swinging anticlockwise around the Belgian, Dutch and Danish coasts. The height of the surge varies considerably from place to place along the coast because of subtle interactions with coastal shape and bottom topography.

Just such a storm surge occurred on 1 February 1953, as the vigorous depression whose gales had sunk the *Princess Victoria* on the previous day moved eastwards toward Norway. The surge concided with the peak of a high spring tide as it moved down the east coast of Britain. The extra 2–3 m of water topped by an angry sea breached the sea defences of East Anglia and the Thames estuary, before going on to inundate an appreciable fraction of Holland, drowning several thousand people. The devastation and loss, and the realization that it could easily happen again, spurred great improvement to sea defences, including thirty years later an adjustable barrage in the Thames to protect the heart of London, and a whole range of investigation which has made the February 1953 storm surge the most intensively studied event of its type, though much less destructive than the cyclone surges of the Bay of Bengal (section 12.4).

11.4 Three-dimensional airflow

Let us consider the paths (*trajectories*) traced out by air parcels as they move in large-scale weather systems. Horizontal winds are so much stronger than vertical ones that you might suppose that motion is effectively horizontal, and that we can find the trajectories of moving air parcels simply by examining conventional horizontal or isobaric weather maps and using the isobars or contours as streamlines of flow. For example, air moving in the low troposphere around the equatorial side of a mature depression would be expected to describe a cyclonically curved path determined by the combination of the instantaneous pattern of isobars (as in Fig. 11.2) and the rate of eastward translation of that pattern. However, two points should make us wary of such an approach: air approaching a front near the surface obviously cannot pass through it as isobaric motion apparently requires, since no available process is able to convert cold air into the warm moist air or vice versa; and in the vicinity of fronts especially, wind speed and direction vary so strongly with height that quite small height variations could transform the shape of air trajectories (compare the north-westerlies near the surface just behind the cold front in Fig. 11.2 with the south-westerly jet stream only about 10 km above the same location).

Large-scale vertical motion would indeed be insignificant if the atmosphere had comparable vertical and horizontal scales, but in fact the atmosphere is so grossly flattened that the weak vertical movements are just about as important in relation to the shallow depth of the troposphere as are the relatively strong horizontal winds in relation to its considerable breadth. For example, a typical updraught in a front is estimated below to be about 10 cm s^{-1}. At this rate it would take an air parcel about 28 h to rise 10 km (i.e. through most of the depth of the mid-latitude troposphere), which compares very closely with the time taken for an air parcel moving at 20 ms^{-1} to traverse a system with horizontal dimension 2000 km. It seems that large-scale weather systems arrange themselves so as to equalize the time periods involved in vertical and horizontal traverse by constituent air parcels, which is a feature which distinguishes the genuinely large-scale system from a large conglomeration of small-scale systems, such as an extensive field of cumulus and cumulonimbus for example. In the present context this means that relatively slight vertical movements are just as important as the much more obvious horizontal ones if we wish to examine the three-dimensional motion of air in large-scale weather systems.

The large-scale vertical motions (remember there are embedded small-scale movements as well — section 10.7) in large weather systems are very small indeed, and we must consider briefly how they can be determined. It is quite inconceivable that a rate of rise of a few centimetres per second could be measured by a vertical anemometer several kilometres above the surface in the disturbed conditions typical of an active front. In fact such values have to be estimated by indirect methods, one of the most physically obvious of which uses observed background rates of rainfall. Warm fronts in particular produce fairly steady precipitation rates for long periods over extensive areas, in which much more vapour is condensed and precipitated than is instantaneously present in the full depth of the troposphere. It follows that there must be a reasonable balance between the upward flow of water vapour in the rising air and the downward flow of condensed water in the precipitation. As shown in appendix 11.1, a typical background rate of rainfall (3 mm per hour) is consistent with an uplift speed of just under 10 cm s^{-1}, assuming a realistic vapour content in the low troposphere. This is a useful order-of-magnitude estimation, but it assumes too much to be used accurately in any particular case.

Dynamic reasoning can be applied to observations of large-scale convergence and divergence etc. to calculate the associated large-scale rates of rise or sink of air (implicit in section 7.13 and Fig. 7.29), with similar results. However, this requires a subtle approach if it is not to demand impossibly high precision from the basic observations. Such methods underlie the forecasting of rates and amounts of precipitation, though it is only quite recently that numerical forecasts have become sophisticated enough to provide usefully accurate predictions of rainfall by such dynamic methods. However, a thermodynamic method has been used from time to time in recent decades to gain glimpses of large-scale vertical air motion, and with it a picture of the three-dimensional flow in large systems. Though used for research rather than routine forecasting, it is relatively straightforward, at least in principle, and depends on the assumption introduced in section 5.6, that vertical motion in large-scale weather systems is sufficiently fast for the procession of air parcel states to be nearly adiabatic (often termed *isentropic* in this context since entropy is conserved in an adiabatic process). When the air is unsaturated the process is nearly dry adiabatic, and when the air is cloud-filled the process approximates to a saturated adiabatic. These assumptions have already been used in the analysis of cumulus convection, where the air trajectories are effectively vertical, but now we are applying them to air motions which are very slightly but extensively sloped.

In the procedure of *isentropic analysis* an array of radiosonde data from the weather system under examination is analysed to pick out winds and other observations on each of several surfaces of constant potential temperature (isentropic surfaces). The most generally conservative label to use is the wet-bulb potential temperature (section 5.11), though in practice the ordinary potential temperature is used in cloudless parts. Figure 11.14 outlines a simple vertical cross-section through isentropic surfaces. Higher values of potential temperature correspond to greater heights at any location, otherwise the air would have a lapse of potential temperature and be explosively unstable (section 5.12), and isentropic surfaces slope upwards towards the cold air in the presence of a horizontal temperature gradient, since potential and actual temperatures on an intersecting isobaric surface must fall towards the cold air.

Fig. 11.14 Idealized vertical section showing how sloping synoptic-scale flow can produce a layered vertical profile at X, as observed by radiosonde or equivalent. In reality, such flows are always three-dimensional, as shown in Fig. 11.15.

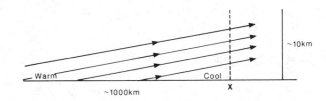

The horizontal flow (the winds) on an isentropic surface can be assumed to represent the flow in the tilted plane of the surface because the angles of tilt of such surfaces are observed to be very small ($< 1°$) in large-scale weather systems, even in the vicinity of fronts. However, it would usually be misleading to use the instantaneous winds directly as streamlines of flow since the isentropic surfaces in most weather systems move with much the same speeds as the systems themselves ($\sim$ 10 m s^{-1} in the case of many depressions). Just as in the analysis of flow in a supercell storm (section 10.6), the system velocity is therefore subtracted vectorially from each observed wind velocity to obtain the wind velocities relative to the moving weather system, and provided the system is not changing intensity or shape too rapidly, these can be used to draw streamlines of airflow in the travelling system. The provisos are obviously met best in mature systems, and as a result it is mature systems which are usually investigated by such isentropic analysis.

Figure 11.15 depicts the typical airflow in a mature depression, as revealed by isentropic analysis of many systems in North America and the north Atlantic [77]. The major feature is the extensive flow of air upwards and polewards in and ahead of the advancing cold front, rising from the low troposphere in the warm air deep in the warm sector, and turning gradually eastwards in the high troposphere in the forward and poleward flange of the warm front. This has become known as the *warm conveyor belt* because the air seems to be rising along an invisible ramp fixed in the moving system, and because this is the warmest of a number of such flows. Since the air entering the conveyor belt is quite moist, it does not travel far up the slope before it reaches saturation and fills with stratiform cloud. The air continues rising, but rather more steeply now that it is moving on a saturated adiabat, and it soon becomes part of the nimbostratus deck of the cold front. The crowding of the

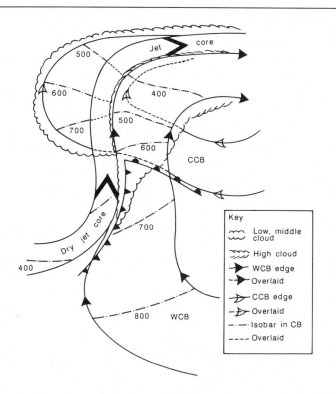

Key
- ∿∿ Low, middle cloud
- ⟩⟩⟩ High cloud
- ▶ WCB edge
- ▷ Overlaid
- ▷ CCB edge
- ▷ Overlaid
- —·— Isobar in CB
- ---- Overlaid

Fig. 11.15 Major sloping airflows and cloud decks associated with a mature depression. WCB and CCB are respectively the warm and cold conveyor belts. Intersections of climbing flows with isobaric surfaces are labelled in millibars. The coldest air in the system is descending equatorwards off the western edge of the diagram. (After [76])

isobars indicates its relatively rapid ascent through the lower middle troposphere (around the 750 mbar level), where the endless production of cloud maintains the background precipitation of the cold front. Ascent continues into the upper parts of the warm front, contributing to some of the precipitation from upper levels and maintaining the anticyclonically curving plume of cirrostratus noted in Fig. 11.6. Notice that the updraught speed of air at any position on the trajectory is given by the product of the relative wind speed and the trajectory slope. Typical values in the middle troposphere are 20 m s^{-1} and $1/100$, giving an updraught of 20 cm s^{-1}, which is consistent with values found by other methods already mentioned, bearing

Fig. 11.16 Haze top at the base of the subsidence inversion of a summer anticyclone, made especially stark by urban pollution (Montreal is virtually invisible under the haze) and by being viewed horizontally from a descending aircraft.

in mind that in the present method we are concentrating on just one slice of the rising air, and probably the most rapidly rising one.

In the low troposphere ahead of the warm front, air ascends in the *cold conveyor belt*, initially parallel to the warm front, but then swinging round in a grand anti-cyclonic arc which sometimes produces a pronounced flange of middle-level cloud poleward and westward of the apex of the warm sector. The north-westward flow of the cold conveyor belt along the warm front might seem to disobey the thermal wind relation, which requires that air flows south-eastwards there, but remember that the flow in Fig. 11.15 is relative to the eastward-moving depression, and that relative westward flow will occur whenever the air ahead of the warm front moves eastwards more slowly than the front itself. Since the front and the whole system tends to move more or less in step with the air in the middle troposphere, the slower-moving air in parts of the low troposphere is being overtaken and appears as relative easterly flow in the pattern of relative flow. The cold conveyor belt contributes particularly to the production of cloud and precipitation in the lower-middle troposphere of the warm front. Other features of the flow in Fig. 11.15 are the largely horizontal flow of air in the high troposphere, catching up rapidly with the system and sinking slightly as it flows from the north-west, and then swinging cyclonically and rising a little to join the north-western flank of the south-westerly jet stream associated with the cold front. Notice that the air in the middle troposphere well behind the cold front is sinking and turning cyclonically as it flows to lower latitudes. This is part of the equatorward sink of cool air which, together with the poleward rise of warm air, accomplishes the meridional and vertical exchange of air which is required for hemispheric energy balance (section 8.12).

It is clear from this brief examination that the flow pattern in a depression is quite different from what might have been deduced from a series of isobaric or horizontal charts. In particular it is more complex and asymmetric, with the cold front looking simpler and more fundamental to the basic scheme (the meridional and vertical exchange of air) than the warm front. In fact the pattern is inescapably three dimensional, which greatly complicates any attempt to portray it on a two-dimensional page. The presence of gently sloping flows is obviously consistent with the term *slope convection* used in Chapter 8 to describe this vast and structured combination of convection and advection.

11.5 Anticyclones

Even casual barometric observation in mid-latitudes shows that fine weather is associated with high (especially rising) pressure, just as wet and windy weather is associated with low (especially falling) pressure. By the early nineteenth century it had become clear that high surface pressure occurs in the middle of anticyclonically rotating air masses (section 7.17), and the name *anticyclone* began to be used. Although much less intense than cyclonic weather systems, anticyclones are nevertheless distinctive weather systems in their own right, with characteristic structure and behaviour, and several variants of the basic type. The basic features of most types of anticyclone, including the slowly subsiding air which fills its broad core, and the subsidence inversion which separates this from the local convecting boundary layer, are most clearly seen in the subtropical anticyclone, and are fully described in section 12.1.

In middle latitudes the *ridges* of high pressure which separate members of a family of depressions can be regarded as relatively narrow poleward extensions of the subtropical high along whose flanks they apparently slide (Fig. 12.1). As seen from a fixed location, the poleward limb of subsidence stunts and then suppresses the showers in the cool air behind the previous cold front, often giving a few hours of cloudless sky before the advance of hooked cirrus from the west heralds the approach of the next warm front (Fig. 11.3). In the summer in particular, the associated brief sunny spell, combined with the fresh, clean low level air, can produce some of the most pleasant weather experienced in the industrial areas of Britain and Western Europe, now that more persistent anticyclonic weather is so often marred by thick industrial haze in the surface layer. But the effect is always short-lived, as the ridge moves on in step with the preceding and following members of the family of cyclonic systems.

BLOCKS

Sometimes an anticyclone may form in the mid-latitudes well to poleward of the normal position of the subtropical highs. The development is associated with a temporary halt to the normal eastward procession of depressions in middle latitudes, with the result that it has become known as a *blocking high* or block.

Once established, a block may persist virtually motionless for days and even weeks, causing large departures from the seasonally typical climate over a wide area, as mentioned in section 4.9. In winter in particular, the presence of a pronounced, low-subsidence inversion can trap smoke and other pollution from large industrial regions and increase surface concentrations of noxious materials to the point of irritation and even danger. Combined with a deck of stratocumulus thick enough to reduce considerably the weak winter sun, the result has been appropriately nicknamed *anticyclonic gloom*. The situation can be considerably worsened if fog forms, as discussed in section 9.12. In summer the enhanced convection tends to maintain deeper convecting layers, but encourages the photochemical reactions underlying the Los Angeles type of smog (Fig. 11.16 and section 12.1).

Climatic anomalies can be very marked in the vicinity of a block. For example, the whole of the British Isles lay under and slightly to the east of a block throughout January, February and part of March 1963, and the combination of loss of warm air advection, absence of thick cloud cover (which normally shields the surface from cooling through the long winter nights), and the advection of very cold air from the interior of Europe, reduced the monthly mean temperatures over much of southern Britain by about 4 °C. Populations of small birds were decimated, transport and industry were disrupted, and the eastern sides of evergreen shrubs bore brown scorch marks for a decade afterwards. In summer the presence of a block is likely to lead to unusually high temperatures, since solar warming through the long, bright days far exceeds the net long-wave cooling. In July and August 1976 the British Isles lay under a block which came at the end of a period of more than a year in which rainfall had been consistently well below the normal because of an unusually high incidence of blocking. With reservoirs not fully restocked by the rather dry winter, and the prolonged excess of evaporation over precipitation in the hot, dry summer, Britain's normally very heavy domestic and industrial consumption of water could not be sustained, and an unusually severe drought ensued. Variations in the frequency of blocking are believed to underlie some of the climatic variations observed in recent centuries and decades (section 4.9).

CONTINENTAL HIGHS

Anticyclones develop regularly in winter over continental interiors, the most pronounced being the Siberian High which extends over Siberia and northern asiatic Russia, and influence a considerable portion of the Asian continent in winter. These are known as *continental highs*, but though they exhibit the usual anticyclonic circulation in the low troposphere (as they must to be geostrophically consistent with the pressure pattern) they differ otherwise from the other types of anticyclone. In particular they are relatively shallow systems, rarely extending their anticyclonic circulation above the middle troposphere. Radiosonde profiles through the centres of continental highs show that air temperatures are substantially lower than those of the surrounding air masses throughout the low troposphere, rising only slowly from the very low temperatures of the snow-covered land surface. It follows that, in contrast to the majority of warm-cored types of anticyclone (page 147), the pressure excess in the centre of the cold-cored type falls rapidly with increasing height.

In early winter, the cooling of the land is greatly accelerated by the first snow falls, which practically eliminate the solar warming while maintaining the net loss by terrestrial radiation. The surface and overlying air cools progressively day by day, as in an extended night. The whole mass of overlying air sinks with the shrinking layer of cooling air near the surface, and it seems that the air aloft is able to converge faster than the air near the surface is able to diverge, with the result that the surface pressure rises. After being set up, the surface pressure remains high despite a dynamic equilibrium between convergence and divergence, and as in the case of the other types of anticyclone, this is not easily explained. Despite persistent subsidence, the absence of an underlying convective boundary layer prevents the appearance of anything like a subsidence inversion. In fact there is simply a relatively smooth inversion which can extend from the surface for kilometres into the troposphere, once the high has become well established (Fig. 11.17). The subsidence is sufficient to keep the low troposphere free of cloud, which has the effect of enhancing the surface cooling and making the anticyclone self-maintaining. Cloud and precipitation can occur in the unaffected middle and upper tropo-

Fig. 11.17 A vertical temperature profile in the Siberian winter anticyclone (Yakutsk, 62 °N, 129 °E), showing a very strong and deep surface-based temperature inversion. The standard British tephigram has had to be extended to include the very low temperatures observed in the low troposphere. (After [78])

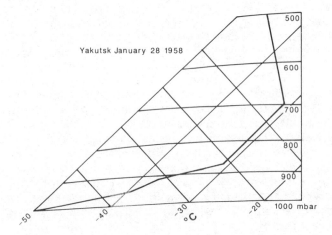

Yakutsk January 28 1958

sphere, but the cyclonic weather systems which would be the primary source of cloud are discouraged by the inert mass of cold air replacing the normal boundary layer, so that continental interiors tend to be largely cut off from the westerlies for much of the winter. The persistence of continental highs throughout much of the middle and latter parts of the winter in Siberia gives rise to extremely low mean surface temperatures, which are lower even than Antarctic values when accentuated by local topography (section 9.11).

11.6   Waves in the westerlies

The great mid-latitude weather systems look like flat waves on weather maps, waves which almost always travel eastwards as they develop. The Norwegian school recognized this, and tried to look for their origins in instability of the Polar Front, similar to the Kelvin–Helmholtz instability which makes a flag flap in response to a difference in wind speed on either side of its fabric (section 10.8). That particular theoretical approach made little headway against the many subtleties which make large-scale atmospheric waves among the most complex phenomena to be treated by fluid dynamics (probably second only to turbulence in this respect). It is beyond the scope of an introductory text to dwell on such matters but, for example, careful analyses of weather maps show that the wave-like disturbances of the great midlatitude weather systems usually move at different speeds at different altitudes, so that it is not strictly possible to describe any one system as a single wave. However, the general idea of treating such weather systems as waves (though not waves on a narrow frontal zone) has proved very fruitful since the late 1930s, as perusal of modern texts on dynamical meteorology will confirm [44]. In this section we will briefly consider some of the most obvious properties of large-scale waves in the westerlies which have emerged from such studies.

CYCLONE WAVES

Previous treatment of the mid-latitude depression has given many examples of its wavy structure, especially in its formative stages (Fig. 11.5). This is especially obvious when isobaric contours are plotted in the middle and upper troposphere, and indeed a sequence of such pictures of a developing depression creates an irresistible impression of an amplifying travelling wave (Fig. 11.18). Theoretical treatment of the dynamics of the baroclinic westerlies by J. Charney and E.T. Eady [80] in the late 1940s showed that they are unstable in a way which leads to the formation of waves which, at least in their early stages, strongly resemble those seen in weather maps of the middle and upper troposphere.

Fig. 11.18 A sequence of simplified pictures of flow at the level of the polar-front jet stream during the development of a mid-latitude depression. Sequences at all levels down to the lower troposphere similarly show an amplifying wavy disturbance of mainly westerly airflow. (After [79])

1   2   3

395

The theoretical methods used are subtle and complex in mathematical detail, but in essence they explore the stability of the initial unwaved flow by imposing very small-amplitude meridional perturbations and looking for subsequent development. They show that wavy perturbations grow exponentially, and at an initially realistic rate, in a range of wavelengths (i.e. west-to-east separations of adjacent troughs or crests) centred on the few thousand kilometres confirmed by inspection of weather maps. One of the most important destabilizing factors is the vertical shear of zonal wind shear in the initially undisturbed flow. When this exceeds about 1 m s⁻¹ per kilometre, a range of unstable wavelengths opens up and widens with increasing shear. Now according to the thermal-wind relation (section 7.10) such a shear is associated geostrophically with a layer-deep temperature gradient across the flow, i.e. with the north–south baroclinity of the zonal flow, and for this reason this type of instability is known as *baroclinic instability*. In fact you can quickly show that a shear of the critical value is associated with a horizontal temperature gradient of about 1 °C per 400 km in middle latitudes, or about 5 °C across the middle 20° of latitude. This is considerably less than the temperature gradient maintained by the meridional gradient of net radiative warming, so that it appears that the troposphere in its normal state is baroclinically unstable in middle latitudes.

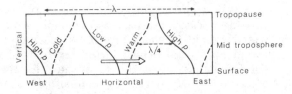

Fig. 11.19 A vertical zonal section through the zonally displaced axes of waves of temperature and pressure (and therefore flow) in Eady's model of the troposphere of a developing depression. The arrow shows the direction of wave propagation. (After [80])

As already mentioned, the details of the theoretically produced waves have certain observably realistic features, one of the most important of which is the offset or phase shift between the waves in the patterns of temperature and flow. Figure 11.19 is taken from Eady's treatment, and clearly shows that the crest of the thermal wave (i.e. the north–south axis of the warmest air) lags behind the crest of the wave in the mid-tropospheric isobaric contours (which represents the flow pattern there through the geostrophic relation), by one quarter of a wavelength. This has the important consequence that warmer air is being carried polewards and cooler air is being carried equatorwards, so that the net effect of the system is to contribute to the poleward advection of heat required to balance the meridional heat budget. This phase shift is observable in many ways in the real atmosphere. It is directly observable in pictures such as Fig. 11.6, where the great plume of cloud indicating the presence of the warm conveyor belt pours north-eastward on the western side of the wave in the middle tropospheric flow. It is also indirectly observable in the westward shift of wave axes with increasing height (Fig. 11.20) which appears in upper air charts of isobaric contours. For example, the presence of a deep layer of warm air to the west of the sea-level ridge of high pressure means that the isobaric surfaces aloft are domed upward over the warm layer so that the axis of the ridge in the middle and upper troposphere lies hundreds of kilometres westward. And the deep layer of cold air to the west of the low-pressure centre at sea level similarly means that the axis of low pressure is shifted westward aloft. This has been a familiar feature of upper air charts since they were first drawn routinely, and was noticed but not recognized by observers in the nineteenth century who reported that barographs from stations on mountain tops (which were then quite popular) lagged behind those nearby at sea level in indicating the passage of a pressure trough. The westward tilt of the axes of both theoretical and observed

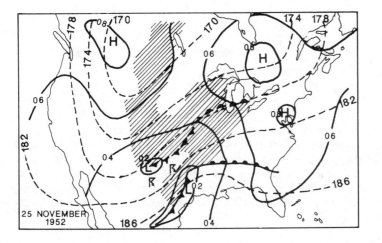

Fig. 11.20 Westward shift of
low-pressure axis with
increasing height. The surface
low pressure is centred at the
apex of the warm centre
while the axis of the
500 mbar trough is located
about 1200 km further west.
(After [72])

systems is very pronounced, the axes making angles of only a degree or so to the
horizontal.

LONG WAVES

Another factor which greatly complicates the identification of weather systems
with individual waves is the simultaneous presence of several identifiable wavelengths in any particular observed situation. There are indications that there are
smaller wavelengths associated with the baroclinic waves mentioned so far, which
may relate to some meso-scale structures observed in frontal cloud and precipitation, and may even represent a frontal instability of the type suspected by the
Norwegians. However, there has been no doubt for several decades that there are
waves which are considerably longer in zonal wavelength than those associated
most directly with individual depressions and ridges. These are known generally as
long waves or planetary waves, but are sometimes named after Rossby who first
discussed their significance and origin, though strictly speaking the term Rossby
wave is now reserved for a particular kind of long wave.

Long waves do not depend on the presence of baroclinity, although they may be
modified by it. But they do depend crucially on the tendency of absolute vorticity
to be conserved in very large-scale flow in the middle troposphere (where there is
little horizontal divergence for reasons discussed in section 7.13 and apparent in
Fig. 7.29 and 12.5). On such vast scales there is little relative vorticity about the
local vertical, so that the main component is the planetary vorticity, which is the
Coriolis parameter f (section 7.14). If we now imagine the very large-scale flow to
be perturbed so that it is angled slightly poleward, the air will move to higher and
higher latitudes, where f is larger and larger. To conserve absolute vorticity the air
develops balancing negative relative vorticity, which at its simplest means anticyclonic curvature (Fig. 11.21). The flow therefore curves back towards its original
latitude, overshoots it and then curves cyclonically back again as it develops
positive relative vorticity. The oscillation about the original latitude continues
indefinitely in the absence of any damping, and we can identify the zonal separation of adjacent crossing of the initial latitude as the wavelength L. It seems clear
that the meridional variation of the Coriolis parameter must play an important
part in determining the wavelength and other features of these long waves. The rate

397

Fig. 11.21 Idealized plan of a Rossby wave with wavelength L in westerly flow.

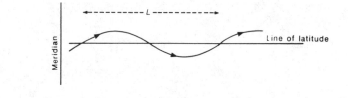

of change of f with northward distance along a meridian is represented by β, and is shown in appendix 11.2 to have values of about 1.6×10^{-11} m^{-1} s^{-1} in middle latitudes. In the specially simple case analysed by Rossby, there are no other important factors, and relatively straightforward analysis shows that such Rossby waves propagate westwards relative to the prevailing air flow with wave speed (i.e. speed of propagation of wave shape) $\beta L^2/2\pi$. This means that the wave speed relative to the Earth's surface is given by

$$C = U - \frac{\beta L^2}{4\pi^2}$$

where U is the speed of the initially uniform westerly flow of air. The form of this relationship shows that Rossby waves propagate eastwards over the Earth's surface at speeds which decrease with increasing wavelength, and that there is a certain wavelength at which the waves are stationary, given by

$$L = 2\pi \{ U/\beta \}^{1/2}$$

If the zonal flow speed is assumed to be 20 m s^{-1}, then it follows that in middle latitudes the wavelength for stationary waves is about 7000 km, four of which would girdle the Earth there. Even longer wavelengths will propagate westward.

Observation is not straightforward because of the somewhat shorter baroclinic waves with which middle latitudes abound, but when the flow in the middle or upper troposphere is averaged over a few days, the fast-moving baroclinic waves are largely smoothed away and the slow-moving long-wave pattern is exposed (Fig. 11.22). These shift majestically with time, usually moving eastwards at a few

Fig. 11.22 Contours of the 500 mbar surface averaged over seven days to reveal the underlying pattern of long waves. The surface fronts at an instant in the middle of the period comprise a series of cyclone waves associated with the long-wave trough. (After [42])

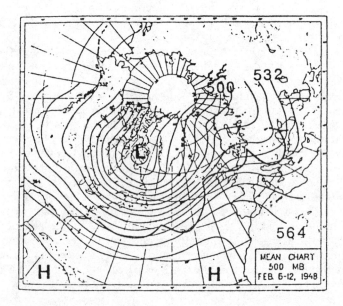

degrees of longitude per day, but transforming from time to time by extensive decay and regrowth in different meridians.

TOPOGRAPHICAL LOCKING

There is clear evidence that there are preferred positions for long waves in relation to large-scale dispositions of land masses and topography, this being most obvious in the northern hemisphere. One mechanism for such linkage can be seen by considering the behaviour of large-scale zonal air flow as it impinges on a long meridional barrier like the Rocky Mountains (Fig. 11.23). As the air reaches the western (upwind) side of the mountains, it is progressively squeezed into a thinner layer by the rising land surface beneath. The squeezing effect is reduced by bodily ascent of the whole atmosphere, and by ponding on the upwind side of the barrier, but these cannot eliminate the effect completely. Since the air is incompressible on this large scale, the vertical squeezing is compensated by horizontal divergence (section 7.14), which in turn acts on the underlying planetary vorticity to produce anticyclonic curvature (i.e. southward in the northern hemisphere) in the previously westerly flow. When the air passes over the crest of the mountain ridges, the falling topography enforces large-scale horizontal convergence which in turn produces cyclonic curvature, tending to re-establish the initial absolute vorticity. However, the air is now at a somewhat lower latitude, where the planetary vorticity f is smaller, so that the air is left with residual cyclonic relative vorticity in the form of cyclonic curvature. There is consequently a *lee trough* on the downwind side of the mountain barrier (Fig. 11.23). Beyond this the air moves polewards, losing cyclonic curvature as it moves to regions with larger f, and beginning to curve anti-cyclonically as it passes the latitude of its original flow west of the barrier. The flow is now in exactly the situation considered at the beginning of the treatment of Rossby waves above, and will behave accordingly, with long-wavelength undulations apparently anchored to the mountain barrier. In practice such anchorage appears only statistically, there being a somewhat greater than average concentration of long-wave troughs just east of the Rockies.

Very significantly, all long waves seem to exert a controlling influence over the faster-moving baroclinic waves both by steering and amplitude control. For example, there is a tendency for baroclinic waves to be suppressed to the west of a long-wave trough, and enhanced to the east, the latter probably corresponding to a zone of repeated cyclogenesis. And once formed, the depressions tend to move as if steered by the contours of long-wave patterns in the middle troposphere.

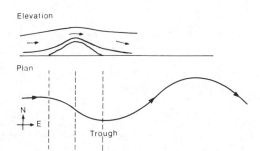

Elevation

Plan

N

E

Trough

Fig. 11.23 Formation of a large-scale trough in the lee of a meridionally extensive mountain range.

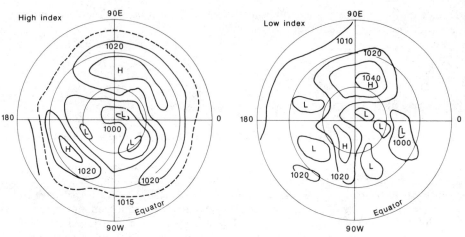

Fig. 11.24 Extremes of index cycle as shown by two examples of seven-day averages of sea-level pressure in the northern hemisphere. In the low-index example, the high pressures are centred over North America and Central Russia. (After [81])

INDEX CYCLE

On an even grander scale, the long-wave pattern round the whole hemisphere seems to *vacillate* (i.e. oscillate irregularly) between two extremes: one of maximum meridional amplitude of waviness, and the other of minimum amplitude (Fig. 11.24). The maximum sinuosity is usually associated with one or more well-developed blocks under the nearly cut-off long-wave ridges, and the minimum corresponds to unusually straight zonal flow over extensive regions, with fast-developing baroclinic waves moving quickly east in the vigorous westerly flow. The long-wave set-up is quantified by the *zonal index* which is defined to be the zonally averaged pressure difference across an agreed meridional segment of middle latitudes at some particular level — for example between latitudes 35° and 55° at the 500 mbar level. The index corresponds to an average geostrophic wind in the zone and is sometimes expressed as such. Regardless of units used, it is small in contorted flow, and large in zonal flow, and the vacillation from one extreme to the next (say from minimum to minimum) is called an *index cycle*. Such cycles have larger amplitudes in winter than in summer, and generally last between 3 and 8 weeks. They clearly correspond to significant events on the very largest possible scale of tropospheric flow — variations in the huge circumpolar vortex. And they can extend their influence far above the tropopause by complex vertical transmission of wave energy, but are not yet susceptible to forecasting of even the most rudimentary kind. Since the periodicities of the zonal index, and the adjustments of very large-scale flow in middle latitudes which they represent, often occupy a significant fraction of a season, they probably account for a considerable part of the year-to-year variations which are such a feature of short-term climatic variations.

11.7 Polar lows and heat lows

It seems likely that there are several different types of low-pressure system at work in mid-latitudes in addition to the extratropical cyclone described by the

Norwegian cyclone model, though the latter is undoubtedly the most common and significant system. As an example let us consider one quite dynamic type, the polar low of the eastern Atlantic, and the relatively static and ephemeral heat low.

POLAR LOWS

In the eastern Atlantic and western European region there is a type of weak low-pressure system which occurs occasionally in the winter and is well-known in Britain for its ability to produce disruptive snow falls. A classic example appears in Fig. 11.25, which shows the typical synoptic situation in which they occur. A fairly gentle northerly flow, to the west of an old stagnant complex low over Scandinavia, is bringing very cold air off the Greenland ice cap and over Iceland to reach the British Isles from the north-north-west. Weak surface lows form in the vicinity of Iceland and move south-eastwards with the air in the low troposphere, bringing substantial amounts of nimbostratus, cumulonimbus and quite heavy falls of snow to the British Isles, before moving on into northern France and weakening.

It was once thought that these systems were essentially conglomerations of fairly shallow cumulonimbus, but fairly recent analysis [25] of this particular situation, amongst others, suggests that they are shallow baroclinic waves forming in strongly baroclinic northerly flow. Though there are snow showers from cumulonimbus in the wake of the system, just as there are showers in the wake (i.e. to the west) of a normal depression, these seem to be secondary to the mass of nimbostratus associated with the shallow low. A sophisticated radar, capable of measuring the motion of precipitation from the Doppler shift of reflected radar waves, indicated the presence of meso-scale convergence (and therefore ascent — see section 7.13) in the low troposphere, but failed to detect any updraughts strong enough to suggest the presence of cumulonimbus embedded in the nimbostratus. Isentropic analysis of the flow revealed a flow structure rather like that found in normal depressions, but confined to the lower half of the troposphere, limited to little more than 500 km in horizontal extent, and of course moving southwards instead of eastwards. If there were fronts they were too weak and on marginally too small a scale to show on the radar or in the standard synoptic observations which were also used.

Theoretical analysis [83] confirms that the shallow but strongly baroclinic zone between the cold airflow off the Greenland ice cap and the much warmer air to the west is capable of generating baroclinic waves of the observed dimensions and speeds of propagation, and at the observed rates, provided that wind speeds are not strong enough to encourage strong damping by turbulent friction (> 10 m s^{-1}). Analysis of more subtle effects, arising from heating by the relatively warm sea surface, suggests that development is inhibited unless the airflow is almost perpendicular to the sea-surface isotherms. Since local sea-surface isotherms are largely zonal, this may well be the reason why polar lows develop only in northerly flows. Between them these restrictions ensure that polar lows develop in only a small fraction of northerly air streams in the eastern Atlantic in winter.

The two lows apparent in Fig. 11.25 produced substantial snow falls in the British Isles, including some exceeding 250 mm in average depth (i.e. about 25 mm equivalent rainfall) which almost totally disrupted traffic in parts of southern England. Although some of the disruption arose because of that area's combination of very high population density and very low preparedness for snow, it is worthwhile noting that there are two quite obvious features of snowfalls which make them especially disruptive of road traffic and other activities. One is the low

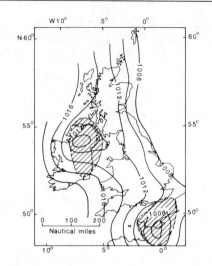

Fig. 11.25 Two polar lows
near the British Isles at 1800 Z
on 8 December 1967. Sea-
level isobars show that the
shallow lows were embedded
in geostrophic northerlies.
Areas of falling snow are
hatched. (After [82])

density of fresh snow, which arises because the spiky arms of the snow crystals
prevent close packing. Fresh snow lies to an average depth which is at least ten
times the equivalent depth of rainfall. The other feature is the tendency for fallen
snow to drift in even quite moderate winds, moving from exposed surfaces and
gathering in great depth in eddies forming in the lee of walls and hedges. Fenced or
hedged roads abound in lowland Britain and are often blocked by several feet of
snow in windy conditions while surrounding meadows have only a skim of snow.
The disruption caused on the occasion of Fig. 11.25 served one useful purpose,
however: it attracted the attention of British meteorologists and led to a
considerable increase in understanding of these weather systems. However, forecasting of polar lows is still very difficult because they usually form in the observational desert north of Iceland and quickly traverse the fairly short distance down to
the British Isles.

HEAT LOWS

In warm summer weather in middle latitudes it is common to see weak puddles of
low pressure developing daily over islands and peninsulas in the absence of any
strong synoptic-scale pressure gradients. Sometimes they may be associated with
showery activity in the afternoon and evening, and sometimes the skies may remain
largely clear. These lows are a meso-scale, even synoptic-scale, response to the
rapid development of strong temperature contrasts between land and adjacent sea.

Figure 11.26 represents a typical heat-low situation over England. During the
previous night the land cooled somewhat below the temperature of the surrounding
sea, but only a few hours of strong sunshine are enough to reverse the situation
completely, because the effective heat capacity of land is so much smaller than that
of the sea (section 9.4). The land surface may soon be 20 °C or more warmer than
the sea, and the temperature difference in the overlying layer of air may be half as
large. This layer expands as it warms, raising the overlying air so that isobaric
surfaces aloft dome upwards slightly overland. For example if a layer of air 300 m
deep warms isobarically by 10 °C in roughly 300 K, its density must fall by about
3% according to the equation of state for a perfect gas. Assuming that the layer is
prevented by its surroundings from expanding laterally, it must expand upwards,

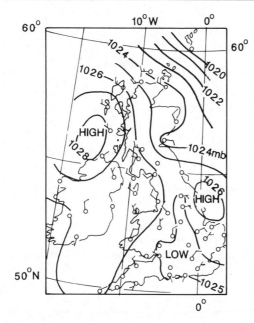

Fig. 11.26 Sea-level isobars showing a heat low over south-east England at 1500 Z on 20 June 1960. Temperatures were just over 25 °C throughout the area of low pressure. Although not plotted here, a succession of earlier maps shows that the low developed as the land heated strongly in the preceding few hours. (From [84])

its upper surface rising by about 3% — i.e. about 10 m in the example chosen.

The position is now as sketched in Fig. 11.27(a), and there is obviously a tendency for the overlying air (virtually the whole troposphere since only a shallow layer is warmed substantially) to flow outwards off the domed boundary layer. The air movement needed to accomplish this on a reasonable time scale (say 3 h) is very small indeed: appendix 11.3 shows that radial speeds ~ 1 cm s^{-1} will suffice for a land surface of radius 100 km. If we assume that the movement is complete when the isobars at the top of the heated layer become horizontal again (Fig. 11.27b), then it is clear that air equivalent in mass to a 10 m layer of the low troposphere has been removed, which corresponds to a loss of pressure of a little more than 1 mbar at the surface and throughout the heated layer, as is typically observed. The sea-level pressure over the warming land has now fallen a millibar or so below the surrounding values, and the weak heat low has appeared on the synoptic charts.

Notice that the upward movement of the top of the warming layer is slow ($\sim$ 1 mm s^{-1} in the worked example) and spread over the whole land surface, and is not to be confused with the updraughts of the thermals embedded in the layer. The latter are strong (~ 1 m s^{-1}), localized and almost completely offset by intervening sinks (section 10.5), and are the means whereby sensible heat is pumped into the warming layer from the underlying surface. Even after the reduction of pressure by shedding of overlying air, the warm layer is strongly baroclinic around the edges of the warmed land, since the isopycnals dip down inland much more steeply than the isobars (Fig. 11.27b). The air in and immediately above the shallow baroclinic layer responds by developing the sea-breeze circulation described in section 9.11. The inflow of cool, dense sea air and outflow of warm land air tends to fill the heat low, but the tendency is impeded by friction in the shallow layer, and by Coriolis deflection of the relatively fast flows, so that the low is depleted only slightly before evening cooling begins to destroy it by reversing the process by which it was created in the first place.

If the warming is not fully offset by the subsequent nocturnal cooling, because of a change in the sypnotic situation, for example, then the shallow residual low may develop further with the next day's heating, and so on for several days. This applies better to larger-scale land masses, such as France or Spain, where the scale is too

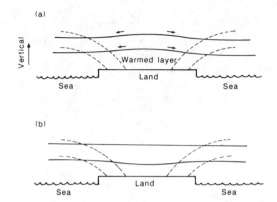

Fig. 11.27 Idealized vertical sections showing the formation of a heat low. In (a) the isobars (solid lines) dome slightly upwards in and above the warmed layer (isotherms dashed), shedding air aloft. In (b) the shedding is complete, leaving isobars dished downwards in the warmed layer.

large for the offsetting sea-breeze circulation to reach more than a small proportion of the total area. Coriolis deflection is the crucial factor here; if the sea breeze persists for more than a few hours, it suffers Coriolis deflection (to the right in the northern hemisphere) and tends to reach quasi-geostrophic equilibrium with the pressure field and friction. The flow in the low troposphere begins to move cyclonically round the shallow low, forming a dynamic barrier to further filling, and confirming the feature as a consistent low-pressure system on the weather map.

Meanwhile the warmed layer inland may be deepening as convection reaches higher above the warming surface, so that the pressure deficit in the heat low increases. However, the vertical development of convection will depend heavily on how convective stability is being influenced by other factors. If it is being reduced, by synoptic-scale vertical stretching associated with the approach of a weak front for example (section 10.3), then the convection may quickly become quite deep, producing widespread outbreaks of heavy showers and thunder. The associated cloud reduces the surface heating, but the deep, warm, cloudy layer may persist for some days as baroclinic adjustment encourages it to rise further in relation to its cooler surroundings. Such *thundery lows* often develop over France in the summer and give spells of humid, thundery weather in Britain as conglomerations of showers swing north-west in the weak circulation. If however the convective stability is being maintained or increased, by subsidence associated with an overall anticyclone, for example, the convection from the warmed surface may remain confined in a fairly shallow layer, and the depth of the heat low be limited by the temperature excess of the heated layer over adjacent cooler layers. Something like this happens in summer over land masses influenced by subtropical anticyclones. Figure 12.6(b) shows the effects of this on sea-level pressures over North Africa, where a semi-permanent heat low over the western Sahara maintains a pressure deficit of about 20 mbar relative to the Azores High to the west, in obvious association with very high surface temperatures over land. Nothing comparable is apparent on the winter maps (Fig. 12.6a). Comparison of summer and winter over the Indian subcontinent shows that there are even more dramatic effects associated with the monsoon there.

11.8 Middle-latitude climates

We have seen that the middle-latitude troposphere is a chronically active zone: opposing radiative imbalances at lower and higher latitudes (section 8.7) maintain

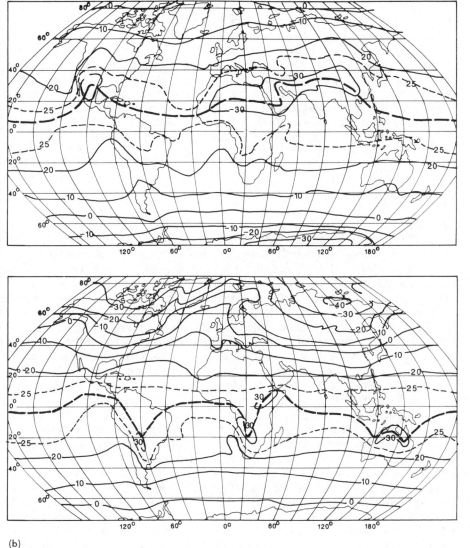

Fig. 11.28 Mean surface temperatures in (a) January and (b) July. (After [85])

(b)

a strong meridionally baroclinic zone whose deep westerly flow is intrinsically unstable (section 11.6), generating long waves and cyclone waves, and through them the whole gamut of smaller-scale disturbances from fronts down to showery and even smaller-scale convection; the seasonal march of the sun produces large seasonal variations in solar and terrestrial radiative fluxes, and their net balance, driving seasonal rhythms of mean temperature and diurnal temperature range which are especially significant overland (Fig. 11.28), an account of the small effective heat capacities there.

The climatic potential in the seasonal and meridional variations of solar elevation is very great: at latitude 65° the noon solar elevation (Fig. 8.8) ranges from only 1.5° in midwinter to 48.5° in midsummer, and at latitude 40° the range is from 26.5 to 73.5. Daylength [51] varies from 2.7 h to 20.2 h at latitude 65°, and from 8.7 h to 14.2 h at latitude 40°, severely limiting the daily insolation and enhancing the net radiative output in winter, and reversing the effects in summer (section 9.3).

These constraints would impose a very sharp meridional gradient of climate, if they were not partly offset by the smoothing actions of the tropospheric disturbances mentioned above, and of the heat capacities of the underlying surfaces (section 9.4). The vast oceanic heat capacity in particular has a very large smoothing effect which, together with the poleward heat flows in major ocean currents like the Gulf Stream and Kuroshio (section 8.13), and ready supply of water vapour, maintains much greater and steadier cyclonic activity over the oceans and western margins than in continental interiors. This maintains a marked west–east zonation of climate from maritime to continental (Fig. 11.28) which dominates the following discussion.

WESTERN MARGINS

Prevailing westerlies and the weather systems they bear allow only very modest seasonal variations of temperature and rainfall (Figs 11.28 and 12.15); they also raise annual average temperatures significantly above the zonal average (in the northern hemisphere, because it contains almost all middle-latitude lands), mainly by raising winter temperatures. Families of depressions are normal throughout the year, interspersed with irregular periods of blocking anticyclones, whose distribution varies only slightly with location and season. Contrasts between tropical and polar air masses are held in check by rapid accommodation to the local sea-surface temperature: for example, showers very rapidly warm the cool polar flows in which they flourish, by pumping latent heat aloft. Summer heat lows may develop inland in slack flows, giving thundery weather; but their warmth is tempered by cloudy loss of sunlight. However, what might seem almost boring uniformity to inhabitants of more extreme climates in fact contains a wealth of day-to-day variation which excites endless comment in native populations and complex climatological commentary which draws attention to subtle *singularities* [86], such as the over-dramatically named European Monsoon — a frequent increase in cyclonic activity in late June, highlighted nowadays by its effects on tennis at Wimbledon (London).

Temperature variations arising from the onset of blocking (warming in summer, cooling in winter — section 11.5), or vigorous cyclonic activity (the reverse), can be so nearly comparable with the small seasonal variations that the warmest days of winter may be warmer than the coolest days of summer. The virtual absence of rainfall in the blocks gives rise to much of the relatively small rainfall variation between corresponding seasons and years at any location. However, the prevailing westerlies maintain strong concentrations of rainfall on western coasts and hills, where there is a fine structure of relief rainfall and rainshadows, with overall reductions inland on scales of 10–100 km (section 9.11). If the relief is modest, as in much of Europe and European Russia, these eastward reductions continue on a larger scale reflecting the inland reduction in cyclonic activity already mentioned; but strong meridional barriers like the Rockies and Andes can compress the transition from maritime to continental climates into much narrower corridors (Fig. 11.28).

Prevailing westerly winds sweep inland, weakening with the increase of surface roughness (section 9.9), and fading patchily in the lee of north–south ridges; and they decrease over longer distances too as the cyclonic activity which drives them decreases inland. On western-facing coasts daily runs of wind (section 2.4) of more than 600 km at the 10 m standard level (an average of Force 4 — appendix 2.2) are commonplace, offering the opportunity of reliable generation of electrical power.

Diurnal temperature variations at screen level increase quite quickly inland, as the decreasing thermal capacity of the upwind fetch allows larger and larger response to the net radiative cycle. This in turn encourages a diurnal cycle in convection and associated cloud and precipitation which reverses seasonally: in winter, showery air masses (like the polar maritime air which sweeps in behind cold fronts) quickly lose their showers as the air flows over progressively colder land, whereas in summer marginally stable flows are progressively destabilized as they pass over warmer and warmer land. The former adds to the normal relief rainfall shadow inland from the western margins. Inland cooling sharply increases the incidence of frost and lying snow in winter.

Many of the climatic features of western margins appear in more limited form on smaller scales, ranging from the North Mediterranean in winter (where cyclogenesis is encouraged by the strong baroclinity between the warm sea and the cold southern European land mass) to the eastern margins of the North American Great Lakes, and the Black and Aral Seas.

CONTINENTAL INTERIORS

Just as western margins have maritime climates because they lie only a short fetch from the sea down the prevailing westerly wind, so the rest of the continental areas have continental climates on account of their much longer fetches from the sea. Even eastern margins are largely continental in climate (Fig. 11.28), since cyclogenesis in the adjacent seas (the the East) only briefly affects the margins before the young depressions eastward move out of range.

For the reasons given above, continental climates are typically much more extreme, meridionally, seasonally and diurnally, than maritime climates. For example, Fig. 11.28 shows that, at latitude 50°, the winter–summer swing in seasonal mean temperatures is from about 5 to 15 °C on the west coasts of North America and Europe, while it is from about -10 to $+20$ °C in north-eastern America (longitude 80 °W) and -15 to $+25$ °C in Asiatic Russia (longitude 80 °E). The seasonal swings in the continental interiors are broadly associated with a transition from cold winter anticyclones to summer heat lows, but considerable differences between North America and Eurasia arise from their particular geographies and are best dealt with separately.

The smaller seasonal temperature range over the *North American* continent arises from its smaller size, a tendency for weak cyclogenesis in the lee of the Rockies (section 11.6), and the presence of a large warm water surface to the south (the Gulf of Mexico). The cold anticyclone in winter is a weak, unstable feature in comparison with its great Asiatic counterpart.

Much of the land north of latitude 60° has a near-Arctic climate, with widespread *permafrost* (permanently frozen ground under a shallow top layer which thaws in summer). In winter, northerlies in the lee of the Rockies bring bitterly cold conditions from this Arctic continental source, which contrast strongly with conditions associated with the southerly flows from the Gulf at other times. Strong outbreaks of the very cold air (*cold waves*) are associated with temperature falls of 10 °C or more in 24 hours. Even sharper contrasts occur when unusually strong westerlies force a very warm, dry Chinook down the eastern slopes of the Rockies (section 9.11). Winter precipitation (snow) is sparse in the central interior, but increases sharply to the east of the Great Lakes (and Hudson Bay, before it freezes over in early winter). Unlike its great Asian counterpart, the cold anticyclone of the

407

North American winter is therefore a statistical feature arising from tendencies underlying considerable day-to-day variability.

In summer, southerly flows of very warm, humid air from the Gulf are asociated with very vigorous outbreaks of severe local storms along sharply defined cold fronts (section 10.6). Less vigorous cyclonic activity draws vapour from the same source to water the east of the continent, but much summer precipitation falls in heavy showers not associated with any air mass contrast. Though driven ultimately by the strong daytime solar heating of the land, these often occur at night, apparently in response to subtle changes in low-level airflow set off by initial nocturnal cooling. The balance between precipitation and evaporation in some of the great grain-growing regions is rather fine, apparently leaving them vulnerable to quite small changes in global climate.

The much greater zonal extent of the *Eurasian* continent (over twice that of North America however the limits are defined) allows the effects of oceanic remoteness full play, despite the lack of any equivalent of the Rocky Mountain barrier to the west. A maximum winter–summer swing of seasonal mean temperature from -40 to $+15$ °C is reached at around longitude 130°.

Winters are extremely harsh over a wide area. Permafrost reaches south of the Arctic circle from about 100 °E to the eastern extremity of Siberia (about 180 °E). Maritime influence from the Arctic ocean is lost when polar pack ice extends to the north coast, and the barriers of the Tibetan and Himalayan massifs to the south prevent any northward intrusion of warm air from the Indian ocean. Snow settles permanently in late autumn over a wide area and remains until spring, raising the ground surface albedo and reflecting what little potential solar input there is (section 9.3). Below the developing winter inversion (Fig. 11.17), falling temperatures reduce the maximum vapour content and with it the local greenhouse effect, allowing the surface to cool even faster by long-wave emission to space (section 8.5). Though the shallow layer of cold air produced in this way is readily channelled by valleys and dammed by ridges, it spills out over a wide area throughout the winter, producing intensely cold waves over North China and Korea. Precipitation is minimal over a wide area.

After a rapid transition in spring, summer conditions arrive with little of the lag apparent in maritime climates. Temperature gradients are modest over a wide area (Fig. 11.28), and a reasonable precipitation total falls mainly as showers (though the northern flanks of the Tibetan massif are semi-desert). China, however, experiences substantial maritime influence, with shallow depressions advancing north-east in late spring (part of the Asian Monsoon), and occasional typhoons in late summer). In the autumn the positive net radiation balance quickly dies and reverses, and rain and snow alternate unpredictably (especially in the west, as both Napoleon and Hitler found to their cost), before winter proper sets in.

Appendix 11.1 Uplift and rainfall

Consider a steady-state weather system in which the upward flux of water substance in the form of vapour is exactly balanced by the downward flux in the form of precipitation. Let us examine this equality in a vertical column of unit horizontal cross-section. The fact that air may be rising in a gentle slope rather than vertically up the column is irrelevant provided there is no significant net horizontal divergence of water substance as a result. Note that we need only the vapour flux into the base of the cloud, since the details of updraught and vapour content at higher levels make no difference to the overall balance.

Suppose that the vapour rises into the base of the cloud in air of density ρ, specific humidity q (g kg^{-1}) and updraught w. Then the upward mass flux of water vapour in kg m^{-2} s^{-1} is

$$10^{-3} \, \rho \, q \, w$$

Now if the equivalent rainfall rate of the precipitation is r mm h^{-1} and the density of water is taken to be 1000 kg m^{-3}, the downward mass flux of precipitation in kg m^{-2} s^{-1} is

$$2.78 \times 10^{-4} \, r$$

Equating these fluxes, and rearranging to isolate the updraught w, we find

$$w = 0.278 \, \frac{r}{\rho \, q}$$

Air enters most cloudy updraughts in the low troposphere where the air density is close to 1 kg m^{-3}. To this approximation therefore, and removing the superfluous accuracy of the numerical factor,

$$w = 0.3 \, \frac{r}{q}$$

where w is in m s^{-1} when r is in mm h^{-1} and q is in g kg^{-1}.

In middle latitudes the specific humidity of air rising into the base of a cloudy mass is $\sim$ 10 g kg^{-1}. It follows that a typical frontal rainfall rate of 3 mm h^{-1} implies an updraught $\sim$ 8 cm s^{-1}. A rainfall rate more typical of a heavy shower is 30 mm h^{-1}, and this implies $w \sim 0.8$ m s^{-1}. However, the presence of evaporative loss from the sides of cumulonimbus, and significant efflux of cloud in the anvil, mean that the simple estimation of w is likely to be considerably too small.

Appendix 11.2 Variation of f with latitude

Consider the rate of change of the Coriolis parameter f with y, the poleward distance along a meridian. According to Fig. 11.29 and the definition of radian measure a very small increment $\mathrm{d}y$ corresponds to a similarly small increment in latitude ϕ through

$$\mathrm{d}y = R \, \mathrm{d}\phi$$

where R is the radius of the Earth. If follows that we can rewrite

$$\beta = \frac{\partial f}{\partial y}$$

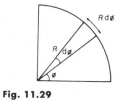

Fig. 11.29

$$= \frac{1}{R} \frac{df}{d\phi}$$

Substituting $2\Omega \sin \phi$ for f and differentiating we find

$$\beta = \frac{2\Omega \cos \phi}{R}$$

$$= 1.6 \times 10^{-11} \text{ m}^{-1} \text{ s}^{-1} \text{ at about latitude } 45°.$$

Appendix 11.3 Radial winds and heat lows

Consider a horizontal circular area A over which the surface pressure falls by $\triangle p$ in time $\triangle t$. Assuming hydrostatic balance, the implied reduction in air mass resting on the area is $(A \triangle p)/g$, so that the average rate of reduction is

$$\frac{\pi R^2}{g} \frac{\triangle p}{\triangle t} \tag{12.4}$$

where R is the radius of A.

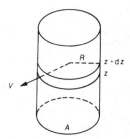

Fig. 11.30

In the simplest model of a heat low this is achieved by air moving radially outwards at all levels above the warming and expanding boundary layer. Figure 11.30 depicts a cylinder of air standing on A. If the radial flow has speed V between heights z and $z + dz$, it is apparent that the contribution to outward mass flux in this range of heights is given by

$$V\rho 2\pi R \, dz$$

Integrating through the full depth of the atmosphere (ignoring the relatively shallow boundary layer) and replacing $\rho \, dz$ by dp/g through the hydrostatic approximation, we find that the rate of export of mass is

$$\frac{2\pi R}{g} \int_0^{p_s} V \, dp$$

where p_s is the surface pressure. If V is uniform, or is the appropriately weighted average radial speed, this expression becomes

$$\frac{2\pi R}{g} V p_s \tag{11.2}$$

The balance between observed depletion rate and rate of export by radial flow is then given by equating expressions (11.1) and (11.2). After rearrangement we have

$$V = \frac{R}{2 p_s} \frac{\triangle p}{\triangle t}$$

If we substitute for the case of a rather large and rapid pressure fall (3 mbar in 3 h) we will find an upper limit to the required radial flow speeds. For a value of R corresponding to a small but substantial island (100 km) we find V to be a little over 1 cm s^{-1}. Notice that this corresponds to a horizontal divergence of a little over 10^{-7} s^{-1}.

The whole analysis can be applied, with V and $\triangle p/\triangle t$ reversed, to the formation of a cold continental high-pressure area.

Problems

Problems for Chapter 11 are included at the end of Chapter 12.

Large-scale weather systems in low latitudes

Middle- and high-latitude weather and climate are strongly affected by the seasonal march of the sun, even over the oceans. We now consider large-scale weather and climate in the rest of the troposphere — the lower latitudes where the seasons, though often very pronounced, are not so directly related to the sun's progress from solstice to solstice. Here seasonal temperature variations are often small, or out of step with the solar calendar, or even show two or more maxima per year; and they may in any case be subordinate to seasonal variations in rainfall or wind direction, or diurnal or irregular day-to-day variations.

More than half of the Earth's surface is included in these lower latitudes (noting that exactly half lies between latitudes 30 °N and S), but the historical growth of meteorology in middle latitudes, the inherent complexity of lower-latitude weather and climate, and the extensive oceans there, have combined to slow our progress in understanding this major part of the Earth's atmosphere. Fortunately the satellite-observation network (section 2.11) is redressing the balance, and the global comprehensiveness required of modern numerical forecasting models is forcing attention on outstanding problems.

12.1 Subtropical anticyclones

These archetypal anticyclones nearly girdle the Earth in the tropics and maintain the subtropical high-pressure zones which are such a major feature of large-scale atmospheric structure (section 4.7). They are very large, persistent, zonally elongated areas of high sea-level pressure, and are particularly well-developed over the oceans, especially in winter. Fig. 12.1 depicts a typical situation in the North Atlantic, where the local segment is known as the Azores High in Europe and the Bermuda High in North America, because it often straddles one or both of those islands. Observations show that the weather is mostly fine, with few reports of precipitation, and those mostly very light apart from some showers which are usually confined to the western flanks of the high. Where skies are not clear, clouds are generally reported to be low, with stratocumulus being the most widespread type. There is very little cloud in the middle troposphere, as is particularly obvious

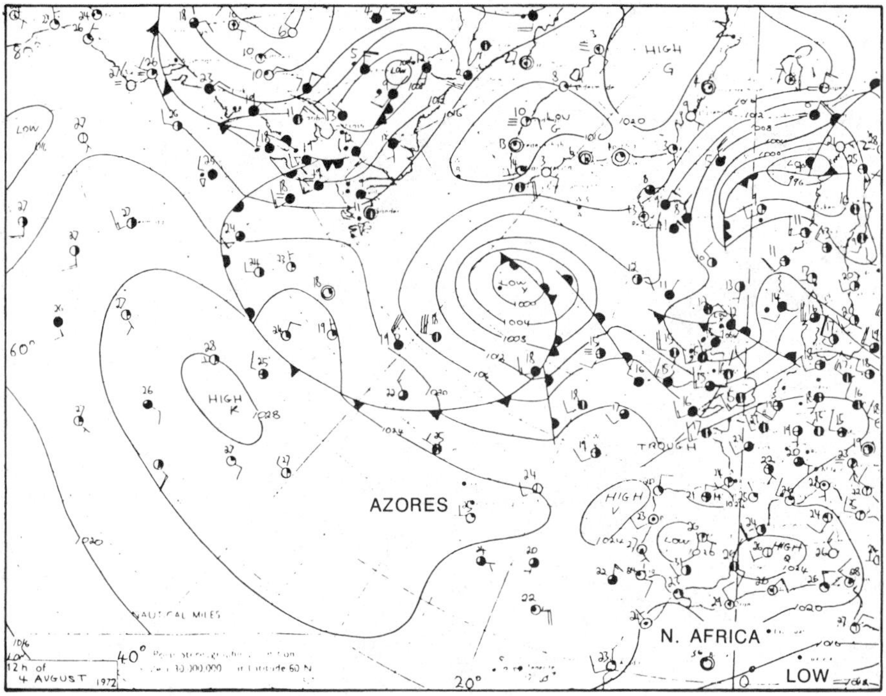

Fig. 12.1 A surface map
showing the Azores High in a
well-developed state.
Depressions and fronts are
moving eastward on its
northern flank, and a ridge of
high pressure is extending
northwards at about 50 °W.
Notice the lowering of sea-
level pressures over north-
west Africa in association
with the summer solar
heating there. (From the
Daily Weather Report of the
Meteorological Office, by
permission)

when observations are made from aircraft, when the sky is seen to be quite empty
above whatever low cloud there may be (Fig. 12.2).

Radiosondes rising through such an anticyclone show clearly why the middle
troposphere is so free of cloud. Above the layer occupied by surface-based
convection (Fig. 12.3), there is a relatively shallow layer which is either nearly
isothermal or contains a temperature inversion (temperature increasing with
height). In either case the potential temperature increases sharply with height, and
the layer therefore represents a convectively very stable interface separating the low
troposphere from potentially much warmer air aloft. The warmer air has very low
relative humidities, often lower than 20%, so that cloud formation or maintenance

Fig. 12.2 A deck of
stratocumulus under a
cloudless middle troposphere
in an anticyclone. Wisps of
high cloud suggest that air in
the high troposphere is
relatively moist.

413

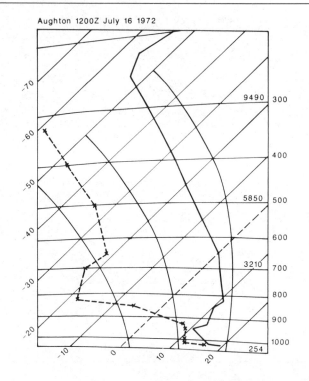

Aughton 1200Z July 16 1972

Fig. 12.3 A radiosonde
ascent through the interior of
an anticyclone, showing a
well-mixed convecting
boundary layer surmounted
by a subsidence inversion and
a deep layer of very dry
subsiding air.

is out of the question. Indeed even liquid aerosol cannot escape evaporation, with
the result that the air in middle levels is remarkably clear of haze. This contrasts
sharply with the air in the low troposphere, below the temperature inversion, which
is often quite hazy, especially over or downwind of continents. Indeed observations from mountains or aircraft often show a very sharp haze top which coincides
with the base of the inversion (Fig. 11.16 shows the same feature in a blocking
anticyclone at higher latitudes) and marks the upper limit of the relatively humid
and hazy air of the convecting layer.

SUBSIDENCE

The potential warmth and the low humidity of air in the middle troposphere of an
anticyclone are both consequences of the gentle but persistent downward motion
which is known from a variety of evidence to be present over large areas of the
system. As air sinks down from high levels it warms, and because there is virtually
no cloud present (usually there is nothing more than a few wisps of cirrus or aircraft
condensation trail), the warming tends to be dry-adiabatic. And because there is no
cloud or precipitation to evaporate in the warming air, the specific humidity in the
air is conserved, with the result that the relative humidity falls as the air warms.
Pretending for a moment that actual descent in time corresponds approximately to
instantaneous descent on Fig. 12.3, you can check the magnitude of this effect by
sliding a dew-point temperature down an isopleth of saturation humidity mixing
ratio (Appendix 12.1) from assumed saturation in the high troposphere, noting the
very large dew-point depression which has developed by the time the air has sunk to
lower middle levels. If the descent of air were truly dry-adiabatic, the temperature
would similarly slide down a dry adiabat from the high troposphere in Fig. 12.3.

The actual temperature profile tends more or less steadily to lower potential temperatures at lower altitudes, which suggests that the sinking air cools diabatically. This cooling is maintained by the net effects of radiation (the only mechanisms capable of reaching air isolated from convection by its own convective stability), in which weak direct solar warming is consistently exceeded by net terrestrial cooling (section 8.6). As noted in the previous section, it is dangerous to assume that three-dimensional air motion in large-scale weather systems can be approximated by horizontal or vertical motion, but in fact more thorough investigations largely confirm the above interpretation. This is largely because the winds in the upper troposphere of an anticyclone are relatively weak, so that air parcels remain in its vicinity for days or longer.

The stable layer in the low troposphere of an anticyclone (Fig. 12.3) is called the *subsidence inversion* because it is maintained by the subsidence of the warm, dry air which makes up the bulk of the air in the system. It forms at the interface between the large-scale descent and the small-scale convection which is maintained by the solar heating of the underlying surface, and is often sufficiently intense to contain a temperature inversion. However, the title subsidence inversion is often used even when the layer is not quite stable enough to produce an inversion. This is unfortunate but not too misleading, since the physically important factor is the presence of strong convective stability, rather than a temperature inversion as such.

CLOUD AND SMOG

If the air in the convective boundary layer is normally humid, the lifting condensation level (section 5.11) will be a few hundred metres below the base of the subsidence inversion, and stunted cumulus will form whenever there is convection from the surface. When such convection is persistent, a layer of stratocumulus forms, which may be thick enough to reduce the sunlight considerably, and even produce a little drizzle. Seen from above, the backscatter of sunlight makes the cloud layer look quite bright by contrast (Fig. 12.2). Once formed, such a cloud layer tends to maintain itself by convection between its top surface, which is cooled more by terrestrial radiation than it is warmed by sunlight, and its base, which is kept warm by the exchange of terrestrial radiation with the underlying surface. The consequent overturning also maintains the globular appearance of the cloud, even when there is no convection below cloud base.

If the air in the convecting boundary layer is dry enough, then the lifting condensation level will be so high that convecting air will still be unsaturated when its upward motion is stopped by the base of the subsidence inversion, and there will be no low cloud. The presence of the subsidence inversion may still be seen from the obvious haze top which is often visible, even from the oblique viewpoint of a surface observer (Fig. 11.16). In these conditions the solar warming of the surface may become intense in the middle of the day, warming the convective boundary layer as a whole and tending to eat into the base of the subsidence inversion. However, the large-scale subsidence is often observed to be able to contain such convection below the lifting condensation level for days on end, and is sometimes strong enough to lower the subsidence inversion to within a few hundred metres of the surface. In such circumstances, industrial pollution of the confined layer may produce significant enhancement of concentrations there. Indeed the so-called *Los Angeles smog* is produced there and in many other cities regularly or occasionally under the influence of anticyclonic subsidence (Fig. 12.4). This type of smog is

Fig. 12.4 Los Angeles Smog in late summer. The stunted cumulus above the distant hills are trapped below the anticyclonic subsidence inversion, as are the pollutant gases and particles and the resulting smog over the city basin. However a much lower inversion (produced by nocturnal cooling or an invasion of cool sea air) is trapping the thickest smog in a much shallower layer and obscuring the city details in the centre of the picture.

especially marked in summer because the ultraviolet component of the strong sunlight encourages photochemical production of ozone and other biologically damaging substances in the confined boundary layer.

On other occasions the convective boundary layer may be several kilometres deep, and yet still not reach the lifting condensation level, because of the dryness of the air as a whole; this can happen over continental subtropical deserts such as the Sahara in North Africa.

The systematic absence of substantial cloud means that rainfall in the subtropical highs is very low. In continental areas the result is arid desert, and indeed there is a belt of hot deserts in the subtropics of each hemisphere, including for example the interior of Australia in the southern hemisphere. Rainfall is sparse, intermittent and unreliable, coming in deep convection during periods of weakness of the anticyclonic circulation. The rain when it comes may fall in torrential showers, running off the baked landscape in flash floods which rush down the gullies (*wadhis* to the arabic nomads who have long eked a living in such harsh conditions in North Africa and Arabia) maintained by similar sporadic events in the past. Though the brief, inefficient moistening of the ground surface is enough to trigger the ecological miracle of the flowering desert, the water soon evaporates and the desert returns. Over the oceans the persistent excess of evaporation over precipitation in the subtropics keeps the surface waters significantly more saline than elsewhere.

AIRFLOW

The anticyclonic circulation which is so pronounced in the lower troposphere, and which gives the whole type of system its title, can be related to the simplified vertical section in Fig. 12.5. Since air is persistently subsiding in the middle troposphere it must be persistently converging in the high troposphere and diverging in the low troposphere. The convergence aloft concentrates the vorticity there, making it more cyclonic (section 7.14), but it seems that the initial vorticity is often very feeble, with the result that the circulation in the upper parts of a deep anticyclone is often only weakly cyclonic. As this air sinks into the low troposphere, however, it diverges strongly and its absolute vorticity is reduced so sharply that it

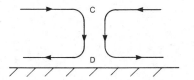

Fig. 12.5 A simplified model of vertical and radial airflow in the troposphere in an anticyclone. Regions of synoptic-scale horizontal convergence and divergence are labelled C and D respectively. Compare with a similar picture of a cyclonic system (Fig. 7.29).

becomes anticyclonic relative to the weather map. Although pronounced, the magnitude of this anticyclonic vorticity is limited as described in section 7.14, so that winds in anticyclones never reach the strengths which are quite common in cyclonic systems.

Near the surface, turbulent friction encourages divergence by enabling air to flow with a substantial component of motion outward across the isobars toward lower pressure. On the western and polar flanks of the subtropical highs such diverging flow feeds air into the warm sectors of the mid-latitude depressions as they sweep eastward, maintaining their warmth and supplying air to the warm conveyor belts there (section 11.4). On the equatorial flanks the diverging flow feeds the trade winds which angle westwards down to the subtropical convergence zone, fuelling the deep convection there (section 4.7).

During the development of a particular subtropical high-pressure cell, convergence aloft obviously exceeds divergence below, in a reversal of what happens during the formation of a depression, where divergence above exceeds convergence below. However, during the quite long periods in which the intensity of an anticyclone changes little, divergence and convergence must balance closely in a dynamic equilibrium. No simple argument explains the continuing presence of the high surface pressure, despite the unwise claims of some descriptive texts. For example, it is not a direct dynamic consequence of the downward motion of air in the main mass of air (like the pressure of a hose squirting water down onto a floor), since this would imply a gross failure of the hydrostatic approximation (section 4.4), which actually must be especially accurate in the quiet conditions prevailing there. It seems that dynamic constraints such as the Coriolis effect maintain the excess of atmospheric mass in the central parts of the anticyclone, once it has been established by the initial excess of convergence. The relative warmth of the subsiding air means that air pressure falls more slowly with increasing height than is the case in the surrounding cooler air masses (section 4.3), with the result that the upward doming of isobaric surfaces (or equivalently the concentration of high pressure on horizontal surfaces) increases with height, often into the upper troposphere. The system is therefore warm and deep, unlike the shallow, cold highs which develop over continental interiors in winter (section 11.7). Over the Sahara in summer, the subsiding warm core is superimposed on a shallow heat low at the very hot surface (section 11.5), with the result that the depression of isobaric surface near the land surface gives way to upward doming in the middle and upper troposphere; a deep high therefore overlies some of the depressed sea-level pressures apparent over North Africa in Fig. 12.6(b).

12.2 Monsoons

The great northward migration of cloud and rain over the Indian subcontinent which takes place early each summer is known to virtually everyone through direct

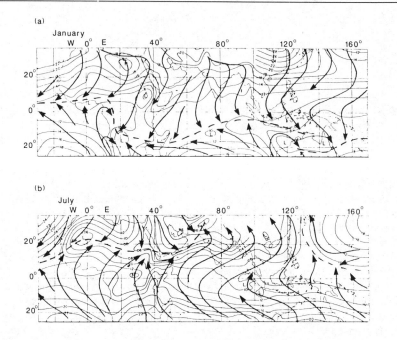

Fig. 12.6 Mean pressure and airflow at the base of the troposphere in low latitudes in January and July. Isobars of mean sea-level pressure (from [86]) are labelled in tens and units of millibars. Arrowheaded thick lines are streamlines of flow near the surface (from [26]). Dashed lines mark zones of confluence (also from [26]), some of which correspond to the showery intertropical convergence zone described in section 12.2.

experience, school geography lessons or news reports. It is often simply called the Indian monsoon, which correctly implies that there are monsoons in other regions, but tends to underplay the scale and interconnection of the huge seasonal events which affect Indo-China, Indonesia and China as well as India. Popular discussion also tends to focus on the wet summer, whereas this is only one part (very important because of the water it brings) of an annual cycle.

The word *monsoon* means season in Arabic, and dramatic seasonal change is a common thread which links a wide variety of climatic events occurring annually in low latitudes. Figure 12.6 shows how the airflow in the low troposphere shifts between January and July. In most places there is some meridional migration of the convergence zones which are the local components of the intertropical convergence zone (section 4.7). Since these zones are dotted with cloud and rain maintained by the convergence of moisture-laden winds (section 8.13), their presence in any area indicates the rainy season there. One of the largest meridional shifts occurs in the vicinity of the Indian subcontinent. In January the nearest convergence zone lies about 10° south of the equator (i.e. in the summer hemisphere), whereas in July, the nearest equivalent feature is the heat low and associated *monsoon trough* extending along the axis of the Ganges and Indus rivers, just south of the great Himalayan mountain barrier. There is another very large shift from northern Australia to an ill-defined location in northern China, and a substantial shift in eastern Africa. By contrast, the convergence zones over the Atlantic and Pacific oceans migrate comparatively short distances, and the convergence zone in the eastern Pacific remains in the northern hemisphere and close to the equator throughout the year.

MECHANISMS

The much larger seasonal shifts of airflow over land as compared with ocean clearly must be related to the relatively very small effective heat capacities of land surfaces (section 9.4): as the zenith sun moves from the tropic of Capricorn in late December to the tropic of Cancer in late June, so the zone of maximum thermal response by surface and atmosphere tries to march with it, and that march must be faster where the total heat capacity is smaller, as it is over land. But the response is made complex and indirect by the role of water vapour and cloud: solar input to moist land surfaces is used largely to evaporate water (section 9.4), which may travel some way before giving up its latent heat in cloud formation. And the atmosphere tends to organize convergent flows of moistened air which gather latent heat from extensive ventilated areas before releasing it in narrow zones of cloudy uplift. Such convergence happens on a range of scales from the Hadley circulation (Fig. 8.13) downwards, all of which are too subtly sensitive to distributions of topography, land and sea to be easily understood.

On the smallest scale, atmospheric response is localized in cumulonimbus, even the largest of which would be individually much too small to appear on maps like Fig. 12.6 (even if they were not smoothed by seasonal averaging). The ascent which constitutes the rising branch of the Hadley circulation actually takes place in a large number of *hot towers* (section 10.5) — narrow, vigorous updraughts embedded in much larger volumes of relatively static or sinking air. And these are organized in a variety of types of large tropical weather system, few of which are observed or understood as clearly as the large-scale weather systems of middle latitudes.

RAINY SEASONS

In terms of local climate, the result of all this is that regions within the range of seasonal excursions of the ITCZ tend to have rainy seasons when it is close by, and dry seasons otherwise. In the simplest picture, places between the extremes of seasonal migration of the convergence zone would have two rainy seasons per year, but there are so many special factors at work in any particular location, especially overland, that the northward and southward migrations are often very different, and only one may be effective as a rainy season.

A rainy season is not simply a period of unceasing rain, or an unrelieved succession of showers: the rains come in bursts as one or other of the low-latitude weather systems develops or moves by, and a particular rainy season may be unusually wet or dry depending on the number, type and intensity of systems occurring. Indeed the combination of the rather short length of many rainy seasons, and the fact that individual weather systems last at least as long, at any location, as mid-latitude systems, mean that year-to-year variability of rainfall is higher than in middle latitudes. This is especially obvious in regions with short rainy seasons, but appears even in well-watered low latitude regions (compare Tables 12.1 and 4.1).

Table 12.1 Annual rainfall at Canton Island 2° 48′ S, 171° 43′ W

Year	Rainfall	Year	Rainfall
1957	1269 mm	1962	402 mm
1958	1597	1963	713
1959	759	1964	519
1960	492	1965	1433
1961	508	1966	1101
mean 879		range	+ 718
			− 477
		standard deviation	414 ie. 47%

THE INDIAN MONSOON

Although the rains enter Burma in April and May, they do not reach India until late May or early June. Then they often sweep across the southern half of the sub-continent in a spectacular *burst*, whose mechanism has intrigued generations of meteorologists, and which may owe more to re-arrangements of flow in the high troposphere than to the prior establishment of the heat low in the north west (Fig. 12.6) which used to be regarded as a prime cause.

The monsoon usually becomes established over the Indian subcontinent by late June, by which time there is a more or less continuous southerly flow of warm, moist surface air toward, but not effectively into, the monsoon trough lying across the north of the peninsula. (Northerly flow to this zone is blocked by the highlands of Tibet and Afghanistan.) The southerly flow is in fact a continuation of the south-east trades of the southern hemisphere. As these cross the equator north-wards, the Coriolis deflection switches from leftward to rightward, and the flow veers to become south-westerly. This flow of already moist air picks up further large quantities of water vapour from the Arabian Sea (warmed to 28 or 29 °C by very strong sunshine before the onset of the monsoon) and reaches the west coast of India as the south-west monsoon. Large quantities of rain fall as a result of ascent forced or triggered along the western edges of the great plateau, and substantial flows continue northwards and merge with the southerly flow from the Bay of Bengal. In the middle and lower middle troposphere there are important westerly flows of air probably originating in Africa and even the Mediterranean.

Slow-moving synoptic-scale low-pressure systems, such as *subtropical cyclones* and *monsoon depressions* (the latter being one of the few types to bring rain to the monsoon trough) form in association with these flows and can produce large areas of rain and quite strong winds even though surface pressures are depressed by only a few millibars.

With the surface thermal equator established locally just a little north of latitude 25 °N, and substantial volumes of air rising in weather systems on its equatorial flank, the normal baroclinity is reversed through almost the full depth of the tropo-sphere between this warmth and the relatively cool troposphere nearer the equator. The thermal wind relationship (section 7.10) then ensures that the *tropical easterly jet stream* is a semi-permanent feature of that region in the summer months at latitudes about 15 °N (Fig. 12.7). The weather systems associated with the monsoon therefore form and act in the context of south-westerly flow at low levels and easterly flow aloft, and it is not surprising that their movements are often sluggish and easterly, and their mechanisms complex.

The south-westerly monsoon begins to retreat in September and usually completes its withdrawal from the Indian subcontinent by the end of November. Well before the end of this period, the easterly jet aloft is replaced by a narrow branch of the subtropical (westerly) jet skirting the southern flank of the

Fig. 12.7 Streamlines (arrowed solid) and isotachs (dashed, labelled in knots) on the 200 mbar surface at 0300 Z on 25 July 1955, showing a strong easterly jet stream over southern India. The zero isotach marks the boundary between westerlies and easterlies, and a south-westerly jet stream lies south-east of the Black Sea, in air flowing towards the north of the Tibetan massif. (After [27] and [88])

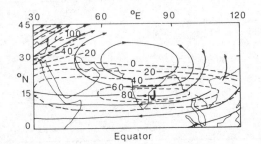

Himalayas. In the north of the subcontinent the ensuing drier seasons are also much cooler because of the moderately high latitudes and flows of cold air from neighbouring high ground. Weak low-pressure systems move south-eastward along what was the axis of the monsoon trough, and later in the winter these may be followed by north-westerly surges of cold air which move right across the peninsula from the great reservoir which is building to the north and east of the Tibetan massif. But this barrier largely protects India from the fierce surges of the bitterly cold north-eastern monsoon which affects China in winter (section 11.8). Before the onset of winter proper, coastal areas of the Indian subcontinent may have been influenced by an important type of low-latitude weather system which we will consider shortly – the tropical cyclone.

12.3 Tropical weather systems

The baroclinity of the tropical troposphere is weak and unstructured by comparison with higher latitudes. Solar heating of surface and troposphere reaches a broad maximum between the tropics, and terrestrial radiation partly offsets this by an amount which leaves a radiative excess varying little with latitude (Fig. 8.7). The result is that the tropical troposphere (meaning the whole troposphere between the tropics, including the equatorial zone) is nearly barotropic in comparison with the essentially baroclinic troposphere of higher latitudes. Horizontal temperature gradients at any level are usually very small, and differences of more than ~ 5 °C are comparatively rare away from the immediate influence of land and sea surfaces. By the same token, seasonal changes in temperature are so small that the seasons in low latitudes are measured more by their contrasts in weather, especially rainfall, than they are by their temperature contrasts.

  Although the tropical troposphere is relatively barotropic, the small horizontal temperature gradients which do arise are associated with very significant patterns of cloud and associated weather. Tall cumulonimbus (hot towers) may be only a couple of degrees warmer than the mid-troposphere they penetrate, which may in turn be a couple of degrees warmer or cooler than its counterpart outside the conglomeration of cumulonimbus which constitutes any particular weather system. In addition, the weakness of the Coriolis effect at low latitudes means that most of the disturbances of the tropical troposphere are not even quasi-geostrophic in their mutual adjustment of wind and pressure fields. This means that the pressure field cannot be used to bolster the observed wind field in the way which is so useful at higher latitudes. (More than a few degrees of latitude away from the equator however, the undisturbed environment is still geostrophic to the extent that for example the thermal-wind relationship holds at least qualitatively.) As a result, quite a vigorous synoptic-scale weather system producing tens of centimetres of rainfall per day and winds of nearly gale force may be associated with a depression in sea-level pressure of only a few millibars, and with horizontal temperature differences aloft of only a very few degrees. In operational meteorology such systems are known as tropical depressions unless surface winds exceed gale force (section 12.4). This sensitivity of linkage between behaviour and ambient conditions poses severe problems for observational, operational and theoretical meteorology — problems which are compounded by the sparseness of observations enforced because so little of the tropical zone is land, and even less is

421

populated. Fortunately, the development of the meteorological satellite in recent decades, especially the geosynchronous type (section 2.11), has improved things considerably. Analysis techniques have been developed to extract wind profiles from the movements of large clouds between consecutive pictures, and cloud and air temperatures from simultaneous measurements of terrestrial radiation at several wavelengths.

In the zones poleward of the equatorial confluence zones, the trade winds maintain an easterly component of flow in the low troposphere, and there is clear evidence from series of consecutive satellite pictures of these zones that much of the cloud and rain over the oceans occurs in large clusters of cloud several thousands of kilometres apart, which remain observably coherent while moving westwards at between 5 and 10 m s^{-1} — a speed which is usually a little slower than the airflow in the low troposphere. Over land, the picture is confused by topography and a strong diurnal rhythm, so that many clusters develop and decay without obvious motion. So although land-based weather systems are the more immediately relevant to people, the oceanic systems are the more obviously coherent and better understood. In addition, many weather systems which form over the tropical oceans (which account for about 75% of the Earth's surface between the tropics) significantly affect adjacent land areas, especially eastern margins.

Many of the cloudy systems over the sea seem to move in association with weak troughs in sea-level pressure (Fig. 12.8) whose wave-like appearance on a weather map gives them their name — *easterly waves*. Note that the reversed pressure gradient compared with the mid-latitude norm of northward pressure lapse gives a superficial impression of a pressure ridge. In fact the trough represents a weak poleward extension of the convergence zone trough whose zonal axis lies some 10° of latitude equatorward of the heart of the wave.

In some cases there is marked asymmetry of cloud distribution, cumulonimbus and some altostratus being concentrated with rain and fairly strong winds to the

Fig. 12.8 An easterly wave in the central Pacific in plan and vertical section. On the map, sea-level pressures and surface winds show a wave with trough line approaching Kwajalein (the Marshall Islands), followed by another wave about 3000 km to the east. Synoptic surface winds are plotted as arrows with a full barb representing each 10 kt. Column totals of water vapour as a proportion of atmospheric mass are plotted as dashed lines and labelled in parts per thousand, with dry and moist zones marked D and M respectively. The dotted line was traversed by instrumented aircraft whose observations of cloud height are represented on the vertical section — average heights by hatched cloud towers and extremes by clear extensions. (Horizontal extents of towers are purely diagrammatic.) (After [27] and [89])

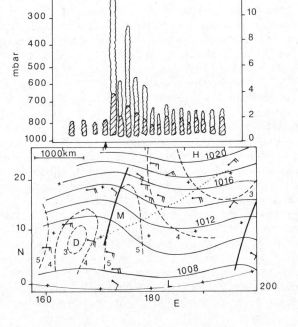

east of (i.e. behind) the trough axis. Such concentration seems to be associated with gentle convergence in the low troposphere there and corresponding divergence to the west of the trough axis, as follows. The convergence implies quite substantial vertical stretching of the lower troposphere which serves to destabilize the layer (sections 7.13 and 10.3) and thereby encourage the growth of cumulonimbus. It also helps concentrate the fuel for cloudy convection, the water vapour evaporating into the boundary layer from the sea surface. The stretching and convergence also deepen the layer with easterly wind components, so that the westerlies are confined to the upper troposphere (Fig. 12.8), and increase the absolute vorticity of the lower troposphere (section 7.14). The latter shows up in the poleward excursion (increasing f) and cyclonic curvature of the airflow as it overtakes the wave crest. A nearly balancing slight divergence seems to be concentrated in the high troposphere, giving a much more asymmetrical vertical distribution of vorticity than is typical in middle latitude systems (Fig. 7.29). To the west of the wave axis each of these effects is reversed to maintain relatively clear skies in the more northerly flow.

In other cases the cloud is banded zonally across the trough axis, and in others again the cloud is distributed fairly homogeneously. There is conflicting evidence about whether or not all these cloudy systems are warmer or cooler on aggregate than their surroundings, i.e. whether the systems are *warm-* or *cold-cored*, but it is clear that they are not conspicuously warm-cored like their much more energetic relatives, the tropical cyclones. The westward movement of such systems across the great oceans at intervals of a few days provides much of the rainfall in these zones, and in the coastal and island areas they affect. They have been studied relatively intensively in the north central Atlantic by American meteorologists interested in their effects on the Carribean and surrounding regions and their capacity to develop occasionally into tropical cyclones (next section). It is clear that a substantial number of the systems reaching the Carribean originate in the lines of cumulonimbus which move westwards across north-west Africa in the summer. Others form in the Carribean convergence zone and on fragments of fronts trailing down from mid-latitudes. The annual total observed in the Atlantic remains fairly constant at about 100, of which about 70 reach the Carribean.

The subtropical cyclones mentioned in the previous section are not limited to monsoon zones, and can give rise to several closed isobars (at the standard 2 mbar interval) at sea level, even though they appear to originate in the upper troposphere. They are cold-cored systems, though the temperature depressions are quite small and they can occasionally become slightly warm-cored through unusually large release of latent heat in the embedded clouds. There is evidence of banded cloud structure, and they are often individually very persistent.

12.4 Tropical cyclones [91]

Of the 70 tropical disturbances (mostly easterly waves) reaching the Caribbean each year, about 25 deepen to the point where there are several closed isobars on surface charts and strong cyclonic circulation, cloud and rain over a substantial area. When winds exceed gale force (19 m s^{-1} at 10 m) they are called *tropical storms*, and are equivalent in intensity to rather weak middle-latitude depressions, though they have no fronts. In summer and autumn, a variable number of these storms, averaging about eight in the Caribbean, intensify to the point where winds exceed hurricane force (33 m s^{-1}) at the 10 m level, whereupon they are known as *tropical cyclones* (popularly *hurricanes*). As is well known by news reports, and even fiction [90], hurricanes are most intense in a ring surrounding a much calmer, very low-pressure *eye*. In the week or so in which they remain in the mature, violent state, they usually travel westwards and then polewards. They are found in many sea areas other than the central west Atlantic and Caribbean (Fig. 12.9), being called *cyclones* in the Indian Ocean and around Australia, and *typhoons* in the rest of the Pacific Ocean. Regardless of local nomenclature, these are all the same type of weather system, which in its combination of intensity and extent is the most powerful disturbance of the Earth's atmosphere.

Tropical cyclones form only over oceans where surface temperatures exceed about 26.5 °C over a substantial area at latitudes of at least 5°, but they occur in every such eligible zone except the south Atlantic. The annual total of about 80 tropical cyclones is small in comparison with totals of extratropical cyclones, and yet hurricanes (as we will now call them all for simplicity) continue to attract the attentions of meteorologists to an extent which is out of all proportion to their number. There are probably several reasons for this, but two will suffice. First, they can be almost unbelievably destructive if they pass over vulnerable areas: buildings and crops suffer severely in winds exceeding hurricane force, and coastal areas risk inundation by the huge waves and exaggerated tides which can be generated (over 200 000 people were drowned in 1970 when a cyclone from the Bay of Bengal made landfall in the mouths of the Ganges, raising the local sea and estuary levels by several metres for a few hours). Second, they are almost unique among tropical weather systems in having a very sharply defined structure which positively invites investigation.

Although hurricanes have been studied intensively for many decades now, their mode of formation from relatively very weak tropical disturbances is still far from clearly understood. Since they form in effectively barotropic conditions there is no obvious source of available potential energy (section 13.2) to compare with the baroclinity of mid-latitudes. However, there is some evidence that residual baroclinity leaking down from higher latitudes in the form of a weak upper trough may help trigger their dramatic development. In the space of a couple of days the central sea-level pressure then falls by 20 mbar or more (normal enough in middle latitudes but unusual in the tropics), and the stage is set for the ultimate intensification. This may still be prevented by factors such as making premature landfall and thereby losing the crucial supply of water vapour from the warm sea surface, or entraining unusually cold air into the centre of the vortex in the high troposphere (which douses the developing warm core of the storm), but if it is not, then the mature hurricane develops rapidly by formation of a relatively narrow ring of extremely strong winds and updraughts around a central, relatively inactive eye.

STRUCTURE

Figure 12.10 is a typical satellite view of a mature hurricane. The dense white ring extending a couple of hundred kilometres out from the centre is the shield of cirrus fanning out in the high troposphere above the ring of updraught surrounding the central eye. Often a small dull spot in the centre and a thin dull ring mark the inner and outer limits of this dense shield, whose appearance amid the more diffuse mass of high cloud is a sure sign that the hurricane has tightened into full maturity. Penetration by special reconnaissance aircraft then reveals the full majesty of the storm.

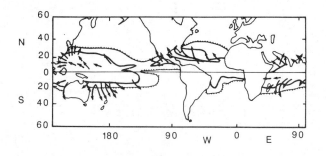

Fig. 12.9 Representative tracks of tropical cyclones (arrowed thick lines) compared with mean sea-surface isotherms (25 °C dotted, 27 °C solid) for September in the northern hemisphere and March in the southern hemisphere. Note that the Mercator projection seriously underestimates the fraction of the Earth's surface area affected by such storms. (After [27] and [70])

The spiral arms of cloud stretching hundreds of kilometres outward from the central structure, often mainly equatorwards and eastwards, are seen to be long curving lines of cumulonimbus. The inner zone contains a dense mass of cumulonimbus and sheets of cloud spreading from them at all levels, circling faster and faster as the centre is approached (Fig. 12.11), until the fastest flow is reached in the vicinity of a nearly solid ring of giant cumulonimbus which often completely encircles the eye, and produces a ring of particularly strong echo on weather radars

Fig. 12.10 Typhoon Clara at 0600 Z on 19 September 1981, centred east of Luzon (The Philippines), as seen in the visible by the Japanese Meteorological Agency's geostationary satellite. The eye and surrounding dense cirrus shield are particularly marked, as is the anticyclonic diverging swirl of thinner high cloud further out on the equatorial flank.

Fig. 12.11 Vertical radial section through hurricane Helene on 26 September 1958. On the right-hand side the solid lines are isotherms labelled in °C, and the dashed lines are isotachs of tangential wind speed labelled in metres per second. The left-hand side is a mirror image of the right showing distributions of cloud, precipitation, the plan position of the strongest radar echo, and the boundaries of the eye wall as defined by the zone of maximum temperature gradient. (Simplified from [27])

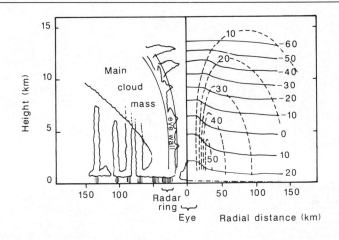

(Fig. 12.12). The eye itself is relatively cloud-free: a quiet amphitheatre a few tens of kilometres across surrounded by the massive, rotating *eye wall* of encircling cumulonimbus. The temperature rises sharply inward at all levels below the very highest troposphere, reaching a warm plateau in the vicinity of the eye wall. The sea-level pressure falls toward the centre, slowly at first and then more and more rapidly to the minimum value which applies across the eye and is a useful measure of the severity of the storm. Most severe hurricanes have minimum sea-level pressures of 950 mbar or less. Such values are not very different from the minima in vigorous extratropical cyclones, although a few hurricanes have much lower minima, but the low pressures in hurricanes are concentrated in very much smaller areas, and are associated with maximum pressure gradients (in the ring of maximum winds) which can be ~ 1 mbar km^{-1} — nearly an order of magnitude larger than the largest synoptic-scale pressure gradients associated with extratropical cyclones. In fact the vigorous central parts of a hurricane are structured on the meso-scale rather than on the synoptic scale, which rendered their detailed observation very problematic before there were meteorological aircraft or satellites. On a conventional surface-weather map, with sea-level isobars drawn at 2 mbar intervals, the innermost 10 or so isobars are not resolvable by synoptic observation, and are interpolated using a simple pattern of concentric circles spaced to match estimates of maximum winds.

WINDS

The very high winds of the hurricane are quite literally its most striking feature, and they exceed hurricane force at the surface in an annulus usually a few tens of kilometres wide. Highest winds occur to the right of the centre, facing in the direction of the storm's overall motion, where the speeds of rotation and translation are added (Fig. 12.13). Sustained speeds as high as 75 m s^{-1} at the 10 m level have been recorded occasionally, though the chances of an anemometer being in such fairly limited zones (and surviving) are quite small.

The effect of wind speeds on surfaces and structures increases roughly in proportion to the kinetic energy of the flow (i.e. to the square of the sustained wind speed), though there is selective sensitivity to embedded gusts depending on their frequency. Wind speeds above 50 m s^{-1} have enormous effects. Over the sea they raise waves ~ 10 m high from trough to crest, and drive sheets of spray and foam

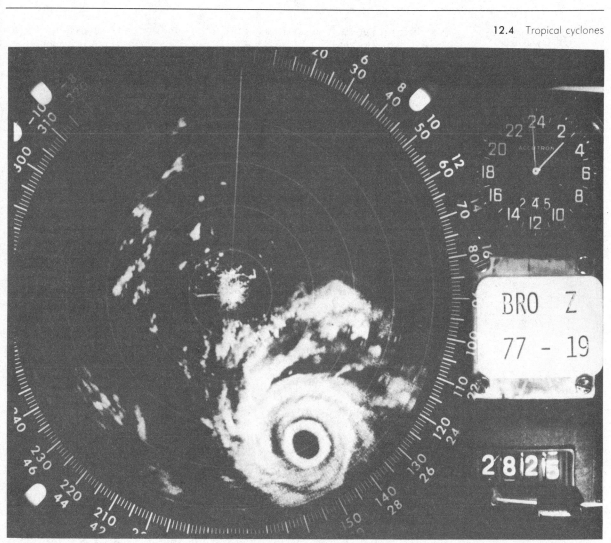

Fig. 12.12 Radar PPI display from Brownsville, Texas of hurricane Anita centred off the Mexican coast on 2 September 1977. Note the echo-free eye surrounded by the solid ring of echo in and beyond the eye wall. Range markers are at 45 km intervals.

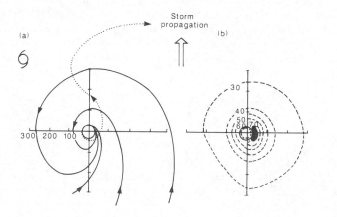

Fig. 12.13 Winds in a very severe hurricane in the northern hemisphere. Solid streamlines in (a) show air spiralling inwards in the low troposphere, while the dotted streamline shows air spiralling outwards in the high troposphere — initially cyclonically like the low-level flow, but then anticyclonically. (b) The associated pattern of isotachs at the 10 m level, labelled in m s⁻¹. The black zone has winds in excess of 80 m s⁻¹ (nearly 160 kt). (Mostly from [91])

over their broken surfaces to the extent that the interface between air and water is blurred. Even modern well-found ships do not brave such conditions by choice, and an accidental encounter between a United States battle fleet and a typhoon in the Philippine Sea in 1944 left 3 ships sunk, 146 aircraft destroyed aboard carriers and 790 men dead.

As in the case of mid-latitude depressions, the areas of strong winds produce storm surges (section 11.3) in coastal waters, and sea levels can be raised or lowered by as much as 4 m compared with the predicted astronomic tide for a few hours, and by half as much again in bays and estuaries. The actual response of the water is a very complex function of storm size, strength and movement, as well as coast and sea-bed topography, but the net result in storms making perpendicular landfall on a coast is a substantial mound of water to the right of the storm centre (in the northern hemisphere). In some situations resonance by surge reflection can lead to a dwindling series of resurgences, the first couple of which may be high enough to cause trouble. If a high surge happens to coincide with a high astronomic tide there is obviously a serious risk of sea defences being breached in vulnerable areas, especially since walls, dunes, etc. are then unusually exposed to battering by large waves.

Over land, hurricane winds flatten crops, uproot trees, severely damage or destroy weakly constructed buildings, and smash most unshuttered windows. Exposed livestock and people obviously risk death and injury by falling debris. Unfortunately in a way, hurricanes are sufficiently rare events at any particular location for it not to be worthwhile for societies to be fully adapted and protected against visitation. So the emphasis is on reliable forecasting to allow time for shuttering of windows and evacuation of areas likely to be flooded. Unfortunately this touches on another aspect of hurricane behaviour which is not properly understood. As the outline storm tracks in Fig. 12.9 suggest, most hurricanes travel westward initially in the easterlies which give them birth, but later tend to *recurve* polewards. Individual paths can be quite contorted in detail for reasons which are not well understood, making it difficult to forecast future positions.

How are such winds produced and sustained for the considerable life of a hurricane? The concentration of strong updraughts around the eye obviously requires a substantial inflow of suitable air, which means considerable horizontal convergence. This is concentrated particularly in the lowest kilometre where friction encourages the flow to spiral in across the nearly circular isobars, but it is known that convergence must persist throughout the layer with significant cyclonic rotation, which is usually at least 10 km deep (Fig. 12.11). In this layer convergence continually concentrates pre-existing vorticity, which in the absence of any synoptic-scale residue is the planetary vorticity f. It is simpler to go back to basics at this point and consider the conservation of angular momentum about a vertical axis in the centre of the eye. Then according to appendix 12.2, the tangential wind speed U reached after a ring of air has contracted from radius R_0 (at which it was stationary on the weather map) to a much smaller radius R, is given to a good approximation by

$$U = \frac{f R_0^2}{2 R} \tag{12.1}$$

At latitude 20° a contraction from nearly 145 km would produce a wind speed of 50 m s^{-1} at radius 30 km (a typical value for the radius of the ring of maximum winds), whereas at latitude 5° the contraction would have to be from nearly 500 km for the same result. In fact both of these contractions are considerable underestimates because some surface friction is communicated throughout the rotating mass by deep convection, eroding original angular momentum as the air converges.

It seems that the convergence required to generate sufficient wind speeds at latitudes less than about 5° is larger than the tropical troposphere can manage, so that hurricanes are not found in this equatorial zone.

The effect of friction is to reduce the radial dependence of tangential winds from the R^{-1} predicted by the conservation of angular momentum to about $R^{-0.6}$. However, the diverging flow at the very top of the troposphere (which spreads the canopy of anvil seen by satellite) is effectively above the reach of this friction, because it originates largely from the updraughts in the eye wall, and tends to preserve the angular momentum of the ring of maximum winds more nearly as it diverges. The result is that, in zones outside the ring of maximum winds, wind speeds aloft are systematically much lower than at lower levels (Fig. 12.11). In fact when air spreads out to radii ~ 200 km it begins to rotate more slowly than the Earth's surface and hence turn anticyclonically on the weather map. The resulting anticyclonic curvature of upper-troposphere flow beyond the dense, central canopy makes the diverging high cloud in the outer parts conform to much the same spiral formation as the converging cyclonically curved lines of cumulonimbus in the lower parts of the troposphere, so that together they contribute to the characteristic spiral 'galactic arm' appearance which is used as the symbol for a mature hurricane on large weather maps (Fig. 12.13).

BALANCE

The reduction in tangential wind speeds in most of the upper parts of the central annulus of the storm (Fig. 12.11), maintained by the above mechanism, plays a vital role in enabling the storm to balance its wind and thermal structures. In this very active zone the horizontal accelerations arising from rotation are so large that they must match the horizontal pressure gradient forces as closely as is the norm in middle latitudes. However, as a rough estimate will show, the centripetal accelerations arising from rapid motion around the storm's centre on the weather map far outweigh the hidden Coriolis component, with the result that the balance is cyclostrophic rather than geostrophic (section 7.12). Assuming for simplicity that all the air at a radial distance R from the storm's centre is actually moving around that circle (actual trajectories must obviously be complicated in detail by the radial motion and the translation of the storm), then the cyclostrophic balance between pressure gradient and centrifugal force is given by

$$\frac{V^2}{R} = \frac{1}{\rho} \frac{\partial p}{\partial R} \tag{12.2}$$

This balance reconciles the extremely strong horizontal pressure gradients and wind speeds found in the inner parts of the storm, but it also implies a cyclostrophic equivalent of the thermal wind equation (section 7.10 and appendix 12.3):

$$\frac{\partial (V^2)}{\partial z} = \frac{g R}{T} \frac{\partial T}{\partial R} \tag{12.3}$$

relating the vertical gradient of the square of the wind speed to the radial temperature gradient. The sense of this relation is the same as for the normal thermal wind: positive shear is associated with low temperatures to the left of the shear in the northern hemisphere. In the hurricane this means that the distinctively warm core on the left requires negative shear (upward decrease of cyclonic winds) for balanced flow — exactly what is maintained by the combination of surface friction and conservation of angular momentum. Inserting realistic wind speeds from Fig. 12.11 into eqn (12.3) shows that the radially inward rise in temperature through the region of maximum winds should be 3–4 °C, in reasonable agreement with observation. The warm core of the hurricane is therefore dynamically confined by the deep vortex of high winds whose strength decreases upwards, in much the same way as adjacent warm and cold air in middle latitudes is confined by the dynamic barrier whose elevated core is the associated jet stream. In each case the warm air would spill out over the cold air, destroying the baroclinity in a few hours, if the dynamic barrier were removed. However, in the case of the mid-latitude front, the dynamic balance involves the Earth's rotation quite directly, through geostrophic balance, whereas in the hurricane the cyclostrophic balance has been constructed from the weak Coriolis effect of low latitudes by powerful convergence. Moreover, the sense of the balanced flow around the warm core of the hurricane requires a kind of deep, inverted jet stream with a circular core in the lower troposphere, with obvious consequences in terms of violent surface winds.

WARM CORE

The very warm core of the hurricane is a conspicuous feature, and one which distinguishes it from other types of tropical disturbance, most of which are cool-cored on the synoptic scale (though shot through by many hot towers which are warm-cored on their much smaller scale). Though less spectacular than the winds, or the torrential rain and electrical activity which mark the inner parts of a hurricane, the warm core is quite noticeable without specialized observations, and many factual and fictional accounts mention the oppressive warmth of the eye, where the more violent distractions are temporarily reduced. The evolution and maintenance of the warm core is vital for the birth and prolonged maturity of the hurricane, but at first glance it is not easy to see how it is achieved. Hurricanes develop only over extensive warm sea surfaces, where deep cumulonimbus convection is endemic anyway, and the troposphere is consequently maintained in a nearly neutrally stable state for such convection (section 5.12). How is it possible for hurricane cumulonimbus to achieve the observed substantial rise in core temperature in the absence of any hotter surface? To try to answer this question we must consider the mechanisms of hurricane development.

Figure 12.14 contains several temperature soundings in and around hurricanes.

Fig. 12.14 Temperature soundings in and around hurricanes, plotted on a tephigram. Curve M is the September Caribbean average. Curve A represents saturated adiabatic ascent from mean surface conditions, whereas B represents ascent after isothermal depressurization to 950 mbar (XB in inset). Curve E represents typical conditions in a hurricane eye. The difference between B and E in the upper troposphere is largely removed by allowing for liberation of latent heat of fusion in glaciating cloud. (Based on [27])

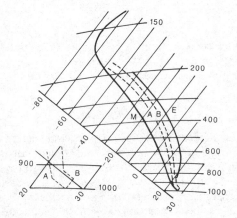

The average sounding for the hurricane season is conditionally unstable throughout most of the depth of the troposphere, so that when convection is triggered locally, a deep cumulonimbus will grow. However, if many showers break out in a smallish region, the effect is for the local troposphere to become less encouraging to deep convection if other factors do not intervene. One such factor which has been closely studied since it was proposed by Charney and Eliassen in 1964 involves positive feedback arising from frictional convergence in the boundary layer supplying convective fuel to the bases of the cumulonimbus. Synoptic-scale convergence concentrates and deepens this particularly moist layer with the result that more cumulonimbus convection is triggered and so on in an exponential development which has realistic properties when it is investigated in sophisticated detail. The process is called conditional instability of the second kind (CISK).

Another much simpler effect becomes important when the central surface pressure has already fallen significantly (say by at least 20 mbar) below values outside the developing storm. The air spiralling in towards lower pressures in the boundary layer is then made potentially warmer through isothermal warming by contact with the uniformly warm sea surface in the presence of falling pressure, so that its buoyancy relative to the distant external atmosphere is significantly increased. The process is depicted in Fig. 12.14, where you can see the advantage gained by having the warming take place at a pressure significantly lower than the surrounding sea-level value. Air rising in deep convection then follows a profile which is warmer at all levels, just as if it had come from a significantly warmer sea surface at 1000 mbar. Cumulonimbus are therefore encouraged and more heat is pumped into the active core of the storm. This too becomes a mechanism for development through positive feedback if the increased heat supply causes a further fall in surface pressure, as will be seen to be the case in a moment. Meteorologists are therefore especially keen to identify the cause of the initial deepening which allows this mechanism to become effective, but the puzzle is still not clearly resolved. In mature hurricanes the very low surface pressures ensure that the effect is very large, equivalent to several °C, and contributes to the relatively high efficiency of the hurricane as a thermodynamic engine.

PRESSURE

Although vertical accelerations in some of the strong updraughts (and downdraughts) are just about significant in comparison with g, the resultant deviations from hydrostatic balance are localized and relatively small. Thus despite the storm's fury, hydrostatic equilibrium still prevails quite accurately on the meso and synoptic scales, and the pressure at any level is simply a measure of the weight and therefore of the mass of air above that level. The warm core is significantly less dense than the surrounding air, so that volume for volume it is significantly less massive. Since no compensating cold air or extra mass (for example in the stratosphere) is observed aloft, the surface pressure below the warm core must therefore be considerably depressed. If for example the core is warmer on average by 6 K in 300 K, the decadal and other scale heights are increased by 2%, and it is readily shown (Appendix 12.4) that this produces a 4.6% fall in surface pressure — nearly 50 mbar. The observed relation between severity and minimum pressure now becomes intrinsically reasonable, to the extent that the core warmth is obviously a measure of the energy of the storm.

THE CENTRAL ZONE

To summarize the spectacular energy and relatively fine structure of the inner zones of a mature hurricane, imagine that you are observing the westward approach of the centre of a hurricane in the northern hemisphere, from the safe vantage point of a well-found lighthouse for example. The north-north-westerly winds (crossing the isobars at more than 20° because of surface friction) have been well above hurricane force for more than an hour, and the very low cloud base is only just visible through the driving, torrential rain and the blizzard of spray whipping off the very heavy seas. Lightning flickers frequently but indistinctly through the murk, but thunder is inaudible above the roar of wind and sea. A tornado writhes briefly by, leaning toward the south in the strong wind shear, and marking a region of small-scale intense convergence below a particularly strong updraught. The barometer stands at 945 mbar and has fallen 15 mbar in the last hour. Suddenly the sky begins to lighten to the east, and the winds begin to ease. Five minutes later they have fallen to gale force, the rain has largely stopped and through breaks in the low cloud you can see that the main mass of cloud is retreating to the west, giving glimpses of an enormous wall of dense cloud (the eye wall) down which some cloud fragments are falling like a waterfall in slow motion. A few blinks of sun show the sea running high and confused, as wave trains from different parts of the eye wall cross and interfere, but the water surface is now largely free of spray and foam under the dying wind. The barometer has steadied at 944 mbar, but the still well in the lighthouse basement shows that the sea level is half a metre above the expected tide and rising, as part of the storm surge in the local coastal waters, including an inverse barometer effect of the very low atmospheric pressure. Some exhausted birds are perching on the balcony of the lighthouse. The wind dies further and begins to change direction uncertainly. After nearly an hour of abatement the wind begins to freshen again, this time from the south, and half an hour later the opposite eye wall is glimpsed approaching from the east and promising a virtual repeat of the earlier experience with winds raging from the south-south-east.

12.5 Low-latitude climates

The Hadley circulation (Fig. 8.4) supplies a framework in which to summarize significant features of low-latitude weather and climate. This covers the largely barotropic region across which the sun is within about 10° of the zenith at least once per year. Air-mass contrasts are weak by comparison with higher latitudes, and there are no fronts, although thermal contrasts between land and sea can be significant, especially when sharpened by sea breezes or monsoon flows on the equatorial sides of the subtropical anticyclones. There are very large meridional gradients of water content, cloud and precipitation between the arid subtropics and the wet equatorial zone, and comparable zonal gradients reflecting distributions of land and sea, many of which show large seasonal variation (Fig. 12.15).

The subtropical high-pressure zones include all of the hot land deserts of the Earth. Widespread subsidence inhibits cloud and precipitation throughout the year, with the result that the land is permanently arid and what little atmospheric water vapour there is has to be advected from adjacent moist areas. The virtual

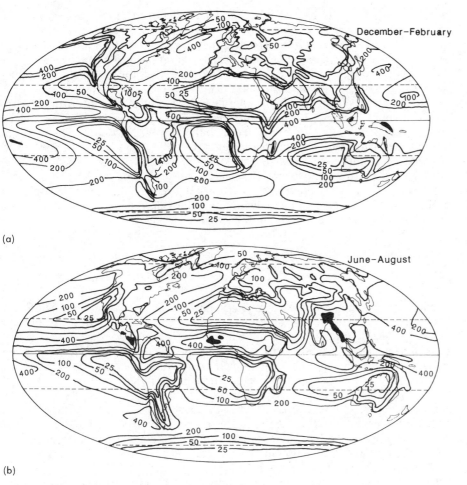

(a)

(b)

Fig. 12.15 Mean annual precipitation in (a) December–February and (b) June–August. (After [92])

absence of evaporative cooling during the day, and the strong radiative cooling through the wide open window (little cloud or vapour) at night can maintain a diurnal surface temperature range of 50° or more (section 9.4).

On the equatorial flanks there are brief, unreliable rainy seasons when the ITCZ reaches its poleward extent. In the North African Sahel, gross inter-annual rainfall variation and largely unexplained longer-term variations have contributed to human misery on an awesome scale in recent decades (section 4.10). In India, the disposition of sea, land and mountain barriers encourage a vast monsoonal invasion of cloud and rain in summer, which despite producing some of the highest rainfall totals in the world on exposed hills, is still sufficiently unreliable to cause large-scale human distress from time to time, especially near its northern extremity.

The trade winds are fed by air flowing equatorwards from the subtropical highs. As the air is transformed by injection of water vapour from beneath (especially over the oceans), it becomes populated by cloudy weather systems which drift westwards providing much of the rainfall in those zones, and imposing substantial rainfall enhancement on eastern sides of islands and larger land masses. The northward flow across the Equator in the Indian summer monsoon is turned westwards however, giving rainfall enhancement on western coasts and slopes (Fig. 9.11).

Over the warmest western sides of oceans at latitudes of at least 5° (i.e. with a large enough Coriolis parameter — section 12.4), some of these disturbances develop into tropical cyclones. In addition to their more notorious behaviour, these storms can deliver very large (sometimes catastrophic) rainfalls to the islands and coasts they cross. Indeed some recurve polewards right out of low latitudes still raining intensely (producing notable floods in the north-eastern USA, for example).

The trade winds from each hemisphere's Hadley cell feed the intertropical convergence zone, supplying air and water vapour to the updraughts of organized populations of tall cumulonimbus. These billow upwards to the high equatorial tropopause, emptying torrential showers onto the underlying surfaces, and incidentally maintaining most of the several thousand thunderstorms usually active in the Earth's atmosphere. The organization of such hot towers is subtly sensitive to local geography and the diurnal cycle, giving each location its distinctive rainy season, and producing an endless display of slowly blossoming canopies of high ice cloud on satellite pictures like Fig. 1.3.

ENSO

The structuring along zonal axes obvious in Fig. 12.15 reflects slight but significant zonal patterning of uplift and subsidence which was first proposed by Sir Gilbert Walker in 1923, and confirmed and elaborated by more recent observation and computation (Fig. 12.16(a)). This *Walker circulation* has been found to undergo irregular reversals every few years which were called the *Southern Oscillation* because they influence weather over a large part of the south central Pacific (Fig. 12.16). More recently the Southern Oscillation has been linked to the notorious El Nino (the Christ Child — a local name for a hesitation in the Humboldt Current which occurs each December and only occasionally develops into a major breakdown), the temporary failure of the cold, Humboldt Current off Peru, which blights the fishing industry there every few years. In fact the whole system is now known as El Nino–Southern Oscillation (ENSO).

In the normal Walker circulation, uplift and rainfall are encouraged over Indonesia and suppressed over the Humboldt current by a slight but persistent vertical circulation superimposed along the centre of the Hadley circulation. Subsidence along the Equator can split the local ITCZ by a cloud-free lane centred along the Equator, which is clearly visible in satellite pictures, and can reach as far west as Canton Island (central Pacific — Table 12.1). In the ENSO, subsidence

Fig. 12.16 Vertical equatorial section of the Walker circulation in (a) normal and (b) ENSO modes, according to computations by Tourre (1984). (After [85])

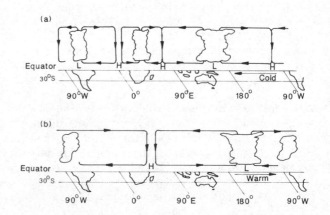

weakens over the failed Humboldt current, and uplift and rainfall are encouraged in a large zone centred on Canton Island. Major ENSOs occurred in 1957–8, 1972–3 and 1982–3, the first of these being associated with markedly increased rainfall in Table 12.1. In fact Fig. 12.16 shows that ENSO affects the vertical circulation along the full length of the Equator, and there is statistical evidence that it is related to shifts in climatic patterns at higher latitudes as well.

Schemes of *teleconnections* are proposed to explain these distant harmonies in tropospheric activity, many involving interactions between atmosphere and ocean. Though largely speculative so far, they point to a degree of interconnectedness in both space and time hardly guessed until 20 years ago, and begin to encourage hope that some forecasting of short-term climatic fluctuations may become possible eventually. In particular the crucial role of air–sea interactions in imposing longer-term order on shorter-term atmospheric chaos is attracting close scrutiny.

Appendix 12.1 Drying by descent

Figure 12.17 shows the upper part of a tephigram, as relevant to the middle and upper troposphere. The point P represents the state of air which has risen to the 300 mbar level (just over 9 km above sea level) by saturated adiabatic ascent with wet-bulb potential temperature 20 °C (equivalent potential temperature 61 °C). It is saturated by 0.53 g kg^{-1} water vapour and its temperature and dew point are − 37 °C. Its potential temperature is 60 °C, and the closeness of this to the equivalent potential temperature is a measure of how little vapour there is left to condense at such low temperatures.

If this air sinks dry-adiabatically to the 500 mbar level (about 5.5 km above sea level) it arrives at the point Q, found by moving down the dry adiabat through P. In fact all the potential temperatures are unaltered in the process, though the simple temperature has risen to 0 °C. Since the vapour content of the parcel too is unaltered by the dry adiabatic process, the dew point at the 500 mbar level is found by sliding down the saturated specific humidity isopleth through P to R. This shows that the dew point rises by only a little over 5 K in the process, so that the dew-point depression at 500 mbar is over 31 K. From the ratio of saturation

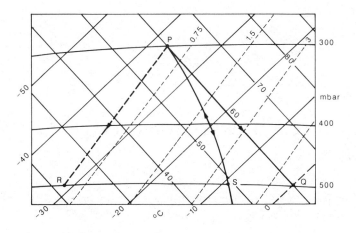

Fig. 12.17 Drying by dry adiabatic descent in the high troposphere (on a tephigram).

435

specific humidities at temperature and dew point it is apparent that the relative humidity of the subsided air is only about 7%. In fact such subsidence is generally accompanied (indeed made possible) by radiative cooling. If this were such as to cool the sinking air along the 20 °C saturated adiabat (but notice this is dry rather than saturated descent), the air would reach the 500 mbar level at S with a temperature of −9 °C, and a relative humidity which, as you should check, is about 13%.

Appendix 12.2 Convergence and conservation of angular momentum

Consider a narrow circular horizontal ring of air which begins at rest with radius R_0 on the Earth's surface, and then shrinks to radius R while conserving angular momentum about the local vertical. Initially it has only the angular velocity of the surface itself, which is $f/2$, where f is the Coriolis parameter. Arbitrarily assuming unit mass for the ring, the initial angular momentum about the centre of the circle is $R_0^2 f/2$. After shrinking to radius R, suppose that the ring is turning with tangential speed U relative to the map. This contributes an additional angular velocity U/R to the planetary $f/2$, so that the total angular momentum is now $R^2 (U/R + f/2)$. Equating initial and final expressions for angular momentum and rearranging we find

$$U = \frac{f}{2R}(R_0^2 - R^2)$$

When the first term in brackets is much larger than the second we have

$$U \cong \frac{fR_0^2}{2R}$$

We can rearrange the latter to find the initial radius needed to produce a required U at a given R, 50 m s^{-1} at 30 km being reasonable for an intense hurricane.

$$R_0^2 \cong \frac{2RU}{f}$$

Notice that R_0 increases with decreasing f, i.e. decreasing latitude.

Appendix 12.3 Thermal wind in cyclostrophic balance

The isobaric form of the equation for cyclostrophically balanced flow at speed V round a circle of radius R is

$$V^2 = R\,g\,\frac{\partial Z_p}{\partial R}$$

where $\partial Z_p/\partial R$ is the upward and outward slope of the isobaric surface which supplies the inward force to maintain the centripetal acceleration of the whirling air parcels. Proceeding as in appendix 7.9 we apply this basic equation to two isobaric surfaces and subtract to find

$$V_2{}^2 - V_1{}^2 = Rg\,\frac{\partial}{\partial R}\,(Z_{p_2} - Z_{p_1})$$

where the right-hand side contains the radial thickness gradient. As in appendix 7.9 we replace the thickness gradient by the gradient of the column mean temperature T.

$$\frac{V_2{}^2 - V_1{}^2}{Z_{p_2} - Z_{p_1}} = \frac{R\,g}{T}\,\frac{\partial T_p}{\partial R}$$

If we consider an extremely thin layer then the differential form appears

$$\frac{\partial(V^2)}{\partial z} = \frac{R\,g}{T}\,\frac{\partial T_p}{\partial R}$$

The strongly negative radial temperature gradient in a hurricane therefore requires an upward reduction of wind speed.

Appendix 12.4 Low pressure and warm core

Suppose that the full height of the uniformly warm core of a hurricane is z. If the scale height of the isothermal core is H, then according to eqn (4.8) the surface pressure p_s is related to the pressure p_t at the top of the core by

$$p_s = p_t \exp\left(\frac{z}{H}\right)$$

For a fixed z and p_t we can find the sensitivity of p_s to H by differentiating

$$\frac{\mathrm{d}p_s}{\mathrm{d}H} = -p_s\frac{z}{H^2}$$

or

$$\frac{\mathrm{d}p_s}{p_s} = -\left(\frac{z}{H}\right)\frac{\mathrm{d}H}{H}$$

The latter expression shows that the fractional (or percentage) change in surface pressure is proportional to the fractional change in scale height, the constant of proportionality being z/H. The minus sign ensures that a rise in scale height is associated with a fall in surface pressure. Since H is directly proportional to the core temperature T (eqn (4.7)), we have established the sensitivity of surface pressure to core temperature. In the case of a hurricane z may be as large as 16 km, so that z/H may approach 2.3, the ratio of decadal to exponential scale heights.

Note that the above is a computational procedure to find the sensitivity of p_s to H and therefore T, and incidentally to demonstrate another usage of scale height. The relationships above do not show how the surface pressure in a particular core may develop during the life of a hurricane, except in so far as z and p_t do actually remain constant, which is a sweeping oversimplification.

Problems on Chapters 11 and 12

Note that the northern hemisphere is assumed unless otherwise stated.

LEVEL 1

1. In not more than one sentence in each case, cite observational evidence for the following large-scale properties of developing extratropical cyclones: broad-scale uplift; convergence in the low troposphere; divergence in the upper troposphere.
2. Whereabouts is the observer of the following situation most likely to be in relation to the classical model of a family of depressions? Pressure is rising and falling by about 20 mbar every couple of days; winds are strongly westerly, backing a little as cloud and rain move in during each fall of pressure, and veering in the subsequent clearances.
3. A square-rigged ship is making northwards in strong westerly winds just to the east of an advancing cold front. If no action is taken by the crew, what is likely to be the situation of the ship after the passage of the surface front?
4. To the north of an old centre of low pressure a mass of cloud is observed advancing from the east, with the edge of its high cloud moving quickly northwards. Decide whether the front is a warm or cold front, and outline your reasoning.
5. Given that air at two different levels 2 km apart on a vertical sounding have risen in parallel conveyor belts with a gradient of about 1 in 100, estimate the horizontal separation of their surface sources.
6. In the conveyor-belt picture of airflows in fronts, a large region of heavy rainfall is likely to be associated with what properties of the flow?
7. Cold damage to evergreen foliage was concentrated on the eastern sides of plants in the severe winter of 1963 in the British Isles. What can you say about the position of the centre of the blocking high?
8. Outline the reason for saying that the rapid onset of continental winters in middle and high latitudes is often associated with the radiative behaviour of snow.

LEVEL 2

9. A low-pressure centre with sea-level pressure 980 mbar lies not far from a ridge with pressure 1012 mbar. Find the difference in sea level associated hydrostatically with this pressure difference.
10. If air has been risen to the 500 mbar level along a sloping warm conveyor belt

with a gradient of about 1 in 75, estimate the geographical position of the region in which it was in the atmospheric boundary layer.

11. Air enters the upper troposphere of an anticyclone (at the 400 mbar level) with a temperature of $-30\ °C$ and is saturated with respect to supercooled water. If it sinks to the 700 mbar level in 3 days, cooling by net radiation at an equivalent isobaric rate of 2 °C per day, find its temperature and relative humidity at the 700 mbar level. Use the tephigram in Fig. 5.5.

12. Using the method of problems 7.23 and 24, find the average horizontal divergence which is needed in the lower troposphere of an anticyclone to accommodate a rate of subsidence of 1 cm s^{-1} in the middle troposphere there.

13. Using the expression for the speed of eastward propagation of a Rossby wave, estimate the speed of a cyclone wave of wavelength 2000 km in middle latitudes. Comment briefly on the inconsistency of the implied assumption.

14. Estimate the depth of a seasonal heat low, such as that over north Africa, associated with a layer of air 2 km deep which is on average 10 °C warmer than comparable air over adjacent seas.

15. Using the method of appendix 12.2, follow air at latitude 20° as it flows outwards from a position in the high troposphere just inside the eye wall of a hurricane, where it was moving cyclonically at speed 30 m s^{-1} round a circle of radius 30 km. Assuming conservation of absolute angular momentum about the vertical in the centre of the eye, find the radius at which the outflow will stop turning cyclonically relative to the surface.

LEVEL 3

16. Compare and contrast the isobaric and isentropic views of air flow in a mature depression, citing the observations which the latter fits, as well as those it does not.

17. Compare and contrast the synoptic-scale structures of a depression and a blocking anticyclone in middle latitudes.

18. Explain carefully how a subsidence inversion is maintained in the low troposphere of an anticyclone.

19. In problem 12.12, the subsidence associated with radiative cooling has been ignored. Assuming that the low troposphere as a whole is cooling at 2 °C per day in the anticyclone, estimate the associated rate of subsidence in the mid-troposphere. The original omission is therefore justified.

20. Draw up a list of all meteorological effects which involve the meridional variation of the Coriolis parameter, including a brief description of the involvement in each case.

21. What differences in regional climate would you expect if the Earth were entirely covered by ocean? (Note that climatologists might then be dolphins!)

22. Discuss the validity of saying that the troposphere in low latitudes is largely but not exactly barotropic.

23. How can tropical cyclones develop conspicuously warm cores over nearly isothermal ocean surfaces, and how are they maintained?

The atmospheric engine

13.1 Introduction

It was mentioned in Chapter 1 that when flows of energy are considered, the atmosphere is seen to be acting like a very large, self-regulating heat engine, whose winds and weather systems correspond to the moving parts (pistons, flywheels etc.) of man-made engines. It is true that the atmospheric engine differs from the ones we construct in many important respects: it is entirely fluid, and has the enviable ability to maintain and regenerate its structures in the course of normal operation; it is very complex, and its activity ranges in intensity from the barely measurable to the overwhelmingly powerful on the human scale; and of course it operates on an enormous range of space and time scales, as detailed throughout this book. But despite such apparent differences, the comparison between the atmosphere (and other natural heat engines) and man-made heat engines is quite strict, and in particular they all obey the same physical laws. Two which constrain them overall and in every part are amongst the most fundamental physical laws known to mankind — the first and second laws of thermodynamics — and we conclude this introduction to meteorology by applying these two laws to highly simplified models of the troposphere and stratosphere as connected heat engines. Figure 13.1 shows the two engines in schematic form: there is a powerful primary engine in the troposphere and a weak secondary one in the stratosphere. We will trace the production and subsequent life history of the energy which passes through the atmospheric engines, keeping them working and maintaining the activity and structure of the atmosphere in the process.

Energy comes in a variety of apparently very different forms, such as heat, kinetic energy, potential energy and chemical energy, but it is the great strength of the concept of energy that each of these is seen to be a different form of the same fundamental thing, and that they can transform according to fixed rates of exchange which have been established by theory and experiment. In fact a common currency (the joule) is now in use in all but the most specialized areas. The ultimate unification is expressed in the principle of conservation of energy which states that if all forms of energy relevant to any particular system are accounted for, including fluxes into and out of the system, then the energy total is conserved. When heat is one of the forms of energy involved, the principle is known as the *first law of thermodynamics*. The first law (as we will simply call it) is fundamental to all

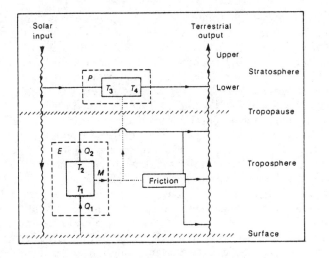

Fig. 13.1 Schematic outline of the tropospheric heat engine (E) and the stratospheric heat pump (P). Radiant-energy fluxes are represented by wavy lines, heat fluxes by straight solid lines, and fluxes of mechanical energy by dotted lines.

energy budgets, since it constrains them to balance in whole and in part. In fact it can be used to determine a component which cannot be measured well enough in its own right, provided all the other components of the energy budget for that particular situation are known. This approach has already been used at various points in the book, especially in Chapter 8, where the imbalance in radiative energy fluxes was used to point to the presence and importance of the convective and advective heat fluxes. We will now use the concept of energy more generally, including the mechanical forms (potential and kinetic energy), to establish an overall framework for the whole range of dynamic atmospheric behaviour, much of it weather-related.

The first law is powerful but undiscriminating. It requires that energy budgets balance, but usually does not determine how the total should be divided between the relevant components in any particular situation. Towards the end of this chapter we will be able to make a little use of the second law of thermodynamics, which was developed out of experience with early steam engines but which is now seen to be one of the cornerstones of the natural sciences, largely because of its ability to describe the direction of energy flow in systems and to discriminate between the very many apparently possible types of mechanism and the much fewer types which actually occur.

The atmospheric engine has basic features in common with all heat engines, however different they may be in detail. It has heat sources, where heat energy effectively enters the system, locations of energy transformation (and our interest is especially in the zones of production of mechanical energy), and heat sinks, where heat leaves to go to some other system. For present purposes all the input Q_1 of solar heat to the tropospheric engine is supposed to occur at a single temperature T_1 which is presumably close to the average temperature of the base of the troposphere. This ignores the substantial minority of solar input which is distributed through the depth of the troposphere (Fig. 8.6). However, this fraction is a secondary feature since it is usually more than offset by the net output of terrestrial radiation from the same zone.

Using this surface heat input, the tropospheric engine generates mechanical energy M by a wide variety of mechanisms, examples of which have appeared throughout this book, but which obviously include the Hadley circulation in low latitudes, the monsoons, the depressions and anticyclones of middle and higher latitudes, and the sporadic cumulus convection found at all latitudes. However,

441

any particular package of mechanical energy is present only temporarily in the troubled state of atmosphere and ocean, since it is eventually transformed to heat by friction of various types. Some of it then joins the much larger flux of heat which has passed through the tropospheric engine in the convective and advective heat fluxes (labelled Q_2 in Fig. 13.1), before being exhausted to space in the form of terrestrial radiation. The rest joins the cascade of terrestrial radiation at lower altitudes, including the surfaces boundary layer with its concentration of turbulent friction. For simplicity the output too is assumed to occur at a single temperature T_2, though it must in principle involve a wide range of tropospheric temperatures because of the variation of tropospheric transparency with radiative wavelength. However, it is reasonable to assume that the value of T_2 should be close to the temperature of the highest opaque layer of the atmosphere (section 8.4). Above this level, which is in the upper troposphere on account of the vertical distribution of water vapour, cloud and carbon dioxide, convection and advection are insignificant and the outward energy flux is carried mainly in the form of terrestrial radiation with no further dynamical significance. Notice that the heat flux through the tropospheric engine in Fig. 13.1 does not contain the considerable but purely radiative fluxes passing through the window and in the cascade of absorption and emission (section 8.4), since they pass through the atmosphere without requiring or encouraging any atmospheric motion (a point considered further in section 13.4).

The oceans interact thermally and dynamically with the troposphere, and so in principle should be represented in Fig. 13.1. But, as mentioned in Chapter 9, the oceans are dynamically passive in that they are driven by the troposphere to pump heat polewards in surface currents like the Gulf Stream. In addition their huge thermal capacity has a powerful smoothing effect on seasonal variations in surface temperature. Hence the main effects of the oceans are to assist the atmosphere in the advection of heat and to smooth the seasonal and latitudinal variations of the temperature T_1 of the heat input to the tropospheric engine. There is a small, intrinsically oceanic engine which transports heat polewards, and which is driven by the descent of cold, dense polar waters, but the heat flux it carries is trivial in comparison with the wind-driven flux.

In comparison with the troposphere, the stratosphere is relatively inactive, as mentioned in Chapters 3 and 8. In the radiative budget of the stratosphere (Fig. 8.6) the absorption of solar ultraviolet is quite closely balanced by the output of terrestrial radiation, according to the scheme outlined in section 3.2. Actually the upper stratosphere in particular is far from inactive, but the amounts of mechanical energy corresponding to the motion of such a tenuous gas are very small in comparison with tropospheric values. In fact there must be slight imbalances of the stratospheric budget because the low stratosphere in particular is known to act as a heat pump (a heat engine working in reverse), forcing heat from the relatively cool and high equatorial zone (at temperature T_3) to the warmer and lower polar zone at T_4. The mechanism is believed to involve synoptic-scale eddies driven by a small leakage of mechanical energy from tropospheric long waves [44]. For simplicity friction has been ignored in the stratosphere, and the temperatures of source and sink assumed to have single values as usual. After transformation, the heat pumped to the polar regions contributes a small component to the outward flux of terrestrial radiation there.

13.2 Forms of energy

Energy exists in the atmosphere as mechanical energy and as heat. Mechanical energy comes in two forms: kinetic energy, which is the energy which a body has an account of its motion, and potential energy, which is the energy it has on account of its position in a field of force. Since gravity is the only force field which is significant in the lower atmosphere (electrostatic and electromagnetic forces are relatively very weak even in the immediate vicinity of thunderstorms), the only type of potential energy we have to consider is gravitational potential energy. This will be called simply potential energy from now on, the 'gravitational' being understood. (In the upper atmosphere by contrast electromagnetic forces can be very important [11].) An expression for the rate of change of mechanical energy of an air parcel of unit mass can be derived from the equation of motion, as shown in Appendix 13.1.

$$\frac{d}{dt} (\tfrac{1}{2} V^2 + g\,z) = PGW - FW \tag{13.1}$$

The terms in brackets on the left-hand side are the familiar expressions for the kinetic and potential energies of a body moving with speed V at height z above an arbitrary datum level, which is usually mean sea level in the atmospheric case. The terms on the right-hand side are written as convenient initials to avoid needless complication. PGW represents the rate of working of ambient pressure gradients on the parcel, and FW represents the rate of working of external friction. Notice that there are no Coriolis terms in the equation. As shown in appendix 13.1 these disappear because their associated apparent forces do not work and hence have no effect on any type of energy. If the right-hand side is zero, then the sum of the kinetic and potential energies of the parcel is fixed. However, the apportionment between the two energy forms varies with parcel height, kinetic increasing at the expense of potential when the parcel falls and vice versa. This is the familiar behaviour of a freely falling body, but the tendency remains even when the right-hand side is not zero, and is an example of the conversion from one type of energy to another which results from the action of a force, in this case the force of gravity. In a neutrally stable atmosphere (section 5.10) the conversion is exactly opposed by the working of the vertical pressure gradient, with the result that no kinetic energy is evolved as the parcel moves up or down.

Heat energy exists in the atmosphere as both sensible and latent heat. The first is measured by the internal energy of the air parcel (section 5.2), which is the kinetic energy of its constituent molecules in their chaotic thermal motion, and the second is the energy which is latent in the presence of water vapour, and can be realized (i.e. converted to sensible heat) by condensation. For simplicity we will retain only the sensible heat terms in the following. This omission will not matter too much when we consider large masses of air rather than small parcels because the role of latent heat is then largely internal, for example tending to raise the altitude of effective solar warming towards cloud level without altering the total rate of input. We can derive an expression for the rate of change of internal energy very quickly from the first law of thermodynamics, as mentioned in Appendix 13.1:

$$\frac{d}{dt} (C_v\,T) = \frac{dQ}{dt} + PC \tag{13.2}$$

The first term on the right-hand side is the diabatic heating term, which in the present context is the parcel's net rate of gain (or loss) of heat by radiation, either by direct atmospheric absorption and emission, or by convective contact with a radiatively warmed or cooled surface, or both. The term PC represents the compressing effect of ambient pressure in raising internal energy. When dQ/dt is zero, eqn (13.2) describes the adiabatic warming or cooling of an air parcel as discussed in Chapter 5.

We find the expression for the rate of change of the total energy (kinetic plus potential plus internal) of a parcel simply by adding eqns (13.1) and (13.2).

$$\frac{d}{dt}(\tfrac{1}{2}V^2 + g\,z + C_v\,T) = \frac{dQ}{dt} + PW - FW \tag{13.3}$$

The term PW is the sum of PGW and PC, and represents the total rate of working by the pressure field on the parcel.

In dealing with the atmosphere on a large scale we are not so much interested in the instantaneous behaviour of individual air parcels as with the average behaviour of large air masses. This might seem likely to complicate matters, since we will have to sum each of the very variable terms in eqn (13.3) throughout extensive volumes and over considerable time periods. But though it is true that a full treatment leads to involved algebra, there are some very important ways in which the aggregated form of eqn (13.3) is simpler than a full aggregation of parcel expressions.

First let us consider a column of air extending from the Earth's surface to the top of the atmosphere. The column total of internal energy per unit horizontal area is simply

$$IE = C_v \int_0^m T\,dm \tag{13.4}$$

where dm is the mass per unit area of a very thin horizontal slice of the column, and m is the total mass of the atmosphere resting on unit horizontal area. A similar expression gives the column total kinetic energy:

$$KE = \tfrac{1}{2} \int_0^m V^2\,dm$$

However, if we assume hydrostatic equilibrium, as we can to a high level of accuracy in most atmospheric behaviour, we find (Appendix 13.2) that the column total of potential energy can be transformed from its obvious form

$$PE = g \int_0^m z\,dm$$

into the expression

$$PE = R \int_0^m T\,dm \tag{13.5}$$

where R is the specific gas constant for air. Comparison of eqns (13.4) and (13.5) shows that PE is directly proportional to IE, the constant of proportionality being R/C_v. The physical reason for this is fairly obvious: in hydrostatic equilibrium the vertical distribution of atmospheric mass, and hence of potential energy, is determined by the air density, which is inversely proportional to absolute air temperature according to the equation of state. If air temperature rises, air density falls and the column expands vertically, raising its potential energy in step with its internal energy. It follows from eqns (13.4) and (13.5) and the definition of C_p that we can define a *total potential energy* (TPE) which is the sum of PE and IE and is given by

$$TPE = C_p \int_0^m T\,dm \tag{13.6}$$

From one viewpoint this describes the effective heat capacity of the column of air (another example of the general relevance of C_p as distinct from C_v — see section

5.3 and appendix 13.3). For example, in the simple case of a column extending to the top of the atmosphere with uniform temperature T and mass m per unit area, the heat capacity per unit area is $C_p m$, and the temperature will rise by $Q/(C_p m)$ if there is a heat input of Q. From another viewpoint the relation

$$TPE = IE + PE = \frac{C_v}{C_p} TPE + \frac{R}{C_p} TPE$$

describes the apportionment of heat input between IE and PE, and suggests that both are reservoirs of energy potentially capable of conversion into kinetic energy (KE). So whereas all fluids and mobile material can convert PE into KE, a readily compressible fluid like air can in addition convert IE in ways which we will not consider in detail but which obviously include the loss of IE in adiabatic decompression.

Now let us now extend the aggregation of the column-totalled terms in eqn (13.3) to cover a wide horizontal area bounded by vertical walls thousands of kilometres apart. The pressure working term effectively vanishes since there is no pressure working across the top or bottom surfaces, and the import and export of energy across the relatively very shallow walls is very small. (The constituent PW terms act overwhelmingly to redistribute energy within the very large bounded volume.) To this extent the equivalent version of eqn (13.3) for the very large volume simplifies to

$$\frac{\mathrm{d}}{\mathrm{d}t}(KE + TPE) = \frac{\mathrm{d}Q}{\mathrm{d}t} - FW \tag{13.7}$$

where the terms now represent horizontal aggregations of the column totals mentioned above. If the atmospheric volume is extended to cover the entire globe, or one hemisphere averaged over one year, then the observed nearly steady state implies that the aggregated total energy $KE + TPE$ is nearly constant and the left-hand side of eqn (13.7) is as nearly zero. In an adiabatic and frictionless atmosphere this is clearly consistent with the fact that each term on the right-hand side is permanently zero. In the real atmosphere it implies that the diabatic and frictional terms effectively cancel. At first sight such a balance seems unlikely, since we have assumed at many previous points that the Earth's surface and atmosphere are in thermal equilibrium in the long term, implying zero $\mathrm{d}Q/\mathrm{d}t$. However, the $\mathrm{d}Q/\mathrm{d}t$ in eqn (13.7) is the portion of the heat input to the atmospheric system which is converted into mechanical energy. After this and subsequent conversions associated with details of the dynamical behaviour of the atmosphere, it is reconverted back to heat by frictional working FW, producing heat at a rate which, at least on the large space and time scales assumed here, exactly balances the diabatic heat input $\mathrm{d}Q/\mathrm{d}t$. The significance of the negative frictional term should now be clear, as should its role in restraining the atmospheric motion maintained by the diabatic heat input. The result is that although heat has been temporarily extracted to drive the atmospheric system, overall thermal equilibrium is maintained by its return through frictional degradation of the kinetic energy of atmospheric motion.

Let us now consider the flow of energy through the global atmosphere assuming a steady, dynamic state (Fig. 13.2). The flow follows the scheme outlined verbally in the previous section: energy enters in the form of net radiative input Q which warms the air in a variety of direct and indirect ways. As the air warms it converts the input into TPE. The warmer, taller columns of air may then give rise to atmospheric motion in ways to be described, converting some of the TPE further into KE. Once in kinetic form the energy begins to be degraded by friction, ending

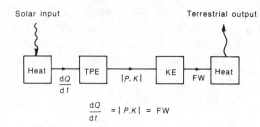

Fig. 13.2 Schematic outline of the flow of energy through the atmosphere in a thermally steady state. Terms are defined in the text.

up as heat which is exhausted to space as part of the terrestrial radiation. According to the steady-state assumption, the rate of input dQ/dt, the rate of conversion from *TPE* to *KE* (represented by $\{P.K\}$), and the rate of frictional degradation of kinetic energy are all equal.

The boxes in Fig. 13.2 represent the reservoirs of *TPE* and *KE* which are maintained by the dynamic equilibrium of the atmosphere, and we can easily estimate their approximate values. Assuming a mean atmospheric temperature of 260 K and mass per square metre of 10 000 kg, the *TPE* in a column resting on unit horizontal area is 2.6×10^3 MJ m^{-2}. However, typical wind speeds of about 15 m s^{-1}, together with the same mass, are equivalent to a *KE* value of just over 1 MJ m^{-2}. This is very much smaller than the pool of *TPE*, but a little thought shows that any surprise we feel at this inequality arises from a quite unwarranted assumption that *TPE* should be completely converted to *KE* by atmospheric systems. Zero *TPE* would be reached only if the atmosphere were reduced to absolute-zero temperature and zero volume at sea level. In reality of course the temperature of any substantial part of the atmosphere is unlikely to fall far below the appropriate average value, which means that a large part of its *TPE* is unavailable for conversion to *KE*. It is therefore more meaningful to replace *TPE* by the very small portion of it which is *available potential energy* (*APE*), i.e. realistically available for conversion to kinetic energy. Though the large unavailable portion corresponds to the warmth and other vital static aspects of the atmosphere, it plays no role in the chain of energy transformations.

We can get an impression of just how little *TPE* is available by considering a very simple incompressible model of mid-latitude baroclinity. Two columns of incompressible 'air' stand side by side, and are initially identical. Then one is heated so that it expands vertically, increasing its *PE* by raising its centre of gravity. The columns are now as depicted in Fig. 13.3(a) — a set-up which is clearly unstable in the sense that the columns can be rearranged to lower their common centre of gravity, freeing some of their *PE* (*IE* is irrelevant in an incompressible fluid) for conversion to *KE*. A very simple re-arrangement is depicted in Fig. 13.3(b), in which the denser air spreads horizontally under the heated air, which spreads evenly over the top without mixing. Calculation of *PE* before and after rearrangement (appendix 13.3) shows that the fall in *PE* as a fraction of the total *PE* is given by

Fig. 13.3 Vertical columns of more and less dense fluid (a) before and (b) after rearrangement to minimize their gravitational potential energy.

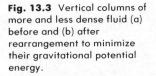

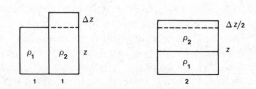

$$\frac{\triangle PE}{PE} = \frac{\triangle \rho}{4\rho}$$

if the fractional difference $\triangle \rho/\rho$ between initial densities is small. Being in fact a perfect gas, the thermal expansion of air is such that a small fractional change in density is equal to the small fractional change in temperature causing it, so that the fractional drop in PE can be written in the following form to relate to atmospheric conditions:

$$\frac{\triangle PE}{PE} = \frac{\triangle T}{4T} \tag{13.8}$$

If we imagine that the model represents adjacent columns of polar and tropical air, then $\triangle T/T$ is about 20/300, and it follows that in this very simple case only about one sixtieth of the potential energy is made available by rearrangement. Figure 13.4 also depicts the even more obvious instability which results when denser air overlies less dense air. When the layers are vertically interchanged without mixing, the fall in PE is twice that in eqn (13.8), but the corresponding atmospheric value is smaller because convection limits $\triangle T$ to a very few degrees by preventing accumulations of convective instability.

(a)   (b)

Fig. 13.4 Horizontal layers of more and less dense fluid (a) before and (b) after rearrangement to minimize their gravitational potential energy. Condition (a) is convectively unstable.

Calculations of this sort have been made for a wide variety of model atmospheres since they were first attempted by the Ukrainian pioneer dynamicist Margules in 1903 when he invented the concept of APE. The values of APE/TPE which we have just found are in fact rather large: considerably smaller values are found when continuous gradients of density are allowed, and when the slight but systematic vertical stability of most of the real atmosphere is included in the baroclinic case (convective instability being confined to a small fraction of the atmosphere at any instant). Regardless of detail, all models show that an essential requirement for the release of any potential energy is that the initial warming should be inhomogeneous. If the two columns in Fig. 13.3(a) had been heated equally there would have been no APE at all, as would have been the case in the convective model if the upper layer had been heated as much as the lower. Even the simplest examination of APE therefore points to the very important fact that not all heat inputs will lead to the production of KE — spatial inequality of heat input is crucial. And even when there is production of KE, the conversion from PE (TPE in the compressible case) is very small.

In the actual compressible atmosphere the available potential energy of any particular configuration can be defined as the maximum TPE which can be made available by adiabatic rearrangement (being compressible there are temperature changes as the air rearranges but no heat is gained or lost overall if the movements are adiabatic). The necessarily detailed and complex calculations needed to find the APE of typical dispositions of atmospheric densities (in fact potential density, by analogy with temperature and potential temperature) indicate that in the real atmosphere rather less than 0.5% of the TPE is available at any instant. This means that the APE corresponding to the initial estimation of typical atmospheric TPE is probably less than 13 MJ m^{-2}, much nearer the estimated KE value, though still somewhat larger. In Fig. 13.2 therefore the *TPE* box should be relabelled *APE*, and its much smaller value understood, but the rate of transformation $\{P.K\}$ to *KE* is unchanged in both physical significance and numerical value. If we regard the value of *APE* as a measure of the degree to which the atmosphere is kept off balance by the continual throughput of solar energy, then it seems that this imbalance is very small, presumably because the solar input is fairly gentle, its gradients quite small, and the mobility and rapidity of response of the atmosphere are sufficient to prevent any sizeable accumulation of available potential energy.

13.3 Energy paths

The chain of energy transformations leading to atmospheric motion begins in the unequal heating of the atmosphere by imbalance or warming by solar radiation and cooling by terrestrial radiation. Of the two main inequalities, the concentration of net warming in low latitudes leads to the motion systems associated with large areas of substantial horizontal winds such as the trades and the mid-latitude westerlies, including the much smaller but still quite extensive areas of very strong horizontal winds such as jet streams. By contrast, the concentration of net warming in the base of the troposphere leads to widespread but highly localized and frictionally limited vertical motion associated with the whole range of vertical convection, from the shimmering dance of air over heated land surfaces to towering cumulonimbus. Because observation and theory of vertical convection is still relatively incomplete, we will concentrate in the following on the large-scale systems of horizontal winds, using the energy types introduced in the previous section to outline cycles of energy transformation and, in the process, summarizing many aspects of the general circulation of the troposphere.

We must expect the variation of net radiative warming with latitude to maintain APE in the form of warmer and taller air columns in lower latitudes which will continually tend to over-run their shorter and cooler neighbours at higher latitudes (like Fig. 13.3 in principle), producing *KE* in the form of horizontal winds in the process. The thermal-wind relation associated with the meridional temperature gradient through the quasi-geostrophic nature of flow in all but the lowest latitudes would lead us to expect these winds to be westerlies. In conventional jargon this would be called conversion from *zonal APE* to *zonal KE*, written as Az and Kz respectively in Fig. 13.5, where the rate of this conversion is represented by $\{Az.Kz\}$. The actual algebra of the terms represented by such symbols is quite complex, and varies with the precise nature of the averaging used and the simplifications made in the theoretical framework, but extensive work in recent decades has produced a fair consensus of data and interpretation. Data from the northern hemisphere still predominate, and behaviour is more clearly observed and understood in middle latitudes than it is close to the equator, but the global annual picture is agreed in broad outline, together with seasonal and gross meridional deviations. It seems clear from this that the expected direct conversion from zonal *APE* to zonal *KE* is relatively unimportant in the annual picture.

The explanation for this surprising result depends largely on the presence of the long (planetary) waves which are such a basic feature of the middle and upper troposphere in middle latitudes (Fig. 11.22). Though the zonal *APE* is certainly the first stage in the energy cycle after radiative heat input, it is apparently followed by a radical redistribution of *APE* in which huge fingers of warm air from low latitudes are interleaved around the hemisphere with equally large fingers of cold air from high latitudes. In technical terms there is a transformation from zonal *APE* to *eddy APE* (*Ae* on Fig. 13.5) at a rate symbolized by {*Az.Ae*}, the symbols again representing complex terms arising from the aggregated energy equations. The term 'eddy' here refers to any deviation from a hemispheric average around a zone of latitude, and therefore includes the effects of monsoons and other geographically tied variations, as well as the mobile cyclone waves and planetary waves of middle latitudes. Descriptively, the initial contrast between hemispheric zonal rings of warmed air in low latitudes and cooled air in high latitudes has been localized and sharpened in the contrasts between adjacent warm and cold fingers.

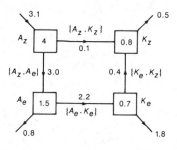

Fig. 13.5 Energy reservoirs and rates of conversion from data covering 12 months in the northern-hemisphere troposphere. Fluxes and conversion rates are expressed in W m^{-2} and reservoir contents (within box) are expressed in MJ m^{-2}. (Adapted from [44])

The stage is now set for the conversion of *APE* to *KE*. The weather systems which feed on and maintain the baroclinity of the middle latitudes convert eddy *APE* to *KE* by sporadic, repeated eruptions of baroclinic instability, producing families of depressions and their associated fronts and jet streams. The conversion is mainly to *eddy KE* in the first place almost by definition, since planetary and cyclone waves are associated with localizations of *KE*, such as jet stream cores and areas of surface gales. The rate of transformation is symbolized by {*Ae.Ke*}.

The crucial feature of all systems converting *PE* to *KE* is that less dense (i.e. warmer) air rises while more dense air sinks, so that the overall centre of gravity falls, just as in the simplest model (Fig. 13.3). This is known as a *thermally direct* circulation, and all self-driving motion systems ranging from centimetric-scale convection to extratropical cyclones and the Hadley circulation are of this type. It is sometimes said that all systems which produce *KE* must therefore be warm-cored, which is ultimately true, but may not be very helpful when the system has a complex structure. For example, on a superficial view we might be tempted to regard the core of a mature depression as the axis of the pressure minima at successively higher levels which slopes upwards and westwards from the surface low, especially since the latter is often called the centre of the low-pressure system. Since this 'core' contains relatively cool air (in fact the pressure trough aloft is enhanced by the consequent depression of thickness), and the presence of some precipitation shows that there is some rising air, it would therefore seem that the depression is a thermally indirect system, consuming *KE* by converting it into *PE*. However, more perceptive observation suggests that the warm conveyor belt (Fig. 11.15) is the true warm core, together with any cooler conveyor belts there may be which are still warmer than the area average, and measurement shows that their *KE* production far outweighs *KE* consumption in rising cold air. In fact any complicated system is likely to have parts in which local energy conversions run counter to the overall trend, i.e. localized thermally indirect parts which are driven by the main thermally direct circulations nearby.

The sporadic and extensive production of eddy *KE* in the large weather systems of middle latitudes maintains the endless westerly flow there by subsequent conversion to zonal *KE*. In fact westerly zonal motion in a ring of air bounded by a lower latitude is enhanced by large-scale eddies if poleward motion across that latitude is systematically associated with stronger westerly flow, and equatorward flow with weaker westerlies. In the northern hemisphere this means that southerly and westerly winds should be positively correlated, which is generally observed to be the case, though it does not apply in every instance. For example such correlation is apparent in the fairly obvious tendency for the south-westerly jet associated with the cold front (which includes the warm conveyor belt in its lower parts) to be stronger than the north-westerly jet lying ahead of the warm front. It is also apparent in the tendency for the axes of cyclone waves to be angled somewhat from south-west to north-east (Fig. 11.10). More fundamentally, both of these correlations result from the tendency to conserve angular momentum as air moves polewards and reduces its perpendicular distance from the Earth's axis of rotation (section 7.14). The resulting rate of conversion from eddy *KE* to zonal *KE* is represented by the symbol $\{Ke.Kz\}$.

These conversions and others to be mentioned below are depicted in Fig. 13.5, together with numerical values taken from one of many detailed budgets calculated from annual data from the northern hemisphere [44]. The production of *APE* from net radiative warming of low latitudes and cooling of high latitudes starts at the top left, and proceeds at a rate of just over 3 W m^{-2}, when expressed as an energy flux per unit area of the hemispheric surface. The precise numerical value varies considerably between analyses, but is never more than a few watts per square metre, which is only a small fraction of the vertical heat flux borne by large-scale motions of the atmospheric engine (~ 50 W m^{-2}, according to section 9.5). This rate of production maintains a reservoir of *APE* which is only 4 MJ m^{-2}, compared with the figure of nearly 3000 MJ m^{-2} in the earlier estimation of *TPE*. The ratio of reservoir content to rate of throughput defines a residence time of just over two weeks for energy in the form of zonal *APE*, with a further five days in the form of eddy *APE*. Most but not all of this *APE* is in turn converted into eddy *KE* in large-scale weather systems. The importance of middle latitudes can be gauged from energy budgets for individual vigorous depressions, which show that conversion rates approach 20 W m^{-2} over the affected areas. It seems that four or five of these systems in action around a hemisphere at any one time can account for a substantial proportion of the total conversion. Notice that the reservoir of atmospheric *KE* is almost equally divided between the eddy and zonal components, although in terms of rates of conversion, only one fifth of the eddy *KE* is passed on to zonal *KE*. The implied residence times for energy are nearly four days in the form of eddy *KE*, which is consistent with the significant role assigned to synoptic-scale weather systems. The much longer residence time of over 20 days in the form of zonal *KE* is consistent with the view that the reservoir of zonal *KE* acts as a big zonal flywheel kept going by the rather small flux of energy reaching the end of the sequence of transformations.

Energy is lost from three corners of the array in Fig. 13.5. The first of these in order of energy flow corresponds to radiative wastage of eddy *APE*, in which warm fingers of air lose some of their warmth and height by net radiative cooling before their contribution to the total *APE* can be converted into *KE*. Some of the original input has therefore been wasted, from the point of view of production of atmospheric *KE*, by what is termed *radiative dissipation*. No doubt a more detailed picture of the earliest stages in the energy cycle would reveal temporary production of even more zonal *APE* than is depicted in Fig. 13.5, followed quickly by wastage through radiative dissipation. It therefore appears that not only is most of the atmospheric *TPE* unavailable for conversion into kinetic energy, but in addition a substantial fraction of what little is available is not in fact converted. The atmospheric engine is in this sense rather inefficient — a fact which is confirmed by a more fundamental criterion in the next section.

Warm anticyclones such as those in the subtropics provide clear examples of radiative dissipation. Their domes of warm air represent sizeable contributions to both zonal and eddy APE. However, these largely static warm cores are particularly prone to radiative cooling because of the near absence of cloud above the subsidence inversion (section 12.1), so that they are continually collapsing, wasting APE which could otherwise have been converted into KE, and maintaining their cloudless cores by the associated subsidence.

The frictional dissipation of KE appears on the right-hand side of the array in Fig. 13.5. The relatively large value of the destruction of eddy KE as compared with zonal KE arises because the highest wind speeds in both the low and high troposphere appear in the eddy category almost by definition. In particular, the strong wind shears in the boundary layers of vigorous depressions act as powerful concentrations of frictional loss. On a simple view, the rate of destruction of KE in the boundary layer is proportional to the cube of the geostrophic wind speed (appendix 13.4), biasing the loss heavily toward regions of strong winds. The presence of large-scale horizontal convergence in the lower parts of cyclonic disturbances maintains strong winds and associated frictional dissipation by conservation of angular momentum (section 7.14). There is some significant frictional dissipation at higher altitudes, particularly in the regions of strong vertical shear below and above the cores of jet streams, where clear-air turbulence (section 10.8) is evidence of turbulent dissipation of larger-scale flow and associated KE.

By contrast, the gentler shears associated with large-scale zonal motion, both in and above the boundary layer, maintain a relatively small rate of frictional dissipation per unit area and allow the reservoir of zonal KE to remain comparable with the eddy KE despite the much smaller throughput of kinetic energy. The zonal 'flywheel' therefore runs relatively freely on a small conversion from eddy KE and a very small leakage from zonal APE.

The values of sequences of Fig. 13.5 correspond to the annual average picture of the general circulation of the hemisphere. If the seasonal extremes are examined separately (and this complicates the analysis considerably) the direct conversion from zonal APE to zonal KE, expected at the beginning of this section, reappears, especially in winter. Examination shows that this arises mainly from the action of the Hadley circulation of low latitudes (section 8.13) — the huge thermally direct circulation powered by the liberation of latent heat in the intertropical convergence zone. The conversion rate $\{Az.Kz\}$ is affected by the conservation of angular momentum as air in the upper troposphere diverges toward the subtropics (section 4.7). Some of the energy flow therefore tends to cut across the top of the equivalent of Fig. 13.5, partly short-circuiting the eddy forms of energy. Analysis by latitude zones confirms this interpretation, and shows that in middle latitudes there is conversion in the opposite sense (from zonal KE to zonal APE) in the thermally indirect Ferrel cell (section 4.7) which in simpler data such as average winds is all but swamped by the unsteady eruption of planetary and cyclone waves. In the northern-hemisphere summer the Asiatic monsoon blurs the effect of the Hadley circulation on the energy sequence by introducing eddy components at lower latitudes. The smallness of $\{Az.Kz\}$ in the hemispheric annual picture (Fig. 13.5) therefore arises because of opposing conversions in different latitude zones, as well as the partial disruption of the Hadley circulation by the summer monsoons.

As mentioned at the beginning of this section, we have deliberately concentrated on the energy pathways in large-scale weather systems rather than the pathways in convective (cumulus-scale) systems, because the latter are so poorly understood. In fact the energy fluxes through the two types of system are believed to be comparable on a global scale, as suggested by the comparability of their associated heat fluxes (Chapter 8). The two types are similar in that they represent the temporary conversion of APE into KE, though the APE for convection arises from the convectively unstable vertical stratification of density, whereas the APE for large-scale weather systems arises from the baroclinically unstable horizontal zoning of density. However, unlike the case in large-scale systems, the APE in convection is converted into KE very quickly and locally, and is similarly reduced to heat by the overwhelming combination of turbulent form drag and entrainment drag (section 7.12), with the result that motion is transient and quickly stifled in all but the most vigorous storms. The effect of all this friction is to keep the reservoir of associated kinetic energy very small indeed although the energy throughput is comparable with that of large-scale motions. This is of course consistent with observation — vertical wind speeds are an order of magnitude down on horizontal wind speeds even in the active cores of cumulus, and drop by another order on averaging through large volumes of the atmosphere. (In principle it is not correct to assume that all KE associated with cumulus is associated with vertical motion, but in terms of contribution to global budget, gales and tornadoes in cumulonimbus are exceptions which prove the general rule.) The smallness of the ratio of reservoir to throughput is simply a measure of the shortness of the residence time of energy in the form of KE, which is consistent with the observed brevity of cumulus lifetimes. The weakness, transience and turbulent complexity of cumulus systems severely hinder our observation and understanding of them and their energy pathways, so that nothing comparable in measured detail with the scheme for large-scale motion has been deduced so far. Some of the most important energetic differences between large and convective scale types of motion are summarized in the following graphic analogy drawn by the writers of a recent textbook [8]: large-scale motion systems are like a person coasting down a certain hill on a racing bicycle, whereas convective systems are like a person coming down the same hill on rusty roller skates. Obviously the cyclist reaches much higher speeds than the skater, although the same initial potential energy is available to each.

Consider finally a descriptive outline of the energy pathways and transformations in the stratosphere (Fig. 13.6). The temperature profile is so stable convectively that the stratosphere cannot support any self-driven activity — even baroclinic instability is suppressed. In fact it seems likely from detailed studies that such muted activity as it shows is driven by eddy kinetic energy propagating up from the great planetary waves of the troposphere (bottom right in Fig. 13.6). Strictly speaking this leakage from the troposphere should appear as a component in the tropospheric cycle in Fig. 13.5, but is omitted because it is smaller than the uncertainties in the numerical values there. Because there is so little friction, this input to the stratosphere maintains a very considerable zonal circulation and eddy and zonal APE by a thermally indirect circulation in which warmer air sinks and

Fig. 13.6 Energy reservoirs and flows in the stratosphere. (After [44])

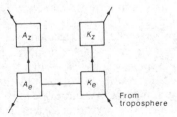

cooler air rises. This sinuous zonation of *APE* then leaks heat to space by weak net radiative cooling which is concentrated in the relatively warm, low stratosphere of higher latitudes. The reversed flow of energy in Fig. 13.6 as compared with Fig. 13.5 in fact corresponds to the action of a heat pump rather than a heat engine, the heat being taken from weak net radiative warming in the cold, high base of the equatorial stratosphere, and pumped to high latitudes by the input of mechanical energy from beneath. It is a kind of gentle refrigerator, keeping the equatorial stratosphere cool by pumping the small radiative excess of heat to higher latitudes.

The life cycle of atmospheric mechanical energy is now complete in outline: produced largely in the low troposphere from a portion of the heat input to the sun-warmed surface, most is degraded to heat again in the troposphere by friction of various types, while a very small part is used to drive the stratospheric heat pump.

The prominent role of friction in the tropospheric engine apparently differs from that in man-made engines, which are of course designed to have minimum friction. But the proper comparison is with man-made engines including the machines that they drive. When this comparison is made, then in man-made engines too, all mechanical energy is eventually degraded to heat, mostly by the action of friction, though this may happen hundreds of kilometres from the engine proper if the power has been transmitted electrically to the point of use. Together with the large throughput of heat enforced by their necessarily limited mechanical efficiency (see next section), this inevitable degradation of mechanical energy to heat means that all energy passing through man-made engines is converted to heat sooner or later, either at source or elsewhere. Man's engines are now so numerous and powerful that unintentional heating is having significant effects in localized areas such as cities, rivers and estuaries. The effects may soon become globally significant too: if our economic, social and industrial systems permit a modestly increased population of five thousand million to produce energy at the *per capita* rate of citizens of the United States in 1950 (Appendix 13.5), then the total rate of artificial heat production will exceed 1% of the flow of mechanical energy through the atmosphere. If the necessary power were to be drawn from sunshine, wind, waves or tides, then it would be little more than a temporary redirection of part of the natural energy flow, but of course at the moment the overwhelming proportion of the power comes from coal, oil, nuclear fission and the irreversible destruction of trees, and is therefore in addition to the natural flow.

13.4   Efficiency of the atmospheric engine

The first law of thermodynamics is a budgeting law, requiring that energy accounts always balance: the second law of thermodynamics shows that not all balancing budgets are permissible. For example, in terms of Fig. 13.1, the first law requires of the tropospheric engine that

$$Q_1 = Q_2 + M \tag{13.9}$$

The second law states that heat must flow from a warm source to a cool sink if a positive quantity of mechanical energy M is to be produced by conversion, however temporarily. In terms of Fig. 13.1, the second law requires that T_1 be higher than T_2. Applied to the stratospheric heat pump, the second law requires that mechanical energy must be supplied if heat is to be taken from a cooler source and delivered to a warmer sink. The stratosphere could not for example pump heat from T_3 to a higher T_4 and produce mechanical energy. More specifically, the first and second laws together set an upper limit to the fraction of the heat input to an engine which can be converted into mechanical energy, i.e. to the *mechanical efficiency* of the engine. This limit is reached in an ideally efficient engine in which there is no friction and in which processes are so smoothly ordered as to be reversible by a minute alteration of conditions. In the terminology of Fig. 13.1, this reversible limit is given by

$$\frac{M}{Q_1} = \frac{(T_1 - T_2)}{T_1} \qquad (13.10)$$

as shown in appendix 13.6. Actual engines must be less efficient than this to the extent that they include friction, which of course they all do, and involve processes which are not in equilibrium, which is also always the case, but there are no simple criteria for estimating this shortfall.

From eqn (13.10) it is clear that we can estimate the optimum mechanical efficiency of the tropospheric engine provided that we can define the source and sink temperatures T_1 and T_2. If the value of the heat input Q_1 is known in addition, then the optimum production of mechanical energy can be found at once, and this places an inescapable upper limit on the mechanical activity of the atmosphere. Note that the terminology of Fig. 13.1 does not specify whether the Qs and M are quantities of energy or rates of energy (powers), since the first and second laws of thermodynamics apply in either case. For comparison with the analysis of the last section it is natural to consider powers.

The simplifications of section 13.1 will be accepted to identify reasonable single values for T_1 and T_2. The global annual average temperature of the Earth's surface is 288 K, and this will be assumed to be the value for T_1. The effective radiative temperature of the planet Earth is close to 255 K (section 8.2), and this will be accepted as the value for T_2, though it must in principle be a little above the temperature of the highest opaque layer (the most obvious temperature to choose for the heat sink) on account of the small output coming through the atmospheric window from the relatively warm surface. It follows that to the extent that these values are realistic, the mechanical efficiency of the tropospheric engine cannot possibly exceed 33/288, or approximately 11%. Because of the relatively small temperature difference between source and sink, therefore, at least 89% of the heat taken into the tropospheric heat engine must pass through without conversion into mechanical energy, and be wasted to the heat sink. (That is wasted from the point of view of the mechanical energy of the engine — not of course from the viewpoint of the comfort of the inhabitants of middle latitudes who bask in the warm draught.) This is apparent in the convective and advective currents of warm and cold air and water described in Chapter 8 and elsewhere and which are associated with the whole range of weather systems large and small. It is likely that a fuller allowance for the complex multiplicity of realistic values for T_1 and T_2 would result in an even smaller value for the maximum mechanical efficiency and an even greater heat transfer without mechanical effect. And of course the irreversible aspects of actual atmospheric behaviour will depress the mechanical efficiency still further.

It is difficult to identify precisely the value of the heat input Q_1 to the tropospheric engine. Considering the scheme of Fig. 8.6, it seems reasonable to focus on the convective heat flux from the underlying surface, since many of the radiative fluxes are not necessarily associated with the existence of a heat engine at all. For example, the sunlight scattered out to space plays no role, and the net flux of terrestrial radiation between the surface and the base of the troposphere could continue as well in a completely static atmosphere. (You cannot be too dogmatic about this though, since the distributions of cloud, water vapour, carbon dioxide and temperature which determine this flux are all themselves maintained by atmospheric motion.) The vertical convective flux which balances the radiative budget corresponds to a global annual average of 30 units or just over 100 W m^{-2}. If we choose this for Q_1 we may be neglecting advection between latitude zones which is implicit in the solar input of 19 units (about 65 W m^{-2}) to the troposphere, since some of this is likely to be offset by net long-wave cooling at higher latitudes rather than locally. However, the nearly parallel gap between the meridional profiles of radiative input and output in Fig. 8.7(b) is very close to 100 W m^{-2} at all latitudes, suggesting that this is a value which fairly represents the combination of convective and advective heat fluxes. So we will identify this value with Q_1 for the purposes of estimating the maximum possible mechanical efficiency for the atmospheric engine.

Actually a more detailed examination would almost certainly reduce the value for Q_1, since a sizeable fraction of it must correspond to heat fluxes passing across temperature differences considerably less than the 33 °C assumed above. You can practically see this effect in the vertical distribution of cloud tops in a field of cumulus, tall clouds being so much rarer than shallow clouds. And it is confirmed by the rapid decrease of convective heat flux with increasing height apparent in Fig. 8.11. But since we have no good way of judging the size of the overestimation, let us accept the 100 W m^{-2} as a round figure, knowing that it is likely to be a substantial overestimate.

We have now established that the maximum mechanical efficiency of the atmospheric engine is 11% and that the input of heat from the heat source is 100 W m^{-2} at most. It follows from eqn 13.10 that the maximum rate of production of mechanical energy by the engine is 11 W m^{-2}, which is quite reasonably consistent with the value of 3 W m^{-2} indicated by the more detailed studies quoted in the previous section, bearing in mind the simplicity of the argument, and the fact that it establishes an upper limit rather than a likely value. Some of the difference between the two values no doubt arises from the difficulty of determining T_1, T_2 and Q_1, especially the last, but some must also arise because the atmospheric engine runs substantially below the thermodynamic optimum because of irreversible processes. The value of this simple analysis is that it shows that, as well as any incidental inefficiencies arising from particular details of atmospheric behaviour, there is an inescapable limitation to its efficiency overall, which is actually quite severe.

13.5 Epilogue

So in the end we find that the incessant, often impressive, and occasionally very destructive activity of the atmosphere is maintained by a conversion to kinetic energy of only a small fraction of the total flux of energy driven through the atmosphere by the Sun, which in turn is only an infinitesimal part of the Sun's prodigal output. The enormous power of the hurricane, the trade winds which dominate over one third of the Earth's surface, the gales which so frequently trouble the seas and western continental margins in middle latitudes — all are manifestations of an average rate of production of kinetic energy in an atmospheric column resting on an area the size of a domestic table which could barely light a small light bulb if it were entirely converted to electrical power. However, this is not as unlikely as it might seem at first glance. As mentioned in section 13.3, the atmosphere is in some respects like a large flywheel which is maintained in nearly steady rotation by the balance between power input from a weak engine and power loss by friction. The balance achieved is such that there is a substantial reservoir of kinetic energy in the flywheel at any time — a quantity representing about one week's production of kinetic energy according to Fig. 13.5. Moreover all the more spectacular motion systems of the atmosphere represent considerable concentrations of kinetic energy: depressions, jet streams, cumulonimbus, tornadoes and so on, all occupy only a very small fraction of the available atmospheric volume, and the most intense (tornadoes on this list) generally occupy the smallest fraction. In these small volumes kinetic energies rise far above the global average. Note that for the most part these motion systems represent concentrations of production, rather than concentrations of pre-existing kinetic energy (though the tornado may be an exception here). According to Fig. 13.5 the flywheel is kept spinning mainly by continual nudging from little, transient, energetic wheels which represent individual weather systems. Even the Hadley circulation is kept turning by the continual eruption of hundreds of cumulonimbus in the intertropical convergence zone.

Because the atmosphere is in a state of nearly steady activity, the rate of destruction of atmospheric kinetic energy by friction must equal its rate of production by solar heating. Therefore to say that the reservoir of kinetic energy corresponds to about a week's production is to say equally that it corresponds to about a week's frictional destruction. If that production were suddenly to cease by some unimaginable solar catastrophe, the flywheel which has been spinning for aeons would grind to a halt in a period of a few days or weeks. The precise length of the period depends on how we imagine the largely surface-based friction to reach to the middle and upper troposphere, and on the effects of the large amounts of heat stored in the surface layers of the ocean, but whatever the details it is bound to be almost infinitely short compared with the timescale on which the Sun has been keeping our atmosphere chronically active. The builders of pre-history who sited their megaliths to mark the annual cycle of the Sun, and the religious devotees, poets and artists of those and subsequent years who have striven to do it homage in many and varied ways, can hardly in their wildest dreams have conceived how absolutely and continually crucial is that 'sovereign eye' to the maintenance of the Earth's surface and atmosphere in their familiar condition.

Appendix 13.1 The energy equation

Multiply the vertical component of the equation of motion (7.12(c)) by the vertica wind speed w to obtain

$$w \frac{dw}{dt} = - \frac{w}{\rho} \frac{\partial p}{\partial z} - g w - w F_z \qquad (13.11$$

Since $w = dz/dt$ and g is effectively constant we can rewrite $g w$ as $d(gz)/dt$, and by elementary differentiation $w \, dw/dt$ can be rewritten as $d(\frac{1}{2} w^2)/dt$. Equation (13.11) can therefore be rewritten in the form

$$\frac{d}{dt}(\tfrac{1}{2}w^2 + g\,z) = -\frac{w}{\rho}\frac{\partial p}{\partial z} - w\,F_z \tag{13.12}$$

Applying the same procedure in turn to the x and y components of the equation of motion (eqn 7.12(a and (b)), we find

$$\frac{d}{dt}(\tfrac{1}{2}u^2) = f\,u\,v - \frac{u}{\rho}\frac{\partial p}{\partial x} - u\,F_x$$

and

$$\frac{d}{dt}(\tfrac{1}{2}v^2) = -f\,v\,u - \frac{v}{\rho}\frac{\partial p}{\partial y} - v\,F_y$$

When these two equations are added the Coriolis terms cancel, showing that the Coriolis effect leaves kinetic energy unchanged. This can be shown to apply very generally to the full array of Coriolis terms mentioned in section 7.7, for the basically simple reason that Coriolis forces, being perpendicular to the relative velocities from which they arise, can perform no work on an air parcel as it moves, and cannot therefore change its kinetic energy. If we add the sum of these two equations to eqn (13.12) we obtain the full expression for the rate of increase of kinetic and potential energy

$$\frac{d}{dt}(\tfrac{1}{2}V^2 + g\,z) = PGW - FW \tag{13.13}$$

where $V^2 = u^2 + v^2 + w^2$, and PGW and FW represent the total rates of working by the pressure gradients and friction respectively, whose components are readily identified in the above equations.

The rate of working represented by PGW arises from the work done on a parcel as it moves down a pressure gradient. Another kind of working PC occurs when ambient pressure squeezes a parcel and increases its internal energy, producing a rise in temperature. The effect is described quantitatively by a rearrangement of eqn (5.1):

$$PC = -p\frac{d\,Vol}{dt} = C_v\frac{dT}{dt} - \frac{dQ}{dt} \tag{13.14}$$

where small changes have been converted into rates of change for compatibility with eqn (13.13). The sum of PGW and PC represents the total rate of working PW by pressure on the air parcel, and eqns (13.13) and (13.14) can now be combined to form an expression for the rate of increase of the total energy (kinetic, potential and internal) of the air parcel

$$\frac{d}{dt}(\tfrac{1}{2}V^2 + g\,z + C_vT) = \frac{dQ}{dt} + PW - FW \tag{13.15}$$

Appendix 13.2 Potential and internal energies of an air column

The total gravitational potential energy of a column of air resting on unit horizontal area of surface is given by

$$PE = g \int_0^m z \, dm \tag{13.16}$$

where m is the total mass of the column. According to the hydrostatic approximation, the mass dm of a thin horizontal slice of the column is related to the pressure difference dp between its top and bottom by $dm = -dp/g$, where the minus sign shows that we are arbitrarily assuming that dm is positive upwards (i.e. that the total mass of the atmosphere is accounted from bottom to top), unlike pressure increments dp. Hence

$$PE = - \int_{p_s}^0 z \, dp = \int_0^{p_s} z \, dp$$

where p_s is the pressure at the base of the atmospheric column (zero height). Integration by parts gives

$$PE = [p \, z]_0^{p_s} - \int_\infty^0 p \, dz$$

The first term on the right-hand side is zero because the product pz is zero both at the top and bottom of the atmosphere: at the top because pressure is zero at finite height, and at the bottom because height is zero at finite pressure. Hence

$$PE = \int_0^\infty p \, dz$$

We can use the equation of state to substitute for p. The resulting specific gas constant can be moved outside the integral because it is a constant, leaving

$$PE = R \int_0^\infty T\rho \, dz$$

Since the product $\rho \, dz$ is simply dm we have

$$PE = R \int_0^m T \, dm \tag{13.17}$$

showing that the gravitational potential energy of the column is directly proportional to its internal energy

$$IE = C_v \int_0^m T \, dm$$

A slight extension of this analysis confirms the special usefulness of the specific heat capacity at constant pressure C_p. If we consider the potential energy of an air column extending from the surface to a height z at which the pressure has non-zero value p, it follows from the above that

$$PE = R \int_0^m T \, dm - pz$$

and that the total potential energy TPE (i.e. $PE + IE$) is given by

$$TPE = C_p \int_0^m T \, dm - pz \tag{13.18}$$

If this layer is now warmed by the input of a quantity of heat $\triangle Q$ while the pressure p at the column top remains constant, the heat energy must be used both to increase the total potential energy of the layer and to do work by expanding vertically and raising the overlying air:

$$\Delta Q = \Delta TPE + p \, \Delta z$$

However,

$$\Delta TPE = C_p \int_0^m \Delta T \, dm - p \, \Delta z$$

and it follows that

$$\Delta Q = C_p \int_0^m \Delta T \, dm$$

showing that the apparent complications arising from the finite depth of the layer and the presence of the overlying air have zero net effect, and that the whole specific energy capacity of the layer is fully described by C_p.

Appendix 13.3 Available potential energy

Figure 13.3(a) represents two adjacent vertical columns of unit horizontal cross-section, one of which is slightly taller and less dense than the other. Assuming the material of column 1 to have uniform density, its centre of gravity is at height $z/2$ above its base. Multiplying by the column weight $g\rho_1 z$, we find its gravitational potential energy to be $g\rho_1 z^2/2$. In the same way the potential energy of column 2 is found to be $g\rho_2 (z + \Delta z)^2/2$. The sum of these expressions is the aggregate potential energy of the columns in their initial configuration.

Now suppose that the columns are rearranged so as to minimize their aggregate potential energy. If we assume their material to be incompressible and require that their horizontal area (two units in total) is unchanged, then the minimum is achieved as in Fig. 13.3(b), with the denser material underlying the less dense. The aggregate potential energy in this final configuration is found by the above procedure to be

$$g\rho_1 z^2/4 + g\rho_2 3z^2/4 + g\rho_2 \Delta z \, (z + \Delta z/4)$$

Subtraction from the initial aggregate potential energy and simplification shows that the reduction ΔPE in potential energy effected by the rearrangement of the columns is given by

$$\Delta PE = \frac{g}{4} [z^2(\rho_1 - \rho_2) + \Delta z^2 \rho_2] \tag{13.19}$$

Since the fractional differences in density and initial column height are assumed to be small, the initial aggregate potential energy PE is closely approximated by $g\rho z^2$, where ρ is the average column density. Writing $\rho_1 - \rho_2$ as $\Delta \rho$, it follows that

$$\frac{\Delta PE}{PE} \cong \frac{1}{4} \left[\frac{\Delta \rho}{\rho} + \left(\frac{\Delta z}{z} \right)^2 \right]$$

If we consider the atmospherically realistic case that the fractional differences in density and initial column height are proportional and very small, the second term in the brackets must be much smaller than the first. The fraction of the initial potential energy made available by feasible rearrangement of the columns is therefore given approximately by $\Delta\rho/4\rho$.

Note that any calculation of available potential energy requires precise prescription of the re-arrangement process, since the fraction available is otherwise undefined. In this simple case density and horizontal area are conserved. In more realistic compressible models, the rearrangements are usually assumed to be adiabatic, but no single model applies to all atmospheric situations.

Appendix 13.4 Frictional dissipation

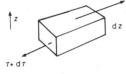

Fig. 13.7 Frictional stresses on the top and bottom faces of a slab of air in the atmospheric boundary layer.

Figure 13.7 represents a slab of air with unit horizontal area being dragged forward (i.e. in the direction of the ambient wind) by the turbulent frictional stress exerted by the faster-moving air above, and being dragged backward by friction with the slower-moving air beneath. According to Fig. 9.21 and the discussions of sections 9.9 and 9.10, the stress at the lower level is slightly larger throughout the depth of frictional influence, increasing from nearly zero at the gradient level to a maximum of τ_s in the surface boundary layer (SBL), in which τ_s is effectively constant. The net frictional force on the slab in Fig. 13.7 therefore opposes the air motion and is given by the difference $d\tau$. If the speed of the slab is V, the rate of working by friction against the wind is $V\,d\tau$, and the aggregated rate of dissipation of kinetic energy in a column extending through the planetary boundary layer is given by

$$\int_0^{\tau_s} V\,d\tau \tag{13.20}$$

This must account for nearly all of the FW term in eqn (13.7) since most of the frictional dissipation is concentrated within the planetary boundary layer.

Figure 13.8 contains a plot of V against τ which is consistent with the above features. The geostrophic wind speed V_g is reached at the gradient level, where τ is zero; wind speeds fall slowly through the Ekman layer (EL) as τ builds up towards the maximum value τ_s which occurs in the SBL, where V falls almost vertically to zero. The aggregated rate of dissipation is the area under the curve, which can be seen to be equal to $V_g\tau_s$ nearly enough to be used in an order-of-magnitude estimation.

Fig. 13.8 Graph of horizontal wind speed V against horizontal frictional stress from the surface S through the atmospheric boundary layer to the gradient level G.

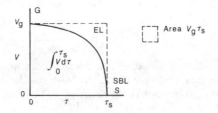

We can use eqn (9.24) to express the frictional stress in the SBL in terms of the square of the 10 m wind speed V_{10} and the appropriate drag coefficient C_{10}:

$$FW \cong \rho\,C_{10}\,V_{10}^2\,V_g \tag{13.21}$$

Substituting a reasonable value of 1 kg m^{-3} for the air density ρ, and 3×10^{-3} for C_{10} (which corresponds to a roughness length of 1 cm and a 10 m observation height), and using the reasonable rule of thumb that $V_{10}/V_g = 0.4$, we have finally

$$FW \cong 5 \times 10^{-4} V_g^3$$

Although only very approximate, this relation emphasizes the great sensitivity of frictional dissipation rate to wind speed. For example the rate implied is 0.06 W m^{-2} for a geostrophic wind speed of 5 m s^{-1}, but is 4 W m^{-2} for a geostrophic wind speed of 20 m s^{-1} (equivalent to about 8 m s^{-1} at 10 m).

Appendix 13.5 Human energy production

According to detailed studies of fuel total industrial and domestic fuel consumption in the United States [93], the *per capita* energy production in 1952 was 62.1 MW h, which is the energy corresponding to a power of 62.1 MW maintained for one hour. Dividing by the number of hours in a year, we find that this corresponds to a power of 7.1 kW continued throughout the year. The total power consumed by 5000 million people at this *per capita* rate would be 3.6×10^{13} W. Assuming realistically that almost all of this is degraded to heat after use (in principle a little must be stored in intermolecular binding energies of long-lived manufactured materials), the implied rate of heat production must almost equal this value. Dividing by the surface area of the Earth, we find the corresponding heat flux in the same units (W m^{-2}) as are used in section 13.3. The value is 7×10^{-2} W m^{-2}, which is a little over 2% of the value quoted for heat production by frictional degradation at the end of the atmospheric mechanical energy cycle. Since the uncertainty in the latter value is about 50%, it is safest to quote the percentage as of order 1.

Appendix 13.6 Optimum mechanical efficiency of a heat engine

In standard thermodynamics textbooks it is shown that there are several equivalent statements of the second law of thermodynamics, one of which (known as Clausius' inequality) can be expressed in terms of the tropospheric engine of Fig. 13.1 as follows:

$$\frac{Q_1}{T_1} \leqslant \frac{Q_2}{T_2} \tag{13.22}$$

The temperatures T_1 and T_2 must be expressed in the absolute scale. The equality describes the ideal reversible limit, while the inequality describes the behaviour of all realistic engines, natural as well as man-made. Equation (13.22) can be simply rearranged:

461

$$\frac{T_2}{T_1} \leqslant \frac{Q_2}{Q_1}$$

From the first law of thermodynamics we can rewrite the expression for the mechanical efficiency of the engine as follows:

$$\frac{M}{Q_1} = \frac{Q_1 - Q_2}{Q_1} = 1 - \frac{Q_2}{Q_1}$$

Combining this with Clausius' inequality we find

$$\frac{M}{Q_1} \leqslant \frac{T_1 - T_2}{T_1}$$

where the equality again describes the relationship at the reversible limit, which is now seen to be associated with maximum efficiency. Since T_2 is always greater than absolute zero (very much greater in the case of the atmospheric engine), it follows that the optimum efficiency of even the most perfect engine is less than unity. In addition, all real engines are irreversible to some extent and therefore fall short of the optimum for their particular source and sink temperatures T_1 and T_2.

Although irreversibility appears as a kind of imperfection in classical thermodynamics (which concentrates on the reversible ideal), recent advances in the study of irreversible systems of many types [94] suggest that many properties of real systems are associated with their intrinsic irreversibility and non-linearity. The irreversible nature of weather systems remains largely unexplored, but is being opened up for scrutiny by chaos theory (section 1.7).

Problems

LEVEL 1

1. In a heat engine, what happens to the heat which is not converted into mechanical energy?
2. In the style of Fig. 1.6 sketch (a) a heat engine and (b) a heat pump operating between a temperature T_1 and a lower temperature T_2, showing the direction of the flows of heat and mechanical energy in each case.
3. Write down the first law of thermodynamics for case (b) in problem 2, where heat Q_2 is absorbed from the cool heat source, Q_1 is delivered to the warm heat sink, and mechanical energy M is used.
4. By considering the case of a domestic refrigerator running in a steady thermal state, show that the resultant effect must be to warm the room in which it operates. Where has the heat come from?
5. Present a non-technical argument to explain why so little of the atmosphere's considerable total potential energy is available for conversion to kinetic energy.
6. Consider the likely dynamic behaviour of an atmosphere which is being heated uniformly throughout its bulk by net radiation.
7. Make a verbal summary of the sequence of inputs, transformations and outputs summarized by Fig. 13.5.
8. Why is the tropospheric engine so inefficient?

9. Find the ratio of available potential energy to potential energy for the case of a two-layer fluid (Fig. 13.4) in which the upper layer is 1% denser than the lower. Assume the relationship mentioned in section 13.2.

10. Find the maximum mechanical efficiency of the intrinsic oceanic heat engine (i.e. ignoring the considerable activity driven by the atmosphere) given that its heat source is in tropical surface waters (temperature about 25 °C) and its heat sink is in polar surface waters (about 0 °C).

11. By assuming conservation of the sum of potential and kinetic energies (i.e. ignoring the right-hand side of eqn (13.1)), find the speed of a rain drop after falling 1 km from rest. Since actual fall speeds are ~ 5 m s^{-1}, estimate the amount of energy per unit mass of raindrop which seems to have disappeared at this point, and discuss where it has gone.

12. If 10 MJ m^{-2} of heat energy is pumped into a vertical column of air extending through the full height of the atmosphere, find the resulting rise in air temperature, and find also the increases in internal and potential energies. (Assume that the column maintains its horizontal area throughout.)

13. Using the approximate relationships in appendix 13.4, find the frictional rates of dissipation of kinetic energy in the boundary layer corresponding to Beaufort wind forces 2, 4, 8 and 12.

14. Find the column total of *KE* in the active part of a mid-latitude depression, assuming wind speeds of 20 m s^{-1} from the 1000 to the 500 mbar levels, 60 m s^{-1} from 500 to 200 mbar, and 20 m s^{-1} in the rest of the atmosphere. (*Hint:* convert the pressure slabs into equivalent masses per unit area assuming hydrostatic balance.)

LEVEL 3

15. Sketch and discuss qualitative modifications of Fig. 13.5 needed to describe (a) convective activity in the troposphere, and (b) large-scale activity in the Hadley circulation.

16. Using the method of appendix 13.3, show that the ratio of available *PE* to actual *PE* (relative to a datum level at the base of the lower layer) for the overturning depicted in Fig. 13.4 is given approximately by $\Delta\rho/2\rho$.

17. Consider eqn (13.1) applied to a slightly buoyant air parcel, and discuss the relative sizes of the terms in a discussion of the parcel's behaviour, firstly in the absence of and then in the presence of drag.

18. Show from the identity of eqn (13.5) and its immediate predecessor that the altitude of the effective centre of gravity of an isothermal atmosphere is one exponential scale height above its base. (Note that this altitude Z is defined to be such that $mgZ = PE$, where m is the atmospheric mass.)

19. Outline the sequence of energy types and conversions for large-scale atmospheric activity in the northern hemisphere on the timescale of a year, qualitatively defining all terms used.

20. Speculate on the sequence of events which would occur in the Earth's surface and atmosphere over a period of a year if the unthinkable happened and the Sun's output suddenly and completely ceased.

Bibliography, references and other sources

Textbooks

The following is a selection of books which I have found to be useful in ways mentioned. The brief opinions are, of course, personal and each of the following deserves close scrutiny by anyone considering their use. The order is alphabetical.

Atkinson, B.W. (Ed.) (1981) *Dynamical meteorology: an introductory selection*. Methuen, London. Although the wide range of authors is responsible for a variety of levels of treatment, the overall coverage is excellent, with good discussions and just enough formalism to show the way to more thorough treatments. Graphics are adequate.

Hess, S.L. (1959) *Introduction to Theoretical Meteorology*. Holt, Rinehart and Winston, New York. An old text, famous for its uncluttered clarity of treatment of thermodynamics, radiation and dynamics (without vectors!). Minimal graphics and little explicit comparison with observation detract but not critically.

Holton, J.R. (1972) *An Introduction to Dynamic Meteorology*. Academic Press, New York. Coverage as implied by the title but pitched at people with considerable experience in applied mathematics. Though the treatment is quite formal throughout, the associated discussion always encourages physical understanding. The treatment follows the conventional bias toward synoptic-scale dynamics.

Lamb, H.H. (1982) *Climate, History and the Modern World*. Methuen, London. A brief outline of the ideas and methods of climatology followed by a fascinating survey of climate, mainly since the last glaciation, with special emphasis on social effects.

Lockwood, J.G. (1979) *Causes of Climate*. Edward Arnold, London. A compact, semi-quantitative survey of climate and climatic change mostly over the last 100 000 years.

Ludlam, F.H. (1980) *Clouds and Storms: the behaviour and effect of water in the atmosphere*. Pennsylvania State University Press, University Park and London. A very large and thorough treatment of many aspects of precipitating and non-precipitating clouds (but not thunderstorm electricity). Unusual for combining cloud physics and small, meso and synoptic-scale meteorology, this book is posthumous testimony to the physical insight and elegant treatment typical of this remarkable scientist, whom I was privileged to have as Ph.D. supervisor. Though not a teaching text, it touches usefully on a wide range of observable meteorological phenomena.

McIntosh, D.H. and Thom, A.S. (1969) *Essentials of Meteorology*. Wykeham Publications Ltd (Taylor & Francis Ltd), London. A small, concise book covering a wide range of meteorological essentials in surprising depth. The treatment of the tephigram is particularly useful. Dynamics are treated using vector notation. Minimal graphics.

Mason, B.J. (1962) *Clouds, Rain and Rainmaking*. Cambridge University Press. A classic introduction to the physics of clouds, precipitation and thunderstorm electricity.

Neiburger, M., Edinger, J.G. and Bonner, W.D. (1982) *Understanding our Atmospheric Environment*. W.H. Freeman and Company, San Francisco. A substantial introductory text, mostly about meteorology but with just enough about pollution and other human interaction to justify the title. The treatment is mostly qualitative but is nevertheless meteorological (rather than physical geographical) in style. Good graphics.

Oke, T.R. (1978) *Boundary Layer Climates*. Methuen, London. A good introduction to boundary-layer meteorology and climatology, with a substantial section on anthropogenic effects. The treatment is mainly semi-quantitative but there are many useful tables of values.

Palmen, E. and Newton, C.W. (1969) *Atmospheric Circulation Systems*. Academic Press, New York. A classic survey of understanding of synoptic- and larger-scale weather systems, with some treatment of organized vigorous small-scale convection. Although the treatment is not at the introductory level, the discussion is full and rewards patient study.

Riehl, H. (1978) *Introduction to the Atmosphere*. McGraw-Hill Kogakusha Ltd, Tokyo. Probably the best purely qualititative introduction to meteorology, with good photographs in the later editions.

Rogers, R.R. (1976) *A Short Course in Cloud Physics*. Pergamon Press, Oxford. A thorough treatment of the whole range of meteorological cloud physics with the curious exception of thunderstorm electricity.

Scorer, R.S. (1978) *Environmental Aerodynamics*. Ellis Horwood Ltd (associated with John Wiley & Sons), Chichester. An unusual and stimulating book, beginning with a formal vector treatment of fluid dynamics which is well beyond introductory level, but then proceeding to a discussion of many smaller-scale aspects of atmospheric behaviour with useful emphasis on physical reality as well as theoretical nicety. It includes related topics such as chimney-plume dispersion and bird soaring.

Simpson, R.H. and Riehl, H. (1981) *The Hurricane and its Impact*. Basil Blackwell, Oxford. A largely qualitative description of these fearsome storms with considerable emphasis on their destructive power.

Wallace, J.M. and Hobbs, P.V. (1977) *Atmospheric Science: an introductory survey*. Academic Press, New York. A modern text on fairly traditional lines, with very solid thermodynamics and radiation, but a much lighter treatment of dynamics. The coverage of cloud, precipitation and electrical processes is particularly detailed. Diagrams and photographs abound.

Wayne, R.P. (1985) *Chemistry of Atmospheres*. Clarendon Press, Oxford. A very full account of the chemistry of the terrestrial and other planetary atmospheres.

Wickham, P.G. (1970) *The Practice of Weather Forecasting*. HMSO, London. An excellent summary of synoptic-scale weather analysis and manual forecasting with lots of maps and soundings from actual cases.

Journals

Most meteorological journals are pitched well beyond the introductory level, though they can be skimmed with profit by looking at diagrams and conclusions before trying to understand the physical meaning of the more formal sections. The popular journal of the Royal Meteorological Society (*Weather*) is an important exception. Though the level of articles is quite variable, there is a good range from the practical to the theoretical in each monthly issue, with a useful policy of public education.

Royal Meteorological Society, 104 Oxford Road, Reading, Berkshire, RG1 7LJ.

The American Meteorological Society publishes a monthly *Bulletin* for its members and others which contains technical reviews and updates which are often informally and accessibly expressed.

45 Beacon Street, Boston, Ma 02108, USA.

The *Scientific American* publishes review articles on climatological and other atmospheric matters (amongst others) in a clear and graphic style by world authorities in the subjects.

Tephigrams

British Meteorological Office blank tephigrams (Metform 2810) from Her Majesty's Stationery Office can be ordered through The Government Bookshop, PO Box 569, London SE1 9NH. Other materials are listed and may be available through the Marketing Services section of the Meteorological Office, London Rd, Bracknell, Berks RG12 2SZ, UK.

References (numbered in the text)

1 J. Gleick, *Chaos* (London, Sphere, 1987).
2 I. Stewart, *Does God Play Dice* (Harmondsworth, Penguin, 1989).
3 Lucretius, *The Nature of the Universe*, trans. R. Latham (Harmondsworth, Penguin, 1966).
4 Meteorological Office, *Handbook of Meteorological Instruments*, Part I (London, HMSO, 1969).
5 Meteorological Office, *The Observer's Handbook* (London, HMSO, 1982).
6 J.S.A. Green, Holiday meteorology: reflections on weather and outdoor comfort, *Weather*, **22**:4 (1956), 128–131.
7 Meteorological Office, *Hygrometric Tables*, Part II screen values, Part III aspirated values (London, HMSO, 1964).
8 J.M. Wallace and P.V. Hobbs, *Atmospheric Science* (New York, Academic Press, 1977).
9 R.C. Ward, *Principles of Hydrology* (London, McGraw-Hill Publishing Company, London).
10 World Meteorological Organization, *International Cloud Atlas*, Vol. I (revised 1975). Simple summary in: Meteorological Office, *Cloud Types for Observers* (London, HMSO, 1972).
11 J.K. Hargreaves, *The Upper Atmosphere and Solar-Terrestrial Relations* (Wokingham, Van Nostrand Reinhold (UK), 1979).
12 C.C. Delwiche, The nitrogen cycle, in *The Biosphere*, A Scientific American Book (San Francisco, W.H. Freeman, 1970). Also: I.M. Campbell, *Energy and the Atmosphere: a physical-chemical approach* (Chichester, Wiley, 1977).
13 P. Cloud and A. Gibor, The oxygen cycle, in *The Biosphere*, A Scientific American Book (San Francisco, W.H. Freeman, 1971). Also: I.M. Campbell (as reference [12]).
14 J.D. Shanklin and B.G. Gardner, *British Antarctic Survey* (Cambridge, UK, 1989).
15 B. Bolin, The carbon cycle, in *The Biosphere*, A Scientific American Book (San Francisco, W.H. Freeman, 1971).
16 *Scientific American, The Biosphere* (San Francisco, W.H. Freeman, 1971).
17 Siegenthaler and Oeschger, *Tellus*, 39b (1987).
18 Department of the Environment, *Global Climate Change* (London, HMSO, 1989).
19 G.E. Likens, R.F. Wright, J.N. Galloway and T.J. Butler, Acid rain, *Scientific American*, **241**:4 (1979), 39–47. See also I.M. Campbell, ref. [12].
20 D.J. Spedding, *Air Pollution* (Oxford, Clarendon Press, 1974). And ref. [42].
21 F. Press and R. Siever, *Earth* (San Francisco, W.H. Freeman, 1974).

22 E.J. Nisbet, *The Young Earth* (London, Allen and Unwin, 1987).

23 Wayne (1985) (see Textbooks).

24 J. Lovelock, *The Ages of Gaia* (Oxford, Oxford University Press, 1988).

25 T.W. Harrold and K.A. Brown, The polar low as a baroclinic disturbance, *Quart. J. Roy. Met. Soc.*, **95**:406 (1969), 710–723.

26 Mintz (1974) (Diagram source).

27 E. Palmen and C.W. Newton, *Atmospheric Circulation Systems, their Structure and Physical Interpretation* (New York, Academic Press, 1969).

28 G. Manley, Central England temperatures: monthly means 1659 to 1973, *Quart. J. Roy. Met. Soc.*, **100**:425 (1974), 389–485.

29 H.H. Lamb, *Climate, History and the Modern World* (London, Methuen, 1982).

30 J.G.L. Lockwood, *The Causes of Climate* (London, Edward Arnold, 1979).

31 A.H. Perry and J.M. Walker, *The Ocean–Atmosphere System* (London, Longman, 1977).

32 Jones and Wigley, University of East Anglia, UK (Diagram source).

33 J.T. Houghton, G.J. Jenkins and J.J. Ephraums (Eds), *Climate Change — the IPCC Assessment* (Cambridge, Cambridge University Press, 1990).

34 B.S. John (Ed.), *The Winters of the World* (North Pomfret (Vt), David & Charles, 1979).

35 F.W. Sears, M.W. Zemansky and H.D. Young, *University Physics*, 6th edn (Reading (Mass.), Addison-Wesley, 1982).

36 Warner and Telford, *Journal of Atmospheric Sciences* (1967).

37 A. Letestu (Ed.), *International Meteorological Tables* (Geneva, Secretariat of the World Meteorological Organization, 1966).

38 Hess (1959) (see Textbooks).

39 F.H. Ludlam and J.F.R. McIlveen, The lag of the humidity sensor in the British radiosonde, *Meteorological Magazine*, **98** (1969), 233–246.

40 F.H. Ludlam, Cumulus and cumulonimbus convection, *Tellus*, **XVIII**:4 (1966), 688–698. And ref. [27].

41 B.J. Mason, *Clouds, Rain and Rainmaking* (Cambridge, Cambridge University Press, 1962).

42 F.H. Ludlam, *Clouds and Storms* (Pennsylvania State University Press, University Park, Pa 16802, 1982).

43 A. Eliassen, Geostrophy, *Quart. J. Roy. Met. Soc.*, **110**:463 (1984), 1–12.

44 J.R. Holton, *An Introduction to Dynamic Meteorology* (New York, Academic Press, 1972).

45 R. Hide, Some laboratory experiments etc., in G.A. Corby (Ed.), *The General Circulation of the Atmosphere* (London, Royal Meteorological Society, 1970).

46 R.S. Scorer, *Environmental Aerodynamics* (Chichester, Ellis Horwood, 1978).

47 S.L. Valley (Ed.), *Handbook of Geophysics and Space Environments* (New York, McGraw-Hill Book Company, 1965).

48 W.S. Von Arx, *An Introduction to Physical Oceanography* (New York, Addison-Wesley, 1962).

49 J. London and T. Sasamori, *Space Research*, **XI** (1971), 639–649.

50 M.I. Budyko, *Climate and Life* (New York, Academic Press, 1974).

51 J.L. Monteith, *Principles of Environmental Physics* (London, Edward Arnold, 1989).

52 G.L. Pickard and W.J. Emery, *Descriptive Physical Oceanography: An Introduction* (Oxford, Pergamon Press, 1982).

53 F. Hoyle, *The Black Cloud* (Harmondsworth, Penguin Books, 1975).

54 T.R. Oke, *Boundary Layer Climates* (London, Methuen, 1978).

55 R.E. Munn, *Descriptive Micrometeorology* (New York/London, Academic Press, 1966).

56 H.H. Lettau and B. Davidson (Eds), *Exploring the Atmosphere's First Mile*, Vols I and II (London, Pergamon Press, 1957).

57 D.J. Tritton, *Physical Fluid Dynamics* (Wokingham, Van Nostrand Reinhold (UK), 1977).

58 S. Pond and G.L. Pickard, *Introductory Dynamic Oceanography* (Oxford, Pergamon Press, 1981).

59 Jarvis *et al.* (1976) (Diagram source).

60 R.G. Barry, *Mountain Weather and Climate* (London/New York, Methuen, 1981).

61 J.E. Simpson, D.A. Mansfield and J.R. Milford, Inland penetration of sea-breeze fronts, *Quart. J. Roy. Met. Soc.*, **103**, (1977), 47–76.

62 C.J.M. Aaneuson (Ed.), Gales in Yorkshire in February 1962, *Meteorological Office Geophys. Mem* (London, HMSO), **14**:108.

63 P.M. Saunders, An observational study of cumulus, *J. Meteorol.*, **18** (1961), 451–467.

64 Newton (1963) (Diagram source).

65 Collinge *et al.* (1990) Radar observations of the Halifax Storm. *Weather*, **45**, 10.

66 J.P. Kuettner, Cloud bands in the Earth's atmosphere: Observations and theory, *Tellus*, **23** (1971) 404–426.

67 E.M. Agee, T.S. Chen and K.E. Dowell, A review of mesoscale cellular convection, *Bull. Amer. Met. Soc.*, **54**:10 (1973), 1004–1012.

68 K.A. Browning, Radar measurements of air motion near fronts (Part II), *Weather*, **27** (1971), 320–340.

69 T.J. Matejka, R.A. Houze and P.V. Hobbs, Microphysics and dynamics of clouds associated with mesoscale rainbands in extratropical cyclones, *Quart. J. Roy. Met. Soc.*, **106** (1980), 29–56.

70 Bergeron (1954) (Diagram source).

71 D. Defoe, *The Storm* (1704). Quoted in S. Brown, *World of the Wind* (London, Alvin Redman, 1962).

72 B. Franklin (1744). Quoted in S. Brown (1962) (see ref. [71]).

73 S. Petterssen, *Weather Analysis and Forecasting*, Vol. 1, *Motion and Motion Systems* (New York, McGraw-Hill Book Company, 1956).

74 Bjerknes and Solberg (1921) (Diagram source).

75 C.L. Godske, T. Bergeron, J. Bjerknes and R.C. Bundgaard, *Dynamic Meteorology and Weather Forecasting* (Boston, Mass., American Met. Soc. and Washington, D.C., Carnegie Inst., 1957).

76 M. Minnaert, *The Nature of Light and Colour in the Open Air* (New York, Dover Publications, 1954).

77 T.N. Carlson, Air flow through midlatitude cyclones and the comma cloud pattern, *Mon. Wea. Rev.*, **108**: Oct (1980), 1498–1509.

78 P.E. Lydolph, *Climates of the Soviet Union*, Vol. 7 of H.E. Landsberg (Ed.), *World Survey of Climatology* (Amsterdam, Elsevier Scientific Publishing Co., 1977).

79 H. Flohn (1959) (Diagram source).

80 E.T. Eady, Long waves and cyclone waves, *Tellus*, **1**:3 (1949), 33–52.

81 Willett (1944) (Diagram source).

82 C.M. Stevenson, The snowfalls of early December 1967, *Weather*, **23** (1968), 156–160.

83 D.A. Mansfield, Polar lows: The development of baroclinic disturbances in cold air outbreaks, *Quart. J. Roy. Met. Soc.*, **100**:427 (1974), 541–554.

84 P.G. Wickham, *The Practice of Weather Forecasting* (London, HMSO, 1982).

85 R.G. Barry and R.J. Chorley, *Atmosphere, Weather and Climate* (Methuen, London, 1989 and earlier editions).

86 G. Manley, *Climate and the British Scene* (London, Collins, 1952).

87 C.S. Ramage, *Monsoon Meteorology* (New York/London, Academic Press, 1971).

88 Koteswaram (1958) (Diagram source).

89 Malkus and Riehl (1981) (Diagram source).

90 R. Hughes, *In Hazard* (various publishers, 1938).

91 R.H. Simpson and H. Riehl, *The Hurricane and its Impact* (Oxford, Basil Blackwell, 1974).

92 Möller (1951) (Diagram source).

93 C.M. Cipolla, *The Economic History of World Population* (Harmondsworth, Penguin Books, 1970).

94 I. Prigogine, *From Being to Becoming: time and complexity in the physical sciences* (San Francisco, W.H. Freeman, 1980).

95 Rowe, M.W. and Meaden, G.T. (1985) Britains Greatest Tornado Outbreak. *Weather*, **40**, 8, 230–234.

Glossary

The Glossary lists terms which are sufficiently important, widespread or peculiar to warrant separate summary definition. Many terms used in the definitions are themselves listed elsewhere in the Glossary to encourage you to refer to several entries to clarify any important definition. Except for very basic terms whose meaning is assumed, all technical terms can be traced through the Index to their definition in the main text, whether or not they appear in the Glossary.

absolute acceleration   Acceleration measured relative to an unaccelerated (inertial) reference frame.

absolute vorticity   Vorticity measured relative to a non-rotating reference frame.

absorptivity   The proportion of incident electromagnetic radiant power absorbed by a surface, expressed as a fraction or percentage.

acceleration   Rate of change of velocity with time.

accretion   Growth of a body of water or ice by the collection of smaller particles or droplets.

adiabatic   In thermal isolation. A thermal process is said to be adiabatic if no heat enters or leaves the system in which it is occurring.

advection   Horizontal transport (of mass, heat etc.) effected by horizontal exchange of air.

advection fog   Fog formed by advection of warm air over a colder surface.

aerodynamic roughness length   The height above the impermeable base of a rough surface at which wind speed falls to zero by extrapolating a log wind profile down through the overlying surface boundary layer.

aerosol   Suspension of very small (maximum diameters of order 1 μm) particles or droplets in the atmosphere, especially concentrated in the low troposphere.

air   The gaseous component of the turbosphere.

air mass   A synoptic-scale segment of troposphere whose temperature and humidity are usefully related to a geographical source region. Usefulness is often marginal in the presence of typically strong wind shears in the vertical.

air parcel   A small body of air which, at least in principle, is coherent and identifiable throughout a useful period, for example a buoyant thermal in a cumulus cloud.

albedo   The proportion of incident electromagnetic radiation (usually solar) reflected or backscattered by a surface, expressed as a fraction or percentage.

alto   Prefix to cloud type, meaning 'in the middle troposphere'; as in altocumulus.

altostratus   Sheet cloud in the middle troposphere (typical of fronts).

anabatic flow or winds   Airflow up sloping terrain.

anemometer   Instrument for measuring wind speed or run of wind.

angular momentum or moment of momentum   The product of the tangential momentum of a body and its radial distance from the axis of rotation.

angular velocity   The rate of turning of a body about an axis expressed in angle turned per unit time (usually radians per second).

anticyclone   Synoptic-scale weather system with winds blowing anticyclonically (clockwise in the northern hemisphere) in the low troposphere.

anvil   A fan of ice cloud at the top of a cumulonimbus.

aphelion   Position or time of maximum separation of planet and sun.

apparent _g_   Gravitational acceleration measured relative to a fixed point on the Earth's surface.

Arctic sea smoke   Steam fog produced when very cold air from polar land masses flows onto the sea.

atmospheric boundary layer or planetary boundary layer   The part of the low troposphere ($\sim$ 500 m deep) most directly influnced by the underlying sea or land surface, including the laminar, surface and Ekman boundary layers.

atmospheric waves   Horizontal or vertical undulations of airflow on scales from the metric to the hemispheric, sometimes in short trains and sometimes solitary (e.g. see _cyclone_ and _gravity waves_).

available potential energy   That part of the atmospheric potential energy which could be converted to kinetic energy by realistic redistribution of air parcels.

azimuth   Wind direction specified by the angle in degrees from which the wind is blowing, counting clockwise from zero at true north.

backing   Wind direction changing so that azimuth is decreasing.

baroclinic   The condition of the atmosphere when large-scale isopycnic and isobaric surfaces are misaligned, usually in association with horizontal temperature gradients.

baroclinic instability   Dynamic instability associated with baroclinic zones and their associated thermal winds, giving rise to cyclogenesis.

barotropic   The condition of the atmosphere when it is not baroclinic.

barometer   An instrument for measuring atmospheric pressure.

barograph   A barometer with a graphical output.

Beaufort scale   A scale of wind strength assessed by effects on surface features such as trees and sea state.

Bergeron–Findeisen   A mechanism for the development of precipitation by preferred growth of ice crystals in clouds consisting largely of supercooled water.

biosphere   The interacting network of organisms living on the Earth.

blackbody   An ideal body which absorbs all incident electromagnetic radiation and emits it with maximum thermodynamic efficiency.

boundary layer   A layer of fluid whose dynamic behaviour is directly influenced by an adjacent solid or liquid surface. Meteorological examples are the atmospheric boundary layer and the sheath of air enveloping a falling raindrop.

Bowen ratio   The vertical flux of sensible heat expressed as a fraction or percentage of the latent heat flux at a particular site or across a geographical region.

Brunt–Väisälä frequency   The frequency of vertical oscillation of an air parcel released after displacement from its equilibrium position in a convectively stable fluid layer.

buoyancy   The net upward force experienced by a body immersed in a fluid of different density in the presence of a hydrostatic pressure gradient.

Buys-Ballot's law   The law defining the relationship between the directions of horizontal pressure gradient and associated geostrophic wind.

cellular convection   Convection organized in quasi-regular cells, as in stratocumulus or the cellular clustering of shower clouds.

Celsius scale   The temperature scale in which the freezing and boiling points of pure water at standard atmospheric pressure are 0 and 100 units ($^\circ$C) respectively.

centripetal   Toward the centre of a circle, as in centripetal acceleration.

centrifugal   Away from the centre of a circle. Ignoring a centripetal acceleration incurs an apparent centrifugal force.

chaos   Apparently random behaviour of a deterministic system.

chinook   A föhn wind near the Rocky Mountains.

cirrus   A fibrous cloud of ice particles.

climate   Weather conditions and their range typical of a region or site.

cloud A dense population of water droplets and/or ice crystals, each of which has diameter of order 10 μm.

cloud seeding Encouragement of precipitation from clouds by freezing some supercooled cloud droplets and speeding the Bergeron–Findeisen mechanism.

cold front See *front*.

collision and coalescence Precipitation development by accretion, in which larger particles or droplets grow as they fall through populations of smaller ones, collecting some by collision and subsequent coalescence.

condensation The growth of water or ice by net diffusion from contiguous vapour — the reverse of evaporation. In the case of ice growth, either process is often called sublimation.

condensation nuclei The components of an aerosol population which act as nuclei for the formation of individual cloud crystals or droplets in a realistically supersaturated vapour.

conditional instability Convective instability which arises when a previously unsaturated layer of air fills wholly or partly with cloud.

conduction of heat Heat transfer through spreading of enhanced thermal agitation by molecular impact.

continuity equation Formal description of the conservation of the mass of a body of moving air, relating the net mass flux into a fixed volume to the rate of accumulation of air mass there.

contour A height contour of an isobaric surface.

convection Usually the small-scale vertical exchange of air parcels, driven either by buoyancy (thermal convection — although density differences do not always depend entirely on temperature differences) or wind shear (mechanical convection). Thermal convection is usually assumed unless mechanical convection is specified. See *slope convection* for application to the large scale.

convective instability Tendency toward buoyant convection arising from disposition of atmospheric density.

convective stability Inhibition of buoyant convection arising from atmospheric density disposition.

convergence Negative divergence.

Coriolis acceleration Component of absolute acceleration arising from the motion of a body relative to the rotating terrestrial reference frame. In meteorology it usually means the component in the local horizontal arising from horizontal flow.

Coriolis parameter The factor $2\,\Omega \sin \phi$, where Ω is the Earth's angular velocity and ϕ is the angle of latitude.

covariance Average value of the product of deviations of matched pairs of values (of vertical and horizontal wind speeds, for example) from their respective mean values.

cryosphere Ice caps, pack ice, glaciers, permafrost — the frozen water substance at and below the Earth's surface.

cumulus The family of hill-shaped clouds ranging from small cumulus to large cumulonimbus. The term on its own usually denotes small cumulus or only slightly larger.

cumulus congestus A moderate or large cumulus cloud, growing conspicuously.

cumulonimbus A precipitating cumulus, often but not necessarily having an anvil.

cyclone Synoptic-scale weather system with winds in the low troposphere rotating about the local vertical in the same sense as the local terrestrial surface but faster (i.e. anticlockwise on a northern hemisphere map). Also a local name for a severe tropical cyclone in the Indian Ocean.

cyclone wave The wave-like deformation of flow in the middle and upper troposphere associated with an extratropical cyclone.

cyclonic shear Horizontal flow sheared in such a sense as to cause an embedded parcel to rotate cyclonically.

cyclogenesis Cyclone formation, usually applied to extratropical cyclones.

cyclostrophic flow Horizontal circular air motion in which centripetal acceleration on the weather map is effectively maintained by centripetal pressure gradient force, Coriolis and friction terms being negligible.

density Mass concentration expressed as mass per unit volume.

depression See *extratropical cyclone*.

detrainment Loss of mass from a rising thermal to surrounding air. The reverse of entrainment.

dew Water condensed onto a solid surface which has cooled to the dew point of the contiguous air.

dew point or dew-point temperature The temperature at which a given parcel of air becomes saturated during isobaric cooling with conservation of vapour content.

dimensions The dependence of a measurable quantity on the fundamental measurables — mass, length, time and temperature. Thus the dimensions of velocity are length divided by time.

dimensionless number Any fundamenal parameter (made up of a dimensionless assembly of observable factors) which usefully describes a regime of dynamic or thermodynamic behaviour, e.g. a Reynolds number.

diurnal With a daily cycle.

divergence (of velocity) The resultant rate of stretching of a fluid as given by the sum of the longitudinal gradients of the three components of flow velocity. In meteorology it usually refers in particular to the stretching components of air in the local horizontal.

drag coefficient The dimensionless ratio of the drag force on a body immersed in a fluid flow to the incident or adjacent fluid momentum flux.

drizzle Precipitation of water droplets with radii $\sim 100~\mu m$.

dry adiabat A line describing a dry adiabatic process on a thermodynamic diagram such as a tephigram.

dry adiabatic Describing an adiabatic movement or process of an air parcel in which there is neither evaporation nor condensation of the water substance — also called isentropic, because entropy is conserved.

dry adiabatic lapse rate Temperature lapse rate equal to the value found in dry adiabatic ascent or descent of an air parcel.

dry air Normal turbospheric air without water vapour.

dynamics Study of forces and associated motional deformation.

dynamical coefficient of viscosity Tangential shearing stress arising from viscosity, divided by lateral shear (spatial velocity gradient). Also known as the Newtonian coefficient of the same.

easterly Flowing or moving from the east.

easterly wave A synoptic-scale wave-like disturbance of the low tropospheric easterlies of low latitudes.

eddy Individual transient element of turbulence — often pictured as a translating and rotating air parcel.

eddy viscosity Apparent viscosity arising from turbulence.

electromagnetic radiation Travelling waves of electric and magnetic disturbance, capable of passing through empty space. Of the full electromagnetic spectrum, solar and terrestrial radiation are the most important components in meteorology.

Ekman layer The topmost layer of the atmospheric boundary layer, between the free atmosphere and the surface boundary layer.

emissivity The actual emittance of a radiating surface expressed as a fraction of the blackbody value for the surface temperature.

emittance The flux of electromagnetic radiant energy emitted by unit area of a radiating surface.

energy The capacity to do work. See *kinetic, gravitational potential, radiant* and *heat* energies for meteorologically important types.

entrainment Incorporation of surrounding air by a rising thermal.

entrainment drag The effective drag on a thermal arising from continual sharing of momentum by entrainment of nearly static ambient air.

equation of motion Equation of relative acceleration with net force acting on an air parcel of unit mass — a form of Newton's second law of motion.

equation of state Equation relating pressure, density and temperature of a parcel of an ideal gas or mixture of ideal gases.

equivalent potential temperature The potential temperature of an air parcel after all water vapour has been condensed by adiabatic decompression.

equinox A time when daylight and night are equally long.

evaporation   Reverse of condensation.

exponential growth (or decay)   Profile of growth (or decay) of a quantity whose gradient is directly proportional to the quantity itself.

extratropical cyclone   A cyclone outside the tropics, distinguished from a tropical cyclone by greater scale, the presence of one or more fronts and the absence of great central intensity.

fall speed   See *terminal velocity*.

fetch   Distance upwind from point of observation to a significant location, as in 'standard synoptic anemometers should be sited to have an unobstructed fetch of at least 100 m'.

flux of energy, etc.   The rate of flow of energy, etc. per unit time through a real or imaginary surface (normal to the flow unless otherwise specified).

flux density of energy, etc.   The flux per unit area of real or imaginary surface.

fog   Dense cloud in contact with a land or water surface, with density specified by visibility.

föhn   A strong, dry, katabatic wind, produced by prior enforced ascent of air over high parts of the European Alps.

free atmosphere   The atmosphere above the atmospheric boundary layer (i.e. above the gradient level).

freezing fog   A fog of supercooled water droplets, freezing on impact with any solid surface.

freezing nuclei   Those solid components of an aerosol population which act as nuclei for the freezing of supercooled cloud droplets.

frequency   The number of events per unit time.

friction velocity   A basic wind speed parameter defined by $\sqrt{\tau/\rho}$, where τ is the turbulent horizontal wind stress in the surface boundary layer and ρ is the air density there.

front   A swath of cloud and precipitation which is synoptic-scale in length and at least large-meso-scale in breadth, and is associated with a significant horizontal temperature gradient in the low troposphere of an extratropical cyclone. A front is called warm or cold depending on whether the warmer or colder air is advancing, and is called occluded when it connects the warm sector to a separated surface pressure minimum.

frontal zone   A region of continually or seasonally preferred cyclogenesis. Also the meso-scale zone of maximum lateral horizontal temperature gradient associated with a front.

frost   Deposition of ice on a land surface by diffusion and sublimation. When thick enough to produce marked whitening of vegetation (especially grass), it is called hoar frost.

gale   A wind whose ten-minute average speed at height 10 equals at least 37 knots.

gas constant   The constant in the equation of state for an ideal gas. In the universal form, a single universal gas constant applies to all gases. In the meteorological form, a different specific gas constant applies to each gas or uniform mixture.

geostrophic wind   A horizontal wind in which the Coriolis acceleration is exactly maintained by the horizontal pressure gradient force (or equivalently the Coriolis and pressure gradient forces exactly balance), i.e. when friction and acceleration relative to the Earth's surface are zero.

glaciation   The process of conversion of supercooled water cloud into ice cloud.

geostationary orbit   An equatorial satellite orbit such that the satellite is stationary relative to the Earth.

gradient   The spatial rate of change of an observable in the direction of maximum rate of increase (as in temperature gradient).

gradient level   The lowest level in the troposphere which is so free of surface drag that actual and gradient winds are indistinguishable.

gradient wind   A horizontal wind in which the combined Coriolis and centripetal accelerations relative to the Earth's surface, arising from cyclonic or anticyclonic flow, are exactly maintained by the horizontal pressure gradient force.

gravitational acceleration or apparent *g*   The downward acceleration of a body falling freely *in vacuo*, as measured from a frame fixed to the Earth's surface. Equivalently it is the gravitational force per unit mass of the body, as measured from the same frame.

gravity waves   Waves of fluid disturbance in which gravity is the predominant restoring force.

greenhouse effect   The elevation of surface and low-troposphere temperatures which is maintained by the atmosphere's transparency to solar radiation and opaqueness to terrestrial radiation.

gust   A short positive departure from the ten-minute average wind speed.

Hadley circulation   The background vertical and meridional circulation of air in low latitudes, consisting of two opposing Hadley cells, each having air rising in the intertropical convergence zone and sinking in a subtropical anticyclone.

hail   Millimetric or larger precipitation particle of ice, formed by the accretion of ice crystals and rapidly freezing supercooled water droplets.

heat   The energy of a material which is stored in the form of sensible heat (kinetic energy of rotational, vibrational and translational thermal motion of constituent atoms and molecules) and/or latent heat.

heat capacity   The amount of heat required to raise the temperature of a body by one degree kelvin. In air or any other highly compressible fluid, its value differs considerably depending on whether heating occurs at constant volume or constant pressure, the latter being the larger.

high pressure zones (highs)   Regions of raised atmospheric pressure at mean sea level. Also known as anticyclones.

hoar frost   See *frost*.

humidity   The vapour content of air. See *relative* and *specific humidity*.

humidity mixing ratio   The mass of vapour as a fraction of the mass of dry air with which it is mixed in a moist air parcel. Numerically indistinguishable from specific humidity in all but the most humid air.

hurricane   See *tropical cyclone*.

hydrologic cycle   The network of pathways of the water substance through the oceans, land surfaces and atmosphere.

hydrosphere   The shell of water which nearly envelopes the Earth in the form of oceans and inland seas. Sometimes includes the *cryosphere*, groundwater and atmospheric water substance.

hydrostatic equation   The formal expression of pure hydrostatic equilibrium.

hydrostatic equilibrium   Balance between gravitational downward pull on an air parcel and upward pressure gradient force arising from vertical pressure lapse. Such equilibrium applies very accurately to all but the most violently disturbed parts of the atmosphere.

hygrograph   A recording hygrometer.

hygrometer   An instrument measuring humidity, often by measuring relative humidity.

hygroscopic   Tending to attract and condense ambient water vapour.

ideal gas   A gas which behaves as if its molecules were infinitely small, interacting only by perfectly elastic collision at the instant of collision, and therefore obeying the equation of state for an ideal gas.

ideal gas constant   See *gas constant*.

inertial reference frame   A reference frame with zero absolute acceleration.

infra-red radiation   Electromagnetic radiation lying beyond the red end of the visible spectrum, but not far beyond.

insolation   Solar irradiance of a surface, sometimes totalled over a finite time period such as a day.

instability   The condition of a body or system which responds to a specified disturbance by increasing the disturbance until an irreversible change has taken place. Sometimes used in context to mean convective instability.

internal energy   The sensible heat capacity of an ideal gas as assessed from its specific heat at constant volume.

intertropical convergence zone (ITCZ)   The zone of persistent convergence of airflow in the low troposphere in very low latitudes.

inversion   An increase of atmospheric temperature with height (an inversion of the normal tropospheric lapse).

ionosphere   The region of the upper atmosphere (usually reckoned above 50 km height) which is chronically in a significantly ionized state.

irradiance   The energy flux density of electromagnetic radiation impinging on a real or imaginary surface.

isentropic   See *dry adiabatic*.

isobar An isopleth of atmospheric pressure.

isopycnal or **isopycnic** An isopleth of air density.

isopleth A line or surface with uniform value of a specified property.

isotach An isopleth of wind speed.

isotherm An isopleth of air temperature.

isotropic Independent of direction; thus microscale turbulence is nearly isotropic while larger scales are vertically squashed.

jet stream A relatively narrow and shallow stream of fast flowing air, usually in the high troposphere (see *polar front* and subtropical jet streams).

joule The unit of energy.

katabatic winds Airflow down sloping terrain.

kelvin The unit or scale of absolute temperature.

kinetic energy The energy which a body has in respect of its motion.

knot A wind speed of one nautical mile per hour.

laminar flow Smooth, viscosity dominated flow.

lapse rate See *temperature lapse rate*.

latent heat The heat which is given out when gases and liquids liquify or solidify, and which is absorbed when solids or liquids melt or evaporate.

lateral acceleration Acceleration across the direction of flow, e.g. centripetal acceleration in cyclonic or anticyclonic flow.

lee trough A synoptic-scale low-pressure system formed or maintained in the lee of a long ridge of high ground.

lifting condensation level The level at which air would become saturated if lifted dry-adiabatically conserving its vapour content (often referring to initial conditions at screen level).

lightning The large electric sparks produced in and around thunderclouds.

linear acceleration Acceleration in the direction of flow.

log wind profile The nearly linear profile of wind speed in the surface boundary layer when plotted against log height.

mass The quantity of material in a body as assessed by its reluctance to accelerate when acted on by a force when free of all other constraints.

mechanical convection Vertical exchange of air parcels accomplished by turbulence driven by vertical wind shear.

melting band The horizontal band of enhanced radar echo produced by snow melting at the 0 °C level in extensive precipitating clouds.

meridional Along a line of meridian.

mesosphere The region of the upper atmosphere lying between the stratosphere and the thermosphere.

mesoscale Spatial scales intermediate between small and synoptic scales of weather systems.

microphysics of clouds Physical processes active on the scale of individual cloud and precipitation droplets and particles.

millibar (mbar) The meteorological unit of pressure; 100 pascals.

mole A mass of pure material whose mass in grams is numerically equal to the material's atomic or molecular weight, and therefore contains Avogadro's number of atoms or molecules.

momentum The product of the mass and velocity of a body — a vector quantity.

monsoon Pronounced seasonal variation of airflow in the lower troposphere in tropical and subtropical regions — for example the Indian monsoon. Sometimes used for similar but weaker behaviour in higher latitudes.

neutral stability The state of a system which accepts a specified disturbance without further response. Often used to mean neutral convective stability: the state of an atmospheric layer which accepts imposed vertical displacement of air parcels without positive or negative response (i.e. which is on the boundary between convective instability and stability).

nimbostratus An extensive layer of cloud precipitating rain and/or snow.

non-inertial reference frame A reference frame with non-zero absolute acceleration.

normalize   To divide a quantity by a more fundamental quantity of the same dimensions to produce a non-dimensional ratio.

Normand's theorem   A particular thermodynamic relation between temperature, thermodynamic wet-bulb temperature and dew-point temperature, expressed in terms of a construction on a tephigram or other thermodynamic diagram.

northerly   Flowing or moving from the north.

occluded front   See *front*.

order of magnitude   The typical magnitude of a quantity to the nearest integral power of 10.

pascal   The SI unit of pressure.

period   The interval between consecutive similar stages in a repeating cycle of events.

planetary boundary layer   See *atmospheric boundary layer*.

photochemistry   Chemical processes encouraged by incident photons of electromagnetic radiation (solar radiation in meterology).

photodissociation   Molecular dissociation caused by absorption of photons of electromagnetic radiation (solar radiation in meteorology).

photoionization   Ionization caused by absorption of photons of electromagnetic radiation (solar radiation in meteorology).

photosynthesis   Synthesis of water and carbon dioxide to form oxygen and sugars in the presence of sunlight and photosynthetic organisms.

Poisson's equation   Equation relating initial and final absolute temperatures and pressures of an ideal gas undergoing dry adiabatic compression or decompression.

polar front   The middle-latitude frontal zone separating air flowing from tropical and polar source regions. A favoured site of cyclogenesis.

potential energy   The energy of a body (or system of bodies) in respect of its position in a field of force. Used in meteorology to mean gravitational potential energy.

potential temperature   The temperature of an air parcel after dry adiabatic compression or decompression from its actual pressure to 1000 mbar.

power   Energy input or output per unit time.

precipitation   Ice particles or water droplets large enough ($\sim$ 100 μm or larger) to fall at least 100 m below cloud base before evaporating. See *drizzle*, *hail*, *rain* and *snow*.

pressure   The apparently continuous and isotropic force exerted on unit area of any real or imaginary surface because of bombardment by molecules of contiguous fluid.

pressure gradient force   The net force on an air parcel arising from its location in a pressure gradient.

pressure tendency   The rate of change of pressure at a fixed location, usually at the Earth's surface.

partial pressure   The pressure which would be exerted by a particular component of a gaseous mixture if all other components were removed without other change.

psychrometer   An instrument measuring vapour content by means of wet- and dry-bulb thermometry.

radian   The angle subtended at its centre by an arc of a circle equal to its radius.

radiometer   An instrument to measure fluxes of electromagnetic radiation.

radiosonde   A balloon-borne package of thermometer, barometer, hygrometer and radar reflector, for sensing the troposphere and low stratosphere during free ascent.

radiation   See *electromagnetic radiation*.

radiant flux   Flow of electromagnetic radiant energy per unit time.

radiant flux density   Radiant flux through unit area of real or imaginary surface, usually perpendicular to the rays.

radiation fog   Fog produced by net radiative cooling of the underlying surface.

rain   Precipitation in the form of millimetric-sized water droplets (as distinct from drizzle).

rain gauge   An instrument to measure totals or rates of rainfall, including drizzle and melted hail and snow.

reference frame   A set of moving or stationary spatial axes used as a basis from which to measure body motion and dynamics.

reference process   See *thermodynamic reference process*.

reflectivity   The proportion of incident radiation reflected by a surface, expressed as a fraction or percentage.

relative acceleration   Acceleration measured relative to a specified reference frame, which may itself have absolute acceleration.

relative humidity   The vapour content of air (measured as vapour density or pressure) as a percentage of the vapour content needed to saturate air at the same temperature.

relative vorticity   Vorticity about the local vertical relative to the local tangential plane of the Earth.

residence time   The average time spent by a particle in one particular component of a system of components in dynamic equilibrium (for example, by a water molecule in the oceans between precipitation and the next evaporation).

reversible   Of a thermodynamic process which is always so close to equilibrium that it can be reversed by a minute change in the set-up.

Reynolds number   The dimensionless ratio of fluid acceleration and accelerations induced by viscosity typical of a particular flow regime.

Richardson number   The dimensionless ratio of velocity shears and buoyancy forces typical of a particular flow regime.

ridge   An elongated zone of high pressure.

Rossby number   The dimensionless ratio of relative accelerations and Coriolis accelerations typical of a particular regime of synoptic-scale flow.

run of wind   The length of airflow registered by an anemometer in a certain time period (often an hour or a day).

saturated adiabat   A line on a tephigram, or other thermodynamic diagram, corresponding to a saturated adiabatic process.

saturated adiabatic lapse rate   *temperature lapse rate* associated with *saturated adiabatic* ascent or descent of an air parcel.

saturated adiabatic process   An adiabatic process in which air is kept in an exactly saturated state by evaporation or condensation of water substance.

saturation   The state of water vapour which is in dynamic equilibrium with a contiguous plane surface of pure water or ice through balanced fluxes of evaporating and condensing water molecules.

scale height   The vertical height interval in which a certain atmospheric property (such as pressure) decreases to a certain proportion (such as 1/10) of its value at the base of the interval.

scale analysis   Analysis of the rough magnitudes of individual terms in dynamic and thermodynamic equations, to determine their relative significance.

scattering   The process of chaotic deflection of electromagnetic radiation by impact on rough surfaces.

screen   See *Stevenson screen*.

seeding of clouds   See *cloud seeding*.

severe local storm   A very severe cumulonimbus.

shear   Cross-flow gradient or difference of the direction and/or speed of fluid flow (speed if direction is unspecified).

shear stress   The tangential force per unit area of sheared fluid which is associated with the shear through molecular or turbulent exchange across the flow.

slope convection   Synoptic-scale ascent and descent associated with extratropical cyclones and anticyclones.

smog   Polluted fog.

snow   Precipitation in the form of well-developed hexagonal ice crystals, either singly or in conglomerate flakes.

solar constant   The average value of normal solar irradiance just outside the Earth's atmosphere. The term can be used for other planets.

solar radiation   Electromagnetic radiation from the Sun, especially in the wavelength range 0.3 to 3 μm which contains nearly all the total irradiance.

solarimeter   An instrument to measure solar irradiance.

specific mass   The mass of a certain material as a proportion of the total mass of mixture in which it is mixed and of which it forms a part.

specific heat capacity   The quantity of heat needed to raise the temperature of unit mass of a particular substance by one degree kelvin.

specific humidity   The mass of water vapour as a proportion of the total mass of moist air of which it forms a part.

spectrum   The distribution of variance or power across wavelength or frequency.

squall   A sudden onset of strong, gusty winds.

squall line or line squall   A linear organization of vigorous cumulonimbus in the United States.

sound   Longitudinal pressure waves audible to the human ear.

stability   The condition of a body or system which responds to a specified disturbance by opposition and suppression. Often used in meteorology to refer to convective stability in particular.

station pressure   Atmospheric pressure measured at station level before correction to sea level.

steam fog   Fog formed by evaporation from a relatively warm water surface into cooler air.

Stevenson screen   A white, wooden ventilated box used to allow thermometers to measure air temperature without radiative influence.

storm   A general term applied to any type of weather system associated with strong surface winds.

stratosphere   The relatively quiet atmospheric layer between the troposphere and mesosphere.

streamline   A line which is instantaneously parallel to fluid flow.

sublimation   See *condensation.*

subsidence inversion   The temperature inversion in the low troposphere of the subsiding core of an anticyclone.

subtropical high pressure   Zones of systematically raised surface pressure between latitudes 20° and 40°.

subtropical jet stream   A westerly jet stream centred near the 200 mbar level at the polar extremity of each *Hadley* cell.

superadiabatic lapse rate   A rate of fall of temperature with height which exceeds the dry adiabatic lapse rate.

supergeostrophic wind   Wind with speed larger than the local geostrophic value.

supersaturation   Greater vapour pressure or density than is required for saturation at the prevailing temperature.

synoptic observations   Observations made simultaneously at the established synoptic grid of positions.

synoptic scale   The minimum horizontal spatial scale of weather system which is well defined by the synoptic observation network.

temperature   The concentration of sensible heat in a body, measured according to an arbitrary scale such as the Celsius and Kelvin scales.

temperature lapse rate   The rate of fall of temperature with increasing height. See *dry* and *saturated adiabiatic lapse rates.*

tephigram   A thermodynamic diagram having temperature and potential temperature as perpendicular axes.

terminal velocity or fall speed   The speed at which a particular body's weight is balanced by its drag as it falls through a particular fluid.

terrestrial radiation   Electromagnetic radiation emitted by materials at terrestrial surface and atmospheric temperatures.

thermal   A buoyant eddy, especially those of size ∼ 100 m and larger.

thermal conductivity   The ratio of heat-flux density to temperature gradient for any particular heat-conducting material.

thermal convection   Buoyant convection driven by temperature differences.

thermal diffusivity   Thermal conductivity divided by the product of density and specific heat capactiy of a heat-conducting material.

thermal wind   The vertical shear of geostrophic wind.

thermocline   A region below a water surface in which temperature increases upward.

thermodynamic diagram   A diagram whose axes are thermodynamic parameters chosen so that area represents energy.

thermodynamic reference process An *adiabatic* (*dry* or *saturated*) series of thermodynamic states of an air parcel used for comparison with complex atmospheric reality.

thermodynamic wet-bulb temperature The lowest temperature to which an air parcel can be cooled by adiabatic evaporation of water into it.

thermograph A recording thermometer.

thickness The vertical depth of a slab of air bounded by chosen isobaric surfaces.

thunder The audible noise made by lightning.

tornado Intense, cloud-cored vortex extending from the base of a severe local storm to the surface.

torque The turning moment of a force.

total potential energy The sum of the internal and gravitational potential energies of an atmospheric column.

trade winds The conspicuously reliable winds blowing obliquely in the low troposphere from a subtropical high to the intertropical convergence zone.

trajectory The path traced by a moving air parcel.

transmissivity The proportion of a flux of electromagnetic radiation which is transmitted through the atmosphere (or part of it).

tropical cyclone A cyclonic disturbance in tropical regions. When surface winds exceed 33 m s^{-1} it is called a hurricane in the Atlantic, a typhoon in the Pacific, and a cyclone in the Indian Ocean.

tropopause The limit of the troposphere (usually well marked and lying between 10 and 15 km above sea level).

troposphere The layer of air continually stirred by weather systems (bounded by the surface and the tropopause).

trough An elongated zone of low atmospheric pressure at a horizontal surface.

turbosphere The atmospheric layer in which diffusive equilibrium is swamped by convective equilibrium (from the surface to about 90 km above sea level).

turbulence Apparently chaotic fluid motion.

typhoon See *tropical cyclone*.

ultraviolet radiation Electromagnetic radiation between the violet end of the visible spectrum and X-radiation.

universal gas constant The constant linking pressure, volume and absolute temperature of a mole of ideal gas.

unstable The condition of a system which amplifies a particular type of imposed disturbance, the type of instability depending on the type of disturbance, e.g. convective, dynamic and baroclinic.

upper-air observations Synoptic observations made by radiosonde or equivalent technique.

vacillation Irregular fluctuation between minimum and maximum sinuosity of middle tropospheric flow in middle latitudes.

vapour The gaseous form of a substance. Often used as an abbreviation for water vapour.

vapour pressure The partial pressure of water vapour.

vector A vector quantity has both magnitude and direction.

veering Wind direction changing so that azimuth is increasing.

velocity The rate of change with time of the position of a body in a specified reference frame.

viscosity The friction in gases and liquids arising from molecular exchange and impact. Sometimes called molecular viscosity to distinguish it from eddy viscosity.

vorticity The measure of fluid rotation. In synoptic scale meteorology the term usually means the relative vorticity about the local vertical.

warm front See *front*.

warm sector The wedge of relatively warm troposphere between a warm front and the following cold front.

water substance Water in any of its physical states — liquid, ice or vapour.

watt One joule per second — the SI unit of power.

wave cyclone An extratropical cyclone (sometimes applied especially to the immature phase).

wavelength   The shortest distance between adjacent wave crests (or any other similar parts) in a train of waves. In a solitary wave it is taken to be the length of the perceptible distortion.

waves   Regular or nearly regular distortions of physical properties which repeat regularly (or nearly so) in space and time.

westerly   Flowing or moving from the west.

wet-bulb depression   The depression of wet-bulb temperature below dry-bulb temperature.

wet-bulb potential temperature   The wet-bulb temperature after an air parcel is taken adiabatically to 1000 mbar. Equivalently it is the temperature at which the saturated adiabat of the parcel crosses 1000 mbar on a tephigram.

wet-bulb temperature   The temperature registered by a wet-bulb thermometer.

wet-bulb thermometer   A thermometer which registers the temperature reached by free evaporation from a saturated wick.

wind   Air motion relative to the Earth's surface, usually its horizontal component.

wind shear   Often the vertical shear of horizontal wind. See *shear*.

wind stress   The horizontal stress exerted on a real or imaginary surface because of adjacent wind shear.

wind vane   A device to measure wind direction.

zonal   Along a line of latitude.

zonal index   A pressure index of sinuosity of large-scale tropospheric flow in middle latitudes.

Notes on selected problems

Chapter 1

5 Pressure is scalar but pressure gradient is vector. **7** About 1/100 of the horizontal. **8** 4. **9** 0.1 N. **10** 1000 m³. **11** Add isotherms to make a subsidiary temperature maximum or minimum or both between A and B. **12** Anticyclone over the British Isles is weakening and belt of rain is moving from Atlantic into NW Ireland. **13** Work out length/timescales ratio for fictional hurricane and compare with Fig. 1.5. **14** Consider atmospheric equivalents of warmed bearing, tyres, ground and slipstream of car.

Chapter 2

9 42.8 units. **10** 78.5 Pa. **11** 12 mm and 4.8 m. **12** 46.1 mbar. **13** 1053 Z, 190.4 km and 53.5 °E of N. **14** 0.11 °C. **15** Glass reduces solar radiation and breezes; warm interior radiates infra-red. **16** For same height rise, pressure fall is directly proportional to g, hence much less on Moon. Barometer fall is same as on Earth (but what about an aneroid barometer?). **17** Drops chill window by evaporation until inside glass surface falls below interior dew point. **18** Snowfall distorted by airflow, like rain but much more so (why?). Unlike rain, after falling, snow is not redistributed by gravity, but is by moderate winds or stronger. **19** Heat flux to bulb directly proportional to bulb surface, but heat capactiy of bulb directly proportional to bulb volume. Hence speed of response proportional to surface/volume, i.e. $1/R$ in sphere radius R.

Chapter 3

1 No, since each type will give uniformity. **8** 1.11 grams, depleted by respiration. **9** 10^{-4} kg l^{-1}, 1.25×10^{21} molecules l^{-1}. **10** 21.8 yr, 403 units. **11** Increase from combustion, further increase by reduction in photosynthesis and increase of decay, slow decrease as chilling ocean absorbs more. **15** Atmospheric source trivial, ice-cap source considerable but very slow. River flooding? Folk memory of inundation of Mediterranean? Coastal inundation after explosion of Thera? **16** Sloping terrain at coasts. Sea level rises, but so does land freed from ice burden. **17** Assume all vapour precipitates and use chain rule.

Chapter 4

9 0.465 kg m^{-3}. **10** 2497, 2355, 3576 m. **11** 246.6 K at 70 °N, 267.3 K at 30 °N. **12** (a) 10 days, (b) 13 days. **13** 51.3 and 95.2 minutes. **14** 0.2 (core speed 60 m s^{-1}), 3.4×10^{-3}. **15** Note $p = \rho R_d T$ where p and ρ are air pressure and density. $T_v = 290.8$ K, cf. $T = 288.2$. **16** 10.9 km for Mars, low atmosphere temperature 220 K. **17** Troposphere much shallower, tropopause and low stratosphere much warmer. Westerlies aloft and easterlies below removed. No rotation, daylight 6 months long. **18** High albedo ice reduces solar input, produces more ice. Baroclinity drives weather systems, precipitates more snow but produces more blanketing cloud.

Chapter 5

2 Note tornado core has low pressure. **8** 4.7 °C. **9** 10.6 and 7.7 g kg^{-1}, 73%. **10** 2550 m. **11** 54%. **12** $-$ 26.6 °C. **13** 55.9 °C, about 18 °C, about 57 °C. **14** 1.3 °C. Cooling could be further reduced. **15** θ 20 °C, q 9 g kg^{-1}, e 14.5 mbar, T_d 12.2 °C, $T_w = \theta_w = 15.2$ °C, LCL pressure 892 mbar, θ_e 46 °C. **16** $-$ 8 °C. Warming by contact with warm interior, heaters etc. **17** Implies 1000 mbar T 55.9 °C, T_w 17 °C, surface T about 19 °C, looks like ocean surface below 30 °C latitude in winter, 40 ° in summer.

Chapter 6

4 Lower RH being lower in the sub-cloud layer, sunshine stronger. **8** Large drips not yet at break-up speed in short available height. **11** 2.14×10^{-3} and 2.25×10^{-2} kg m^{-3}. Surface air grossly subsaturated. **12** 7.3%, 5000 m. **13** 0.36, 3.57 μm. **14** 0.56 s at 11 °C, 5.2 s at -20 °C assuming same D value. **15** 13.4 min to 500 μm, another 116 s to 1 mm. **16** Thunder begins 0.61 s after flash, loud bang centred about 3 s after flash, end of rumble 6.1 s after flash. **17** 9.3×10^{21} molecules $m^{-2} s^{-1}$ deposited, 8.6×10^{26} impacts. **18** 0.62, 6.18 μm; 0.124, 0.012%. **19** Bases — explosive growth, well-mixed layer; edges — rapid evaporation in subsaturated air. Deactivated invisible, activated highly visible. **20** About 830 m.

Chapter 7

1 No vertical contribution. **3b** Reading enhanced. **7** About 6 m s^{-1} compared with 100 m s^{-1} needed if air used. **11** Assuming craft contents are weightless, i.e. it is in ballistic orbit or flight. **14** 300 m s^{-1}! Ignore Coriolis. **15** 0.24 m s^{-1}. **16** Note longitudinal accelerations $\sim 4 \times 10^{-3}$ m s^{-2}. **17** 39.7, 68.8, 394 m s^{-1}. The last is very unrealistic. Pressure gradients like this occur in hurricanes, but balance is then cyclostrophic not geostrophic. **18** Each speed 4 $\times$ greater. 5.1×10^{-4} in all cases (i.e. 51 m per 100 km). **19** 20 m s^{-1}. **21** 3.28×10^{-2} m s^{-2}, 8.1 m s^{-2}. **22** 10^{-2} m s^{-2}, i.e. about 1/3. **23** 4215, 1776, 749 km^2. **24** 1, 3 and 5 cm s^{-1}. **25** 1.98, 3.93, 7.78 absolute and 0.98, 2.93, 6.78 relative vorticities, all in units of 10^{-4} s^{-1}. **26** Note deep tropospheric drag by mountain waves and deep convection. **28** Obviously more isotropic. Effective height of solar heat input might not then be at the surface, giving marine-like inactivity in lower troposphere. **31** Sloping frontal zone gives sharp T gradient in only shallow zone on any vertical. **32** Slope 10 : 1 (not far from vertical — slowly diverging funnel where pressure = LCL value).

Chapter 8

3 Note this is the opposite of the conventional radiometer output, which would make the cat look similar in the IR and visible wavelengths. **9** 6.5 °. **16** 2.696, 1.397, 0.605 kW m^{-2}. **17** 244 K. **18** 448.3, 463.8, 339.6, 233.0, 167.7 W m^{-2}. **19** Sc — uniform darkish field, darker than visible; Cb — white tops in dark field; frontal decks — white stripes amid darker on dark field. **20** Stratosphere/troposphere = 1.43. **21** 699, 295 W m^{-2}. **25** Solar constant 5.6 $\times$ present value, Earth much hotter, terrestrial radiation in near IR, solar radiation red, wavelength overlap between solar and IR. **27** $T_g^4 = 2T_a^4$, $T_g^4 = S/[\sigma (1 - a/2)]$. With $S = 350$ W m^{-2}, T_g is 318 K, T_a is 267 K. **28** E.g. snow-covered low ground not too cold, hence warmer and darker than high cloud in IR though indistinguishable in visible. **29** 439 W m^{-2} above atmosphere, compared with 300 W m^{-2} observed suggests effective albedo 0.32.

10 Polewards and eastwards. **14** ITCZ air cooler and saturated; STH air warm and very unsaturated. **22** Water 19 °C, atmosphere 79 °C, ice depth 0.32 m. **23** About 14 kW per head of 4 billion population. **24** $F = 31.3$ W m^{-2}, reasonable agreement. **33** See problem 22. **34** Find mass of moving water to be 5×10^5 kg m^{-2}. The average poleward heat flux is 31 MJ m^{-1}. Fair agreement, though a little low.

Chapter 9

3 0.125. **5** The transparency since it exposes a huge mass. **8** Values keeping well above 0.25 even in the coldest (pre-dawn) period. **15** 906, 87, 574 W m^{-2}. **16** 428, 41, 271 W m^{-2}. **17** 11.84, 5.59 MJ m^{-2}. **18** 0 °, i.e. on the horizon, 47 °. **19** About 0.3 m s^{-1} depending on choice of θ (note can treat θ as T). **20** -2 m^2 s^{-2}, $\sqrt{2}$ m s^{-1}, 1.7 N m^{-2}. **21** 6.25 cm, 7.32 m s^{-1}, 0.58 m s^{-1}. **22** 2 g kg^{-1} (assuming not far from 1000 mbar at comparison level and apparently assuming saturated rising air, but see next problem). **23** 780 mbar, 2.3 km. Note that the maximum upslope/downslope T difference is preserved throughout the upslope sub-cloud layer. **24** Little cloud, hot summers, cold winters, snow, big range of day length and noon solar elevation. **25** Consider turbulent fluxes of sensible heat and momentum in the vertical. **26** Maximum solar elevation from problem 20. Solar day 24 hr long, solar input by method of problem 19 34.1 MJ m^{-2}. The LW loss depends heavily on humidity and cloud cover of sky. Crudely, ground T is 283 K, sky T 233 K, all day input gives 17 MJ m^{-2}. **27** 1.303.

Chapter 10

2 0.66 m s^{-1}. **4** Consider hexagonal cells crossing observer, some diametrically, some tangentially. 54 km. **7** Are they thrown or are they left? **8** 50 mm hr^{-1}. **9** 2.5, 0.33, 2.83 °C. **10** At 800 mbar T values are 20.2 and 16 °C. Convection inhibited. **11** $\partial\theta/\partial z$ is approximately 1/50 °C m^{-1} and assuming θ 273 K, B–V time period is 3.9 min. **12** 12.8 m s^{-1} using appendix 11.2 replacing Γ_d by $(\Gamma_d + 1/200)$. **13** 83 m s^{-1} (i.e. a jet core). An underestimate. **14** No room for compensating sink. **15** Buoyancy increases on condensation, decreases on evaporation. Locate active zones on simple cloud model. **16** 6.7 °C, surface wet-bulb temperature. **17** Cooling by 1.25 °C. After appendix 11.1 this is partly offset to give effective cooling of 1.05 °C. **19** Need about 25 km height to fall through assuming average air density 0.8 kg m^{-3}. Large stone held while falling through updraught at 10 m s^{-1} for 30 min followed by 12 km fall to ground, gives effective total 24 km. Suggests such growth needs encounter with rich droplet source when large (near base of updraught?). **21** 12 m s^{-2}, i.e. 1.2 g (accelerating faster than free fall) followed by 2.4 g jolt upward. This is why civil aircraft keep clear of big Cb! **22** Sub-cloud layer area reduces to 1/220 in 30 min (vertically stretched to same extent). Some radioactive aerosol in converged and uplifted air was rained and washed to surface in rain so contaminated that public health warnings were broadcast.

Chapters 11 and 12

2 Well to west of old low centre where open waves form and move quickly. **3** Taken aback (i.e. blown backwards) by near headwinds against sails still set NW–SE. **9** 32 cm. **10** About 400 km on simplest straight-line geometry relative to the moving system. Path curvature could reduce this somewhat, but eastward system motion could add considerably to geographical distance of source. **11** 6.8 °C and 8.9% RH. **12** 2×10^{-6} s^{-1}. **13** 1.6 m s^{-1} slower eastward than the airflow in mid troposphere (e.g. if latter is 20 m s^{-1} E'wards then cyclone wave moves eastwards at 18.4 m s^{-1}.) Assume broad airstream and no divergence or convergence. **14** Use method of appendix 12.7 to find 7.6 mbar for average T 300 K and ambient p_s 1000 mbar. **15** 192 km. **16** If use dry adiabatic flow, then cloudy uplifts underestimated. Model in Fig. 11.15 does not describe intense activity in low troposphere near surface cold front. **19** About 0.45 mm s^{-1} as mean temperature falls.

Chapter 13

4 In steady state, inward heat leakage balances outward heat pumped. Additional heat generated by motor and inefficient heat pump is net gain. **5** Do not use technical jargon. **6** No dynamic behaviour. **8** Heat source and sink temperatures quite close, much friction and irreversibility. **9** 0.5%. **10** 8.4%. **11** 140 m s^{-1}. About 9797 J per kg of raindrop has to be converted into heat by friction and shared with ambient air. **12** 0.98 °C. 2.87 MJ m^{-2} into PE, 7.13 MJ m^{-2} into IE. **13** Use $FW \cong 7.5 \times 10^{-3} V_{10}^3$ (appendix 13.4) to find 0.1, 2.3, 51, 262 W m^{-2}. **14** 6.93 MJ m^{-2}. **15** (a) No zonal and eddy categories, but horizontal and vertical possible instead. Some A from A_e — convection in fronts. Turbulent loss dominates. **17** PGW (upwards) slightly exceeds d(gz)/dt (buoyancy), giving relatively small rate of increase of KE $(w^2/2)$. Friction nearly completely offsets buoyancy giving much smaller KE increase.

Appendix

Section equivalents between Fundamentals of Weather and Climate (FWC) and Basic Meteorology (BM). All details are omitted where numbering is identical (though titles may differ), and new sections in FWC are listed without BM equivalence.

	FWC	BM
Section	1.4, 1.5	1.4
	1.6	1.5
	1.7	–
Section	3.1	3.1, 3.2, 3.3
	3.2, 3.3	3.4
	3.4, 3.5	3.5
	3.8	–
Appendix	–	3.1
Section	4.9	4.9
	4.10	–
Appendix	4.3	–
	4.4	4.3
	4.5	4.4
Appendix	5.2	–
	5.3	5.2
	5.4	5.3
Section	7.2	7.2, 7.3, 7.4
	7.3	7.5
	7.4	7.6
	7.5	7.7
	7.6	7.9
	7.7	7.10
	7.8	7.11
	7.9	7.12
	7.10	7.13
	7.11	7.14
	7.12	7.15
	7.13	7.16
	7.14	7.17
Appendix	7.5	Section 7.8

Index